INSTRUCTOR'S ANNOTATED EXERCISES TO ACCOMPANY

Mathematical Ideas

EIGHTH EDITION

CHARLES D. MILLER

VERN E. HEEREN
American River College

E. JOHN HORNSBY, JR.
University of New Orleans

An imprint of Addison Wesley Longman, Inc.

Reading, Massachusetts • Menlo Park, California • New York • Harlow, England
Don Mills, Ontario • Sydney • Mexico City • Madrid • Amsterdam

Sponsoring Editor: Anne Kelly
Developmental Editor: Greg McRill
Project Editor: Cathy Wacaser
Design Administrator: Jess Schaal
Text and Cover Design: Lesiak/Crampton Design Inc: Lucy Lesiak
Cover Photo: Doug Armand/Tony Stone Images
Photo Researcher: Karen Koblik
Production Administrator: Randee Wire
Compositor: Interactive Composition Corporation
Printer and Binder: R.R. Donnelley & Sons Company
Cover Printer: Phoenix Color Corporation

For permission to use copyrighted material, grateful acknowledgment is made to the copyright holders on page 355, which is hereby made part of this copyright page.

Instructor's Annotated Exercises to accompany *Mathematical Ideas,* Eighth Edition

ISBN 0-673-98241-6 (Instructor's Annotated Exercises)

96 97 98 99 —DOW— 9 8 7 6 5 4 3 2 1

To the Instructor

For your convenience, all of the exercises from the student's edition of *Mathematical Ideas,* Eighth Edition are reprinted in the *Instructor's Annotated Exercises.* Answers are printed in place for all exercises except those for which many correct answers are possible (i.e., proofs or writing exercises). Each exercise set is cross-referenced to the pages on which the exercises appear in the student text. A special binding allows the open manual to lie flat on a desk or lectern.

Conceptual and writing exercises are identified for the instructor by the symbols ⊙ and ✎, respectively. Challenging exercises, which will require most students to stretch beyond the concepts discussed in the text, are marked with the symbol ▲. Calculator exercises are indicated by 🖩 or 🖩.

Contents

CHAPTER 1 Approach to Problem Solving

1.1 Exercises (Text Page 7)

Decide whether each of the following is an example of inductive or deductive reasoning in Exercises 1–10.

1. If it is your wish, then it is my command. It is your wish. Therefore, it is my command. **deductive**
2. Automobile repairs always take longer than the mechanic predicts. The mechanic told me that fixing the brakes on my car will take three hours. Therefore, my car will be in the shop for more than three hours. **deductive**
3. The last four governors in our state have been Republicans. Therefore, the next governor will be Republican. **inductive**
4. Joshua had 40 baseball cards. He gave 10 away. Therefore, he has 30 left. **deductive**
5. If the same number is added to both sides of a true equation, the resulting equation is true. $3 + 2 = 5$. Therefore, $(3 + 2) + 9 = 5 + 9$. **deductive**
6. Linda's first four children were girls. Therefore, her next child will be a girl. **inductive**
7. Prudent people never buy \$3,000 television sets. Lisa Wunderle is prudent. Therefore, Lisa will never buy a \$3,000 television set. **deductive**
8. The best player on our girls' team this year attended a basketball clinic to improve her skills. Therefore, in order to be the best player on your team, you must attend a basketball clinic. **inductive**
9. For the past 20 years, a rare plant has bloomed in a South American rain forest each spring, alternating between pink and white flowers. Last spring, it bloomed with pink flowers. Therefore, this spring it will bloom with white flowers. **inductive**
10. All men are mortal. Socrates is a man. Therefore, Socrates is mortal. **deductive**
11. Discuss the differences between inductive and deductive reasoning. Give an example of each.

Determine the most probable next term in each list of numbers.

12. 6, 9, 12, 15, 18 **21**
13. 13, 18, 23, 28, 33 **38**
14. 3, 12, 48, 192, 768 **3,072**
15. 32, 16, 8, 4, 2 **1**
16. 3, 6, 9, 15, 24, 39 **63**
17. 1/3, 3/5, 5/7, 7/9, 9/11 **11/13**
18. 1/2, 3/4, 5/6, 7/8, 9/10 **11/12**
19. 1, 4, 9, 16, 25 **36**
20. 1, 8, 27, 64, 125 **216**
21. 1, 0, 1, 1, 0, 1, 1, 1, 0, 1, 1, 1, 1, 0, 1, 1 **1**
22. 1, 1, 1, 1, 0, 0, 0, 1, 1, 1, 0, 0, 0, 1, 1, 0, 0 **0**
23. Construct a list of numbers similar to those in Exercises 12–22 such that the most probable next number in the list is 60. **There are many possibilities. One such list is 10, 20, 30, 40, 50, ...**
24. Construct a list of numbers similar to those in Exercises 12–22 such that either 20 or 25 may be a true conjecture for the next number in the list. **There are many possibilities. One such list is 5, 10, 15, ...**

In Exercises 25–36, a list of equations is given. Use the list and inductive reasoning to predict the next equation, and then verify your conjecture.

25.
$(1 \times 9) + 2 = 11$
$(12 \times 9) + 3 = 111$
$(123 \times 9) + 4 = 1{,}111$
$(1{,}234 \times 9) + 5 = 11{,}111$
$\mathbf{(12{,}345 \times 9) + 6 = 111{,}111}$

26.
$(9 \times 9) + 7 = 88$
$(98 \times 9) + 6 = 888$
$(987 \times 9) + 5 = 8{,}888$
$(9{,}876 \times 9) + 4 = 88{,}888$
$\mathbf{(98{,}765 \times 9) + 3 = 888{,}888}$

    SCIENTIFIC CALCULATOR  GRAPHING CALCULATOR

27.
$15{,}873 \times 7 = 111{,}111$
$15{,}873 \times 14 = 222{,}222$
$15{,}873 \times 21 = 333{,}333$
$15{,}873 \times 28 = 444{,}444$
$\mathbf{15{,}873 \times 35 = 555{,}555}$

28.
$3{,}367 \times 3 = 10{,}101$
$3{,}367 \times 6 = 20{,}202$
$3{,}367 \times 9 = 30{,}303$
$3{,}367 \times 12 = 40{,}404$
$\mathbf{3{,}367 \times 15 = 50{,}505}$

29.
$11 \times 11 = 121$
$111 \times 111 = 12{,}321$
$1{,}111 \times 1{,}111 = 1{,}234{,}321$
$\mathbf{11{,}111 \times 11{,}111 = 123{,}454{,}321}$

30.
$34 \times 34 = 1{,}156$
$334 \times 334 = 111{,}556$
$3{,}334 \times 3{,}334 = 11{,}115{,}556$
$\mathbf{33{,}334 \times 33{,}334 = 1{,}111{,}155{,}556}$

31.
$2 = 4 - 2$
$2 + 4 = 8 - 2$
$2 + 4 + 8 = 16 - 2$
$2 + 4 + 8 + 16 = 32 - 2$
$\mathbf{2 + 4 + 8 + 16 + 32 = 64 - 2}$

32.
$$3 = \frac{3(2)}{2}$$
$$3 + 6 = \frac{6(3)}{2}$$
$$3 + 6 + 9 = \frac{9(4)}{2}$$
$$3 + 6 + 9 + 12 = \frac{12(5)}{2}$$
$$\mathbf{3 + 6 + 9 + 12 + 15 = \frac{15(6)}{2}}$$

33.
$$3 = \frac{3(3 - 1)}{2}$$
$$3 + 9 = \frac{3(9 - 1)}{2}$$
$$3 + 9 + 27 = \frac{3(27 - 1)}{2}$$
$$3 + 9 + 27 + 81 = \frac{3(81 - 1)}{2}$$
$$\mathbf{3 + 9 + 27 + 81 + 243 = \frac{3(243 - 1)}{2}}$$

34.
$$5(6) = 6(6 - 1)$$
$$5(6) + 5(36) = 6(36 - 1)$$
$$5(6) + 5(36) + 5(216) = 6(216 - 1)$$
$$5(6) + 5(36) + 5(216) + 5(1{,}296) = 6(1{,}296 - 1)$$
$$\mathbf{5(6) + 5(36) + 5(216) + 5(1{,}296) + 5(7{,}776) = 6(7{,}776 - 1)}$$

35.
$$\frac{1}{1 \cdot 2} = \frac{1}{2}$$
$$\frac{1}{1 \cdot 2} + \frac{1}{2 \cdot 3} = \frac{2}{3}$$
$$\frac{1}{1 \cdot 2} + \frac{1}{2 \cdot 3} + \frac{1}{3 \cdot 4} = \frac{3}{4}$$
$$\frac{1}{1 \cdot 2} + \frac{1}{2 \cdot 3} + \frac{1}{3 \cdot 4} + \frac{1}{4 \cdot 5} = \frac{4}{5}$$
$$\mathbf{\frac{1}{1 \cdot 2} + \frac{1}{2 \cdot 3} + \frac{1}{3 \cdot 4} + \frac{1}{4 \cdot 5} + \frac{1}{5 \cdot 6} = \frac{5}{6}}$$

36.
$$\frac{1}{2} = 1 - \frac{1}{2}$$
$$\frac{1}{2} + \frac{1}{4} = 1 - \frac{1}{4}$$
$$\frac{1}{2} + \frac{1}{4} + \frac{1}{8} = 1 - \frac{1}{8}$$
$$\frac{1}{2} + \frac{1}{4} + \frac{1}{8} + \frac{1}{16} = 1 - \frac{1}{16}$$
$$\mathbf{\frac{1}{2} + \frac{1}{4} + \frac{1}{8} + \frac{1}{16} + \frac{1}{32} = 1 - \frac{1}{32}}$$

CONCEPTUAL | WRITING | CHALLENGING | SCIENTIFIC CALCULATOR | GRAPHING CALCULATOR

A story is often told about how the great mathematician Carl Friedrich Gauss (1777–1855) at a very young age was told by his teacher to find the sum of the first 100 *counting numbers. While his classmates toiled at the problem, Carl simply wrote down a single number and handed it in to his teacher. His answer was correct. When asked how he did it, the young Carl explained that he observed that there were* 50 *pairs of numbers that each added up to* 101. *(See below.) So the sum of all the numbers must be* $50 \times 101 = 5{,}050$.

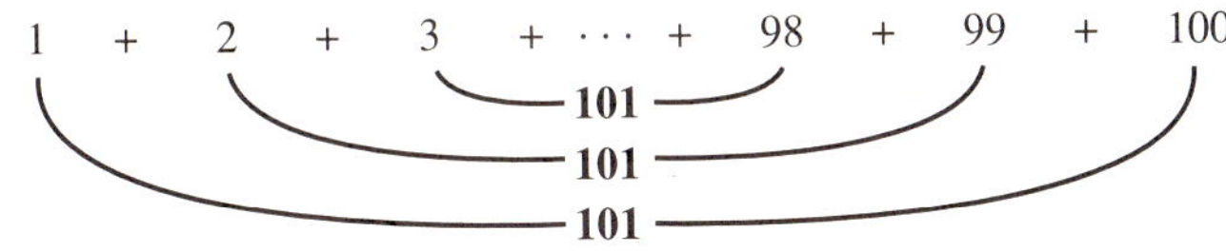

50 sums of 101 = $50 \times 101 = 5{,}050$

Use the method of Gauss to find each of the following sums.

37. $1 + 2 + 3 + \cdots + 150$ **11,325**

38. $1 + 2 + 3 + \cdots + 300$ **45,150**

39. $1 + 2 + 3 + \cdots + 500$ **125,250**

40. $1 + 2 + 3 + \cdots + 1{,}000$ **500,500**

41. Modify the procedure of Gauss to find the sum $1 + 2 + 3 + \cdots + 125$. **7,875**

42. Explain in your own words how the procedure of Gauss can be modified to find the sum $1 + 2 + 3 + \cdots + n$, where n is an odd natural number. (An odd natural number, when divided by 2, leaves a remainder of 1.)

43. Modify the procedure of Gauss to find the sum $2 + 4 + 6 + \cdots + 100$. **2,550**

44. Use the result of Exercise 43 to find the sum $4 + 8 + 12 + \cdots + 200$. **5,100**

---

45. Consider the following table.

0	2	2	2	0	0	0	0	0
0	2	4	6	4	2	0	0	0
0	2	6	12	14	12	6	2	0
0	2	8	20	32	38	32	20	8

Find a pattern and predict the next row of the table.
0 2 10 30 60 90 102 90 60 (To find any term, choose the term directly above it and add to it the two preceding terms. If there are fewer than two terms, add as many as there are.)

46. Find a pattern in the following list of figures and use inductive reasoning to predict the next figure in the list.

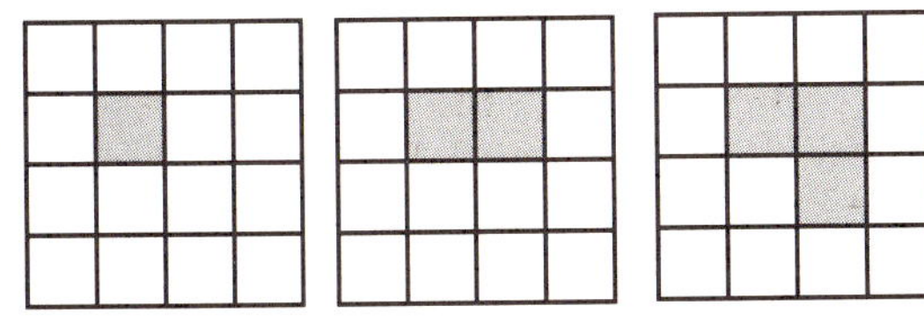

(a) (b) (c)

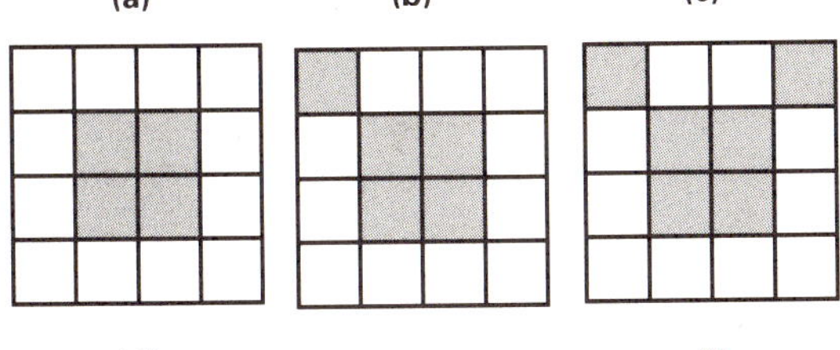

(d) (e) (f)

47. What is the next term in this list? O, T, T, F, F, S, S, E, N, T (*Hint:* Think about words and their relationship to numbers.) **E (One, Two, Three, and so on.)**

48. What is the most probable next number in this list? 12, 1, 1, 1, 2, 1, 3 (*Hint:* Think about a clock.) **1 (These are the numbers of chimes a clock rings, starting with 12 o'clock, if it rings the number of hours on the hour, and 1 chime on the half-hour.)**

49. Choose any number, and follow these steps.
(a) Multiply by 2.
(b) Add 6.
(c) Divide by 2.
(d) Subtract the number you started with.
(e) Record your result.
Repeat the process, except in Step (b), add 8. Record your final result. Repeat the process once more, except in Step (b), add 10. Record your final result.

(f) Observe what you have done; use inductive reasoning to explain how to predict the final result. You may wish to use this exercise as a "number trick" to amuse your friends. **The final result will always be half of the number added in Step (b).**

50. (a) Choose any three-digit number with all different digits. Now reverse the digits, and subtract the smaller from the larger. Record your result. Choose another three-digit number and repeat this process. Do this as many times as it takes for you to see a pattern in the different results you obtain. (*Hint:* What is the middle digit? What is the sum of the first and third digits?) **The middle digit is always 9, and the sum of the first and third digits is always 9 (considering 0 as the first digit if the difference has only two digits).**

(b) Write an explanation of this pattern. You may wish to use this exercise as a "number trick" to amuse your friends.

51. Complete the following.

$12{,}345{,}679 \times 9 =$ _____

$12{,}345{,}679 \times 18 =$ _____

$12{,}345{,}679 \times 27 =$ _____

By what number would you have to multiply 12,345,679 in order to get an answer of 888,888,888? **111,111,111; 222,222,222; 333,333,333; 72**

52. Complete the following.

$142{,}857 \times 1 =$ _____

$142{,}857 \times 2 =$ _____

$142{,}857 \times 3 =$ _____

$142{,}857 \times 4 =$ _____

$142{,}857 \times 5 =$ _____

$142{,}857 \times 6 =$ _____

What pattern exists in the successive answers? Now multiply 142,857 by 7 to obtain an interesting result. **142,857; 285,714; 428,571; 571,428; 714,285; 857,142. Each result consists of the same six digits, but in a different order. 142,857 × 7 = 999,999**

53. Choose two natural numbers. Add 1 to the second and divide by the first to get a third. Add 1 to the third and divide by the second to get a fourth. Add 1 to the fourth and divide by the third to get a fifth. Continue this process until you discover a pattern. What is the pattern? **A cycle will form that has its sixth number equal to the first number chosen. For example, choosing 5 and 6 gives 5, 6, 1.4, .4, 1, 5, 6 as the first seven terms.**

54. Refer to Figures 1(b)–(f). Instead of counting interior regions of the circle, count the chords formed. Use inductive reasoning to predict the number of chords that would be formed if seven points were used. **21 chords**

55. Discuss one example of inductive reasoning that you have used recently in your life. Test your premises and your conjecture. Did your conclusion ultimately prove to be true or false?

56. Give an example of faulty inductive reasoning.

1.2 Exercises (Text Page 17)

Use the method of successive differences to determine the next number in each sequence.

1. 2, 8, 16, 26, . . . **38**

2. 15, 24, 39, 60, . . . **87**

3. 21, 34, 51, 72, 97, . . . **126**

4. 11, 18, 29, 44, 63, . . . **86**

5. 5, 28, 87, 200, 385, . . . **660**

6. 11, 34, 81, 164, 295, . . . **486**

7. 1, 15, 79, 253, 622, 1,296, . . . **2,410**

8. 5, 23, 91, 269, 641, 1,315, . . . **2,423**

9. Suppose that the expression $n^2 + 5$ determines the nth term in a sequence. That is, to find the first term, let $n = 1$, to find the second term, let $n = 2$, and so on.

(a) Find the first four terms of the sequence. **6, 9, 14, 21**

(b) Use the method of successive differences to predict the fifth term of the sequence. **Add 9 to 21 to get 30.**

(c) Find the fifth term of the sequence by letting $n = 5$ in the expression $n^2 + 5$. Does your result agree with the one you found in part (b)? **$5^2 + 5 = 30$. The results agree.**

10. Refer to Figure 1 in the previous section. The method of successive differences can be applied to the sequence of interior regions, 1, 2, 4, 8, 16, 31, to find the number of regions determined by seven points on the circle. What is the next term in this sequence? How many regions would be determined by eight points? Verify this using the formula given at the end of that section. **57; 99**

In each of the following, several equations are given illustrating a suspected number pattern. Determine what the next equation would be, and verify that it is indeed a true statement.

11. $(1 \times 8) + 1 = 9$
$(12 \times 8) + 2 = 98$
$(123 \times 8) + 3 = 987$ **$(1{,}234 \times 8) + 4 = 9{,}876$**

12. $(1 \times 9) - 1 = 8$
$(21 \times 9) - 1 = 188$
$(321 \times 9) - 1 = 2{,}888$
$(4{,}321 \times 9) - 1 = 38{,}888$

13. $101 \times 101 = 10{,}201$
$10{,}101 \times 10{,}101 = 102{,}030{,}201$
$1{,}010{,}101 \times 1{,}010{,}101 = 1{,}020{,}304{,}030{,}201$

14. $999{,}999 \times 2 = 1{,}999{,}998$
$999{,}999 \times 3 = 2{,}999{,}997$
$999{,}999 \times 4 = 3{,}999{,}996$

15. $1 = 1^2$
$1 + 2 + 1 = 2^2$
$1 + 2 + 3 + 2 + 1 = 3^2$
$1 + 2 + 3 + 4 + 3 + 2 + 1 = 4^2$
$1 + 2 + 3 + 4 + 5 + 4 + 3 + 2 + 1 = 5^2$

16. $3^2 - 1^2 = 2^3$
$6^2 - 3^2 = 3^3$
$10^2 - 6^2 = 4^3$
$15^2 - 10^2 = 5^3$
$21^2 - 15^2 = 6^3$

17. $1^2 + 1 = 2^2 - 2$
$2^2 + 2 = 3^2 - 3$
$3^2 + 3 = 4^2 - 4$ **$4^2 + 4 = 5^2 - 5$**

18. $2^2 - 1^2 = 2 + 1$
$3^2 - 2^2 = 3 + 2$
$4^2 - 3^2 = 4 + 3$ **$5^2 - 4^2 = 5 + 4$**

19. $1 + 2 = 3$
$4 + 5 + 6 = 7 + 8$
$9 + 10 + 11 + 12 = 13 + 14 + 15$
$16 + 17 + 18 + 19 + 20 = 21 + 22 + 23 + 24$

20. $1 = 1 \times 1$
$1 + 5 = 2 \times 3$
$1 + 5 + 9 = 3 \times 5$
$1 + 5 + 9 + 13 = 4 \times 7$

Use the formula $S = \dfrac{n(n + 1)}{2}$ *derived in this section to find each of the following sums.*

21. $1 + 2 + 3 + \cdots + 100$ **5,050**

22. $1 + 2 + 3 + \cdots + 400$ **80,200**

23. $1 + 2 + 3 + \cdots + 525$ **138,075**

24. $1 + 2 + 3 + \cdots + 600$ **180,300**

Use the formula $S = n^2$ *discussed in this section to find each of the following sums. (Hint: To find n, add 1 to the last term and divide by 2.)*

25. $1 + 3 + 5 + \cdots + 49$ **625**

26. $1 + 3 + 5 + \cdots + 101$ **2,601**

27. $1 + 3 + 5 + \cdots + 301$ **22,801**

28. $1 + 3 + 5 + \cdots + 999$ **250,000**

29. Use the formula for finding the sum $1 + 2 + 3 + \cdots + n$ to discover a formula for finding the sum $2 + 4 + 6 + \cdots + 2n$. **$S = n(n + 1)$**

30. State in your own words the following formula discussed in this section:
$(1 + 2 + 3 + \cdots + n)^2 = 1^3 + 2^3 + 3^3 + \cdots + n^3$.

31. Explain how the following diagram geometrically illustrates the formula
$1 + 2 + 3 + 4 = \frac{4 \times 5}{2}$.

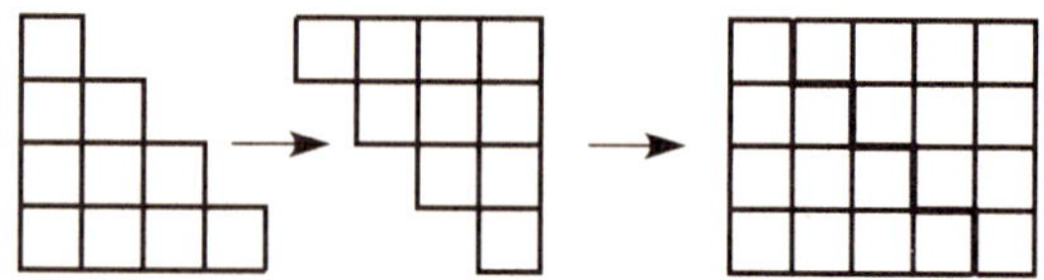

32. Explain how the following diagram geometrically illustrates the formula
$1 + 3 + 5 + 7 + 9 = 5^2$.

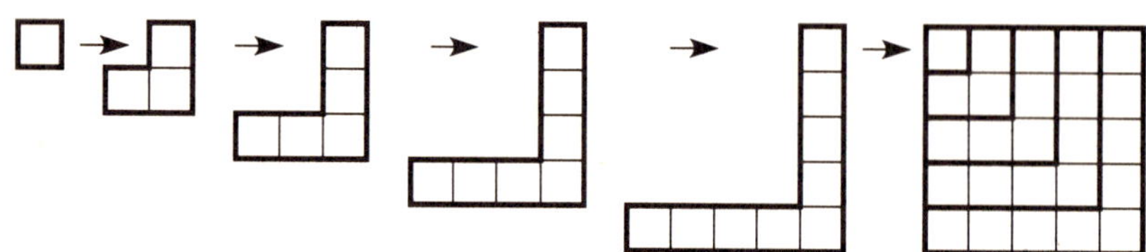

33. Use patterns to complete the table below. row 1: 28, 36; row 2: 36, 49, 64; row 3: 35, 51, 70, 92; row 4: 28, 45, 66, 91, 120; row 5: 18, 34, 55, 81, 112, 148; row 6: 8, 21, 40, 65, 96, 133, 176

Figurate Number	1st	2nd	3rd	4th	5th	6th	7th	8th
Triangular	1	3	6	10	15	21		
Square	1	4	9	16	25			
Pentagonal	1	5	12	22				
Hexagonal	1	6	15					
Heptagonal	1	7						
Octagonal	1							

34. The first five triangular, square, and pentagonal numbers may be obtained using sums of terms of sequences, as shown below.

Triangular	Square	Pentagonal
$1 = 1$	$1 = 1$	$1 = 1$
$3 = 1 + 2$	$4 = 1 + 3$	$5 = 1 + 4$
$6 = 1 + 2 + 3$	$9 = 1 + 3 + 5$	$12 = 1 + 4 + 7$
$10 = 1 + 2 + 3 + 4$	$16 = 1 + 3 + 5 + 7$	$22 = 1 + 4 + 7 + 10$
$15 = 1 + 2 + 3 + 4 + 5$	$25 = 1 + 3 + 5 + 7 + 9$	$35 = 1 + 4 + 7 + 10 + 13$

Notice the successive differences of the added terms on the right sides of the equations. The next type of figurate number is the **hexagonal** number. (A hexagon has six sides.) Use the patterns above to predict the first five hexagonal numbers.
1, 6, 15, 28, 45

35. Eight times any triangular number, plus 1, is a square number. Show that this is true for the first four triangular numbers. $8(1) + 1 = 9 = 3^2$; $8(3) + 1 = 25 = 5^2$; $8(6) + 1 = 49 = 7^2$; $8(10) + 1 = 81 = 9^2$

36. Every square number can be written as the sum of two triangular numbers. For example, 16 = 6 + 10. This can be represented geometrically by dividing a square array of dots with a line as illustrated at the right.

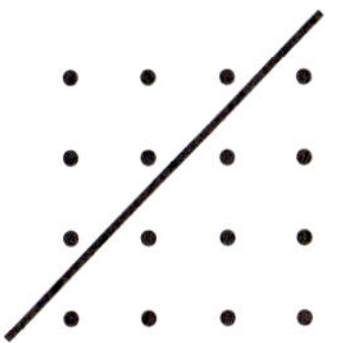

The triangular arrangement above the line represents 6, the one below the line represents 10, and the whole arrangement represents 16. Show how the square numbers 25 and 36 may likewise be geometrically represented as the sum of two triangular numbers.

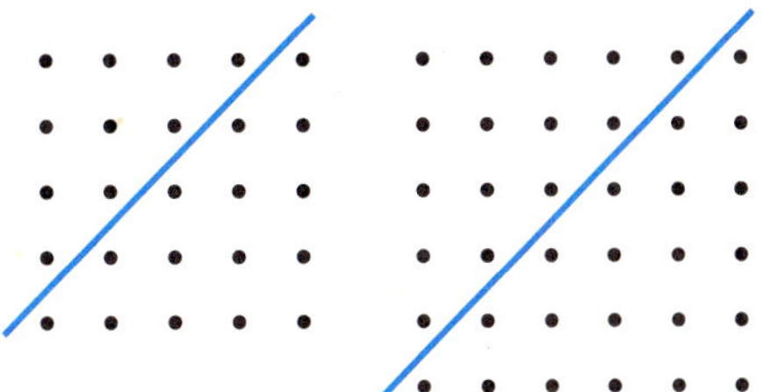

37. Divide the first triangular number by 3 and record the remainder. Divide the second triangular number by 3 and record the remainder. Repeat this procedure several more times. Do you notice a pattern? **The pattern is 1, 0, 0, 1, 0, 0, ...**

38. Repeat Exercise 37, but instead use square numbers and divide by 4. What pattern is determined? **The pattern is 1, 0, 1, 0, 1, 0, ...**

39. Exercises 37 and 38 are specific cases of the following: In general, when the numbers in the sequence of n-agonal numbers are divided by n, the sequence of remainders obtained is a repeating sequence. Verify this for $n = 5$ and $n = 6$. **For $n = 5$, the pattern is 1, 0, 2, 2, 0, 1, 0, 2, 2, 0, ... For $n = 6$, the pattern is 1, 0, 3, 4, 3, 0, 1, 0, 3, 4, 3, 0, ...**

Refer to the table of figurate numbers in Exercise 33. The 2 × 2 *square of entries formed by the third and fourth triangular and square numbers is*

$$\begin{matrix} 6 & 10 \\ 9 & 16. \end{matrix}$$

If we add the terms in the diagonal from upper left to lower right, add the terms in the diagonal from upper right to lower left, and then subtract these sums, we obtain

$$(6 + 16) - (10 + 9) = 3.$$

Notice that this difference, 3, is the column (the third) in which the terms on the left side of the square are located.

Verify this property for the following 2 × 2 *squares from the table.*

40. $\begin{matrix} 3 & 6 \\ 4 & 9 \end{matrix}$ **The difference is 2. 3 and 4 are in column 2.**

41. $\begin{matrix} 10 & 15 \\ 16 & 25 \end{matrix}$ **The difference is 4. 10 and 16 are in column 4.**

42. $\begin{matrix} 9 & 16 \\ 12 & 22 \end{matrix}$ **The difference is 3. 9 and 12 are in column 3.**

43. $\begin{matrix} 5 & 12 \\ 6 & 15 \end{matrix}$ **The difference is 2. 5 and 6 are in column 2.**

44. The 2 × 2 square

$$\begin{matrix} 189 & 235 \\ 225 & 280 \end{matrix}$$

appears in an extended table of figurate numbers, similar to the one in Exercise 33. In what two columns do these numbers appear? (*Hint:* Refer to the explanation preceding Exercise 40.) **columns 9 and 10**

45. Complete the following table.

n	2	3	4	5	6	7	8
A Square of n							
B (Square of n) + n							
C One-half of Row B entry							
D (Row A entry) − n							
E One-half of Row D entry							

Use your results to answer the following, using inductive reasoning.

(a) What kind of figurate number is obtained when you find the average of n^2 and n? (See Row C.) **a triangular number**

(b) If you square n and then subtract n from the result, and then divide by 2, what kind of figurate number is obtained? (See Row E.) **a triangular number**

46. A fraction is reduced to *lowest terms* if the greatest common factor of its numerator and its denominator is 1. For example, 3/8 is reduced to lowest terms, but 4/12 is not.

(a) For $n = 2$ to $n = 8$, form the fractions

$$\frac{n\text{th square number}}{(n+1)\text{th square number}}.$$

4/9, 9/16, 16/25, 25/36, 36/49, 49/64, 64/81

(b) Repeat part (a), but use triangular numbers instead. **3/6, 6/10, 10/15, 15/21, 21/28, 28/36, 36/45**

(c) Use inductive reasoning to make a conjecture based on your results from parts (a) and (b), observing whether the fractions are reduced to lowest terms.
The fraction formed by two consecutive square numbers is always reduced to lowest terms, while the fraction formed by two consecutive triangular numbers is never reduced to lowest terms.

*In addition to the formulas for T_n, S_n, and P_n shown in the text, the following formulas are true for **hexagonal** numbers (H), **heptagonal** numbers (Hp), and **octagonal** numbers (O):*

$$H_n = \frac{n(4n-2)}{2} \qquad Hp_n = \frac{n(5n-3)}{2} \qquad O_n = \frac{n(6n-4)}{2}.$$

Use these formulas to find each of the following.

47. the seventeenth square number **289**

48. the tenth triangular number **55**

49. the eighth pentagonal number **92**

50. the sixth hexagonal number **66**

51. the eighth heptagonal number **148**

52. the sixth octagonal number **96**

53. Observe the formulas given for H_n, Hp_n, and O_n, and use patterns and inductive reasoning to predict the formula for N_n, the nth **nonagonal** number. (A nonagon has 9 sides.) Then use the fact that the sixth nonagonal number is 111 to further confirm your conjecture. $N_n = \frac{n(7n-5)}{2}$

54. Use the result of Exercise 53 to find the seventh nonagonal number. **154**

CONCEPTUAL  WRITING ▲ CHALLENGING SCIENTIFIC CALCULATOR GRAPHING CALCULATOR

Use inductive reasoning to answer each question in Exercises 55–58.

55. If you add two consecutive triangular numbers, what kind of figurate number do you get? **a triangular number**

56. If you add the squares of two consecutive triangular numbers, what kind of figurate number do you get? **a triangular number**

57. Square a triangular number. Square the next triangular number. Subtract the smaller result from the larger. What kind of number do you get? **a perfect cube**

58. Choose a value of n greater than or equal to 2. Find T_{n-1}, multiply it by 3, and add n. What kind of figurate number do you get? **a pentagonal number**

1.3 Exercises (Text Page 27)

Use problem-solving strategies to solve each of the following problems. In many cases there is more than one possible approach, so be creative. Keep in mind that ingenuity is an important approach in problem solving.

1. If you raise 3 to the 328th power, what is the units digit of the result? **1**

2. What is the units digit in 7^{491}? **3**

3. Barbara Burnett bought a book for $10 and then spent half her remaining money on a train ticket. She then bought lunch for $4 and spent half her remaining money at a bazaar. She left the bazaar with $20. How much money did she start with? **$98**

4. I am thinking of a positive number. If I square it, double the result, take half of that result, and then add 12, I get 21. What is my number? **3**

5. A frog is at the bottom of a 20-foot well. Each day it crawls up 4 feet but each night it slips back 3 feet. After how many days will the frog reach the top of the well? **17 days**

6. A drawer contains 20 black socks and 20 white socks. If the light is off and you reach into the drawer to get your socks, what is the minimum number of socks you must pull out in order to be sure that you have a matching pair? **3 socks**

7. How many squares are in the following figure? **14**

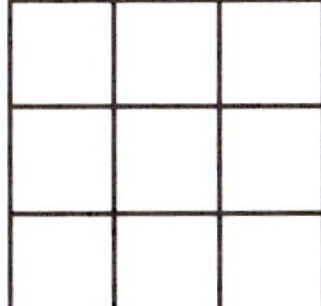

8. How many triangles are in the following figure? **35**

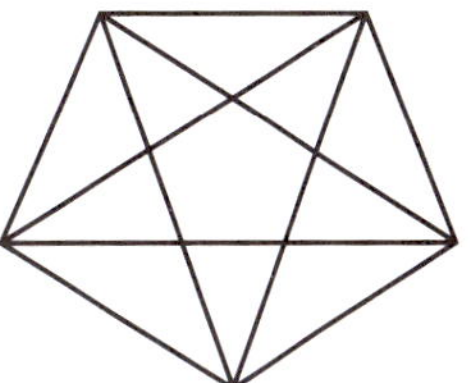

9. Some children are standing in a circular arrangement. They are evenly spaced and marked in numerical order. The fourth child is standing directly opposite the twelfth child. How many children are there in the circle? **16**

10. A *perfect number* is a counting number that is equal to the sum of all its counting number divisors except itself. For example, 28 is a perfect number, since its divisors other than itself are 1, 2, 4, 7, and 14, and $1 + 2 + 4 + 7 + 14 = 28$. What is the smallest perfect number? **6**

11. A *palindromic number* is a number whose digits read the same from left to right as they read from right to left. For example, 14,641 is palindromic. What three-digit palindromic numbers have the sum of their digits equal to the sum of the digits of the largest two-digit palindromic number? **585, 666, 747, 828, 909 (The largest two-digit palindromic number is 99, so the sum must be $9 + 9 = 18$.)**

12. A lily pad grows so that each day it doubles its size. On the fifteenth day of its life, it completely covers a pond. On what day was the pond half covered? **the fourteenth day**

13. Comment on an interesting property of this sentence: "A man, a plan, a canal, Panama." (*Hint:* See Exercise 11.) **It is a palindrome, since it reads the same backward as forward.**

14. Draw a diagram that satisfies the following description, using the minimum number of birds: "Two birds above a bird, two birds below a bird, and a bird between two birds."

15. Assuming that he lives that long, one of the authors of this book will be 76 years old in the year x^2, where x is a counting number. In what year was he born? 1949

16. The same author mentioned in Exercise 15 graduated from high school in the year that satisfies these conditions: (1) The sum of the digits is 23; (2) The hundreds digit is 3 more than the tens digit; (3) No digit is an 8. In what year did he graduate? 1967

17. Donna is taller than David but shorter than Bill. Dan is shorter than Bob. What is the first letter in the name of the tallest person? B, since either Bill or Bob must be the tallest person.

18. There is a two-digit number between 20 and 30 such that the sum of the cubes of its digits is equal to three times the number. What is the number? 24

19. Eve said to Adam, "If you give me one dollar, then we will have the same amount of money." Adam then replied, "Eve, if you give me one dollar I will have double the amount of money you are left with." How much does each have? Eve has $5 and Adam has $7.

20. The number of hens and cows in a barnyard adds up to 11. The number of legs among them is 32. How many hens and how many cows are in the barnyard? 6 hens and 5 cows

21. If Earl Karn weighs 170 pounds standing on one leg, how much does he weigh standing on two legs? 170 pounds

22. If you take 7 bowling pins from 10 bowling pins, what do you have? 7 bowling pins

23. In the addition problem below, some digits are missing as indicated by the blanks. If the problem is done correctly, what is the sum of the missing digits?

```
     _ 3 5
     8 _ 6
 +   1 4 _
 ---------
   _, 4 0 8
```

14 (The correct problem is

```
   435
   826
 + 147
 -----
 1,408.)
```

24. Fill in the blanks so that the multiplication problem below uses all digits 0, 1, 2, 3, . . . , 9 exactly once, and is correctly worked.

```
      _ 0 2
 ×      3 _
 ----------
 _ 5, _ _ _
```

The correct problem is

```
   402
 ×  39
 -------
 15,678.
```

25. A *magic square* is a square array of numbers that has the property that the sum of the numbers in each row, column, and diagonal is the same. Fill in the square below so that it becomes a magic square, and all digits 1, 2, 3, . . . , 9 are used exactly once.

6		8
	5	
		4

6	1	8
7	5	3
2	9	4

26. Refer to Exercise 25. Complete the magic square below so that all digits 1, 2, 3, . . . , 16 are used exactly once, and the sum in each row, column, and diagonal is 34.

6			9
	15		14
11		10	
16		13	

6	12	7	9
1	15	4	14
11	5	10	8
16	2	13	3

27. With a 5-minute sand timer and a 9-minute sand timer, what is the easiest way to time an egg to boil for 13 minutes? **Start the two timers together. When the 5-minute timer runs out, start boiling the egg. When the 9-minute timer runs out, turn it over and time another 9 minutes. When it is finished, the egg will have boiled (9 − 5) + 9 = 13 minutes.**

28. Michael Booth has an unlimited number of cents (pennies), nickels, and dimes. In how many different ways can he pay 15¢ for a chocolate mint? (For example, one way is 1 dime and 5 pennies.) **6 ways**

▲ **29.** What is the minimum number of pitches that a baseball pitcher who pitches a complete game can make in a regulation 9-inning baseball game? **25 pitches (The visiting team's pitcher retires 24 consecutive batters through the first eight innings, using only one pitch per batter. His team does not score either. Going into the bottom of the ninth tied 0–0, the first batter for the home team hits his first pitch for a home run. The pitcher threw 25 pitches and loses the game by a score of 1–0.)**

30. What is the least natural number whose written name in the English language has its letters in alphabetical order? **forty**

31. If it takes 7 1/2 minutes to boil an egg, how long does it take to boil five eggs? **7 1/2 minutes (Boil them all at the same time.)**

32. A hayfield has two haystacks in one corner and four haystacks in another corner. If all the hay is put into the middle of the field, how many haystacks will there be? **1 haystack**

33. Several soldiers must cross a deep river at a point where there is no bridge. The soldiers spot two children playing in a small rowboat. The rowboat can hold only two children or one soldier. All the soldiers get across the river. How? **The two children row across. One stays on the opposite bank and the other returns. One soldier rows across, and the child on the opposite bank then rows back. The two children row across. One stays and the other returns. Now another soldier rows across. This process continues until all the soldiers are across.**

34. A person must take a wolf, a goat, and some cabbage across a river. The rowboat to be used has room for the person plus either the wolf, the goat, or the cabbage. If the person takes the cabbage in the boat, the wolf will eat the goat. While the wolf crosses in the boat, the cabbage will be eaten by the goat. The goat and cabbage are safe only when the person is present. Even so, the person gets everything across the river. Explain how. (This problem dates back to around the year 750.) **The person takes the goat across and returns alone. On the second trip, the person takes the wolf across and returns with the goat. On the third trip, the goat is left on the first side while the person takes the cabbage across. Then the person returns alone and brings the goat back across.**

35. You have eight coins. Seven are genuine and one is a fake, which weighs a little less than the other seven. You have a balance scale, which you may use only three times. Tell how to locate the bad coin in three weighings. (Then show how to detect the bad coin in only *two* weighings.) **For three weighings, first balance four against four. Of the lighter four, balance two against the other two. Finally, of the lighter two, balance them one against the other. To find the bad coin in two weighings, divide the eight coins into groups of 3, 3, 2. Weigh the groups of three against each other on the scale. If the groups weigh the same, the fake is in the two left out and can be found in one additional weighing. If the two groups of three do not weigh the same, pick the lighter group. Choose any two of the coins and weigh them. If one of these is lighter, it is the fake; if they weigh the same, then the third coin is the fake.**

▲ **36.** Three women, Ms. Thompson, Ms. Johnson, and Ms. Andersen, are sitting side by side at a meeting of the neighborhood improvement group. Ms. Thompson always tells the truth, Ms. Johnson sometimes tells the truth, and Ms. Andersen never tells the truth. The woman on the left says, "Ms. Thompson is in the middle." The woman in the middle says, "I'm Ms. Johnson," while the woman on the right says, "Ms. Andersen is in the middle." What are the correct positions of the women? **From left to right, they are Johnson, Andersen, and Thompson.**

37. In the following problem, each letter stands for a specific digit. Assuming the problem is worked correctly, what do the letters stand for?

$$\begin{array}{r} \text{P P P} \\ +\quad\ \text{A} \\ \hline \text{A B B B} \end{array}$$

P represents 9, A represents 1, and B represents 0.

38. Repeat Exercise 37 for the following.

$$\begin{array}{r} \text{X Y} \\ +\ \ \text{Y} \\ \hline \text{Y X} \end{array}$$

X represents 8 and Y represents 9.

39. Draw a square in the following figure so that no two dogs share the same region.

40. When the diagram shown is folded to form a cube, what letter is opposite the face marked Y? **P**

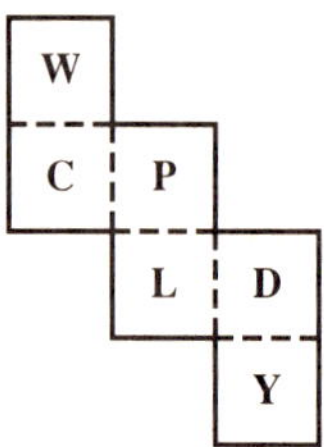

Exercises 41–42 are ancient Hindu mathematical problems.

41. Beautiful maiden with beaming eyes, tell me . . . which is the number that when multiplied by 3, then increased by 3/4 the product, then divided by 7, diminished by 1/3 of the quotient, multiplied by itself, diminished by 52, by the extraction of the square root, addition of 8, and division by 10 gives the number 2? **28**

42. The mixed price of 9 citrons and 7 fragrant wood apples is 107; again, the mixed price of 7 citrons and 9 fragrant wood apples is 101. O you arithmetician, tell me quickly the price of a citron and of a wood apple, having distinctly separated those prices well. **5 for a wood apple; 8 for a citron**

43. How many times can you subtract 10 from 100? **Only once—after you have subtracted 10 from 100, you are no longer subtracting from 100.**

44. At a variety store, 1 costs $.50 while 5082 costs $2.00. What is being sold? **house address numbers**

45. Draw the following figure without picking up your pencil from the paper and without tracing over a line you have already drawn.

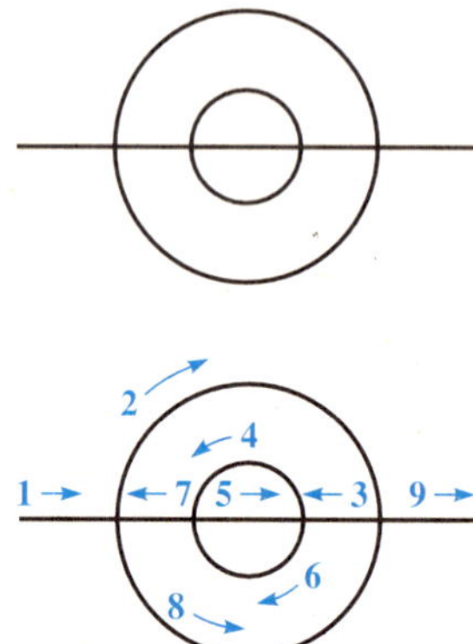

46. Repeat Exercise 45 for the figure shown here.

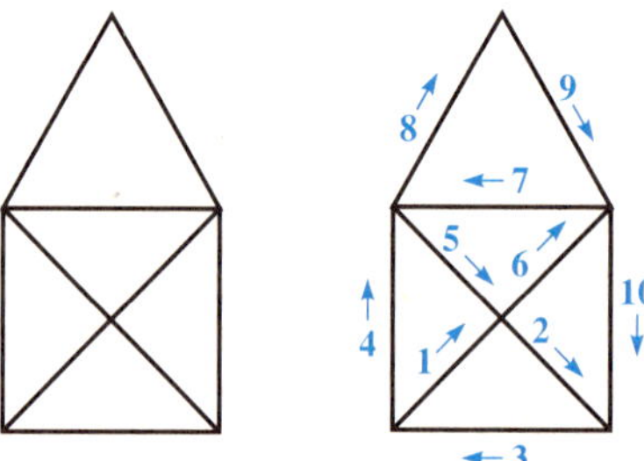

47. Volumes 1 and 2 of *Mathematics Is for Everyone* are standing in numerical order from left to right on your bookshelf. Volume 1 has 300 pages and Volume 2 has 250 pages. Excluding the covers, how many pages are between page 1 of Volume 1 and page 250 of Volume 2? **none**

48. The brother of the chief executive officer (CEO) of a major industrial firm died. The man who died has no brother. How is this possible? **The CEO is a woman.**

49. A teenager's age increased by 2 gives a perfect square. Her age decreased by 10 gives the square root of that perfect square. She is 5 years older than her brother. How old is her brother? **9 years old**

50. James, Dan, Jessica, and Cathy form a pair of married couples. Their ages are 35, 30, 29, and 28. Jessica is married to the oldest person in the group. James is older than Jessica but younger than Cathy. Who is married to whom, and what are their ages? **Dan (35) is married to Jessica (28); James (29) is married to Cathy (30).**

51. Only one of the numbers here is a perfect square. Which one is it? (No extensive calculation is necessary. *Hint:* Consider the units digit.)
(a) 23,784,855,888,784
(b) 49,552,384,781,228
(c) 33,587,988,047,112
(d) 64,987,237,001,667
(a), since perfect squares must have a units digit of 0, 1, 4, 9, 6, or 5. ($23{,}784{,}855{,}888{,}784 = (4{,}876{,}972)^2$)

52. After you have solved the problem in Exercise 51 correctly, write a few sentences explaining the procedure you used.

53. What is the maximum number of small squares in which we may place a cross (×) and not have any row, column, or diagonal completely filled with crosses? Illustrate your answer. **6**

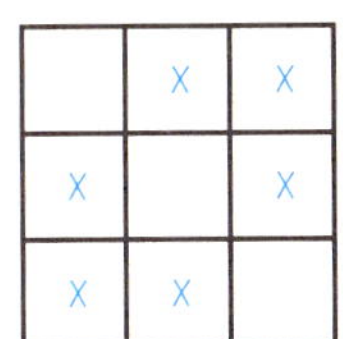

One of several possibilities

54. How much dirt is there in a cubical hole, 4 feet on each side? **None, since there is no dirt in a hole.**

55. Refer to Example 1, and observe the sequence of numbers in color, the Fibonacci sequence. Choose any four consecutive terms. Multiply the first one chosen by the fourth, and then multiply the two middle terms. Repeat this process a few more times. What do you notice when the two products are compared? **The products will always differ by 1.**

56. What is the 100th digit in the decimal representation for $1/7$? **8**

57. In how many different ways can you make change for a half dollar using currently minted U.S. coins, if cents (pennies) are not allowed? **10**

58. Some months have 30 days and some have 31 days. How many months have 28 days? **12; All months have 28 days.**

59. If a year has two consecutive months with Friday the thirteenth, what months must they be? **They must be February and March. This can occur only in a non-leap year. The next time it will happen is in 1998.**

60. **(a)** Consider the following multiplication problems. They each consist of two 2-digit factors that are obtained by reversing the digits. Notice that the products are the same.

$$\begin{array}{r} 84 \\ \times\ 24 \\ \hline 2{,}016 \end{array} \qquad \begin{array}{r} 48 \\ \times\ 42 \\ \hline 2{,}016 \end{array}$$

$$\begin{array}{r} 23 \\ \times\ 64 \\ \hline 1{,}472 \end{array} \qquad \begin{array}{r} 32 \\ \times\ 46 \\ \hline 1{,}472 \end{array}$$

What pattern do you notice? **The product of the tens digits must equal the product of the ones digits.**

(b) Explain the necessary condition for this pattern to exist.

61. The first man introduced himself to the first woman with a brief "palindromic" greeting. What was the greeting? (*Hint:* See Exercises 11, 13, and 19.) **"Madam, I'm Adam."**

62. Refer to Exercises 11 and 15. Another author of this book (who expects to be somewhat older than 76 in the year x^2) has a palindromic, seven-digit phone number, say *abc-defg*, satisfying these conditions:

$b = 2 \times a$

$e = f$

a, b, and c are all multiples of 4

$a + b + c + d + e + f + g = 47.$

Determine his phone number. **488-7884**

1.4 Exercises (Text Page 35)

Exercises 1–18 are designed to give you practice in learning how to do some basic operations on your calculator. Perform the indicated operations and give as many digits in your answer as shown on your calculator display. (The number of displayed digits will vary depending on the model used.)

1. $28.3 + (9.7 - 4.8)$ **33.2**

2. $4.2 \times (3.1 - 1.6)$ **6.3**

3. $-4.1 + (7.8 \times 1.4)$ **6.82**

4. $[1.331 \div (-11)] + 3$ **2.879**

5. $\sqrt{6.036849}$ **2.457**

6. $\sqrt{34{,}774{,}609}$ **5,897**

7. $\sqrt[3]{260{,}917{,}119}$ **639**

8. $\sqrt[3]{109.215352}$ **4.78**

9. 7.4^2 **54.76**

10. 8.2^2 **67.24**

11. 9.1^3 **753.571**

12. 6.55^3 **281.011375**

13. 2.1^5 **40.84101**

14. 1.8^6 **34.01224**

15. $\sqrt[6]{1.29}$ **1.043353839**

16. $\sqrt[5]{9.68}$ **1.574617492**

17. 2π **6.283185307**

18. π^2 **9.869604401**

19. Use your calculator to *square* the following two-digit numbers ending in 5: 15, 25, 35, 45, 55, 65, 75, 85. Write down your results, and examine the pattern that develops. Then use inductive reasoning to predict the value of 95^2. Write an explanation of how you can mentally square a two-digit number ending in 5. **$95^2 = 9{,}025$**

20. Choose any number consisting of five digits. Multiply it by 9 on your calculator. Now add the digits in the answer. If the sum is more than 9, add the digits again, and repeat until the sum is less than 10. Your answer will always be 9. Repeat the exercise with a number consisting of six digits. Does the same result hold? **yes**

By examining several similar computation problems and their answers obtained on a calculator, we can use inductive reasoning to make conjectures about certain rules, laws, properties, and definitions in mathematics. Perform each calculation and observe the answers. Then fill in the blank with the appropriate response. (Justification of these results will be discussed later in the book.)

21. $5 \times (-4)$; -3×8; $2.7 \times (-4.3)$

Multiplying a negative number and a positive number gives a **negative** (negative/positive) product.

22. $(-3) \times (-8)$; $(-5) \times (-4)$; $(-2.7) \times (-4.3)$

Multiplying two negative numbers gives a **positive** (negative/positive) product.

23. 1^2; 1^3; 1^{-3}; 1^0; 1^{13}

Raising 1 to any power gives a result of **1**.

24. 5.6^0; π^0; 2^0; 120^0; $.5^0$

Raising a nonzero number to the power 0 gives a result of **1**.

25. $1/7$; $1/(-9)$; $1/3$; $1/(-8)$

The sign of the reciprocal of a number is **the same as** (the same as/different from) the sign of the number.

26. $0/8$; $0/2$; $0/(-3)$; $0/\pi$

Zero divided by a nonzero number gives a quotient of **0**.

27. $5/0$; $9/0$; $\pi/0$; $-3/0$; $0/0$

Dividing a number by 0 gives a result of ______ on a calculator. (What do you think this indicates?) **an error message; Division by 0 is not allowed.**

CONCEPTUAL | WRITING | CHALLENGING | SCIENTIFIC CALCULATOR | GRAPHING CALCULATOR

28. $\sqrt{-3}$; $\sqrt{-5}$; $\sqrt{-6}$; $\sqrt{-10}$
Taking the square root of a negative number gives a result of _____ on a calculator. (What do you think this indicates?) **an error message; The square root of a negative number is not a real number.**

29. $(-3) \times (-4) \times (-5)$; $(-3) \times (-4) \times (-5) \times (-6) \times (-7)$;
$(-3) \times (-4) \times (-5) \times (-6) \times (-7) \times (-8) \times (-9)$
Multiplying an *odd* number of negative numbers gives a **negative** (positive/negative) product.

30. $(-3) \times (-4)$; $(-3) \times (-4) \times (-5) \times (-6)$;
$(-3) \times (-4) \times (-5) \times (-6) \times (-7) \times (-8)$
Multiplying an *even* number of negative numbers gives a **positive** (positive/negative) product.

31. Find the decimal fraction for 2/3 on your calculator. The display will show perhaps a lead 0, a decimal point, and a string of all 6s or a string of 6s with final digit 7. Does your calculator *truncate* or *round off*? **If the last digit is 6, it truncates. If it is 7, it rounds off.**

32. From algebra we know that we are allowed to raise a negative number to a counting number power. For example,

$$(-3)^5 = (-3) \times (-3) \times (-3) \times (-3) \times (-3) = -243.$$

However, some scientific calculators will not allow a negative number to be raised to a power, despite the fact that this is mathematically valid. Suppose your calculator gives an error message when you try to evaluate $(-3)^5$. Give an explanation of how you would use your calculator to do this computation. (*Hint:* See the result of Exercise 29.) **Raise 3 to the fifth power to get 243. Since a negative number is being raised to an *odd* power, the result must be negative. So $(-3)^5 = -243$.**

33. Choose any three-digit number and enter the digits into a calculator. Then enter them again to get a six-digit number. Divide this six-digit number by 7. Divide the result by 13. Divide the result by 11. What is your answer? Explain why this happens. **The result is the three-digit number you started with. Dividing by 7, then 13, and then 11 is the same as dividing by 1,001. A three-digit number *abc* multiplied by 1,001 gives *abcabc*, so we have just reversed the process.**

34. Choose any digit except 0. Multiply it by 429. Now multiply the result by 259. What is your answer? Explain why this happens. **The result is a six-digit number consisting of only the digit you started with. Multiplying by 429 and 259 is the same as multiplying by 111,111, giving all the same digits.**

For each of the following, perform the indicated calculation on a scientific calculator. (These exercises do not apply to typical graphing calculator display screens.) Then turn your calculator upside down to read the word that belongs in the blank in the accompanying sentence.

35. 11,669 × 3; One of teen idol Fabian's 1959 records was *Turn Me* _____. **LOOSE**

36. 7,531,886.6 ÷ 1.4; Have you ever had a case of the _____? **GIGGLES**

37. 5 × 10,609; Imelda has an obsession for _____. **ShOES**

38. 128,396 − 93,016; When Mrs. Percy Pearl Washington died in 1972, she weighed about 880 pounds. She was very _____. **OBESE**

39. Make up your own exercise similar to Exercises 35–38. When a standard calculator is turned upside down, the digits in the display correspond to letters of the English alphabet as follows: **Answers will vary.**

0 ↔ O	3 ↔ E	7 ↔ L
1 ↔ I	4 ↔ h	8 ↔ B
2 ↔ Z	5 ↔ S	9 ↔ G.

40. Displayed digits on most calculators usually show some or all of the parts in the pattern shown in the figure.

For the digits 0 through 9:

(a) Which part is used most frequently?

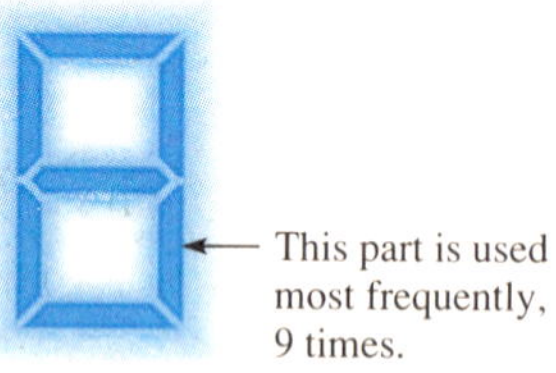

(b) Which part is used the least?

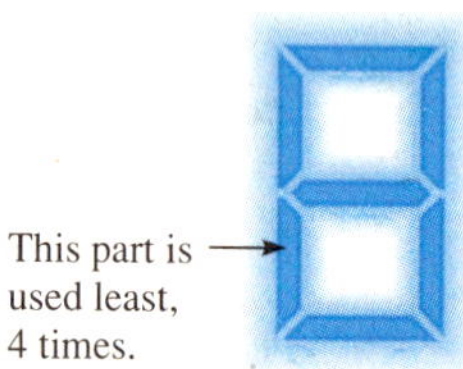

(c) Which digit uses the most parts? **8 (all seven parts)**

(d) Which digit uses the fewest parts? **1 (two parts)**

Give the appropriate counting number answer to each question in Exercises 41–44.

41. A certain type of carrying case will hold a maximum of 48 audio cassettes. If you need to store 490 audio cassettes, how many carrying cases will you require? 11

42. A plastic page designed to store baseball cards will hold up to 16 cards. If you must store your collection of 484 cards, how many pages will you need? 31

43. Each room available for administering a placement test will hold up to 40 students. Two hundred fifty students have signed up for the test. How many rooms will be used? 7

44. A gardener wants to fertilize 2,000 tomato plants. Each bag of fertilizer will supply up to 150 plants. How many bags does she need to do the job? 14

In Exercises 45–50, use estimation to choose the letter of the choice closest to the correct answer.

45. The planet Mercury takes 88.0 Earth days to revolve around the sun. Pluto takes 90,824.2 days to do the same. When Pluto has revolved around the sun once, about how many times will Mercury have revolved around the sun? (b)

(a) 100 **(b)** 1,000 **(c)** 10,000 **(d)** 100,000

46. Hale County in Texas has a population of 34,671 and covers 1,005 square miles. About how many inhabitants per square mile does the county have? (a)

(a) 35 **(b)** 350 **(c)** 3,500 **(d)** 35,000

 CONCEPTUAL  WRITING  CHALLENGING   SCIENTIFIC CALCULATOR   GRAPHING CALCULATOR

47. The 1990 U.S. census showed that the total population of the country was 248,709,873. Of these, 25,223,086 were in the 45–50-year age bracket. On the average, about one in every _____ citizens is in this age bracket. (b)
(a) 5 **(b)** 10 **(c)** 15 **(d)** 20

48. The minimum start-up fee to open a Dairy Queen franchise is $375,000. Suppose that 19 people decide to put up equal amounts toward the fee. About how much would each have to contribute? (c)
(a) $10,000 **(b)** $15,000 **(c)** $20,000 **(d)** $25,000

49. In voting for the portrayal of which age of Elvis that was to be depicted on a U.S. postage stamp, voters cast 851,000 votes for the younger Elvis against 277,723 votes for the more mature Elvis. A newspaper article read "Younger Elvis is Winner by _____ to 1 Margin." What number belongs in the blank? (c)
(a) 5 **(b)** 4 **(c)** 3 **(d)** 2

50. The distance from Springfield, MO, to Seattle, WA, is 2,009 miles. If a bus averages about 50 miles per hour, about how many hours would this trip take? (c)
(a) 30 **(b)** 35 **(c)** 40 **(d)** 45

When the Dawkins family left Hattiesburg to visit their cousins in Columbus, their car's instrument panel looked like this.

When they arrived in Columbus, it looked like this.

51. What is the round-trip distance between Columbus and Hattiesburg? **360 miles**

52. If the car's tank holds 16 gallons of gasoline, how much is left? **12 gallons**

53. How many miles per gallon did the car get? **45 miles per gallon**

54. How many gallons of gas will the car have left after the return trip? **8 gallons**

Use the bar graph to answer the questions in Exercises 55–58.

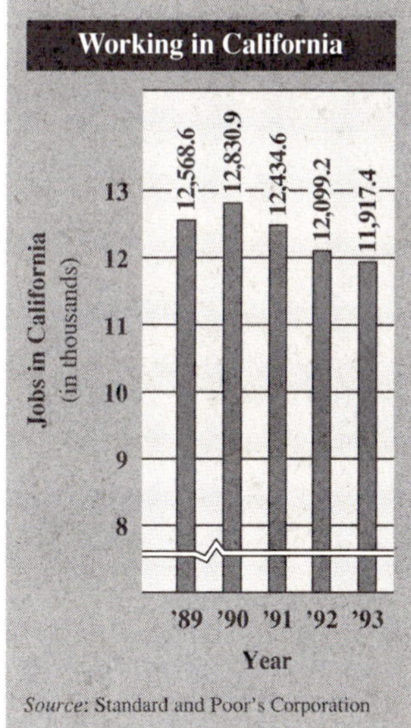

55. Between what two consecutive years shown on the graph did the number of jobs in California increase? How much was the increase? **between 1989 and 1990; 262.3 thousand**

56. How many jobs did California lose between 1990 and 1993? **913.5 thousand**

57. According to Standard & Poor's Corporation, half the losses between 1990 and 1993 were in defense and 40% were in construction. How many were lost in defense? How many were lost in construction? **456.75 thousand in defense; 365.4 thousand in construction**

58. Suppose that the same number of jobs were lost between 1993 and 1994 as were lost between 1992 and 1993. How many jobs were there in California in 1994 based on this assumption? **11,735.6 thousand**

Use the line graph to answer the questions in Exercises 59–62.

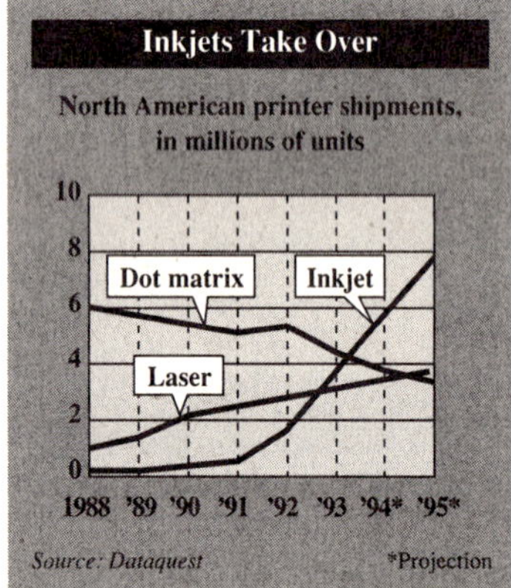

59. Which one of the following would be the best estimate for the total number of units of printer shipments in North America in 1989? **(d)**

(a) 5.8 million **(b)** 1.5 million
(c) .2 million **(d)** 7.5 million

60. In what year did the number of Inkjet printer shipments first exceed the number of Laser printer shipments? **1993**

61. What type of printer has shown the most rapid increase in shipments since 1992? **Inkjet**

⊙ **62.** How has the trend in shipments of Dot Matrix printers differed from the shipments of Inkjet and Laser printers since 1992? **Dot matrix shipments have decreased while the others have increased.**

Use the pie chart to answer the questions in Exercises 63–66.

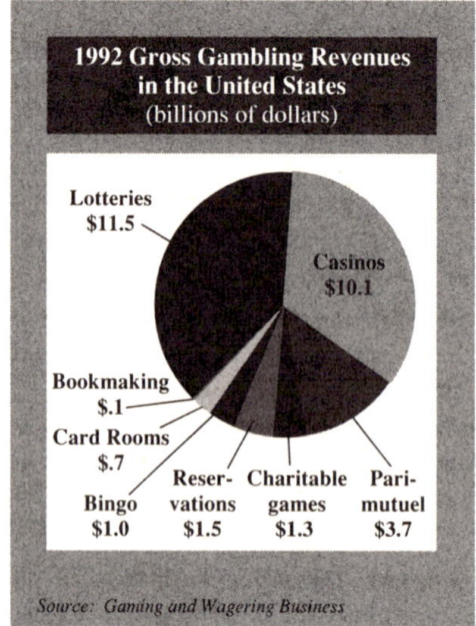

63. In 1992, the gaming industry had $29.9 billion in gross revenues. What percent of this was from lotteries? **38.5%**

64. Suppose that a 4% tax had been levied on gaming revenues in 1992. How much tax would have been collected from casinos? **$404 million**

65. Suppose that casinos paid out 83% of their gross revenues. How much would they have paid out in 1992? **$8.38 billion**

66. How much more in gross revenues did lotteries take in than bookmaking, card rooms, bingo, charitable games, reservations, and pari-mutuel combined? **$3.2 billion**

67. Which type of graph studied in this section would be most suitable to illustrate the percentages of different races in a group of 1,000 people? **pie chart**

68. Why would a pie chart not be suitable to depict the trends in price fluctuations in gasoline over a five-year period?

⊙ CONCEPTUAL  WRITING ▲ CHALLENGING SCIENTIFIC CALCULATOR GRAPHING CALCULATOR

The map shows the predicted weather information for a summer day in July. Use the map and the accompanying legend to answer the following questions.

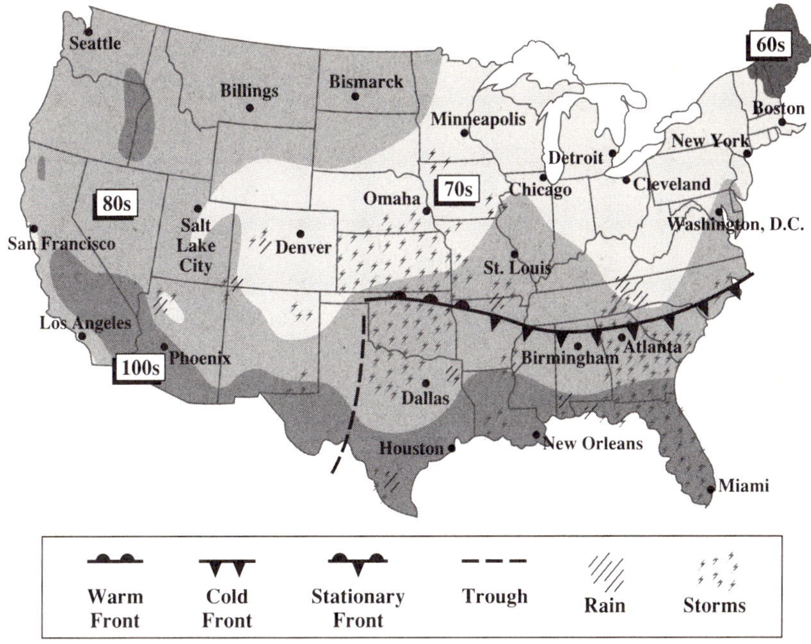

Source: Weather Data Incorporated

69. Of the cities Houston, Minneapolis, St. Louis, and Omaha, which one has a predicted temperature in the 80s? **St. Louis**

70. Of the cities Billings, Detroit, Miami, and Washington, D.C., which one has a prediction of storms? **Miami**

71. What kind of front is just north of Birmingham and Atlanta? **cold front**

72. In what state is a trough located? **Texas**

Chapter 1 Test (Text Page 49)

Decide whether each of Exercises 1 and 2 is an example of inductive or deductive reasoning.

1. Angelica Canales is a salesperson for a major publishing company. She has met her annual sales goal in each of the past four years. Therefore, she will reach her goal this year. **inductive**

2. If you square a positive number, you will get a positive number. The number 5 is positive. Therefore, 5^2 is a positive number. **deductive**

3. What is the most probable next number in the list 1, 1/3, 1/5, 1/7, . . . ? **1/9**

4. Use the list of equations and inductive reasoning to predict the next equation, and then verify your conjecture. **65,359,477,124,183 × 68 = 4,444,444,444,444,444**

$$65{,}359{,}477{,}124{,}183 \times 17 = 1{,}111{,}111{,}111{,}111{,}111$$
$$65{,}359{,}477{,}124{,}183 \times 34 = 2{,}222{,}222{,}222{,}222{,}222$$
$$65{,}359{,}477{,}124{,}183 \times 51 = 3{,}333{,}333{,}333{,}333{,}333$$

5. Use the method of successive differences to find the next term in the sequence

3, 11, 31, 69, 131, 223 . . . . **351**

6. Find the sum $1 + 2 + 3 + \cdots + 1{,}000$. **500,500**

7. Consider the following equations, where the left side of each is an octagonal number.

$$1 = 1$$
$$8 = 1 + 7$$
$$21 = 1 + 7 + 13$$
$$40 = 1 + 7 + 13 + 19$$

Use the pattern established on the right sides to predict the next octagonal number. What is the next equation in the list?
65; $65 = 1 + 7 + 13 + 19 + 25$

8. Use the result of Exercise 7 and the method of successive differences to find the first eight octagonal numbers. Then divide each by 4 and record the remainder. What is the pattern obtained?
1, 8, 21, 40, 65, 96, 133, 176; The pattern is 1, 0, 1, 0, 1, 0, 1, 0, ...

9. Explain the pattern used to obtain the terms of the Fibonacci sequence 1, 1, 2, 3, 5, 8, 13, 21, ...

10. What is the units digit in 2^{35}? **8**

11. If fence posts are placed 6 feet apart, how many posts are necessary to construct a 60-foot fence (with no gates)? **11**

12. How many rectangles are in this figure? **9**

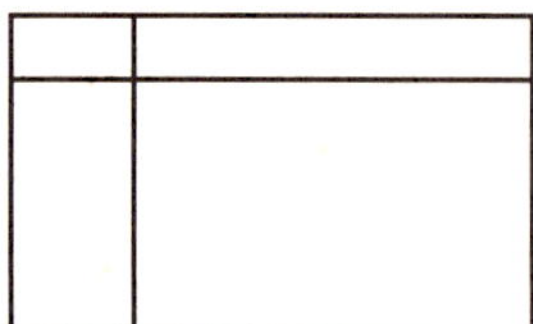

13. I am thinking of a number. If I double it, add 6 to the result, triple that result and then subtract 4, the final result is 50. What is my number? **6**

14. Which is correct? Three cubed *is* nine or three cubed *are* nine? **Neither is correct, since $3^3 = 27$.**

15. What is the 103rd digit in the decimal representation for $1/11$? **0**

16. Based on your knowledge of elementary arithmetic, explain the pattern that can be observed when the following operations are performed: $9 \times 1, 9 \times 2, 9 \times 3, \ldots, 9 \times 9$. (*Hint:* Add the digits in the answers. What do you notice?) **The sum of the digits is always 9.**

Use your calculator to evaluate each of the following. Give as many decimal places as the calculator displays.

17. $\sqrt{17.4}$ **4.171330723**

18. 1.5^3 **3.375**

19. In 1989 Blaise Bryant, a running back for the Iowa State football team, carried the ball 299 times and scored 19 touchdowns. This means that he scored a touchdown once in about every _____ times he carried the ball. **(b)**

(a) 10 **(b)** 15 **(c)** 20 **(d)** 25

20. The line graph depicts how the average National Football League player's salary increased from 1987 to 1993.

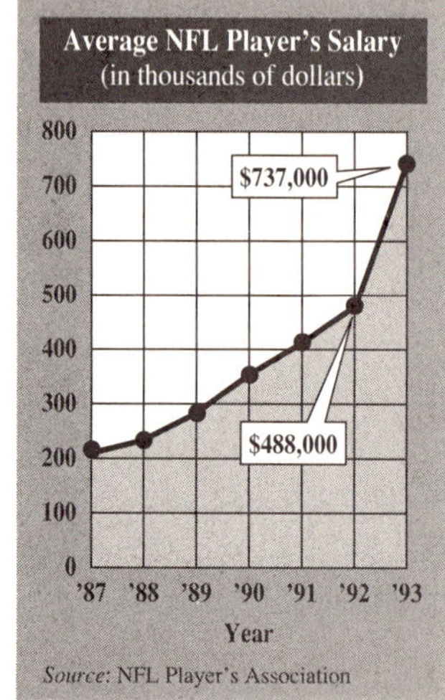

(a) By how much did the salary increase from 1992 to 1993? **$249,000**

(b) Determine whether the following statement is true or false: The average salary in 1993 was less than three times the average salary in 1987. **true**

CONCEPTUAL WRITING CHALLENGING SCIENTIFIC CALCULATOR GRAPHING CALCULATOR

CHAPTER 2 Sets

2.1 Exercises (Text Page 56)

List all the elements of each set.

1. the set of all counting numbers less than or equal to 4 **{1, 2, 3, 4}**
2. the set of all whole numbers greater than 8 and less than 16 **{9, 10, 11, 12, 13, 14, 15}**
3. the set of all whole numbers not greater than 6 **{0, 1, 2, 3, 4, 5, 6}**
4. the set of all counting numbers between 2 and 12 **{3, 4, 5, 6, 7, 8, 9, 10, 11}**
5. {6, 7, 8, . . . , 14} **{6, 7, 8, 9, 10, 11, 12, 13, 14}**
6. {3, 6, 9, 12, . . . , 30} **{3, 6, 9, 12, 15, 18, 21, 24, 27, 30}**
7. {−15, −13, −11, . . . , −1} **{−15, −13, −11, −9, −7, −5, −3, −1}**
8. {−4, −3, −2, . . . , 4} **{−4, −3, −2, −1, 0, 1, 2, 3, 4}**
9. {2, 4, 8, . . . , 256} **{2, 4, 8, 16, 32, 64, 128, 256}**
10. {90, 87, 84, . . . , 54} **{90, 87, 84, 81, 78, 75, 72, 69, 66, 63, 60, 57, 54}**
11. {1, 1/3, 1/9, . . . , 1/243} **{1, 1/3, 1/9, 1/27, 1/81, 1/243}**
12. {1/2, 1/4, 1/6, . . . , 1/20} **{1/2, 1/4, 1/6, 1/8, 1/10, 1/12, 1/14, 1/16, 1/18, 1/20}**
13. $\{x \mid x$ is an even whole number less than 15} **{0, 2, 4, 6, 8, 10, 12, 14}**
14. $\{x \mid x$ is an odd integer between −8 and 5} **{−7, −5, −3, −1, 1, 3}**

Denote each set by the listing method. There may be more than one correct answer.

15. the set of all counting numbers greater than 20 **{21, 22, 23, . . . }**
16. the set of all integers between −200 and 500 **{−199, −198, −197, . . . , 499}**
17. the set of traditional major political parties in the United States **{Democrat, Republican}**
18. the set of all persons living on February 1, 1992 who had been President of the United States. (See the photograph.) **{Bush, Reagan, Carter, Ford, Nixon}**

19. $\{x \mid x$ is a positive multiple of 4} **{4, 8, 12, . . .}**
20. $\{x \mid x$ is a negative multiple of 7} **{−7, −14, −21, . . . }**
21. $\{x \mid x$ is the reciprocal of a natural number} **{1, 1/2, 1/3, 1/4, . . . }**
22. $\{x \mid x$ is a positive integer power of 3} **{3, 9, 27, 81, . . . }**

Denote each set by set-builder notation, using x as the variable. There may be more than one correct answer. **In Exercises 23–28, there may be other acceptable descriptions.**

23. the set of all rational numbers **$\{x \mid x$ is a rational number}**
24. the set of all even counting numbers **$\{x \mid x$ is an even counting number}**

25. the set of all movies released this year $\{x \mid x$ **is a movie released this year**$\}$

26. the set of all multinational corporations $\{x \mid x$ **is a multinational corporation**$\}$

27. $\{1, 3, 5, \ldots, 99\}$ $\{x \mid x$ **is an odd counting number less than 100**$\}$

28. $\{35, 40, 45, \ldots, 995\}$ $\{x \mid x$ **is a multiple of 5 between 30 and 1,000**$\}$

Identify each set as finite *or* infinite.

29. $\{2, 4, 6, \ldots, 28\}$ **finite**

30. $\{6, 12, 18, \ldots\}$ **infinite**

31. $\{1/2, 2/3, 3/4, \ldots, 99/100\}$ **finite**

32. $\{-10, -8, -6, \ldots, 0\}$ **finite**

33. $\{x \mid x$ is a counting number greater than $30\}$ **infinite**

34. $\{x \mid x$ is a counting number less than $30\}$ **finite**

35. $\{x \mid x$ is a rational number$\}$ **infinite**

36. $\{x \mid x$ is a rational number between 0 and $1\}$ **infinite**

Find $n(A)$ for each set.

37. $A = \{0, 1, 2, 3, 4, 5, 6\}$ **7**

38. $A = \{-3, -2, -1, 0, 1, 2\}$ **6**

39. $A = \{2, 4, 6, \ldots, 1{,}000\}$ **500**

40. $A = \{0, 1, 2, 3, \ldots, 3{,}000\}$ **3,001**

41. $A = \{a, b, c, \ldots, z\}$ **26**

42. $A = \{x \mid x$ is a vowel in the English alphabet$\}$ **5**

43. $A =$ the set of integers between -10 and 10 **19**

44. $A =$ the set of current U.S. senators **100**

45. $A = \{1/3, 2/4, 3/5, 4/6, \ldots, 27/29, 28/30\}$ **28**

46. $A = \{1/2, -1/2, 1/3, -1/3, \ldots, 1/10, -1/10\}$ **18**

Identify each set as well defined *or* not well defined.

47. $\{x \mid x$ is a real number$\}$ **well defined**

48. $\{x \mid x$ is a negative number$\}$ **well defined**

49. $\{x \mid x$ is a good singer$\}$ **not well defined**

50. $\{x \mid x$ is a skillful actor$\}$ **not well defined**

51. $\{x \mid x$ is a difficult class$\}$ **not well defined**

52. $\{x \mid x$ is a counting number less than $1\}$ **well defined**

Fill each blank with either $\in$ or $\notin$ to make the following statements true.

53. 5 ____ $\{2, 4, 5, 6\}$ $\in$

54. 8 ____ $\{3, -2, 5, 9, 8\}$ $\in$

55. -4 ____ $\{4, 7, 9, 12\}$ $\notin$

56. -12 ____ $\{3, 8, 12, 16\}$ $\notin$

57. 0 ____ $\{-2, 0, 5, 7\}$ $\in$

58. 0 ____ $\{3, 4, 7, 8, 10\}$ $\notin$

▲ 59. $\{3\}$ ____ $\{2, 3, 4, 5\}$ $\notin$

▲ 60. $\{5\}$ ____ $\{3, 4, 5, 6, 7\}$ $\notin$

Write true *or* false *for each of the following statements.*

61. $3 \in \{2, 5, 6, 8\}$ **false**

62. $6 \in \{-2, 5, 8, 9\}$ **false**

63. $b \in \{h, c, d, a, b\}$ **true**

64. $m \in \{1, m, n, o, p\}$ **true**

65. $9 \notin \{6, 3, 4, 8\}$ **true**

66. $2 \notin \{7, 6, 5, 4\}$ **true**

67. $\{k, c, r, a\} = \{k, c, a, r\}$ **true**

68. $\{e, h, a, n\} = \{a, h, e, n\}$ **true**

69. $\{5, 8, 9\} = \{5, 8, 9, 0\}$ **false**

70. $\{3, 7, 12, 14\} = \{3, 7, 12, 14, 0\}$ **false**

71. $\{d, x, m, x, d\} = \{m, d, x\}$ **true**

72. $\{u, v, u, v\} = \{u, v\}$ **true**

73. $\{x \mid x$ is a counting number less than $3\} = \{1, 2\}$ **true**

74. $\{x \mid x$ is a counting number greater than $10\} = \{11, 12, 13, \ldots\}$ **true**

◉ CONCEPTUAL ✎ WRITING ▲ CHALLENGING ▦ SCIENTIFIC CALCULATOR ▦ GRAPHING CALCULATOR

Write true *or* false *for each of the following statements.*

Let $A = \{2, 4, 6, 8, 10, 12\}$
$B = \{2, 4, 8, 10\}$
$C = \{4, 10, 12\}$.

75. $4 \in A$ **true**

76. $8 \in B$ **true**

77. $4 \notin C$ **false**

78. $8 \notin B$ **false**

79. $10 \notin A$ **false**

80. $6 \notin A$ **false**

81. Every element of C is also an element of A. **true**

82. Every element of C is also an element of B. **false**

83. This chapter opened with the statement, "The human mind likes to create collections." Why do you suppose this is so? In explaining your thoughts, utilize one or more particular "collections," mathematical or otherwise.

84. Explain the difference between a well-defined set and a not well-defined set. Give examples and utilize terms introduced in this section.

Recall that two sets are called equal *if they contain identical elements. On the other hand, two sets are called* equivalent *if they contain the same number of elements (but not necessarily the same elements). For each of the following conditions, give an example or explain why it is impossible.*

85. two sets that are neither equal nor equivalent
{2} and {3, 4} (Other examples are possible.)

86. two sets that are equal but not equivalent
Impossible. If they are equal, then they have the same number of elements.

87. two sets that are equivalent but not equal **$\{a, b\}$ and $\{a, c\}$ (Other examples are possible.)**

88. two sets that are both equal and equivalent **{8} and {5 + 3} (Other examples are possible.)**

89. Michelle Perulli is health conscious, but she does like a certain chocolate bar, each of which contains 220 calories. In order to burn off unwanted calories, Michelle participates in her favorite activities, shown below, in increments of one hour and never repeats a given activity on a given day.

Activity	Symbol	Calories Burned per Hour
Volleyball	v	160
Golf	g	260
Canoeing	c	340
Swimming	s	410
Running	r	680

(a) On Monday, Michelle has time for no more than two hours of activities. List all possible sets of activities that would burn off at least the number of calories obtained from three chocolate bars.
$\{r\}, \{g, s\}, \{c, s\}, \{v, r\}, \{g, r\}, \{c, r\}, \{s, r\}$

(b) Assume that Michelle can afford up to three hours of time for activities on Saturday. List all sets of activities that would burn off at least the number of calories in five chocolate bars.
$\{v, g, r\}, \{v, c, r\}, \{v, s, r\}, \{g, c, r\}, \{g, s, r\}, \{c, s, r\}$

90. The table below categorizes municipal solid waste generated in the United States.

Category	Symbol	Percentage
Food wastes	F	7
Yard wastes	Y	18
Metals	M	8
Glass	G	7
Plastics	L	8
Paper	P	40
Rubber, leather, textile, wood, other	R	12

Using the given symbols, list the elements of the following sets.

(a) $\{x \mid x$ is a category accounting for more than 15% of the wastes$\}$ $\{Y, P\}$

(b) $\{x \mid x$ is a category accounting for less than 10% of the wastes$\}$ $\{F, M, G, L\}$

2.2 Exercises (Text Page 64)

Insert $\subseteq$ or $\not\subseteq$ in each blank so that the resulting statement is true.

1. $\{-2, 0, 2\}$ ____ $\{-2, -1, 1, 2\}$ $\not\subseteq$

2. {Monday, Wednesday, Friday} ____ {Sunday, Monday, Tuesday, Wednesday, Thursday} $\not\subseteq$

3. $\{2, 5\}$ ____ $\{0, 1, 5, 3, 4, 2\}$ $\subseteq$

4. {a, n, d} ____ {r, a, n, d, y} $\subseteq$

5. $\emptyset$ ____ {a, b, c, d, e} $\subseteq$

6. $\emptyset$ ____ $\emptyset$ $\subseteq$

7. $\{-7, 4, 9\}$ ____ $\{x \mid x$ is an odd integer$\}$ $\not\subseteq$

8. $\{2, 1/3, 5/9\}$ ____ the set of rational numbers $\subseteq$

Decide whether $\subset$, or $\subseteq$, or both, *or* neither *can be placed in the blank to make the statement true.*

9. {B, C, D} ____ {B, C, D, F} **both**

10. {red, green, blue, yellow} ____ {green, yellow, blue, red} $\subseteq$

11. $\{9, 1, 7, 3, 5\}$ ____ $\{1, 3, 5, 7, 9\}$ $\subseteq$

12. {S, M, T, W, Th} ____ {M, W, Th, S} **neither**

13. $\emptyset$ ____ $\{0\}$ **both**

14. $\emptyset$ ____ $\emptyset$ $\subseteq$

15. $\{-1, 0, 1, 2, 3\}$ ____ $\{0, 1, 2, 3, 4\}$ **neither**

16. $\{5/6, 9/8\}$ ____ $\{6/5, 8/9\}$ **neither**

For Exercises 17–38, tell whether each statement is true *or* false.

$$\text{Let } U = \{a, b, c, d, e, f, g\}$$
$$A = \{a, e\}$$
$$B = \{a, b, e, f, g\}$$
$$C = \{b, f, g\}$$
$$D = \{d, e\}.$$

17. $A \subset U$ **true**

18. $C \subset U$ **true**

19. $D \subseteq B$ **false**

20. $D \subseteq A$ **false**

21. $A \subset B$ **true**

22. $B \subseteq C$ **false**

23. $\emptyset \subset A$ **true**

24. $\emptyset \subseteq D$ **true**

25. $\emptyset \subseteq \emptyset$ **true**

26. $D \subset B$ **false**

27. $\{g, f, b\} \subset B$ **true**

28. $\{0\} \subset D$ **false**

29. $D \not\subseteq B$ **true**

30. $A \not\subseteq B$ **false**

31. There are exactly 6 subsets of C. **false**

32. There are exactly 31 subsets of B. **false**

33. There are exactly 3 subsets of A. **false**

34. There are exactly 4 subsets of D. **true**

35. There is exactly 1 subset of $\emptyset$. **true**

36. There are exactly 127 proper subsets of U. **true**

37. The drawing below correctly represents the relationship among sets A, C, and U. **false**

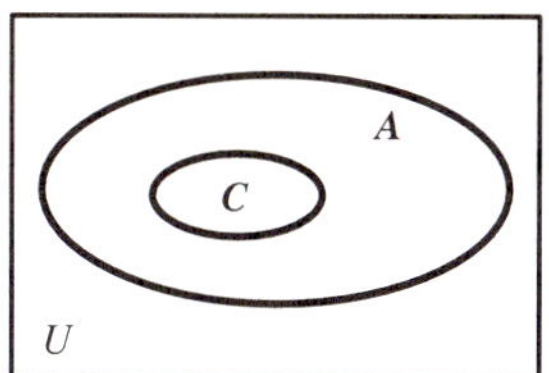

38. The drawing below correctly represents the relationship among sets B, C, and U. **true**

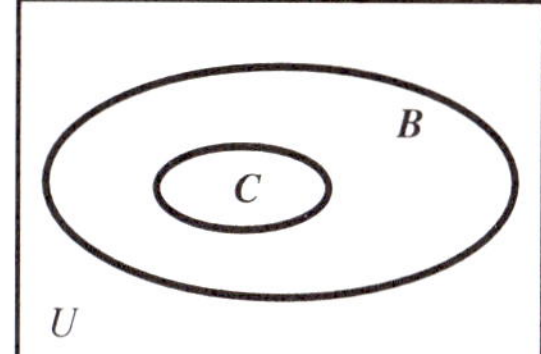

Find the number of subsets and the number of proper subsets of each of the following sets.

39. {1, 3, 5} **8; 7**

40. {−8, −6, −4, −2} **16; 15**

41. {a, b, c, d, e, f} **64; 63**

42. the set of days of the week **128; 127**

43. $\{x \mid x$ is an odd integer between -4 and $6\}$ **32; 31**

44. $\{x \mid x$ is an even whole number less than $3\}$ **4; 3**

Let $U = \{1, 2, 3, 4, 5, 6, 7, 8, 9, 10\}$ *and find the complement of each of the following sets.*

45. {1, 4, 6, 8} **{2, 3, 5, 7, 9, 10}**

46. {2, 5, 7, 9, 10} **{1, 3, 4, 6, 8}**

47. {1, 3, 4, 5, 6, 7, 8, 9, 10} **{2}**

48. {3, 5, 7, 9} **{1, 2, 4, 6, 8, 10}**

49. $\emptyset$ **{1, 2, 3, 4, 5, 6, 7, 8, 9, 10}**

50. U **$\emptyset$**

51. Greg Odjakjian and his family, wishing to see a movie this evening, have made up the following lists of characteristics for their two main options.

Go to a Movie Theater	Rent a Home Video
High cost	Low cost
Entertaining	Entertaining
Fixed schedule	Flexible schedule
Current films	Older films

Find the smallest universal set U that contains all listed characteristics of both options.
{High cost, Entertaining, Fixed schedule, Current films, Low cost, Flexible schedule, Older films}

Let T represent the set of characteristics of the movie theater option, and let V represent the set of characteristics of the home video option. Using the universal set from Exercise 51, find each of the following.

52. T' **{Low cost, Flexible schedule, Older films}**

53. V' **{High cost, Fixed schedule, Current films}**

Find the set of elements common to both of the sets in Exercises 54–57, where T and V are defined as above.

54. T and V **{Entertaining}**

55. T' and V **{Low cost, Flexible schedule, Older films}**

56. T and V' **{High cost, Fixed schedule, Current films}**

57. T' and V' **$\emptyset$**

Anita Bartelson, Brett Sullivan, Curt Reynolds, Duncan MacKinnon, and Emma Price plan to meet in the hospitality suite at a sales convention to compare notes. Denoting these five people by A, B, C, D, and E, list all the possible ways that the given number of them can gather in the suite.

58. 5 **one way (all five present): *ABCDE***

59. 4 ***ABCD, ABCE, ABDE, ACDE, BCDE***

60. 3 ***ABC, ABD, ABE, ACD, ACE, ADE, BCD, BCE, BDE, CDE***

61. 2 ***AB, AC, AD, AE, BC, BD, BE, CD, CE, DE***

62. 1 ***A, B, C, D, E***

63. 0 **one way (none present)**

64. Find the total number of ways that members of this group can gather in the suite. (*Hint:* Add your answers to Exercises 58–63.)
1 + 5 + 10 + 10 + 5 + 1 = 32

▲◉ **65.** How does your answer in Exercise 64 compare with the number of subsets of a set of 5 elements? How can you interpret the answer to Exercise 64 in terms of subsets? **They are the same: $32 = 2^5$. The number of ways that people from a group of five can gather is the same as the number of subsets there are of a set of five elements.**

▲ **66.** In discovering the formula (2^n) for the number of subsets of a set with n elements, we observed that for the first few values of n, increasing the number of elements by one doubles the number of subsets. Here, you can prove the formula in general by showing that the same is true for any value of n. Assume set A has n elements and s subsets. Now add one additional element, say e, to the set A. (We now have a new set, say B, with $n + 1$ elements.) Divide the subsets of B into those that do not contain e and those that do.

(a) How many subsets of B do not contain e? (*Hint:* Each of these is a subset of the original set A.) s

(b) How many subsets of B do contain e? (*Hint:* Each of these would be a subset of the original set A, with the element e thrown in.) s

(c) What is the total number of subsets of B? $2s$

◉ **(d)** What do you conclude? **Adding one more element will always double the number of subsets, so the formula 2^n is true in general.**

◉ *Suppose you have available the bills shown here.*

67. If you must select at least one bill, and you may select up to all of the bills, how many different sums of money could you make? **15**

68. In Exercise 67, remove the condition "you must select at least one bill." Now, how many sums are possible? **16, since it is now possible to select *no* bills.**

2.3 Exercises (Text Page 76)

Perform the indicated operations.

Let $U = \{a, b, c, d, e, f, g\}$
$X = \{a, c, e, g\}$
$Y = \{a, b, c\}$
$Z = \{b, c, d, e, f\}$.

1. $X \cap Y$ **{a, c}**
2. $X \cup Y$ **{a, b, c, e, g}**
3. $Y \cup Z$ **{a, b, c, d, e, f}**
4. $Y \cap Z$ **{b, c}**
5. $X \cup U$ **{a, b, c, d, e, f, g}**
6. $Y \cap U$ **{a, b, c}**
7. X' **{b, d, f}**
8. Y' **{d, e, f, g}**
9. $X' \cap Y'$ **{d, f}**
10. $X' \cap Z$ **{b, d, f}**
11. $Z' \cap \emptyset$ $\boldsymbol{\emptyset}$
12. $Y' \cup \emptyset$ **{d, e, f, g}**
13. $X \cup (Y \cap Z)$ **{a, b, c, e, g}**

14. $Y \cap (X \cup Z)$ {a, b, c}
15. $(Y \cap Z') \cup X$ {a, c, e, g}
16. $(X' \cup Y') \cup Z$ {b, c, d, e, f, g}
17. $(Z \cup X')' \cap Y$ {a}
18. $(Y \cap X')' \cup Z'$ {a, c, d, e, f, g}
19. $X - Y$ {e, g}
20. $Y - X$ {b}
21. $X' - Y$ {d, f}
22. $Y' - X$ {d, f}
23. $X \cap (X - Y)$ {e, g}
24. $Y \cup (Y - X)$ {a, b, c}

Describe each set in words. **In Exercises 25–30, there may be other acceptable descriptions.**

25. $A \cup (B' \cap C')$ **the set of all elements that either are in A, or are not in B and not in C**

26. $(A \cap B') \cup (B \cap A')$ **the set of all elements that are in A and not in B, or are in B and not in A**

27. $(C - B) \cup A$ **the set of all elements that are in C but not in B, or are in A**

28. $B \cap (A' - C)$ **the set of all elements that are in B and are not in A and not in C**

29. $(A - C) \cup (B - C)$ **the set of all elements that are in A but not in C, or in B but not in C**

30. $(A' \cap B') \cup C'$ **the set of all elements that are not in A and not in B, or are not in C**

31. The table lists some common adverse effects of prolonged tobacco and alcohol use.

Tobacco	**Alcohol**
Emphysema, e	Liver damage, l
Heart damage, h	Brain damage, b
Cancer, c	Heart damage, h

Find the smallest possible universal set U that includes all the effects listed.
$U = \{e, h, c, l, b\}$

Let T be the set of listed effects of tobacco and A be the set of listed effects of alcohol. (See Exercise 31.) Find each set.

32. A' $\{e, c\}$
33. T' $\{l, b\}$
34. $T \cap A$ $\{h\}$
35. $T \cup A$ $\{e, h, c, l, b\}$
36. $A \cap T'$ $\{l, b\}$

Describe in words each set in Exercises 37–42.

Let U = the set of all tax returns
A = the set of all tax returns with itemized deductions
B = the set of all tax returns showing business income
C = the set of all tax returns filed in 1996
D = the set of all tax returns selected for audit.

37. $B \cup C$ **the set of all tax returns showing business income or filed in 1996**

38. $A \cap D$ **the set of all tax returns with itemized deductions and selected for audit**

39. $C - A$ **the set of all tax returns filed in 1996 without itemized deductions**

40. $D \cup A'$ **the set of all tax returns selected for audit or without itemized deductions**

41. $(A \cup B) - D$ **the set of all tax returns with itemized deductions or showing business income, but not selected for audit**

42. $(C \cap A) \cap B'$ **the set of all tax returns filed in 1996 and with itemized deductions but not showing business income**

◎ *Assuming that A and B represent* any two sets, *identify each of the following statements as either* always true *or* not always true.

43. $A \subseteq (A \cup B)$ **always true**

44. $A \subseteq (A \cap B)$ **not always true**

45. $(A \cap B) \subseteq A$ **always true**

46. $(A \cup B) \subseteq A$ **not always true**

47. $n(A \cup B) = n(A) + n(B)$ **not always true**

48. $n(A \cap B) = n(A) - n(B)$ **not always true**

49. $n(A \cup B) = n(A) + n(B) - n(A \cap B)$ **always true**

50. $n(A \cap B) = n(A) + n(B) - n(A \cup B)$ **always true**

51. If $B \subseteq A$, $n(A) - n(B) = n(A - B)$ **always true**

52. $n(A - B) = n(B - A)$ **not always true**

For Exercises 53–60,

$$\text{Let } U = \{1, 2, 3, 4, 5\}$$
$$X = \{1, 3, 5\}$$
$$Y = \{1, 2, 3\}$$
$$Z = \{3, 4, 5\}.$$

◎ *In each case, state a general conjecture based on your observation.*

53. **(a)** Find $X \cup Y$. **{1, 3, 5, 2}** **(b)** Find $Y \cup X$. **{1, 2, 3, 5}**
(c) State a conjecture. **For any sets X and Y, $X \cup Y = Y \cup X$.**

54. **(a)** Find $X \cap Y$. **{1, 3}** **(b)** Find $Y \cap X$. **{1, 3}**
(c) State a conjecture. **For any sets X and Y, $X \cap Y = Y \cap X$.**

55. **(a)** Find $X \cup (Y \cup Z)$. **{1, 3, 5, 2, 4}** **(b)** Find $(X \cup Y) \cup Z$. **{1, 3, 5, 2, 4}**
(c) State a conjecture. **For any sets X, Y, and Z, $X \cup (Y \cup Z) = (X \cup Y) \cup Z$.**

56. **(a)** Find $X \cap (Y \cap Z)$. **{3}** **(b)** Find $(X \cap Y) \cap Z$. **{3}**
(c) State a conjecture. **For any sets X, Y, and Z, $X \cap (Y \cap Z) = (X \cap Y) \cap Z$.**

57. **(a)** Find $(X \cup Y)'$. **{4}** **(b)** Find $X' \cap Y'$. **{4}**
(c) State a conjecture. **For any sets X and Y, $(X \cup Y)' = X' \cap Y'$.**

58. **(a)** Find $(X \cap Y)'$. **{2, 4, 5}** **(b)** Find $X' \cup Y'$. **{2, 4, 5}**
(c) State a conjecture. **For any sets X and Y, $(X \cap Y)' = X' \cup Y'$.**

59. **(a)** Find $X \cup \emptyset$. **{1, 3, 5}** **(b)** State a conjecture. **For any set X, $X \cup \emptyset = X$.**

60. **(a)** Find $X \cap \emptyset$. **$\emptyset$** **(b)** State a conjecture. **For any set X, $X \cap \emptyset = \emptyset$.**

Write true *or* false *for each of the following.*

61. $(3, 2) = (5 - 2, 1 + 1)$ **true**

62. $(10, 4) = (7 + 3, 5 - 1)$ **true**

63. $(4, 12) = (4, 3)$ **false**

64. $(5, 9) = (2, 9)$ **false**

65. $(6, 3) = (3, 6)$ **false**

66. $(2, 13) = (13, 2)$ **false**

67. $\{6, 3\} = \{3, 6\}$ **true**

68. $\{2, 13\} = \{13, 2\}$ **true**

69. $\{(1, 2), (3, 4)\} = \{(3, 4), (1, 2)\}$ **true**

70. $\{(5, 9), (4, 8), (4, 2)\} = \{(4, 8), (5, 9), (4, 2)\}$ **true**

Find $A \times B$ and $B \times A$, for A and B defined as follows.

71. $A = \{2, 8, 12\}$, $B = \{4, 9\}$ **$A \times B = \{(2, 4), (2, 9), (8, 4), (8, 9), (12, 4), (12, 9)\}$;**
$B \times A = \{(4, 2), (4, 8), (4, 12), (9, 2), (9, 8), (9, 12)\}$

72. $A = \{3, 6, 9, 12\}$, $B = \{6, 8\}$
$A \times B = \{(3, 6), (3, 8), (6, 6), (6, 8), (9, 6), (9, 8), (12, 6), (12, 8)\}$;
$B \times A = \{(6, 3), (6, 6), (6, 9), (6, 12), (8, 3), (8, 6), (8, 9), (8, 12)\}$

◎ CONCEPTUAL WRITING ▲ CHALLENGING SCIENTIFIC CALCULATOR GRAPHING CALCULATOR

73. $A = \{d, o, g\}$, $\quad B = \{p, i, g\}$
$\mathbf{A \times B = \{(d, p), (d, i), (d, g), (o, p), (o, i), (o, g), (g, p), (g, i), (g, g)\}}$;
$\mathbf{B \times A = \{(p, d), (p, o), (p, g), (i, d), (i, o), (i, g), (g, d), (g, o), (g, g)\}}$

74. $A = \{b, l, u, e\}$, $\quad B = \{r, e, d\}$
$\mathbf{A \times B = \{(b, r), (b, e), (b, d), (l, r), (l, e), (l, d), (u, r), (u, e), (u, d), (e, r), (e, e), (e, d)\}}$;
$\mathbf{B \times A = \{(r, b), (r, l), (r, u), (r, e), (e, b), (e, l), (e, u), (e, e), (d, b), (d, l), (d, u), (d, e)\}}$

Use the given information to find $n(A \times B)$ and $n(B \times A)$ in Exercises 75–78.

75. the sets in Exercise 71 $\quad \mathbf{n(A \times B) = 6; \quad n(B \times A) = 6}$

76. the sets in Exercise 73 $\quad \mathbf{n(A \times B) = 9; \quad n(B \times A) = 9}$

77. $n(A) = 35$ and $n(B) = 6$ $\quad \mathbf{n(A \times B) = 210; \quad n(B \times A) = 210}$

78. $n(A) = 13$ and $n(B) = 5$ $\quad \mathbf{n(A \times B) = 65; \quad n(B \times A) = 65}$

79. If $n(A \times B) = 36$ and $n(A) = 12$, find $n(B)$. **3**

80. If $n(A \times B) = 100$ and $n(B) = 4$, find $n(A)$. **25**

Place the elements of these sets in the proper location on the given Venn diagram.

81. Let $U = \{a, b, c, d, e, f, g\}$
$A = \{b, d, f, g\}$
$B = \{a, b, d, e, g\}$.

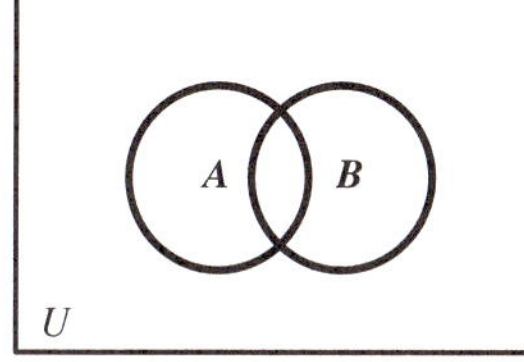

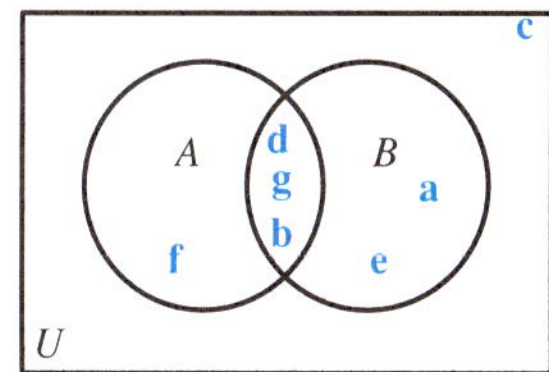

82. Let
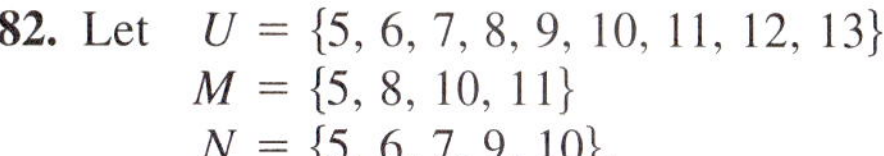
$U = \{5, 6, 7, 8, 9, 10, 11, 12, 13\}$
$M = \{5, 8, 10, 11\}$
$N = \{5, 6, 7, 9, 10\}$.

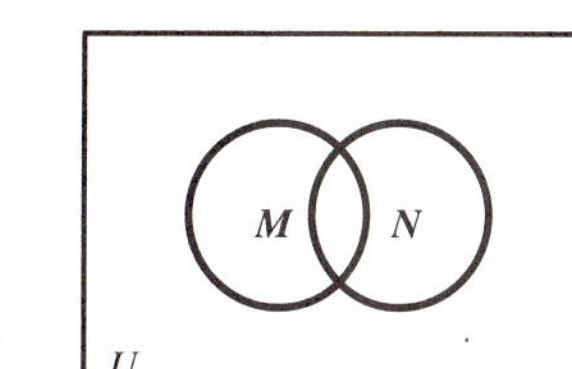

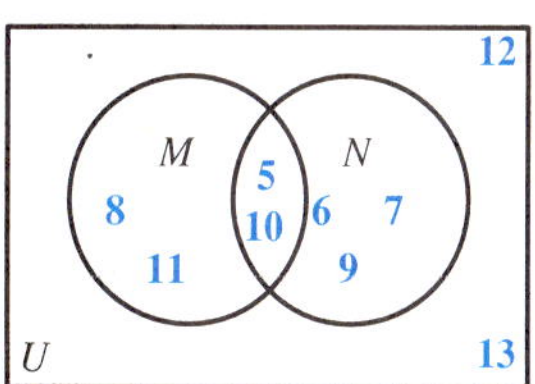

Use a Venn diagram similar to the one shown here to shade each of the following sets.

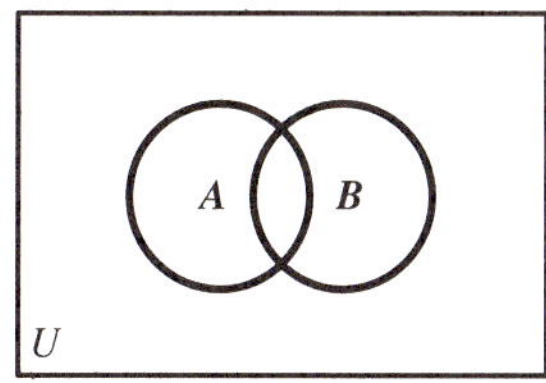

83. $B \cap A'$

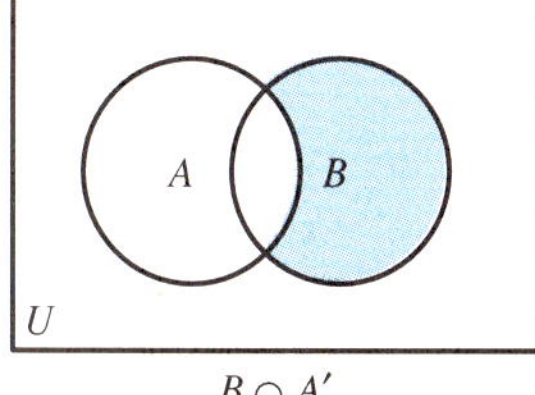

$B \cap A'$

84. $A \cup B$

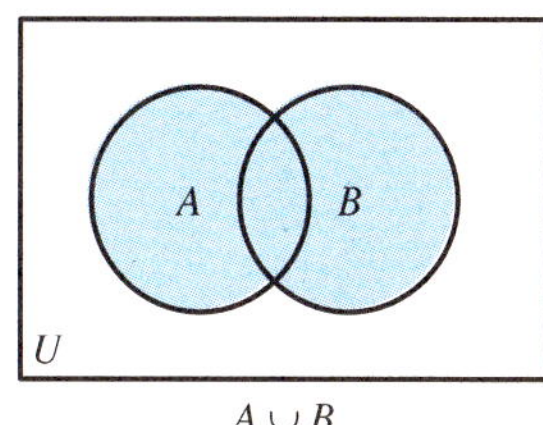

$A \cup B$

85. $A' \cup B$

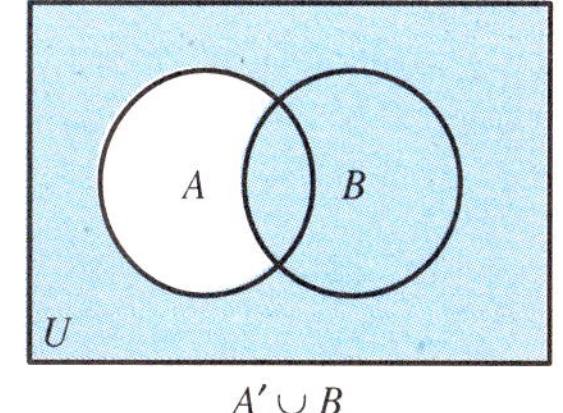

$A' \cup B$

86. $A' \cap B'$

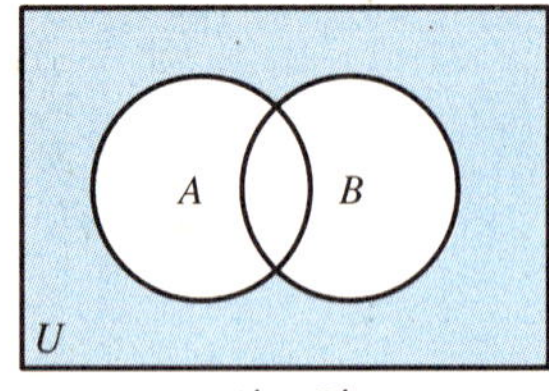
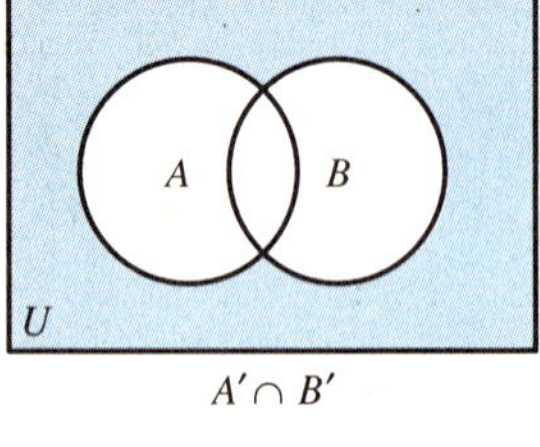

$A' \cap B'$

87. $B' \cup A$

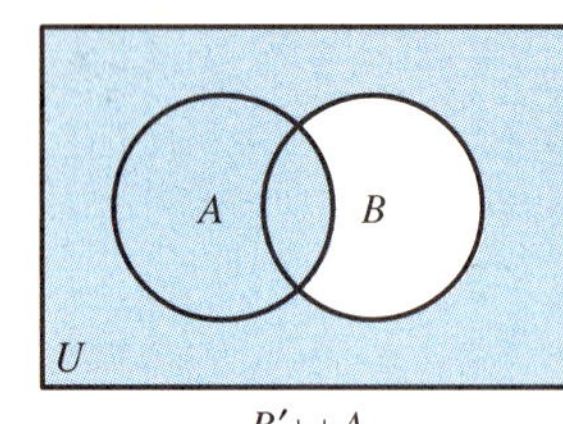

$B' \cup A$

88. $A' \cup A$

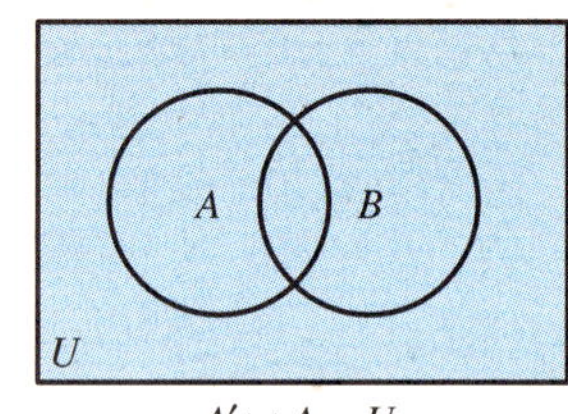

$A' \cup A = U$

89. $B' \cap B$

$B' \cap B = \emptyset$

90. $A \cap B'$

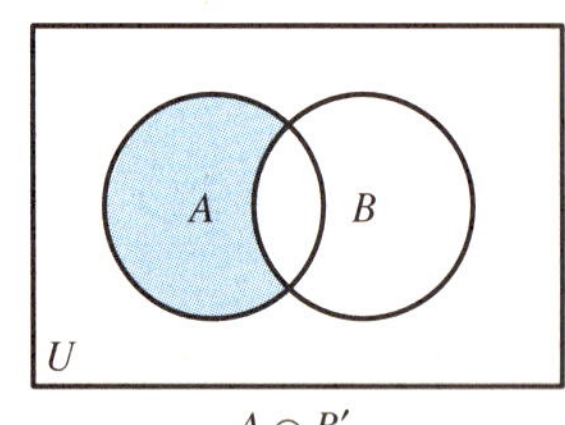

$A \cap B'$

91. $B' \cup (A' \cap B')$

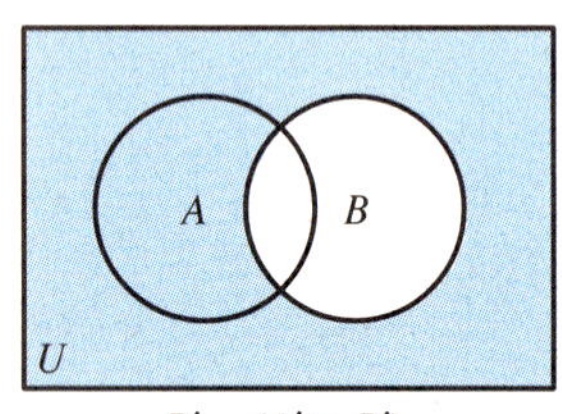

$B' \cup (A' \cap B')$

92. $(A \cap B) \cup B$

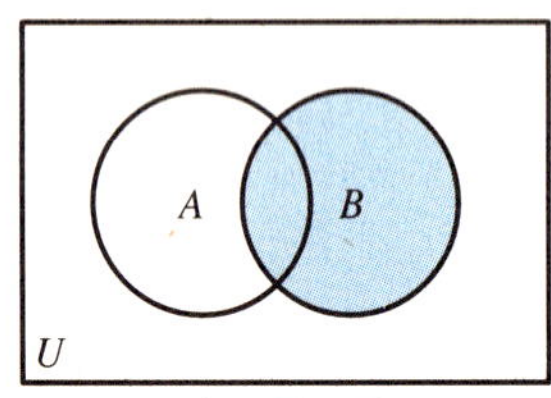

$(A \cap B) \cup B$

93. U'

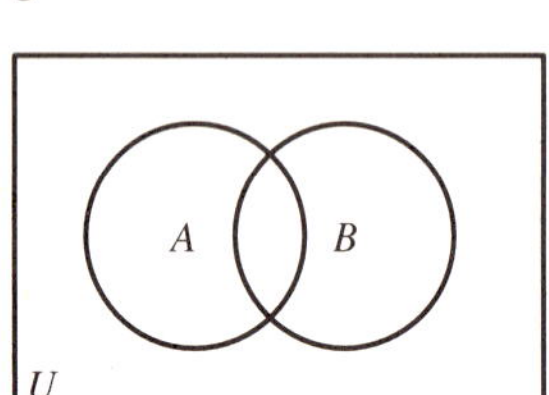

$U' = \emptyset$

94. $\emptyset'$

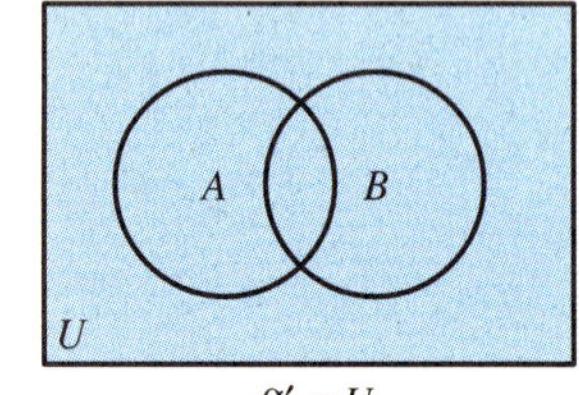

$\emptyset' = U$

95. Let $U = \{m, n, o, p, q, r, s, t, u, v, w\}$
$A = \{m, n, p, q, r, t\}$
$B = \{m, o, p, q, s, u\}$
$C = \{m, o, p, r, s, t, u, v\}$.
Place the elements of these sets in the proper location on a Venn diagram similar to the one shown here.

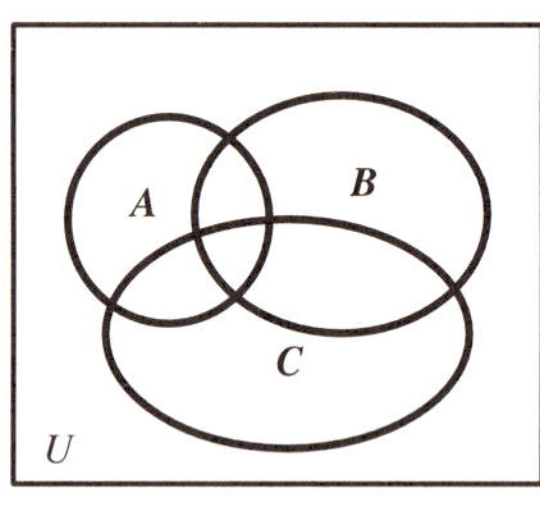

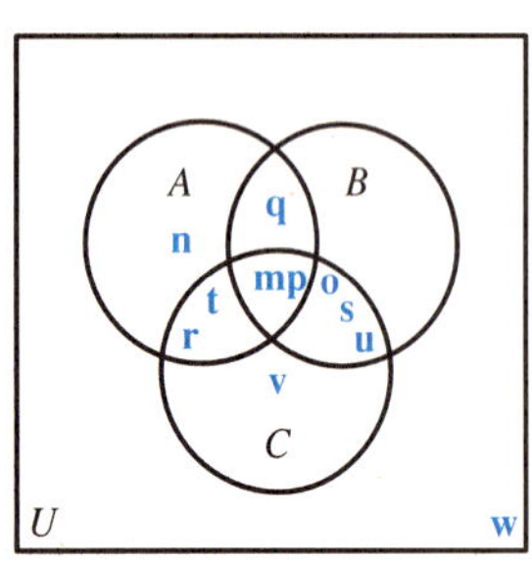

96. Let $U = \{1, 2, 3, 4, 5, 6, 7, 8, 9\}$
$A = \{1, 3, 5, 7\}$
$B = \{1, 3, 4, 6, 8\}$
$C = \{1, 4, 5, 6, 7, 9\}$.

Place the elements of these sets in the proper location on a Venn diagram.

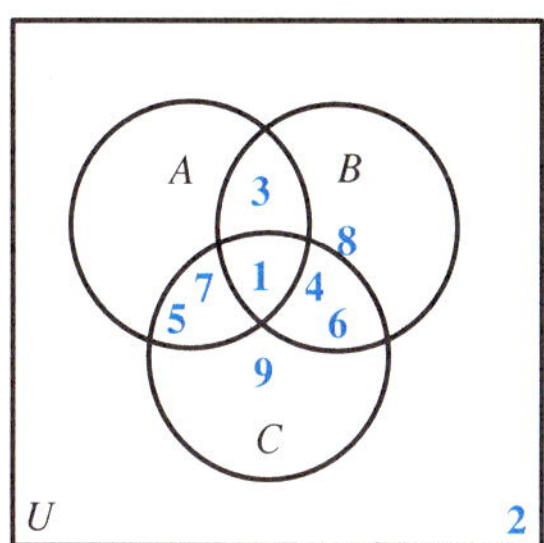

Use a Venn diagram to shade each of the following sets.

97. $(A \cap B) \cap C$

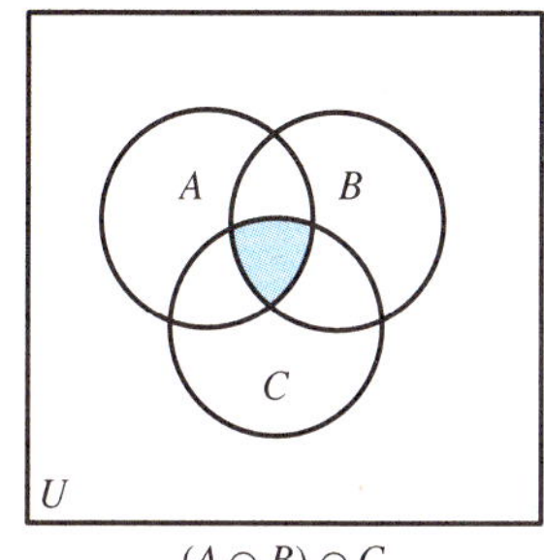

$(A \cap B) \cap C$

98. $(A \cap C') \cup B$

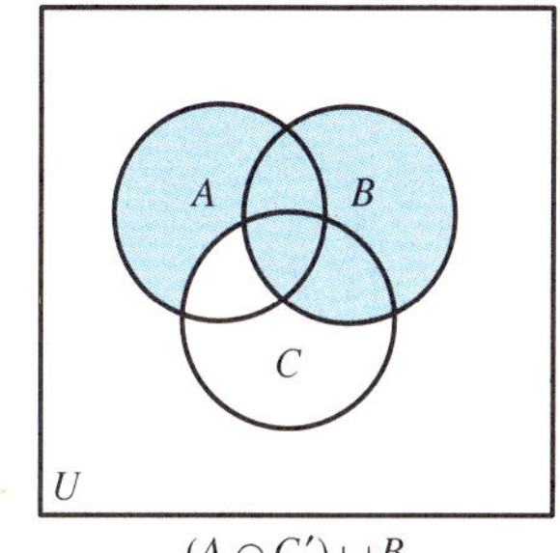

$(A \cap C') \cup B$

99. $(A \cap B) \cup C'$

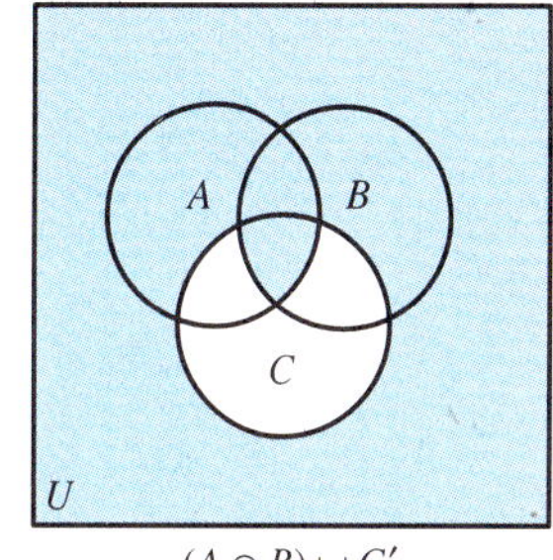

$(A \cap B) \cup C'$

100. $(A' \cap B) \cap C$

$(A' \cap B) \cap C$

101. $(A' \cap B') \cap C$

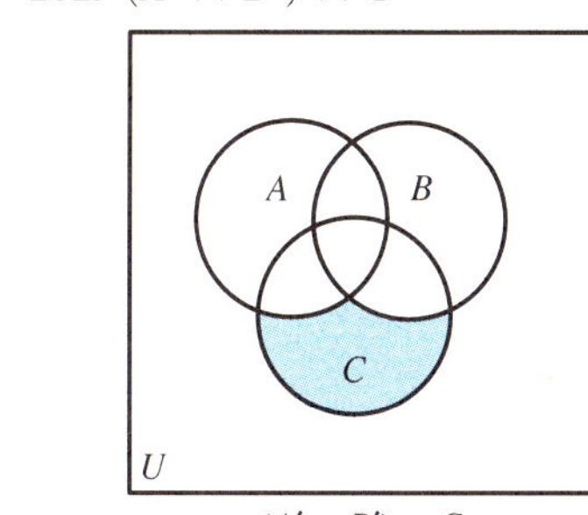

$(A' \cap B') \cap C$

102. $(A \cup B) \cup C$

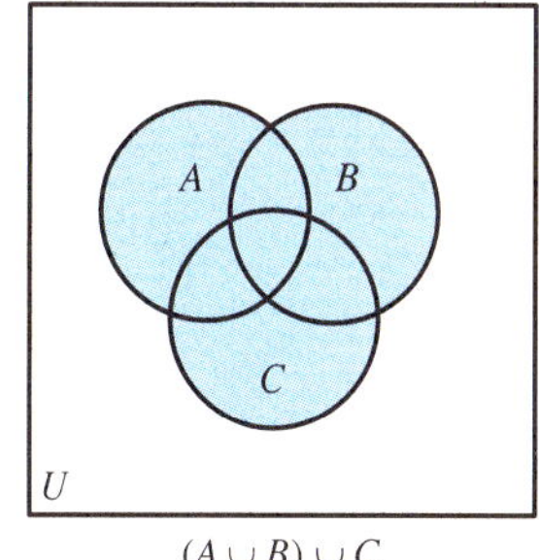

$(A \cup B) \cup C$

103. $(A \cap B') \cup C$

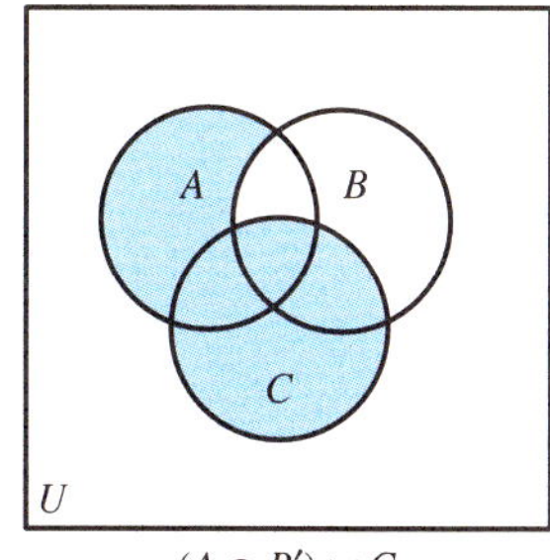

$(A \cap B') \cup C$

104. $(A \cap C') \cap B$

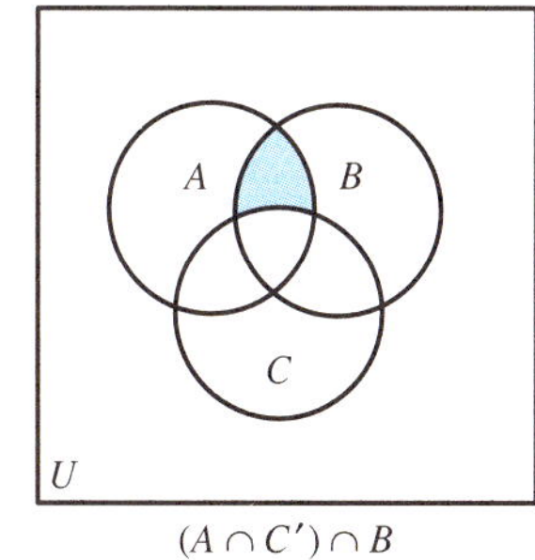

$(A \cap C') \cap B$

105. $(A \cap B') \cap C'$

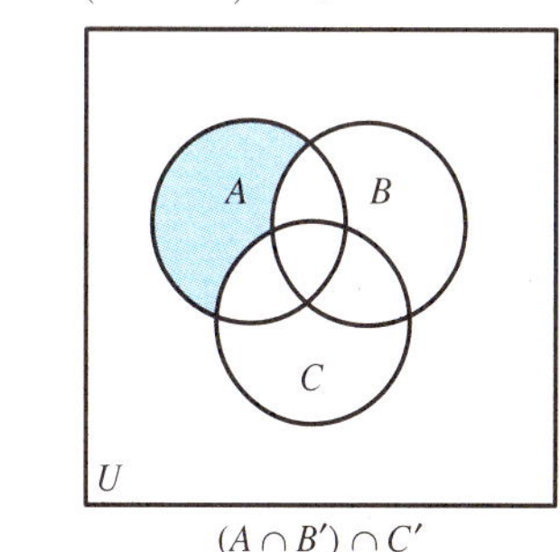

$(A \cap B') \cap C'$

106. $(A' \cap B') \cup C$

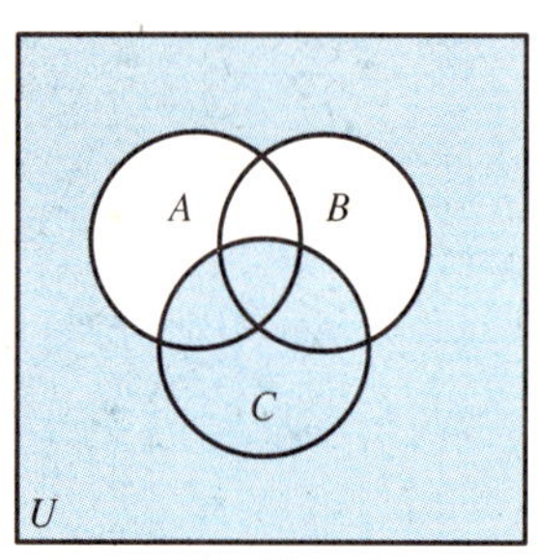

$(A' \cap B') \cup C$

107. $(A' \cap B') \cup C'$

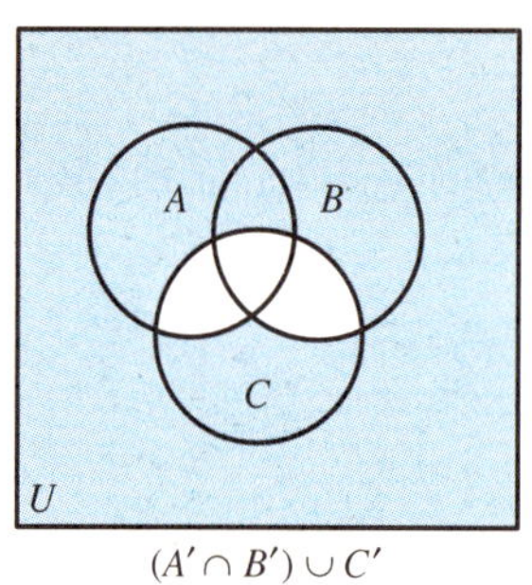

$(A' \cap B') \cup C'$

108. $(A \cap B)' \cup C$

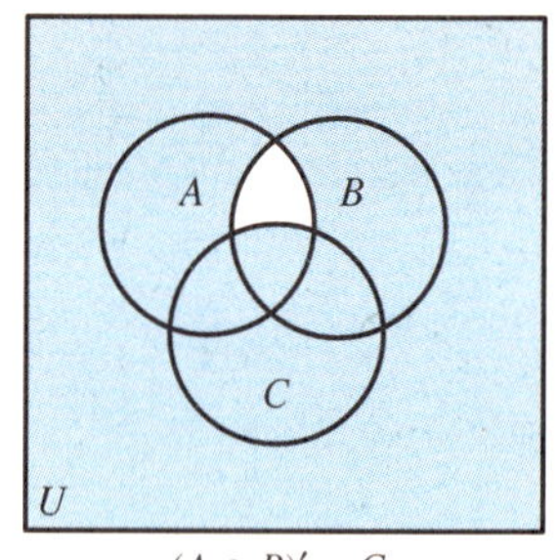

$(A \cap B)' \cup C$

Write a description of each shaded area. Use the symbols A, B, C, ∩, ∪, −, and ′ as necessary. More than one answer may be possible.

109.

A
B
U

$A' \cap B'$ or $(A \cup B)'$

110.

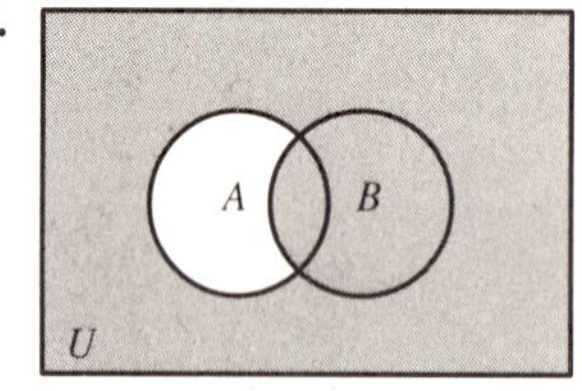

$A' \cup B$ or $(A - B)'$

111.

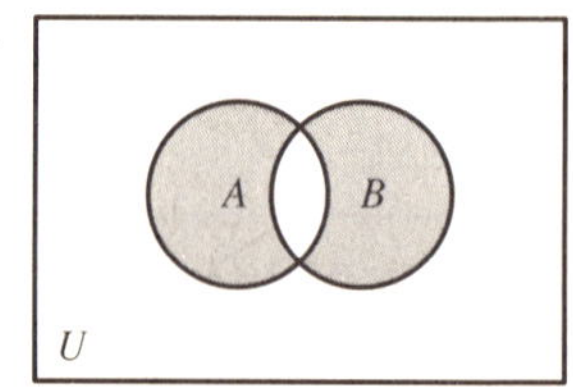

$(A \cup B) \cap [(A \cap B)']$ or $(A \cup B) - (A \cap B)$

112.

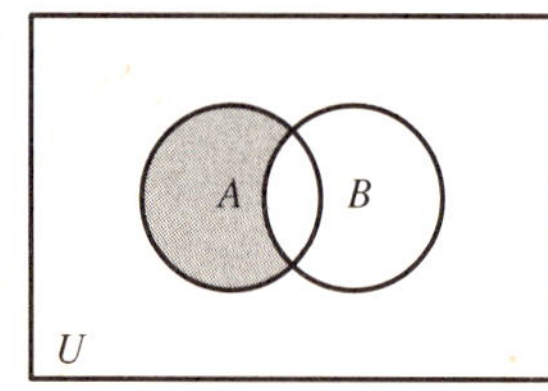

$A \cap B'$ or $A - B$

113.

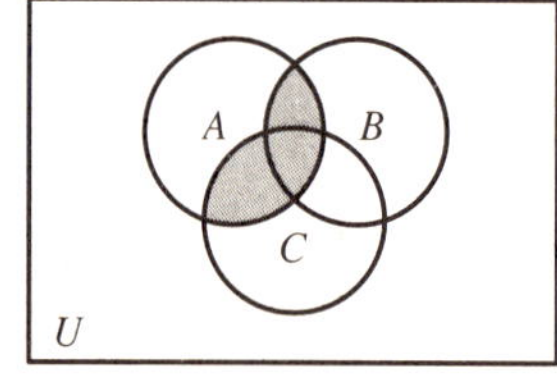

$(A \cap B) \cup (A \cap C)$ or $A \cap (B \cup C)$

114.

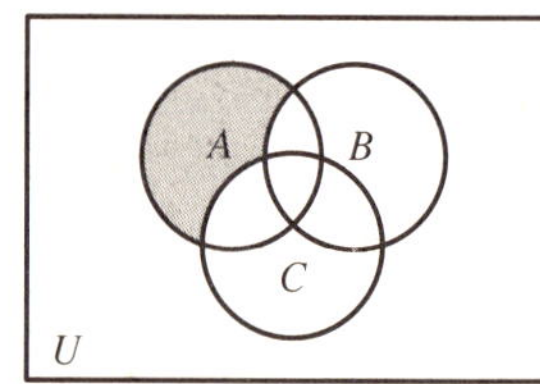

$A \cap (B \cup C)'$ or $A \cap (B' \cap C')$

115.

A
B
C
U

$(A \cap B) \cap C'$ or $(A \cap B) - C$

116.

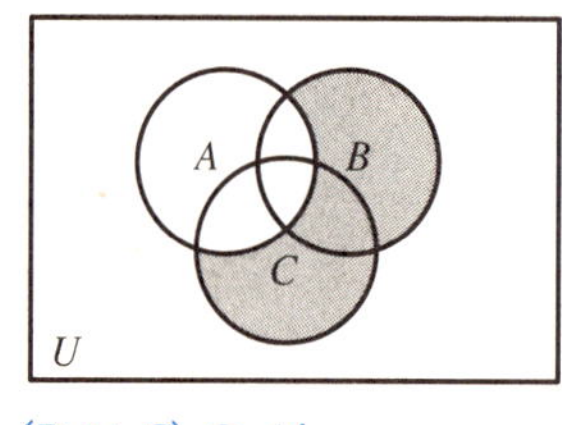

$(B \cup C) \cap A'$ or $(B \cup C) - A$

▲ ◉ *Suppose A and B are sets. Describe the conditions under which each of the following statements would be true.*

117. $A = A - B$ $A \cap B = \emptyset$

118. $A = B - A$ $A = B = \emptyset$

119. $A = A - \emptyset$ true for *any* set A

120. $A = \emptyset - A$ $A = \emptyset$

121. $A \cup \emptyset = \emptyset$ $A = \emptyset$

122. $A \cap \emptyset = \emptyset$ true for *any* set A

◉ CONCEPTUAL  WRITING  CHALLENGING  SCIENTIFIC CALCULATOR GRAPHING CALCULATOR

123. $A \cap \emptyset = A$ **$A = \emptyset$**

124. $A \cup \emptyset = A$ **true for *any* set A**

125. $A \cup A = \emptyset$ **$A = \emptyset$**

126. $A \cap A = \emptyset$ **$A = \emptyset$**

127. Give examples of how the conciseness of the "language of mathematics" can be an advantage.

128. Give examples of how a language such as English, Spanish, Arabic, or Vietnamese can have an advantage over the symbolic language of mathematics.

129. If A and B are sets, is it necessarily true that $n(A - B) = n(A) - n(B)$? Explain.

130. If $Q = \{x \mid x \text{ is a rational number}\}$ and $H = \{x \mid x \text{ is an irrational number}\}$, describe each of the following sets.

(a) $Q \cup H$ **(b)** $Q \cap H$

White light can be viewed as a blending of the three primary *colors red, green, and blue. Or, we can obtain a secondary color by blending any* two *primary colors. (For example, red and blue produce magenta, blue and green produce cyan, and red and green produce yellow.) For the following exercises, refer to the photo shown here.*

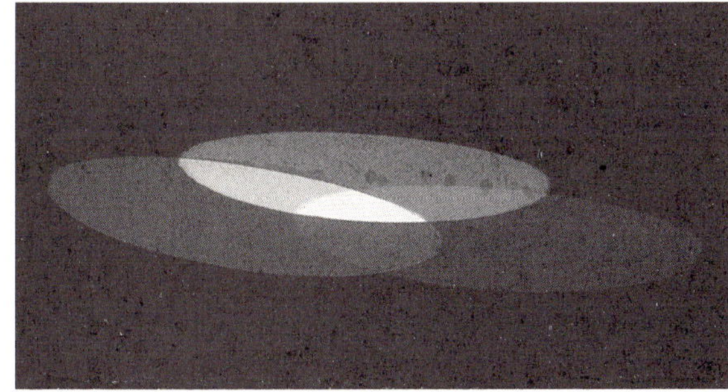

131. Name all the secondary colors of light. For each one, give its primary components. **yellow (components: red and green); magenta (components: red and blue); cyan (components: blue and green)**

132. In terms of set operations, white light is the three-way "intersection" of red, green, and blue. What *other* three-way intersection would also produce white light? **yellow, magenta, and cyan**

133. Explain why scientists sometimes refer to yellow as "minus blue."

134. What color is obtained if red is filtered out of (or subtracted from) white light? **cyan**

135. What must be filtered out of white light to obtain green light? **both red and blue**

For each of the following exercises, draw two appropriate Venn diagrams to decide whether the given statement is always true *or* not always true.

136. $A \cap A' = \emptyset$ **always true**

137. $A \cup A' = U$ **always true**

138. $(A \cap B) \subseteq A$ **always true**

139. $(A \cup B) \subseteq A$ **not always true**

140. If $A \subseteq B$, then $A \cup B = A$. **not always true**

141. If $A \subseteq B$, then $A \cap B = B$. **not always true**

142. $(A \cup B)' = A' \cap B'$ (De Morgan's second law) **always true**

2.4 Exercises (Text Page 83)

Use Venn diagrams to answer each of the following.

1. Ana Pott, a psychology professor at a Southern California college, was planning a study of viewer responses to certain aspects of the movies *Casablanca, High Noon,* and *Vertigo*. Upon surveying her class of 55 students, she determined the following data:

17 had seen *Casablanca*
17 had seen *High Noon*
23 had seen *Vertigo*
6 had seen *Casablanca* and *High Noon*
8 had seen *Casablanca* and *Vertigo*
10 had seen *High Noon* and *Vertigo*
2 had seen all three of these movies.

How many students had seen

(a) exactly two of these movies? **18**
(b) exactly one of these movies? **15**
(c) none of these movies? **20**
(d) *Casablanca* but neither of the others? **5**

2. At a Florida community college, half of the 48 mathematics majors were receiving federal financial aid. Of these:

5 had Pell Grants
14 participated in the College Work Study Program
4 had Stafford Loans
2 had Stafford Loans and participated in Work Study.

Those with Pell Grants had no other federal aid.

How many of the 48 math majors had:
(a) no federal aid? **24**
(b) more than one of these three forms of aid? **2**
(c) federal aid other than these three forms? **3**
(d) a Stafford Loan or Work Study? **16**

3. The following list shows the preferences of 102 people at a wine-tasting party:

99 like Spañada
96 like Ripple
99 like Boone's Farm Apple Wine
95 like Spañada and Ripple
94 like Ripple and Boone's
96 like Spañada and Boone's
93 like all three.

How many people like:
(a) none of the three? **0**
(b) Spañada, but not Ripple? **4**
(c) anything but Boone's Farm? **3**
(d) only Ripple? **0**
(e) exactly two kinds of wine? **6**

4. Bob Carlton (Example 3 in the text) was again reassigned, this time to the home economics department of the electric utility. He interviewed 140 people in a suburban shopping center to find out some of their cooking habits. He obtained the following results. Should he be reassigned yet one more time?

58 use microwave ovens
63 use electric ranges
58 use gas ranges
19 use microwave ovens and electric ranges
17 use microwave ovens and gas ranges
4 use both gas and electric ranges
1 uses all three
2 cook only with solar energy

yes; His data add up to 142 people.

5. A chicken farmer surveyed his flock with the following results. The farmer has:

9 fat red roosters
2 fat red hens
26 fat roosters
37 fat chickens
7 thin brown hens
18 thin brown roosters
6 thin red roosters
5 thin red hens.

Answer the following questions about the flock. [*Hint:* You need a Venn diagram with circles for fat, for male (a rooster is a male, a hen is a female), and for red (assume that brown and red are opposites in the chicken world).] How many chickens are:

(a) fat? 37
(b) red? 22
(c) male? 50
(d) fat, but not male? 11
(e) brown, but not fat? 25
(f) red and fat? 11

6. It was once said that Country-Western songs emphasize three basic themes: love, prison, and trucks. A survey of the local Country-Western radio station produced the following data:

12 songs about a truck driver who is in love while in prison
13 about a prisoner in love
28 about a person in love
18 about a truck driver in love
3 about a truck driver in prison who is not in love
2 about people in prison who are not in love and do not drive trucks
8 about people who are out of prison, are not in love, and do not drive a truck
16 about truck drivers who are not in prison.

(a) How many songs were surveyed? 51
Find the number of songs about:
(b) truck drivers 31
(c) prisoners 18
(d) truck drivers in prison 15
(e) people not in prison 33
(f) people not in love. 23

▲ 7. Nadine Tracy conducted a survey among 75 patients admitted to the cardiac unit of a Massachusetts hospital during a two-week period.

Let B = the set of patients with high blood pressure
C = the set of patients with high cholesterol levels
S = the set of patients who smoke cigarettes.

Nadine's data are as follows:

$n(B) = 47$
$n(C) = 46$
$n(S) = 52$
$n[(B \cap C) \cup (B \cap S) \cup (C \cap S)] = 51$
$n(B \cap S) = 33$
$n(B \cap C) = 31$
$n(B \cap C \cap S) = 21.$

Find the number of these patients who:
(a) had either high blood pressure or high cholesterol levels, but not both 31
(b) had fewer than two of the indications listed 24
(c) were smokers but had neither high blood pressure nor high cholesterol levels 11
(d) did not have exactly two of the indications listed. 45

▲ 8. Stacy Thrash, who sells college textbooks, interviewed freshmen on a west coast campus to find out the main goals of today's students.

Let W = the set of those who want to become wealthy
F = the set of those who want to raise a family
E = the set of those who want to become experts in their field.

Stacy's findings are summarized here:

$$n(W) = 160 \qquad n(E \cap F) = 90$$
$$n(F) = 140 \qquad n(W \cap F \cap E) = 80$$
$$n(E) = 130 \qquad n(E') = 95$$
$$n(W \cap F) = 95 \qquad n[(W \cup F \cup E)'] = 10.$$

Find the total number of students interviewed. **225**

9. Dwaine Tomlinson runs a basketball program in California. On the first day of the season, 60 young men showed up and were categorized by age level and by preferred basketball position, as shown in the following table.

		Position			
		Guard (G)	**Forward (F)**	**Center (N)**	**Totals**
	Junior High (J)	9	6	4	19
Age	**Senior High (S)**	12	5	9	26
	College (C)	5	8	2	15
	Totals	26	19	15	60

Using the set labels (letters) in the table, find the number of players in each of the following sets.

(a) $J \cap G$ **9** **(b)** $S \cap N$ **9** **(c)** $N \cup (S \cap F)$ **20**
(d) $S' \cap (G \cup N)$ **20** **(e)** $(S \cap N') \cup (C \cap G')$ **27** **(f)** $N' \cap (S' \cap C')$ **15**

10. A study of U.S. Army housing trends categorized personnel as commissioned officers (C), warrant officers (W), or enlisted (E), and categorized their living facilities as on-base (B), rented off-base (R), or owned off-base (O). One survey yielded the following data.

		Facilities			
		B	***R***	***O***	**Totals**
	C	12	29	54	95
Personnel	***W***	4	5	6	15
	E	374	71	285	730
Totals		390	105	345	840

Find the number of personnel in each of the following sets.

(a) $W \cap O$ **6** **(b)** $C \cup B$ **473** **(c)** $R' \cup W'$ **835**
(d) $(C \cup W) \cap (B \cup R)$ **50** **(e)** $(C \cap B) \cup (E \cap O)$ **297** **(f)** $B \cap (W \cup R)'$ **386**

In the following exercises, make use of an appropriate formula.

11. Evaluate $n(A \cup B)$ if $n(A) = 8$, $n(B) = 14$, and $n(A \cap B) = 5$. **17**

12. Evaluate $n(A \cap B)$ if $n(A) = 15$, $n(B) = 12$, and $n(A \cup B) = 25$. **2**

13. Evaluate $n(A)$ if $n(B) = 20$, $n(A \cap B) = 6$, and $n(A \cup B) = 30$. **16**

14. Evaluate $n(B)$ if $n(A) = 35$, $n(A \cap B) = 15$, and $n(A \cup B) = 55$. **35**

Draw an appropriate Venn diagram and use the given information to fill in the number of elements in each region.

15. $n(U) = 43, n(A) = 25, n(A \cap B) = 5, n(B') = 30$

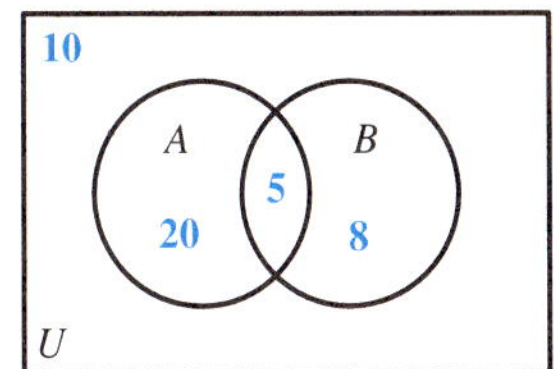

16. $n(A) = 19, n(B) = 13, n(A \cup B) = 25, n(A') = 11$

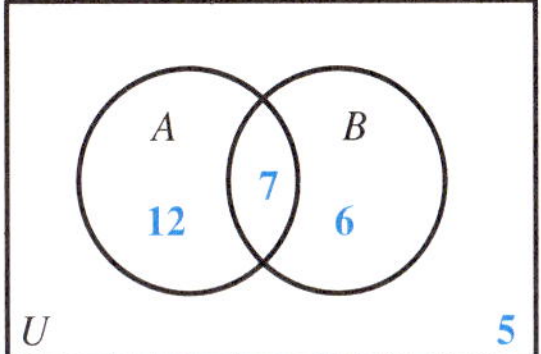

17. $n(A \cup B) = 15, n(A \cap B) = 8, n(A) = 13, n(A' \cup B') = 11$

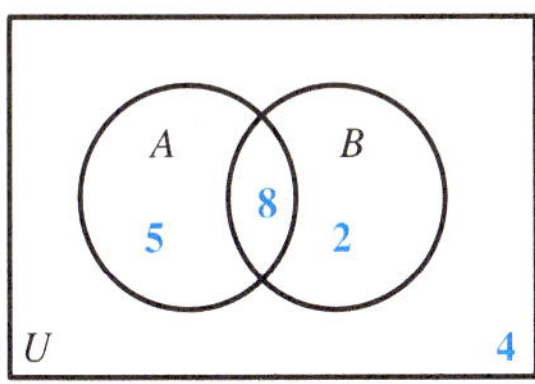

18. $n(A') = 25, n(B) = 28, n(A' \cup B') = 40, n(A \cap B) = 10$

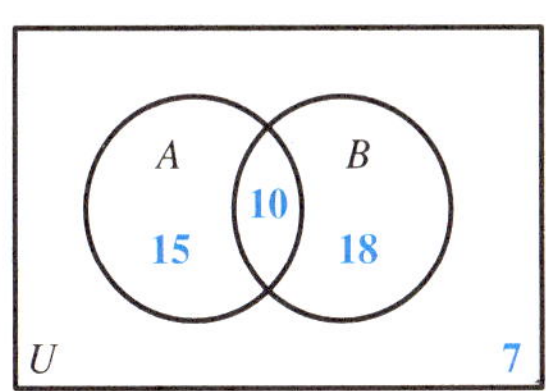

19. $n(A) = 24, n(B) = 24, n(C) = 26, n(A \cap B) = 10, n(B \cap C) = 8, n(A \cap C) = 15,$ $n(A \cap B \cap C) = 6, n(U) = 50$

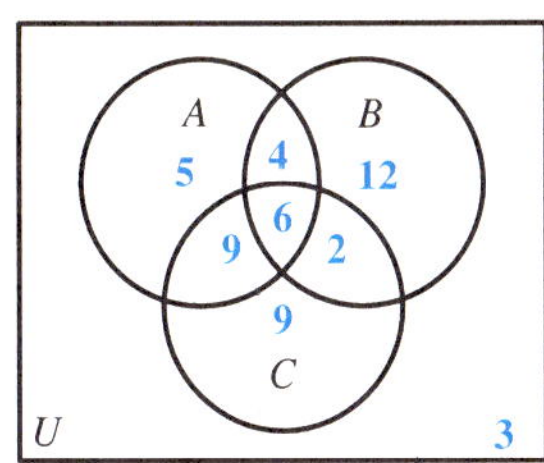

20. $n(A) = 57, n(A \cap B) = 35, n(A \cup B) = 81, n(A \cap B \cap C) = 15, n(A \cap C) = 21,$ $n(B \cap C) = 25, n(C) = 49, n(B') = 52$

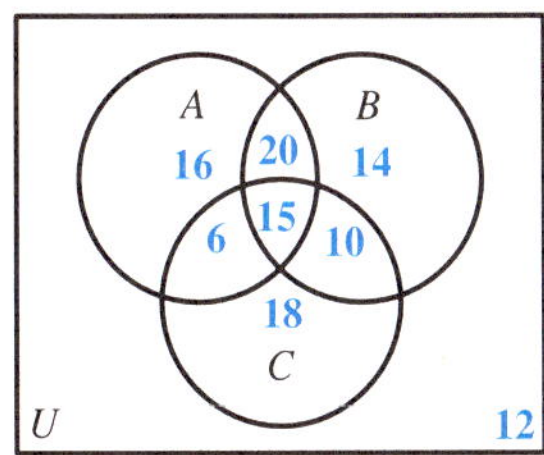

21. $n(A \cap B) = 21$, $n(A \cap B \cap C) = 6$, $n(A \cap C) = 26$, $n(B \cap C) = 7$, $n(A \cap C') = 20$, $n(B \cap C') = 25$, $n(C) = 40$, $n(A' \cap B' \cap C') = 2$

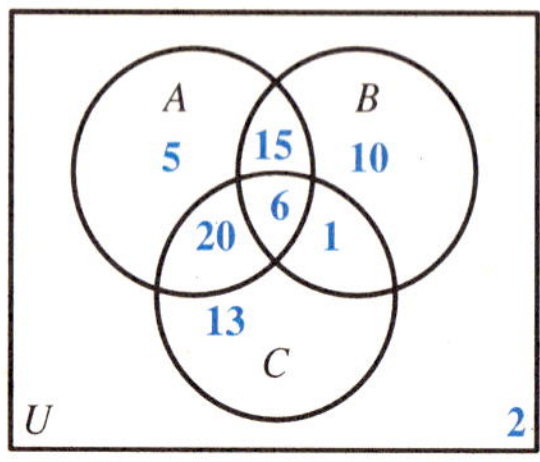

22. $n(A) = 15$, $n(A \cap B \cap C) = 5$, $n(A \cap C) = 13$, $n(A \cap B') = 9$, $n(B \cap C) = 8$, $n(A' \cap B' \cap C') = 21$, $n(B \cap C') = 3$, $n(B \cup C) = 32$

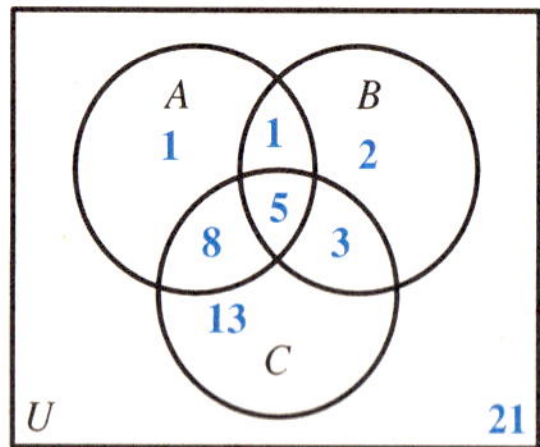

23. Could the information of Example 4 have been presented in a Venn diagram similar to those in Examples 1 and 3? If so, construct such a diagram. Otherwise explain the essential difference of Example 4.

24. Explain how a cardinal number formula can be derived for the case where *three* sets occur. Specifically, give a formula relating $n(A \cup B \cup C)$ to $n(A)$, $n(B)$, $n(C)$, $n(A \cap B)$, $n(A \cap C)$, $n(B \cap C)$, and $n(A \cap B \cap C)$. Illustrate with a Venn diagram.

25. In the previous section, we looked at Venn diagrams containing one, two, or three sets.* Use this information to complete the following table. (*Hint:* Make a prediction for 4 sets from the pattern of 1, 2, and 3 sets.) **4, 8, 16**

Number of sets	1	2	3	4
Number of regions dividing U	2	____	____	____

The figure at the right shows U divided into 16 regions by four sets, A, B, C, and D. Find the numbers of the regions belonging to each set in Exercises 26–29.

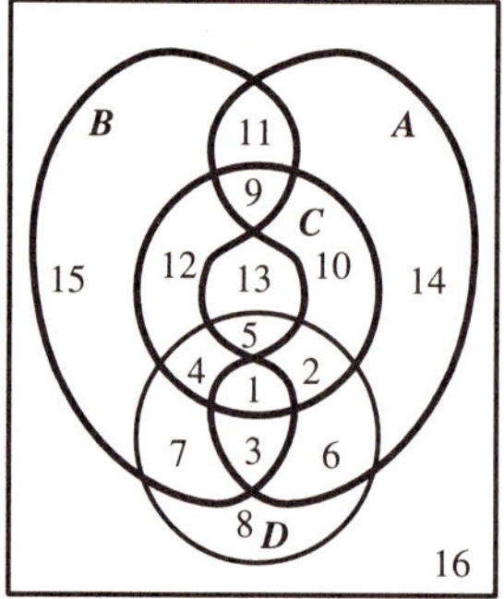

*For information on using more than four sets in a Venn diagram, see "The Construction of Venn Diagrams," by Branko Grünbaum, in *The College Mathematics Journal,* June 1984.

26. $A \cap B \cap C \cap D$ **1**

27. $A \cup B \cup C \cup D$ **1, 2, 3, 4, 5, 6, 7, 8, 9, 10, 11, 12, 13, 14, 15 (all except 16)**

28. $(A \cap B) \cup (C \cap D)$ **1, 2, 3, 4, 5, 9, 11**

29. $(A' \cap B') \cap (C \cup D)$ **5, 8, 13**

◉ 30. If we placed five most generally related sets inside U, how many regions should result? Make a general formula. If n general sets are placed inside U, how many regions would result? **$2^5 = 32$; 2^n**

▲ 31. A survey of 130 television viewers revealed the following facts:

52 watch football
56 watch basketball
62 watch tennis
60 watch golf
21 watch football and basketball
19 watch football and tennis
22 watch basketball and tennis
27 watch football and golf
30 watch basketball and golf
21 watch tennis and golf
3 watch football, basketball, and tennis
15 watch football, basketball, and golf
10 watch football, tennis, and golf
10 watch basketball, tennis, and golf
3 watch all four of these sports
5 don't watch any of these four sports.

Use a Venn diagram with four sets like the one in Exercises 26–29 to answer the following questions.

(a) How many of these viewers watch football, basketball, and tennis, but not golf? **none**

(b) How many watch exactly one of these four sports? **52**

(c) How many watch exactly two of these four sports? **44**

2.5 Exercises (Text Page 93)

Place each pair of sets into a one-to-one correspondence, if possible.

1. {I, II, III} and {x, y, z} **(Other correspondences are possible.)**

{I, II, III}
↕ ↕ ↕
{x, y, z}

2. {a, b, c, d} and {2, 4, 6} **not possible**

3. {a, d, d, i, t, i, o, n} and {a, n, s, w, e, r} **(Other correspondences are possible.)**

{a, d, i, t, o, n}
↕ ↕ ↕ ↕ ↕ ↕
{a, n, s, w, e, r}

4. {Reagan, Bush, Clinton} and {Nancy, Barbara, Hillary} **(Other correspondences are possible. This particular one pairs each president with his wife.)**

{Clinton, Reagan, Bush}
↕ ↕ ↕
{Hillary, Nancy, Barbara}

Give the cardinal number of each set.

5. {a, b, c, d, . . . , k} **11**

6. {9, 12, 15, . . . , 36} **10**

7. ∅ **0**

8. {0} **1**

9. {300, 400, 500, . . .} **$\aleph_0$**

10. {−35, −28, −21, . . . , 56} **14**

11. {−1/4, −1/8, −1/12, . . .} **$\aleph_0$**

12. $\{x \mid x$ is an even integer$\}$ **$\aleph_0$**

13. $\{x \mid x$ is an odd counting number$\}$ **$\aleph_0$**

14. {b, a, l, l, a, d} **4**

15. {Jan, Feb, Mar, . . . , Dec} **12**

16. {Alabama, Alaska, Arizona, . . . , Wisconsin, Wyoming} **50**

17. Lew Lefton of the University of New Orleans has revised the old song "100 Bottles of Beer on the Wall" to illustrate a property of infinite cardinal numbers. Fill in the blank in the first line of Lefton's composition:
"$\aleph_0$ bottles of beer on the wall, $\aleph_0$ bottles of beer, take one down and pass it around, ______ bottles of beer on the wall." $\aleph_0$

▲ ◉ **18.** Two one-to-one correspondences are considered "different" if some elements are paired differently in one than in the other. For example:

$$\begin{matrix}\{a, & b, & c\} \\ \updownarrow & \updownarrow & \updownarrow \\ \{a, & b, & c\}\end{matrix} \text{ and } \begin{matrix}\{a, & b, & c\} \\ \updownarrow & \updownarrow & \updownarrow \\ \{c, & b, & a\}\end{matrix} \text{ are different, while } \begin{matrix}\{a, & b, & c\} \\ \updownarrow & \updownarrow & \updownarrow \\ \{c, & a, & b\}\end{matrix} \text{ and } \begin{matrix}\{b, & c, & a\} \\ \updownarrow & \updownarrow & \updownarrow \\ \{a, & b, & c\}\end{matrix} \text{ are not.}$$

(a) How many *different* correspondences can be set up between the two sets

{George Burns, John Wayne, Chuck Norris}
and {Carlos Ray, Nathan Birnbaum, Marion Morrison}? **6**

(b) Which one of these correspondences pairs each man with himself?

$$\begin{matrix}\{\text{George Burns}, & \text{John Wayne}, & \text{Chuck Norris}\} \\ \updownarrow & \updownarrow & \updownarrow \\ \{\text{Nathan Birnbaum}, & \text{Marion Morrison}, & \text{Carlos Ray}\}\end{matrix}$$

Determine whether the following pairs of sets are equal, equivalent, both, *or* neither.

19. {u, v, w}, {v, u, w} **both**

20. {48, 6}, {4, 86} **equivalent**

21. {X, Y, Z}, {x, y, z} **equivalent**

22. {lea}, {ale} **equivalent**

23. $\{x \mid x$ is a positive real number$\}$, $\{x \mid x$ is a negative real number$\}$ **equivalent**

24. $\{x \mid x$ is a positive rational number$\}$, $\{x \mid x$ is a negative real number$\}$ **neither**

Show that each set has cardinal number $\aleph_0$ by setting up a one-to-one correspondence between the given set and the set of counting numbers.

25. the set of positive even numbers

$$\begin{matrix}\{2, & 4, & 6, & 8, & \ldots, & 2n, & \ldots\} \\ \updownarrow & \updownarrow & \updownarrow & \updownarrow & & \updownarrow & \\ \{1, & 2, & 3, & 4, & \ldots, & n, & \ldots\}\end{matrix}$$

26. $\{-10, -20, -30, -40, \ldots\}$

$$\begin{matrix}\{-10, & -20, & -30, & -40, & \ldots, & -10n, & \ldots\} \\ \updownarrow & \updownarrow & \updownarrow & \updownarrow & & \updownarrow & \\ \{1, & 2, & 3, & 4, & \ldots, & n, & \ldots\}\end{matrix}$$

27. {1,000,000, 2,000,000, 3,000,000, . . .}

$$\begin{matrix}\{1{,}000{,}000, & 2{,}000{,}000, & 3{,}000{,}000, & \ldots, & 1{,}000{,}000n, & \ldots\} \\ \updownarrow & \updownarrow & \updownarrow & & \updownarrow & \\ \{\ 1, & 2, & 3, & \ldots, & n, & \ldots\}\end{matrix}$$

28. the set of odd integers

$$\begin{matrix}\{1, & -1, & 3, & -3, & 5, & -5, & \ldots, & a, & \ldots\} \\ \updownarrow & \updownarrow & \updownarrow & \updownarrow & \updownarrow & \updownarrow & & \updownarrow & \\ \{1, & 2, & 3, & 4, & 5, & 6, & \ldots, & n, & \ldots\}\end{matrix}$$

where $a = n$ if n is odd and $a = 1 - n$ if n is even

29. {2, 4, 8, 16, 32, . . .} (*Hint:* $4 = 2^2$, $8 = 2^3$, $16 = 2^4$, and so on)

$$\begin{matrix}\{2, & 4, & 8, & 16, & 32, & \ldots, & 2^n, & \ldots\} \\ \updownarrow & \updownarrow & \updownarrow & \updownarrow & \updownarrow & & \updownarrow & \\ \{1, & 2, & 3, & 4, & 5, & \ldots, & n, & \ldots\}\end{matrix}$$

▲ **30.** $\{-17, -22, -27, -32, \ldots\}$

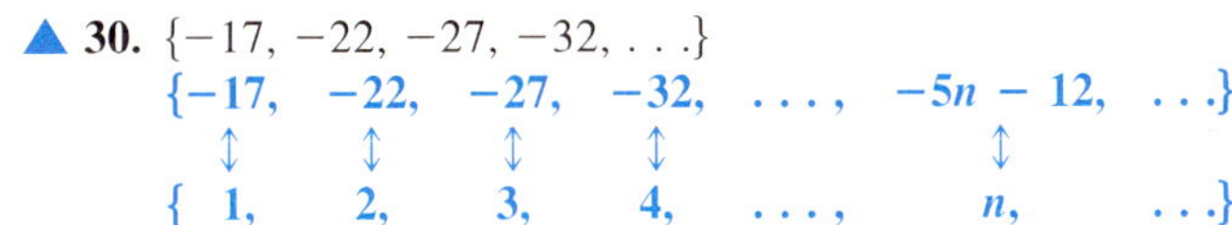

{−17, −22, −27, −32, . . . , −5n − 12, . . .}
↕ ↕ ↕ ↕ ↕
{ 1, 2, 3, 4, . . . , n, . . .}

◎ *In each of Exercises 31–34, identify the given statement as* always true *or* not always true. *If not always true, give an example.*

31. If A and B are infinite sets, then $A \sim B$. **Not always true. For example, let A = the set of counting numbers, B = the set of real numbers.**

32. If set A is an infinite set and set B can be put in a one-to-one correspondence with a proper subset of A, then B must be infinite. **Not always true. For example, let A = the set of counting numbers, $B = \{a, b, c\}$.**

33. If A is an infinite set and A is not equivalent to the set of counting numbers, then $n(A) = c$. **Not always true. For example, A could be the set of all subsets of the set of reals. Then $n(A)$ would be an infinite number *greater* than c.**

34. If A and B are both countably infinite sets, then $n(A \cup B) = \aleph_0$. **Always true.**

▲ ◎ **35.** The set of real numbers can be represented by an infinite line, extending indefinitely in both directions. Each point on the line corresponds to a unique real number, and each real number corresponds to a unique point on the line.

(a) Use the figure below, where the line segment between 0 and 1 has been bent into a semicircle and positioned above the line, to prove that

$$\{x \mid x \text{ is a real number between 0 and 1}\} \sim \{x \mid x \text{ is a real number}\}.$$

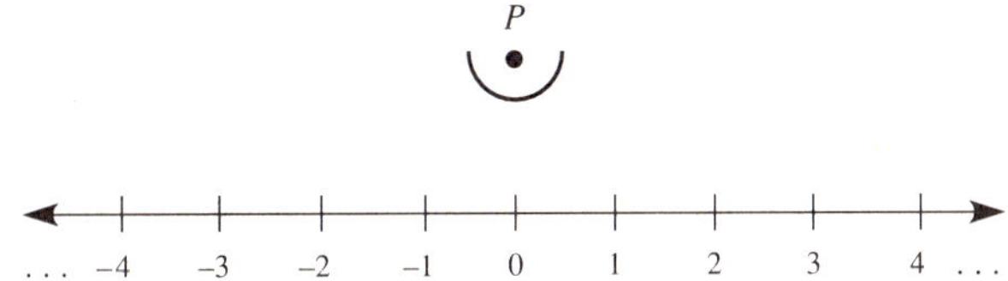

Rays emanating from point P will establish a geometric pairing of the points on the semicircle with the points on the line.

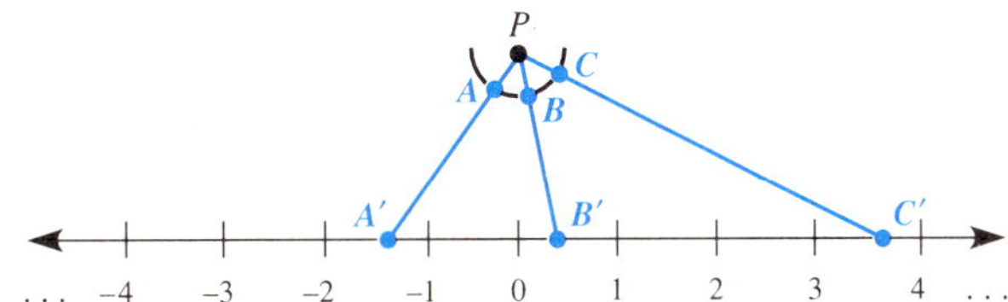

(b) What fact does part (a) establish about the set of real numbers? **The set of real numbers is infinite (having been placed in a one-to-one correspondence with a proper subset of itself).**

▲ ◎ **36.** Show that the two vertical line segments shown here both have the same number of points.

Show that each of the following sets can be placed in a one-to-one correspondence with a proper subset of itself to prove that each set is infinite.

37. {3, 6, 9, 12, . . .}

$\{3, 6, 9, 12, \ldots, 3n, \ldots\}$
$\updownarrow$
$\{6, 9, 12, 15, \ldots, 3n + 3, \ldots\}$

38. {4, 7, 10, 13, 16, . . .}

$\{4, 7, 10, 13, \ldots, 3n + 1, \ldots\}$
$\updownarrow$
$\{7, 10, 13, 16, \ldots, 3n + 4, \ldots\}$

39. {3/4, 3/8, 3/12, 3/16, . . .}

$\{3/4, 3/8, 3/12, 3/16, \ldots, 3/(4n), \ldots\}$
$\updownarrow$
$\{3/8, 3/12, 3/16, 3/20, \ldots, 3/(4n + 4), \ldots\}$

40. {1, 4/3, 5/3, 2, . . .}

$\{1, 4/3, 5/3, 2, \ldots, (n + 2)/3, \ldots\}$
$\updownarrow$
$\{4/3, 5/3, 2, 7/3, \ldots, (n + 3)/3, \ldots\}$

▲ **41.** {1/9, 1/18, 1/27, 1/36, . . .}

$\{1/9, 1/18, 1/27, \ldots, 1/(9n), \ldots\}$
$\updownarrow$
$\{1/18, 1/27, 1/36, \ldots, 1/(9n + 9), \ldots\}$

▲ **42.** {−3, −5, −9, −17, . . .}

$\{-3, -5, -9, -17, \ldots, -(2^n + 1), \ldots\}$
$\updownarrow$
$\{-5, -9, -17, -33, \ldots, -(2^{n+1} + 1), \ldots\}$

▲ ◎ ✍ **43.** Describe the difference between *equal* and *equivalent* sets.

▲ ◎ ✍ **44.** Explain how the correspondence suggested in Example 4 shows that the set of real numbers between 0 and 1 is not countable.

Chapter 2 Test (Text Page 96)

For Exercises 1–14,

Let $U = \{a, b, c, d, e, f, g, h\}$
$A = \{a, b, c, d\}$
$B = \{b, e, a, d\}$
$C = \{a, e\}$.

Find each of the following sets.

1. $A \cup C$ **{a, b, c, d, e}** **2.** $B \cap A$ **{a, b, d}** **3.** B' **{c, f, g, h}** **4.** $A - (B \cap C')$ **{a, c}**

Identify each of the following statements as true *or* false.

5. $b \in A$ **true** **6.** $C \subseteq A$ **false** **7.** $B \subset (A \cup C)$ **true** **8.** $c \notin C$ **true**

9. $n[(A \cup B) - C] = 4$ **false** **10.** $\emptyset \subset C$ **true**

11. $(A \cap B') \sim (B \cap A')$ **true** **12.** $(A \cup B)' = A' \cap B'$ **true**

Find each of the following.

13. $n(A \times C)$ **8** **14.** the number of proper subsets of A **15**

Give a word description for each of the following sets. **Answers may vary for Exercises 15–18.**

15. {−3, −1, 1, 3, 5, 7, 9} **the set of odd integers between −4 and 10**

16. {January, February, March, . . . , December} **the set of months of the year**

Express each of the following sets in set-builder notation.

17. $\{-1, -2, -3, -4, \ldots\}$ **$\{x \mid x$ is a negative integer$\}$**

18. $\{24, 32, 40, 48, \ldots, 88\}$ **$\{x \mid x$ is a multiple of 8 between 20 and 90$\}$**

Place $\subset$, $\subseteq$, both, *or* neither *in each blank to make a true statement.*

19. $\emptyset$ _____ $\{x \mid x$ is a counting number between 17 and 18$\}$ $\subseteq$

20. $\{4, 9, 16\}$ _____ $\{4, 5, 6, 7, 8, 9, 10\}$ **neither**

Shade each of the following sets in an appropriate Venn diagram.

21. $X \cup Y'$

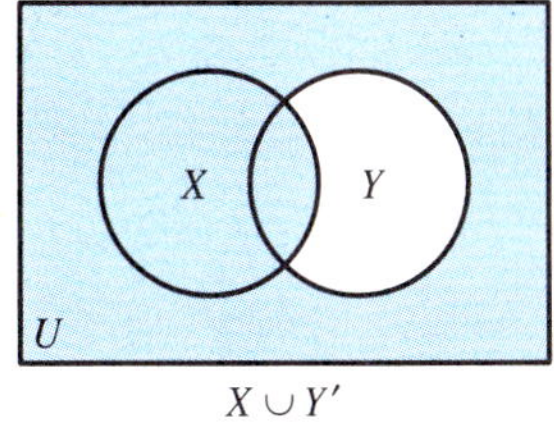

$X \cup Y'$

22. $X' \cap Y'$

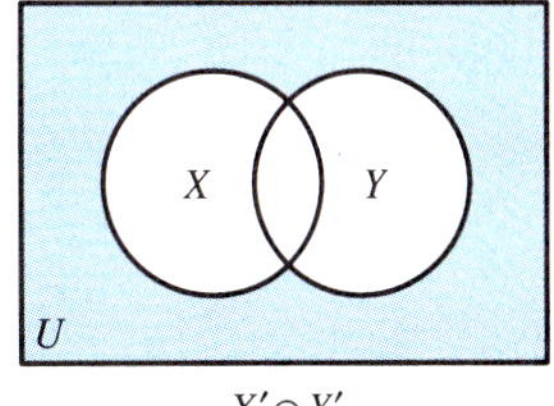

$X' \cap Y'$

23. $(X \cup Y) - Z$

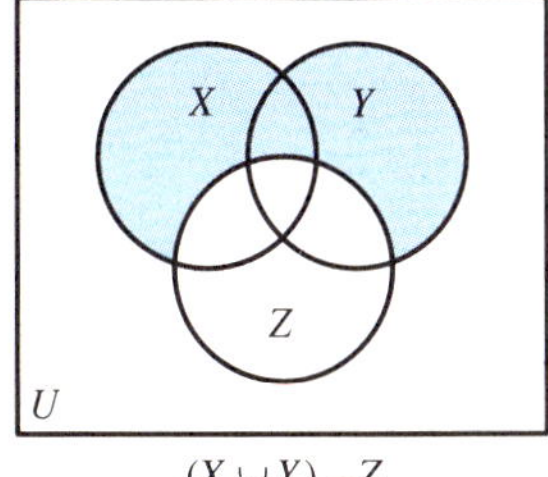

$(X \cup Y) - Z$

24. $[(X \cap Y) \cup (Y \cap Z) \cup (X \cap Z)] - (X \cap Y \cap Z)$

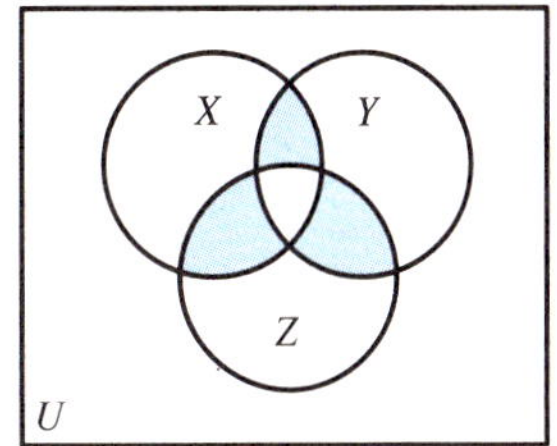

$[(X \cap Y) \cup (Y \cap Z) \cup (X \cap Z)] - (X \cap Y \cap Z)$

The following table lists ten inventions, important directly or indirectly in our lives, together with other pertinent data.

Invention	Date	Inventor	Nation
Adding machine	1642	Pascal	France
Barometer	1643	Torricelli	Italy
Electric razor	1917	Schick	U.S.
Fiber optics	1955	Kapany	England
Geiger counter	1913	Geiger	Germany
Pendulum clock	1657	Huygens	Holland
Radar	1940	Watson-Watt	Scotland
Telegraph	1837	Morse	U.S.
Thermometer	1593	Galileo	Italy
Zipper	1891	Judson	U.S.

Let U = the set of all ten inventions
A = the set of items invented in the United States
T = the set of items invented in the twentieth century.

List the elements of each of the following sets.

25. $A \cap T$ **{Electric razor}**

26. $(A \cup T)'$ **{Adding machine, Barometer, Pendulum clock, Thermometer}**

27. $A - T'$ **{Electric razor}**

28. State De Morgan's laws for sets in words rather than symbols.

29. Explain in your own words why, if A and B are any two nonempty sets, $A \times B \neq B \times A$ but $n(A \times B) = n(B \times A)$.

In one recent year, financial aid available to college students in the United States was nearly 30 billion dollars. (Much of it went unclaimed, mostly because qualified students were not aware of it, did not know how to obtain or fill out the required applications, or did not feel the results would be worth their effort.) The three major sources of aid are government grants, private scholarships, and the colleges themselves.

30. Jill Lynn Yuen, Financial Aid Director of a small private Midwestern college, surveyed the records of 100 sophomores and found the following:

49 receive government grants
55 receive private scholarships
43 receive aid from the college
23 receive government grants and private scholarships
18 receive government grants and aid from the college
28 receive private scholarships and aid from the college
8 receive help from all three sources.

How many of the students in the survey:
(a) have a government grant only? **16**
(b) have a private scholarship but not a government grant? **32**
(c) receive financial aid from only one of these sources? **33**
(d) receive aid from exactly two of these sources? **45**
(e) receive no financial aid from any of these sources? **14**
(f) receive no aid from the college or from the government? **26**

CHAPTER 3 Logic

3.1 Exercises (Text Page 103)

Decide whether or not each of the following is a statement.

1. A ZIP code for New Orleans is 70115. **statement**
2. October 12, 1949 was a Wednesday. **statement**
3. Have a nice day. **not a statement**
4. Stand up and be counted. **not a statement**
5. $8 + 15 = 23$ **statement**
6. $9 - 4 = 5$ and $2 + 1 = 5$ **statement**
7. Chester A. Arthur was president in 1882. **statement**
8. Not all numbers are positive. **statement**
9. Dancing is enjoyable. **not a statement**
10. Since 1950, more people have died in automobile accidents than of cancer. **statement**
11. Toyotas are better cars than Dodges. **not a statement**
12. Sit up and take notice. **not a statement**
13. One gallon of water weighs more than 5 pounds. **statement**
14. Kevin "Catfish" McCarthy once took a prolonged continuous shower for 340 hours, 40 minutes. **statement**

Decide whether each of the following statements is compound.

15. My syster got married in Paris. **not compound**

16. I read novels and I read newspapers. **compound**

17. Denise Clark is younger than 40 years of age, and so is Rodger Klas. **compound**

18. Yesterday was Friday. **not compound**

19. $4 + 2 \neq 8$ **compound**

20. $5 \neq 4 + 2$ **compound**

21. If Buddy is a politician, then Eddie is a crook. **compound**

22. If Earl Karn sells his quota, then Kerry Freeman will be happy. **compound**

Write a negation for each of the following statements.

23. The flowers are to be watered. **The flowers are not to be watered.**

24. Her aunt's name is Lucia. **Her aunt's name is not Lucia.**

25. No rain fell in southern California today. **Some rain fell in southern California today.**

26. Every dog has its day. **At least one dog does not have its day.**

27. All students present will get another chance. **At least one student present will not get another chance.**

28. Some books are longer than this book. **No book is longer than this book.**

29. Some people have all the luck. **No people have all the luck.**

30. No computer repairman can play blackjack. **At least one computer repairman can play blackjack.**

31. Nobody doesn't like Sara Lee. **Someone does not like Sara Lee.**

32. Everybody loves somebody sometime. **Someone does not love somebody sometime.**

Give a negation of each inequality. Do not use a slash symbol.

33. $x > 3$ $x \leq 3$ **34.** $y < -2$ $y \geq -2$

35. $p \geq 4$ $p < 4$ **36.** $q \leq 12$ $q > 12$

37. Explain why the negation of "$r > 4$" is not "$r < 4$."

38. Try to negate the sentence "The exact number of words in this sentence is ten" and see what happens. Explain the problem that arises.

Let p represent the statement "She has blue eyes" *and let q represent the statement* "He is 43 years old." *Translate each symbolic compound statement into words.*

39. $\sim p$ **She does not have blue eyes.**

40. $\sim q$ **He is not 43 years old.**

41. $p \wedge q$ **She has blue eyes and he is 43 years old.**

42. $p \vee q$ **She has blue eyes or he is 43 years old.**

43. $\sim p \vee q$ **She does not have blue eyes or he is 43 years old.**

44. $p \wedge \sim q$ **She has blue eyes and he is not 43 years old.**

45. $\sim p \vee \sim q$ **She does not have blue eyes or he is not 43 years old.**

46. $\sim p \wedge \sim q$ **She does not have blue eyes and he is not 43 years old.**

47. $\sim(\sim p \wedge q)$ **It is not the case that she does not have blue eyes and he is 43 years old.**

48. $\sim(p \vee \sim q)$ **It is not the case that she has blue eyes or he is not 43 years old.**

Let p represent the statement "Chris collects videotapes" *and let q represent the statement* "Jack is a shortstop." *Convert each of the following compound statements into symbols.*

49. Chris collects videotapes and Jack is not a shortstop. $p \wedge \sim q$

50. Chris does not collect videotapes or Jack is not a shortstop. $\sim p \vee \sim q$

51. Chris does not collect videotapes or Jack is a shortstop. $\sim p \vee q$

52. Jack is a shortstop and Chris does not collect videotapes. $q \wedge \sim p$

53. Neither Chris collects videotapes nor Jack is a shortstop. $\sim(p \vee q)$ **or, equivalently,** $\sim p \wedge \sim q$

54. Either Jack is a shortstop or Chris collects videotapes, and it is not the case that both Jack is a shortstop and Chris collects videotapes. $(q \vee p) \wedge [\sim(q \wedge p)]$

55. Incorrect use of quantifiers is often heard in everyday language. Suppose you hear that a local electronics chain is having a 30% off sale, and the radio advertisement states "All items are not available in all stores." Do you think that, literally translated, the ad really means what it says? What do you think is really meant? Explain your answer.

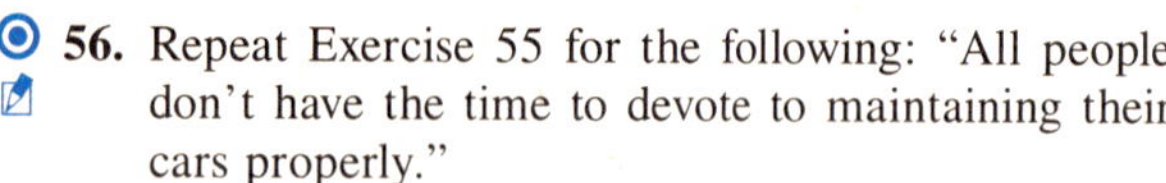

56. Repeat Exercise 55 for the following: "All people don't have the time to devote to maintaining their cars properly."

Decide whether each statement involving a quantifier is true *or* false.

57. Every natural number is an integer. **true**

58. Every whole number is an integer. **true**

59. There exists an integer that is not a natural number. **true**

60. There exists a rational number that is not an integer. **true**

61. All irrational numbers are real numbers. **true**

62. All rational numbers are real numbers. **true**

63. Some whole numbers are not rational numbers. **false**

64. Some rational numbers are not integers. **true**

65. Each rational number is a positive number. **false**

66. Each whole number is a positive number. **false**

Refer to the sketches labeled A, B, and C, and identify the sketch (or sketches) that is (are) satisfied by the given statement involving a quantifier.

A

B

C

67. All plants have a flower. **C**

68. At least one plant has a flower. **A, C**

69. No plant has a flower. **B**

70. All plants do not have a flower. **B**

71. At least one plant does not have a flower. **A, B**

72. No plant does not have a flower. **C**

73. Not every plant has a flower. **A, B**

74. Not every plant does not have a flower. **A, C**

75. Explain the difference between the following statements:

All students did not pass the test.
Not all students passed the test.

76. Write the following statement using "every": There is no one here who has not done that at one time or another. **Everyone here has done that at one time or another.**

77. The statement "For some real number x, $x^2 \geq 0$" is true. However, your friend does not understand why, since he claims that $x^2 \geq 0$ for *all* real numbers x (and not *some*). How would you explain his misconception to him?

78. Only one of the following statements is true. Which one is it? **(a)**

(a) For some real number x, $x \not< 0$.
(b) For all real numbers x, $x^3 > 0$.
(c) For all real numbers x less than 0, x^2 is also less than 0.
(d) For some real number x, $x^2 < 0$.

CONCEPTUAL WRITING CHALLENGING SCIENTIFIC CALCULATOR GRAPHING CALCULATOR

3.2 Exercises (Text Page 115)

Use the concepts introduced in this section to answer Exercises 1–6.

1. If we know that p is true, what do we know about the truth value of $p \vee q$ even if we are not given the truth value of q? **It must be true.**

2. If we know that p is false, what do we know about the truth value of $p \wedge q$ even if we are not given the truth value of q? **It must be false.**

3. If p is false, what is the truth value of $p \wedge (q \vee \sim r)$? **F**

4. If p is true, what is the truth value of $p \vee (q \vee \sim r)$? **T**

5. Explain in your own words the condition that must exist for a conjunction of two component statements to be true.

6. Explain in your own words the condition that must exist for a disjunction of two component statements to be false.

Let p represent a false statement and let q represent a true statement. Find the truth value of the given compound statement.

7. $\sim p$ **T** **8.** $\sim q$ **F** **9.** $p \vee q$ **T** **10.** $p \wedge q$ **F** **11.** $p \vee \sim q$ **F**

12. $\sim p \wedge q$ **T** **13.** $\sim p \vee \sim q$ **T** **14.** $p \wedge \sim q$ **F** **15.** $\sim(p \wedge \sim q)$ **T** **16.** $\sim(\sim p \vee \sim q)$ **F**

17. $\sim[\sim p \wedge (\sim q \vee p)]$ **T** **18.** $\sim[(\sim p \wedge \sim q) \vee \sim q]$ **T**

19. Is the statement $3 \geq 1$ a conjunction or a disjunction? **a disjunction**

20. Why is the statement $6 \geq 2$ true? Why is $6 \geq 6$ true? **It is true because $6 > 2$ is true; it is true because $6 = 6$ is true.**

Let p represent a true statement, and q and r represent false statements. Find the truth value of the given compound statement.

21. $(q \vee \sim r) \wedge p$ **T** **22.** $(p \wedge r) \vee \sim q$ **T** **23.** $(\sim p \wedge q) \vee \sim r$ **T** **24.** $p \wedge (q \vee r)$ **F**

25. $(\sim r \wedge \sim q) \vee (\sim r \wedge q)$ **T** **26.** $\sim(p \wedge q) \wedge (r \vee \sim q)$ **T**

27. $\sim[r \vee (\sim q \wedge \sim p)]$ **T** **28.** $\sim[(\sim p \wedge q) \vee r]$ **T**

Let p represent the statement $1 > 9$, let q represent the statement $7 \not> 5$, and let r represent the statement $13 \geq 13$. Find the truth value of the given compound statement.

29. $p \wedge r$ **F** **30.** $p \vee \sim q$ **T** **31.** $\sim q \vee \sim r$ **T** **32.** $\sim p \wedge \sim r$ **F**

33. $(p \wedge q) \vee r$ **T** **34.** $\sim p \vee (\sim r \vee \sim q)$ **T** **35.** $(\sim r \wedge q) \vee \sim p$ **T** **36.** $\sim(p \vee \sim q) \vee \sim r$ **F**

Give the number of rows in the truth table for each of the following compound statements.

37. $p \vee \sim r$ **4** **38.** $p \wedge (r \wedge \sim s)$ **8**

39. $(\sim p \wedge q) \vee (\sim r \vee \sim s) \wedge r$ **16** **40.** $[(p \vee q) \wedge (r \wedge s)] \wedge (t \vee \sim p)$ **32**

41. $[(\sim p \wedge \sim q) \wedge (\sim r \wedge s \wedge \sim t)] \wedge (\sim u \vee \sim v)$ **128**

42. $[(\sim p \wedge \sim q) \vee (\sim r \vee \sim s)] \vee [(\sim m \wedge \sim n) \wedge (u \wedge \sim v)]$ **256**

43. If the truth table for a certain compound statement has 64 rows, how many distinct component statements does it have? **6**

44. Is it possible for the truth table of a compound statement to have exactly 48 rows? Explain.

Construct a truth table for each compound statement. **In Exercises 45–60, we give the truth values found in the order in which they appear in the final column of the truth table, when the truth table is constructed in the format described in the section.**

45. $\sim p \wedge q$ **FFTF**

46. $\sim p \vee \sim q$ **FTTT**

47. $\sim(p \wedge q)$ **FTTT**

48. $p \vee \sim q$ **TTFT**

49. $(q \vee \sim p) \vee \sim q$ **TTTT**

50. $(p \wedge \sim q) \wedge p$ **FTFF**

51. $\sim q \wedge (\sim p \vee q)$ **FFFT**

52. $\sim p \vee (\sim q \wedge \sim p)$ **FFTT**

53. $(p \vee \sim q) \wedge (p \wedge q)$ **TFFF**

54. $(\sim p \wedge \sim q) \vee (\sim p \vee q)$ **TFTT**

55. $(\sim p \wedge q) \wedge r$ **FFFFTFFF**

56. $r \vee (p \wedge \sim q)$ **TFTTTFTF**

57. $(\sim p \wedge \sim q) \vee (\sim r \vee \sim p)$ **FTFTTTTT**

58. $(\sim r \vee \sim p) \wedge (\sim p \vee \sim q)$ **FFFTTTTT**

▲ **59.** $\sim(\sim p \wedge \sim q) \vee (\sim r \vee \sim s)$ **TTTTTTTTTTTTFTTT**

▲ **60.** $(\sim r \vee s) \wedge (\sim p \wedge q)$ **FFFFFFFFTFTTFFFF**

Use one of De Morgan's laws to write the negation of each statement.

61. You can pay me now or you can pay me later.
You can't pay me now and you can't pay me later.

62. I am not going or she is going. **I am going and she is not going.**

63. It is summer and there is no snow. **It is not summer or there is snow.**

64. 1/2 is a positive number and −12 is less than zero.
1/2 is not a positive number or −12 is not less than zero.

65. I said yes but she said no. **I did not say yes or she did not say no.**

66. Kelly Bell tried to sell the book, but she was unable to do so.
Kelly Bell did not try to sell the book, or she was able to do so.

67. $5 - 1 = 4$ and $9 + 12 \neq 7$ **$5 - 1 \neq 4$ or $9 + 12 = 7$**

68. $3 < 10$ or $7 \neq 2$ **$3 \not< 10$ and $7 = 2$**

▲ **69.** Dasher or Dancer will lead Santa's sleigh next Christmas.
Neither Dasher nor Dancer will lead Santa's sleigh next Christmas.

▲ **70.** The lawyer and the client appeared in court.
The lawyer did not appear in court or the client did not appear in court.

Identify each of the following statements as true *or* false.

71. For every real number y, $y < 12$ or $y > 4$. **T**

72. For every real number t, $t > 3$ or $t < 3$. **F**

73. For some integer p, $p \geq 5$ and $p \leq 5$. **T**

74. There exists an integer n such that $n > 0$ and $n < 0$. **F**

▲ **75.** Complete the truth table for *exclusive disjunction*. The symbol $\underline{\vee}$ represents "one or the other is true, but not both." **FTTF**

p	q	$p \underline{\vee} q$
T	T	
T	F	
F	T	
F	F	

Exclusive disjunction

◉ CONCEPTUAL ✎ WRITING ▲ CHALLENGING ▦ SCIENTIFIC CALCULATOR ▦ GRAPHING CALCULATOR

76. Lawyers sometimes use the phrase "and/or." This phrase corresponds to which usage of the word *or*: inclusive or exclusive? **inclusive disjunction**

▲ *Decide whether the following compound statements are* true *or* false. *Here,* or *is the* exclusive disjunction; *that is, assume "either p or q is true, but not both."*

77. $3 + 1 = 4$ or $2 + 5 = 7$ **F**

78. $3 + 1 = 4$ or $2 + 5 = 9$ **T**

79. $3 + 1 = 7$ or $2 + 5 = 7$ **T**

80. $3 + 1 = 7$ or $2 + 5 = 9$ **F**

3.3 Exercises (Text Page 122)

In Exercises 1–8, decide whether each statement is true *or* false.

1. If the antecedent of a conditional statement is false, the conditional statement is true. **true**
2. If the consequent of a conditional statement is true, the conditional statement is true. **true**
3. If q is true, then $(p \wedge q) \rightarrow q$ is true. **true**
4. If p is true, then $\sim p \rightarrow (q \vee r)$ is true. **true**
5. The negation of "If pigs fly, I'll believe it" is "If pigs don't fly, I won't believe it." **false**
6. The statements "If it flies, then it's a bird" and "It does not fly or it's a bird" are logically equivalent. **true**
7. Given that $\sim p$ is true and q is false, the conditional $p \rightarrow q$ is true. **true**
8. Given that $\sim p$ is false and q is false, the conditional $p \rightarrow q$ is true. **false**
9. In a few sentences, explain how we determine the truth value of a conditional statement.
10. Explain why the statement "If $3 = 5$, then $4 = 6$" is true.

Rewrite each statement using the if . . . then *connective. Rearrange the wording or add words as necessary.*

11. You can believe it if it's in *USA Today.* **If it's in *USA Today,* then you can believe it.**
12. It must be dead if it doesn't move. **If it doesn't move, then it must be dead.**
13. Kathi Callahan's area code is 708. **If the person is Kathi Callahan, then her area code is 708.**
14. Kara Gourley goes to Hawaii every summer. **If it is summer, then Kara Gourley goes to Hawaii.**
15. All soldiers maintain their weapons. **If you're a soldier, then you maintain your weapon.**
16. Every dog has its day. **If it's a dog, then it has its day.**
17. No koalas live in Mississippi. **If it's a koala, then it doesn't live in Mississippi.**
18. No guinea pigs are scholars. **If it's a guinea pig, then it is not a scholar.**
19. An alligator cannot live in these waters. **If it's an alligator, then it cannot live in these waters.**
20. Romeo loves Juliet. **If she is Juliet, then Romeo loves her.**

Tell whether each conditional is true *or* false. *Here* T *represents a true statement and* F *represents a false statement.*

21. $F \rightarrow (4 = 7)$ **T**

22. $T \rightarrow (4 < 2)$ **F**

23. $(6 = 6) \rightarrow F$ **F**

24. $F \rightarrow (3 = 3)$ **T**

25. $(4 = 11 - 7) \rightarrow (3 > 0)$ **T**

26. $(4^2 \neq 16) \rightarrow (4 + 4 = 8)$ **T**

Let s represent "I study in the library," *let p represent the statement* "I pass my psychology course," *and let m represent* "I major in mathematics." *Express each compound statement in words.*

27. $\sim m \rightarrow p$ **If I do not major in mathematics, then I pass my psychology course.**

28. $p \rightarrow \sim m$ **If I pass my psychology course, then I do not major in mathematics.**

29. $s \rightarrow (m \wedge p)$ **If I study in the library, then I major in mathematics and I pass my psychology course.**

30. $(s \wedge p) \rightarrow m$ **If I study in the library and I pass my psychology course, then I major in mathematics.**

31. $\sim p \rightarrow (\sim m \vee s)$ **If I do not pass my psychology course, then I do not major in mathematics or I study in the library.**

32. $(\sim s \vee \sim m) \rightarrow \sim p$ **If I do not study in the library or do not major in mathematics, then I do not pass my psychology course.**

Let d represent "I drive my car," *let s represent* "it snows," *and let c represent* "classes are cancelled." *Write each compound statement in symbols.*

33. If it snows, then I drive my car. $s \rightarrow d$

34. If I drive my car, then classes are cancelled. $d \rightarrow c$

35. If I do not drive my car, then it does not snow. $\sim d \rightarrow \sim s$

36. If classes are cancelled, then it does not snow. $c \rightarrow \sim s$

37. I drive my car, or if classes are cancelled then it snows. $d \vee (c \rightarrow s)$

38. Classes are cancelled, and if it snows then I do not drive my car. $c \wedge (s \rightarrow \sim d)$

39. I'll drive my car if it doesn't snow. $\sim s \rightarrow d$

40. It snows if classes are cancelled. $c \rightarrow s$

Find the truth value of each statement. Assume that p and r are false, and q is true.

41. $\sim r \rightarrow q$ **T**

42. $\sim p \rightarrow \sim r$ **T**

43. $q \rightarrow p$ **F**

44. $\sim r \rightarrow p$ **F**

45. $p \rightarrow q$ **T**

46. $\sim q \rightarrow r$ **T**

47. $\sim p \rightarrow (q \wedge r)$ **F**

48. $(\sim r \vee p) \rightarrow p$ **F**

49. $\sim q \rightarrow (p \wedge r)$ **T**

50. $(\sim p \wedge \sim q) \rightarrow (p \wedge \sim r)$ **T**

51. $(p \rightarrow \sim q) \rightarrow (\sim p \wedge \sim r)$ **T**

52. $(p \rightarrow \sim q) \wedge (p \rightarrow r)$ **T**

53. Explain why, if we know that p is true, we also know that

$$[r \vee (p \vee s)] \rightarrow (p \vee q)$$

is true, even if we are not given the truth values of q, r, and s.

54. Construct a true statement involving a conditional, a conjunction, a disjunction, and a negation (not necessarily in that order), that consists of component statements p, q, and r, with all of these component statements false. **Answers will vary. One example is $p \rightarrow [(\sim q \wedge r) \vee p]$.**

Construct a truth table for each statement. Identify any tautologies. **In Exercises 55–65, we give the truth values found in the order in which they appear in the final column of the truth table, when the truth table is constructed in the format described in Section 3.2.**

55. $\sim q \rightarrow p$ **TTTF**

56. $p \rightarrow \sim q$ **FTTT**

57. $(\sim p \rightarrow q) \rightarrow p$ **TTFT**

58. $(\sim q \rightarrow \sim p) \rightarrow \sim q$ **FTFT**

59. $(p \vee q) \rightarrow (q \vee p)$ **TTTT; tautology**

60. $(p \wedge q) \rightarrow (p \vee q)$ **TTTT; tautology**

61. $(\sim p \rightarrow \sim q) \rightarrow (p \wedge q)$ **TFTF**

62. $r \rightarrow (p \wedge \sim q)$ **FTTTFTFT**

63. $[(r \vee p) \wedge \sim q] \rightarrow p$ **TTTTTTFT**

▲ **64.** $(\sim r \rightarrow s) \vee (p \rightarrow \sim q)$ **TTTFTTTTTTTTTTTT**

▲ **65.** $(\sim p \wedge \sim q) \rightarrow (\sim r \rightarrow \sim s)$ **TTTTTTTTTTTTTTFT**

66. What is the minimum number of Fs that need appear in the final column of a truth table for us to be assured that the statement is not a tautology? **one**

CONCEPTUAL WRITING ▲ CHALLENGING SCIENTIFIC CALCULATOR GRAPHING CALCULATOR

Write the negation of each statement. Remember that the negation of $p \rightarrow q$ is $p \wedge \sim q$.

67. If you give your plants tender, loving care, they flourish. **You give your plants tender, loving care and they do not flourish.**

68. If the check is in the mail, I'll be surprised. **The check is in the mail and I'll not be surprised.**

69. If she doesn't, he will. **She doesn't and he will not.**

70. If I say yes, she says no. **I say yes and she does not say no.**

71. All residents of Boise are residents of Idaho. **The person is a resident of Boise and is not a resident of Idaho.**

72. All men were once boys. **He is a man and was not once a boy.**

Write each statement as an equivalent statement that does not use the if . . . then *connective. Remember that $p \rightarrow q$ is equivalent to $\sim p \vee q$.*

73. If you give your plants tender, loving care, they flourish. **You do not give your plants tender, loving care or they flourish.**

74. If the check is in the mail, I'll be surprised. **The check is not in the mail or I'll be surprised.**

75. If she doesn't, he will. **She does or he will.**

76. If I say yes, she says no. **I do not say yes or she says no.**

77. All residents of Boise are residents of Idaho. **The person is not a resident of Boise or is a resident of Idaho.**

78. All men were once boys. **The person is not a man or was once a boy.**

Use truth tables to decide which of the pairs of statements are equivalent.

79. $p \rightarrow q$; $\sim p \vee q$ **equivalent**

80. $\sim(p \rightarrow q)$; $p \wedge \sim q$ **equivalent**

81. $p \rightarrow q$; $\sim q \rightarrow \sim p$ **equivalent**

82. $q \rightarrow p$; $\sim p \rightarrow \sim q$ **equivalent**

83. $\sim(\sim p)$; p **equivalent**

84. $p \rightarrow q$; $q \rightarrow p$ **not equivalent**

85. $p \wedge \sim q$; $\sim q \rightarrow \sim p$ **not equivalent**

86. $\sim p \wedge q$; $\sim p \rightarrow q$ **not equivalent**

Extension Exercises (Text Page 126)

Write a logical statement representing each of the following circuits. Simplify each circuit when possible.

1.

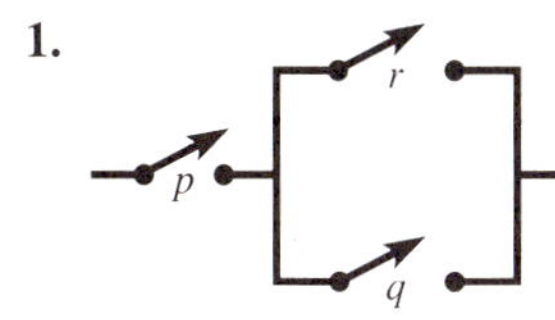

$p \wedge (r \vee q)$

2.

$(p \wedge q) \vee (p \wedge \sim q)$; The statement simplifies to p.

3.

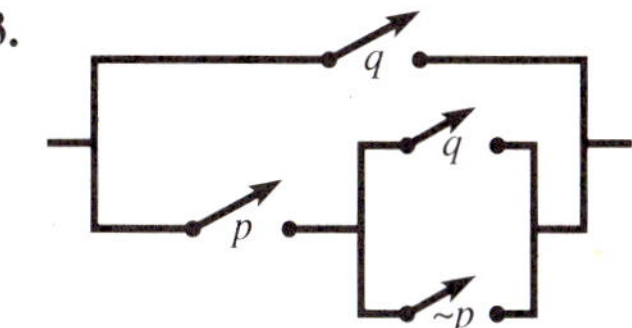

$q \vee [p \wedge (q \vee \sim p)]$; The statement simplifies to q.

4.

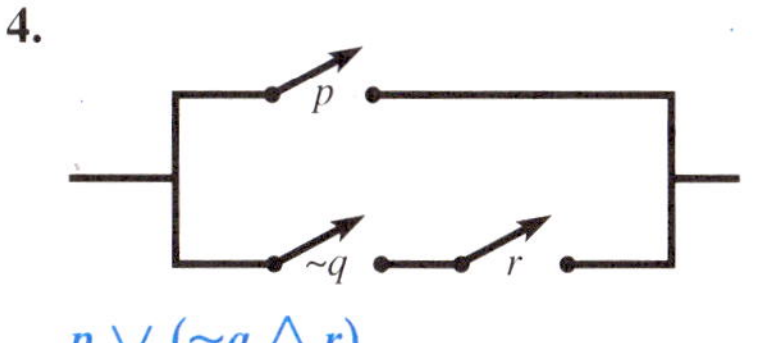

$p \vee (\sim q \wedge r)$

5.

$(\sim p \vee q) \vee (\sim p \vee \sim q)$; The statement simplifies to T.

6.

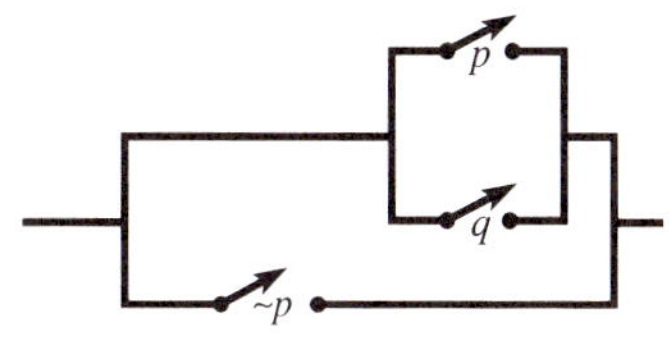

$\sim p \vee (p \vee q)$; The statement simplifies to T.

Draw circuits representing the following statements as they are given. Simplify if possible.

7. $(\sim p \wedge \sim q) \wedge \sim r$

8. $p \wedge (q \vee \sim p)$
The statement simplifies to $p \wedge q$.

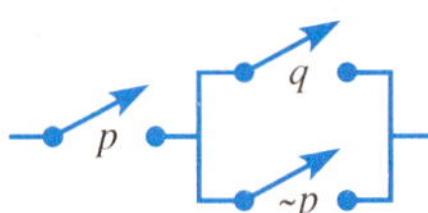

9. $(\sim q \wedge \sim p) \vee (\sim p \vee q)$

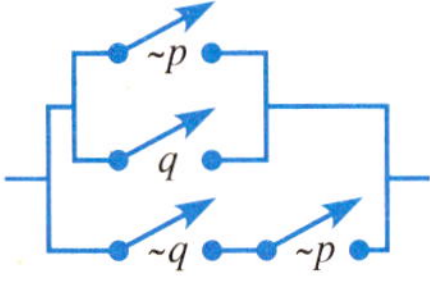

10. $(p \vee q) \wedge (\sim p \wedge \sim q)$
The statement simplifies to F.

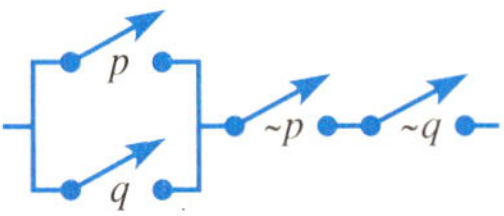

11. $[(\sim p \wedge \sim r) \vee \sim q] \wedge (\sim p \wedge r)$
The statement simplifies to $(\sim p \wedge r) \wedge \sim q$.

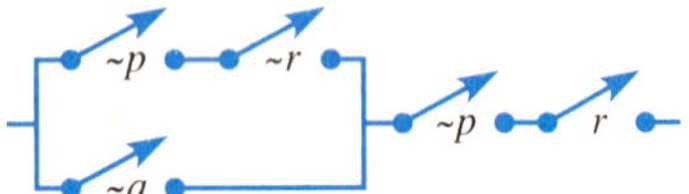

12. $[(p \vee q) \wedge r] \wedge \sim p$
The statement simplifies to $(r \wedge \sim p) \wedge q$.

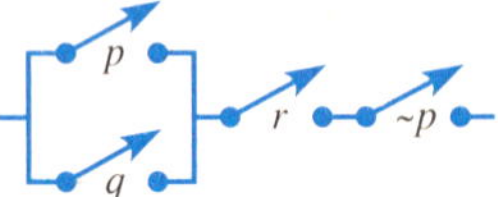

13. $\sim p \rightarrow (\sim p \vee \sim q)$
The statement simplifies to T.

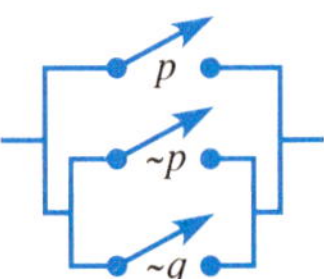

14. $\sim q \rightarrow (\sim p \rightarrow q)$
The statement simplifies to $p \vee q$.

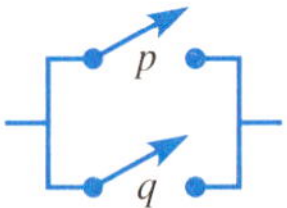

15. Refer to Figures 5 and 6 in Example 1. Suppose the cost of the use of one switch for an hour is 3¢. By using the circuit in Figure 6 rather than the circuit in Figure 5, what is the savings for a year of 365 days, assuming that the circuit is in continuous use? **$262.80**

16. Explain why the circuit

will always have exactly one open switch. What does this circuit simplify to?

CONCEPTUAL  WRITING ▲ CHALLENGING  SCIENTIFIC CALCULATOR GRAPHING CALCULATOR

3.4 Exercises (Text Page 132)

For each given direct statement, write **(a)** *the converse,* **(b)** *the inverse, and* **(c)** *the contrapositive in* if . . . then *form. In Exercises 3–10, it may be helpful to restate the direct statement in* if . . . then *form.*
Wording may vary in the answers to Exercises 1–10.

1. If you lead, then I will follow. **(a) If I follow, then you lead. (b) If you do not lead, then I will not follow. (c) If I do not follow, then you do not lead.**

2. If beauty were a minute, then you would be an hour. **(a) If you were an hour, then beauty would be a minute. (b) If beauty were not a minute, then you would not be an hour. (c) If you were not an hour, then beauty would not be a minute.**

3. If I had a nickel for each time that happened, I would be rich. **(a) If I were rich, then I would have a nickel for each time that happened. (b) If I did not have a nickel for each time that happened, then I would not be rich. (c) If I were not rich, then I would not have a nickel for each time that happened.**

4. If it ain't broke, don't fix it. **(a) If you don't fix it, then it ain't broke. (b) If it's broke, then fix it. (c) If you fix it, then it's broke.**

5. Milk contains calcium. **(a) If it contains calcium, then it's milk. (b) If it's not milk, then it does not contain calcium. (c) If it does not contain calcium, then it's not milk.**

6. Walking in front of a moving car is dangerous to your health. **(a) If it is dangerous to your health, then you walk in front of a moving car. (b) If you do not walk in front of a moving car, then it is not dangerous to your health. (c) If it is not dangerous to your health, then you do not walk in front of a moving car.**

7. A rolling stone gathers no moss. **(a) If it gathers no moss, then it is a rolling stone. (b) If it is not a rolling stone, then it gathers moss. (c) If it gathers moss, then it is not a rolling stone.**

8. Birds of a feather flock together. **(a) If they flock together, then they are birds of a feather. (b) If they are not birds of a feather, then they do not flock together. (c) If they do not flock together, then they are not birds of a feather.**

9. Where there's smoke, there's fire. **(a) If there's fire, then there's smoke. (b) If there's no smoke, then there's no fire. (c) If there's no fire, then there's no smoke.**

10. If you build it, he will come. **(a) If he comes, then you build it. (b) If you don't build it, then he won't come. (c) If he doesn't come, then you won't build it.**

11. $p \rightarrow \sim q$ **(a)** $\sim q \rightarrow p$ **(b)** $\sim p \rightarrow q$ **(c)** $q \rightarrow \sim p$

12. $\sim p \rightarrow q$ **(a)** $q \rightarrow \sim p$ **(b)** $p \rightarrow \sim q$ **(c)** $\sim q \rightarrow p$

13. $\sim p \rightarrow \sim q$ **(a)** $\sim q \rightarrow \sim p$ **(b)** $p \rightarrow q$ **(c)** $q \rightarrow p$

14. $\sim q \rightarrow \sim p$ **(a)** $\sim p \rightarrow \sim q$ **(b)** $q \rightarrow p$ **(c)** $p \rightarrow q$

▲ **15.** $p \rightarrow (q \vee r)$ (*Hint:* Use one of De Morgan's laws as necessary.) **(a)** $(q \vee r) \rightarrow p$ **(b)** $\sim p \rightarrow (\sim q \wedge \sim r)$ **(c)** $(\sim q \wedge \sim r) \rightarrow \sim p$

▲ **16.** $(r \vee \sim q) \rightarrow p$ (*Hint:* Use one of De Morgan's laws as necessary.) **(a)** $p \rightarrow (r \vee \sim q)$ **(b)** $(\sim r \wedge q) \rightarrow \sim p$ **(c)** $\sim p \rightarrow (\sim r \wedge q)$

17. Discuss the equivalences that exist among the direct conditional statement, the converse, the inverse, and the contrapositive.

18. State the contrapositive of "If the square of a natural number is even, then the natural number is even." The two statements must have the same truth value. Use several examples and inductive reasoning to decide whether both are true or both are false.

Write each of the following statements in the form "if *p*, then *q*."

19. If I finish studying, I'll go to the party. **If I finish studying, then I'll go to the party.**

20. If it is muddy, I'll wear my galoshes. **If it is muddy, then I'll wear my galoshes.**

21. "Today is Wednesday" implies that yesterday was Tuesday. **If it is Wednesday, then yesterday was Tuesday.**

22. "17 is positive" implies that 17 + 1 is positive. **If 17 is positive, then 17 + 1 is positive.**

23. All whole numbers are integers. **If a number is a whole number, then it is an integer.**

24. All integers are rational numbers. **If a number is an integer, then it is a rational number.**

25. Being in Fort Lauderdale is sufficient for being in Florida. **If you are in Fort Lauderdale, then you are in Florida.**

26. Doing crossword puzzles is sufficient for driving me crazy. **If I do crossword puzzles, then I am driven crazy.**

27. Being an environmentalist is necessary for being elected. **If one is elected, then one is an environmentalist.**

28. A day's growth of beard is necessary for Greg Odjakjian to shave. **If Greg Odjakjian is to shave, then he must have a day's growth of beard.**

29. The principal will hire more teachers only if the school board approves. **If the principal hires more teachers, then the school board approves.**

30. I can go from Park Place to Baltic Avenue only if I pass GO. **If I go from Park Place to Baltic Avenue, then I pass GO.**

31. No integers are irrational numbers. **If a number is an integer, then it is rational.**

32. No whole numbers are not integers. **If a number is a whole number, then it is an integer.**

33. Newt will be a liberal when pigs fly. **If pigs fly, then Newt will be a liberal.**

34. The Indians will win the pennant when their pitching improves. **If their pitching improves, then the Indians will win the pennant.**

35. A parallelogram is a four-sided figure with opposite sides parallel. **If the figure is a parallelogram, then it is a four-sided figure with opposite sides parallel.**

36. A rectangle is a parallelogram with a right angle. **If the figure is a rectangle, then it is a parallelogram with a right angle.**

37. A square is a rectangle with two adjacent sides equal. **If the figure is a square, then it is a rectangle with two adjacent sides equal.**

38. A triangle with two sides of the same length is isosceles. **If a triangle has two sides of the same length, then it is isosceles.**

39. An integer whose units digit is 0 or 5 is divisible by 5. **If an integer has a units digit of 0 or 5, then it is divisible by 5.**

40. The square of a two-digit number whose units digit is 5 will end in 25. **If a two-digit number whose units digit is 5 is squared, then it will end in 25.**

41. One of the following statements is not equivalent to all the others. Which one is it? **(d)**
(a) r only if s.
(b) r implies s.
(c) If r, then s.
(d) r is necessary for s.

42. Many students have difficulty interpreting *necessary* and *sufficient*. Use the statement "Being in Canada is sufficient for being in North America" to explain why "p is sufficient for q" translates as "if p, then q."

43. Use the statement "To be an integer, it is necessary that a number be rational" to explain why "p is necessary for q" translates as "if q, then p."

44. Explain why the statement "A week has eight days if and only if December has forty days" is true.

Identify each statement as true *or* false.

45. $5 = 9 - 4$ if and only if $8 + 2 = 10$. **true**

46. $3 + 1 \neq 6$ if and only if $9 \neq 8$. **true**

47. $8 + 7 \neq 15$ if and only if $3 \times 5 = 9$. **true**

48. $6 \times 2 = 14$ if and only if $9 + 7 \neq 16$. **true**

49. John F. Kennedy was president if and only if Ronald Reagan was not president. **false**

50. Burger King sells Big Macs if and only if Guess sells jeans. **false**

Two statements that can both be true about the same object are **consistent.** For example, "It is green" *and* "It is small" *are consistent statements. Statements that cannot both be true about the same object are called* **contrary;** "It is a Ford" *and* "It is a Chevrolet" *are contrary.*

Label the following pairs of statements as either contrary *or* consistent.

51. Elvis is alive. Elvis is dead. **contrary**

52. Bill Clinton is a Democrat. Bill Clinton is a Republican. **contrary**

53. That animal has four legs. That animal is a dog. **consistent**

CONCEPTUAL  WRITING ▲ CHALLENGING SCIENTIFIC CALCULATOR GRAPHING CALCULATOR

54. That book is nonfiction. That book costs more than $40. **consistent**

55. This number is an integer. This number is irrational. **contrary**

56. This number is positive. This number is a natural number. **consistent**

57. Make up two statements that are consistent. **For example: That man is Arnold Parker; That man sells books.**

58. Make up two statements that are contrary. **For example: Barbara Burnett is 35 years old; Barbara Burnett is older than 35 years.**

3.5 Exercises (Text Page 136)

Decide whether each argument is valid *or* invalid.

1. All museums exhibit art.
The Louvre is a museum.
———
The Louvre exhibits art. **valid**

2. All children love to swim.
Troy Kroeger is a child.
———
Troy Kroeger loves to swim. **valid**

3. All homeowners have a plumber.
Vonalaine Crowe has a plumber.
———
Vonalaine Crowe is a homeowner. **invalid**

4. All politicians have questionable ethics.
That man has questionable ethics.
———
That man is a politician. **invalid**

5. All residents of St. Tammany parish live on farms.
Jay Beckenstein lives on a farm.
———
Jay Beckenstein is a resident of St. Tammany parish. **invalid**

6. All dogs love to bury bones.
Archie is a dog.
———
Archie loves to bury bones. **valid**

7. All members of the credit union have savings accounts.
Kristyn Wasag does not have a savings account.
———
Kristyn Wasag is not a member of the credit union. **valid**

8. All engineers need mathematics.
Shane Stagg does not need mathematics.
———
Shane Stagg is not an engineer. **valid**

9. All residents of New Orleans have huge utility bills in July.
Erin Kelly has a huge utility bill in July.
———
Erin Kelly lives in New Orleans. **invalid**

10. All people applying for a home loan must provide a down payment.
Cynthia Herring provided a down payment.
———
Cynthia Herring applied for a home loan. **invalid**

11. Some mathematicians are absent-minded.
Diane Gray is a mathematician.
———
Diane Gray is absent-minded. **invalid**

12. Some animals are nocturnal.
Oliver Owl is an animal.
———
Oliver Owl is nocturnal. **invalid**

13. Some cars have automatic door locks.
Some cars are red.
———
Some red cars have automatic door locks. **invalid**

14. Some doctors appreciate classical music.
Kevin Howell is a doctor.
———
Kevin Howell appreciates classical music. **invalid**

15. Refer to Example 4 in this section. Give a different conclusion than the one given there so that the argument is still valid. **One possible conclusion is "All expensive things make you feel good."**

16. Construct a valid argument based on the Euler diagram shown here.

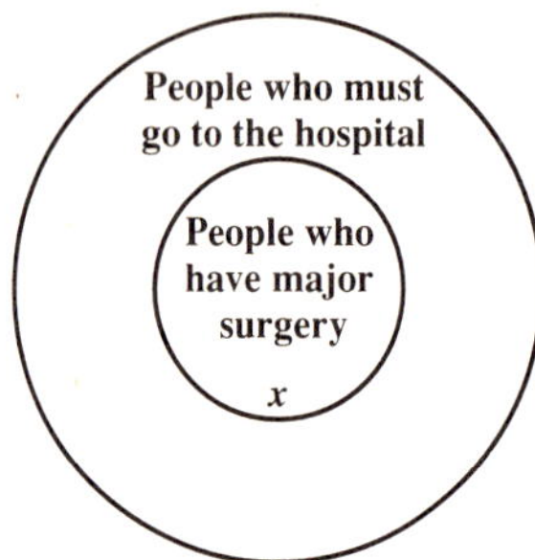

x represents Cynthia Biron.

People who have major surgery must go to the hospital.
Cynthia Biron is having major surgery.
Cynthia Biron must go to the hospital.

As mentioned in the text, an argument can have a true conclusion yet be invalid. In these exercises, each argument has a true *conclusion. Identify each argument as* valid *or* invalid.

17. All birds fly.
All planes fly.
A bird is not a plane. **invalid**

18. All cars have tires.
All tires are rubber.
All cars have rubber. **valid**

19. All chickens have a beak.
All hens are chickens.
All hens have a beak. **valid**

20. All chickens have a beak.
All birds have a beak.
All chickens are birds. **invalid**

21. Quebec is northeast of Ottawa.
Quebec is northeast of Toronto.
Ottawa is northeast of Toronto. **invalid**

22. Veracruz is south of Tampico.
Tampico is south of Monterrey.
Veracruz is south of Monterrey. **valid**

23. No whole numbers are negative.
-1 is negative.
-1 is not a whole number. **valid**

24. A scalene triangle has a longest side.
A scalene triangle has a largest angle.
The largest angle in a scalene triangle is opposite the longest side. **invalid**

In Exercises 25–30, the premises marked A, B, and C are followed by several possible conclusions. Take each conclusion in turn, and check whether the resulting argument is valid *or* invalid.

A. All people who drive contribute to air pollution.
B. All people who contribute to air pollution make life a little worse.
C. Some people who live in a suburb make life a little worse.

25. Some people who live in a suburb drive. **invalid**

26. Some people who live in a suburb contribute to air pollution. **invalid**

27. Some people who contribute to air pollution live in a suburb. **invalid**

28. Suburban residents never drive. **invalid**

29. All people who drive make life a little worse. **valid**

30. Some people who make life a little worse live in a suburb. **valid**

31. Find examples of arguments in magazine ads. Check them for validity.

32. Find examples of arguments on television commercials. Check them for validity.

Logic puzzles of the type found in Exercises 33–34 can be solved by using a "grid" where information is entered in each box. If the situation is impossible, write No *in the grid. If it is possible write* Yes. *For example, consider the following puzzle.*

Norma, Harriet, Betty, and Geneva are married to Don, Bill, John, and Nathan, but not necessarily in the order given. One couple has first names that start with the same letter. Harriet is married to John. Don's wife is neither Geneva nor Norma. Pair up each husband and wife.

The grid that follows from the puzzle is shown below.

	Don	**Bill**	**John**	**Nathan**
Norma	No	No	No	Yes
Harriet	No	No	Yes	No
Betty	Yes	No	No	No
Geneva	No	Yes	No	No

From the grid, we deduce that Harriet is married to John, Norma is married to Nathan, Betty is married to Don, and Geneva is married to Bill.

▲ *Solve each of the following puzzles.*

33. Juanita, Evita, Li, Fred, and Butch arrived at a party at different times. Evita arrived after Juanita but before Butch. Butch was neither the first nor the last to arrive. Juanita and Evita arrived after Fred but all three of them were present when Li got there. In what order did the five arrive? **Fred, Juanita, Evita, Butch, Li**

34. There are five girls whose first names are Cherie, Leann, Chris, Mary, and Monica. (The last two are twins.) The twins have the same last name, and the four last names are Parkerson, Bingle, Plunk, and Waters. Cherie's last name begins with a "P." The twins have never met the girl whose last name is Waters, and Leann has the longest last name. Determine the girls' complete names. **The girls' names are Cherie Plunk, Leann Parkerson, Chris Waters, and Mary and Monica Bingle.**

3.6 Exercises (Text Page 146)

Each of the following arguments is either valid by one of the forms of valid arguments discussed in this section, or a fallacy by one of the forms of invalid arguments discussed. (See the summary boxes.) Decide whether the argument is valid *or a* fallacy, *and give the form that applies.*

1. If you build it, he will come.
If he comes, then you will see your father.
___
If you build it, then you will see your father. **valid by reasoning by transitivity**

2. If Harry Connick, Jr. comes to town, then I will go to the concert.
If I go to the concert, then I'll be broke until payday.
___
If Harry Connick, Jr. comes to town, then I'll be broke until payday.
valid by reasoning by transitivity

3. If Doug Gilbert sells his quota, he'll get a bonus.
Doug Gilbert sold his quota.
___
He got a bonus. **valid by modus ponens**

4. If Cyndi Keen works hard enough, she will get a raise.
Cyndi Keen worked hard enough.
___
She got a raise. **valid by modus ponens**

5. If she buys a dress, then she will buy shoes.
She buys shoes.
___
She buys a dress. **fallacy by fallacy of the converse**

6. If I didn't have to write a term paper, I'd be ecstatic.
I am ecstatic.
___
I don't have to write a term paper. **fallacy by fallacy of the converse**

7. If beggars were choosers, then I could ask for it.
I cannot ask for it.
___
Beggars aren't choosers. **valid by modus tollens**

8. If Roger Clemens pitches, the Red Sox will win.
The Red Sox will not win.
———
Roger Clemens will not pitch. **valid by modus tollens**

9. "If I have seen farther than others, it is because I have stood on the shoulders of giants." (Sir Isaac Newton)
I have not seen farther than others.
———
I have not stood on the shoulders of giants. **fallacy by fallacy of the inverse**

10. "If we evolved a race of Isaac Newtons, that would not be progress." (Aldous Huxley)
We have not evolved a race of Isaac Newtons.
———
That is progress. **fallacy by fallacy of the inverse**

11. Alice Lavin sings or Barbara Lumer dances.
Barbara Lumer does not dance.
———
Alice Lavin sings. **valid by disjunctive syllogism**

12. She charges it on Visa or she orders it C.O.D.
She doesn't charge it on Visa.
———
She orders it C.O.D. **valid by disjunctive syllogism**

Use a truth table to determine whether the argument is valid *or* invalid.

13. $p \wedge \sim q$
p
———
$\sim q$ **valid**

14. $p \vee q$
p
———
$\sim q$ **invalid**

15. $p \vee \sim q$
p
———
$\sim q$ **invalid**

16. $\sim p \rightarrow \sim q$
q
———
p **valid**

17. $\sim p \rightarrow q$
p
———
$\sim q$ **invalid**

18. $p \rightarrow q$
$q \rightarrow p$
———
$p \wedge q$ **invalid**

19. $p \rightarrow \sim q$
$\sim p$
———
$\sim q$ **invalid**

20. $p \rightarrow \sim q$
q
———
$\sim p$ **valid**

21. $(p \rightarrow q) \wedge (q \rightarrow p)$
p
———
$p \vee q$ **valid**

22. $(\sim p \vee q) \wedge (\sim p \rightarrow q)$
p
———
$\sim q$ **invalid**

▲ **23.** $(r \wedge p) \rightarrow (r \vee q)$
$(q \wedge p)$
———
$r \vee p$ **valid**

▲ **24.** $(\sim p \wedge r) \rightarrow (p \vee q)$
$(\sim r \rightarrow p)$
———
$q \rightarrow r$ **invalid**

◉ **25.** Earlier we showed how to analyze arguments using Euler diagrams. Refer to Example 4 in this section, restate each premise and the conclusion using a quantifier, and then draw an Euler diagram to illustrate the relationship.

Every time something squeaks, I use WD-40.
Every time I use WD-40, I go to the hardware store.
———
Every time something squeaks, I go to the hardware store.

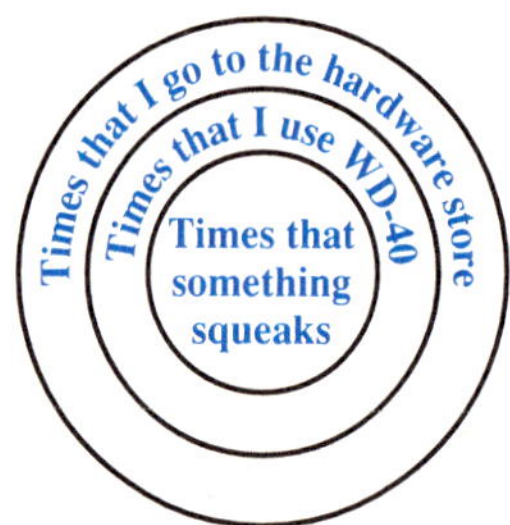

◉ **26.** Explain in a few sentences how to determine the statement for which a truth table will be constructed so that the arguments in Exercises 27–36 can be analyzed for validity.

◉ CONCEPTUAL  WRITING ▲ CHALLENGING  SCIENTIFIC CALCULATOR GRAPHING CALCULATOR

Determine whether the following arguments are valid *or* invalid.

27. Wally's hobby is amateur radio. If his wife likes to read, then Wally's hobby is not amateur radio. If his wife does not like to read, then Nikolas likes cartoons. Therefore, Nikolas likes cartoons. **valid**

28. If you are infected with a virus, then it can be transmitted. The consequences are serious and it cannot be transmitted. Therefore, if the consequences are not serious, then you are not infected with a virus. **valid**

29. Paula Abdul sings or Tom Cruise is not a hunk. If Tom Cruise is not a hunk, then Garth Brooks does not win a Grammy. Garth Brooks wins a Grammy. Therefore, Paula Abdul does not sing. **invalid**

30. If Bill so desires, then Al will be the vice president. Magic is a spokesman or Al will be the vice president. Magic is not a spokesman. Therefore, Bill does not so desire. **invalid**

31. The Falcons will be in the playoffs if and only if Morten is an all-pro. Janet loves the Falcons or Morten is an all-pro. Janet does not love the Falcons. Therefore, the Falcons will not be in the playoffs. **invalid**

32. If you're a big girl, then you don't cry. If you don't cry, then your momma does not say, "Shame on you." You don't cry or your momma says, "Shame on you." Therefore, if you're a big girl, then your momma says, "Shame on you." **invalid**

33. If I were your woman and you were my man, then I'd never stop loving you. I've stopped loving you. Therefore, I am not your woman or you are not my man. **valid**

34. If Charlie is a salesman, then he lives in Hattiesburg. Charlie lives in Hattiesburg and he loves to fish. Therefore, if Charlie does not love to fish, he is not a salesman. **valid**

35. All men are mortal. Socrates is a man. Therefore, Socrates is mortal. **valid**

36. All men are created equal. All people who are created equal are women. Therefore, all men are women. **valid**

37. Mittie Arnold made the following observation: "If I want to determine whether an argument leading to the statement

$$[(p \to q) \wedge \sim q] \to \sim p$$

is valid, I only need to consider the lines of the truth table which lead to T for the column headed $(p \to q) \wedge \sim q$." Mittie was very perceptive. Can you explain why her observation was correct?

38. Suppose that you ask someone for the time and you get the following response:

> "If I tell you the time, then we'll start chatting. If we start chatting, then you'll want to meet me at a truck stop. If we meet at a truck stop, then we'll discuss my family. If we discuss my family, then you'll find out that my daughter is available for marriage. If you find out that she is available for marriage, then you'll want to marry her. If you want to marry her, then my life will be miserable since I don't want my daughter married to some fool who can't afford a $10 watch."

Use reasoning by transitivity to draw a valid conclusion. **If I tell you the time, then my life will be miserable.**

In the arguments used by Lewis Carroll, it is helpful to restate a premise in if . . . then *form in order to more easily lead to a valid conclusion. The following premises come from Lewis Carroll. Write each premise in* if . . . then *form.* **Answers in Exercises 39–46 may be replaced by their contrapositives.**

39. None of your sons can do logic. **If he is your son, then he cannot do logic.**

40. All my poultry are ducks. **If it is my poultry, then it is a duck.**

41. No teetotalers are pawnbrokers. **If the person is a teetotaler, then the person is not a pawnbroker.**

42. Guinea pigs are hopelessly ignorant of music. **If it is a guinea pig, then it is hopelessly ignorant of music.**

43. Opium-eaters have no self-command. **If it is an opium-eater, then it has no self-command.**

44. No teachable kitten has green eyes. **If it is a teachable kitten, then it does not have green eyes.**

45. All of them written on blue paper are filed. **If it is written on blue paper, then it is filed.**

46. I have not filed any of them that I can read. **If I can read it, then I have not filed it.**

The following exercises involve premises from Lewis Carroll. Write each premise in symbols, and then in the final part, give a valid conclusion.

47. Let p be "one is able to do logic," q be "one is fit to serve on a jury," r be "one is sane," and s be "he is your son."

(a) Everyone who is sane can do logic. $r \rightarrow p$

(b) No lunatics are fit to serve on a jury. $\sim r \rightarrow \sim q$

(c) None of your sons can do logic. $s \rightarrow \sim p$

(d) Give a valid conclusion. **Your sons are not fit to serve on a jury.**

48. Let p be "it is a duck," q be "it is my poultry," r be "one is an officer," and s be "one is willing to waltz."

(a) No ducks are willing to waltz. $p \rightarrow \sim s$

(b) No officers ever decline to waltz. $r \rightarrow s$

(c) All my poultry are ducks. $q \rightarrow p$

(d) Give a valid conclusion. **None of my poultry are officers.**

49. Let p be "it is a guinea pig," q be "it is hopelessly ignorant of music," r be "it keeps silent while the *Moonlight Sonata* is being played," and s be "it appreciates Beethoven."

(a) Nobody who really appreciates Beethoven fails to keep silent while the *Moonlight Sonata* is being played. $s \rightarrow r$

(b) Guinea pigs are hopelessly ignorant of music. $p \rightarrow q$

(c) No one who is hopelessly ignorant of music ever keeps silent while the *Moonlight Sonata* is being played. $q \rightarrow \sim r$

(d) Give a valid conclusion. **Guinea pigs don't appreciate Beethoven.**

50. Let p be "one is honest," q be "one is a pawnbroker," r be "one is a promise-breaker," s be "one is trustworthy," t be "one is very communicative," and u be "one is a wine-drinker."

(a) Promise-breakers are untrustworthy. $r \rightarrow \sim s$

(b) Wine-drinkers are very communicative. $u \rightarrow t$

(c) A person who keeps a promise is honest. $\sim r \rightarrow p$

(d) No teetotalers are pawnbrokers. (*Hint:* Assume "teetotaler" is the opposite of "wine-drinker.") $\sim u \rightarrow \sim q$

(e) One can always trust a very communicative person. $t \rightarrow s$

(f) Give a valid conclusion. **All pawnbrokers are honest.**

51. Let p be "he is going to a party," q be "he brushes his hair," r be "he has self-command," s be "he looks fascinating," t be "he is an opium-eater," u be "he is tidy," and v be "he wears white kid gloves."

(a) No one who is going to a party ever fails to brush his hair. $p \rightarrow q$

(b) No one looks fascinating if he is untidy. $\sim u \rightarrow \sim s$

(c) Opium-eaters have no self-command. $t \rightarrow \sim r$

(d) Everyone who has brushed his hair looks fascinating. $q \rightarrow s$

(e) No one wears white kid gloves unless he is going to a party. (*Hint:* "a unless b" $\equiv \sim b \rightarrow a$.) $v \rightarrow p$

(f) A man is always untidy if he has no self-command. $\sim r \rightarrow \sim u$

(g) Give a valid conclusion. **Opium-eaters do not wear white kid gloves.**

52. Let p be "it begins with 'Dear Sir'," q be "it is crossed," r be "it is dated," s be "it is filed," t be "it is in black ink," u be "it is in the third person," v be "I can read it," w be "it is on blue paper," x be "it is on one sheet," and y be "it is written by Brown."

(a) All the dated letters in this room are written on blue paper. $r \rightarrow w$

(b) None of them are in black ink, except those that are written in the third person. $\sim u \rightarrow \sim t$

(c) I have not filed any of them that I can read. $v \rightarrow \sim s$

(d) None of them that are written on one sheet are undated. $x \rightarrow r$

(e) All of them that are not crossed are in black ink. $\sim q \rightarrow t$

(f) All of them written by Brown begin with "Dear Sir." $y \rightarrow p$

(g) All of them written on blue paper are filed. $w \rightarrow s$

(h) None of them written on more than one sheet are crossed. $\sim x \rightarrow \sim q$

(i) None of them that begin with "Dear Sir" are written in the third person. $p \rightarrow \sim u$

(j) Give a valid conclusion. **I can't read any of Brown's letters.**

CONCEPTUAL  WRITING CHALLENGING SCIENTIFIC CALCULATOR GRAPHING CALCULATOR

Chapter 3 Test (Text Page 150)

Write a negation for each of the following statements.

1. $5 + 3 = 9$ **$5 + 3 \neq 9$**
2. Every good boy deserves favour. **There is a good boy who does not deserve favour.**
3. Some people here can't play this game. **All people here can play this game.**
4. If it ever comes to that, I won't be here. **It comes to that and I am here.**
5. My mind is made up and you can't change it. **My mind is not made up or you can change it.**

Let p represent "it is broken" *and let q represent* "you can fix it." *Write each of the following in symbols.*

6. If it isn't broken, then you can fix it. **$\sim p \rightarrow q$**
7. It is broken or you can't fix it. **$p \vee \sim q$**
8. You can't fix anything that is broken. **$p \rightarrow \sim q$**

Using the same directions as for Exercises 6–8, write each of the following in words.

9. $\sim p \wedge q$ **It is not broken and you can fix it.**
10. $p \leftrightarrow \sim q$ **It is broken if and only if you can't fix it.**

In each of the following, assume that p and q are true, with r false. Find the truth value of each statement.

11. $\sim p \wedge \sim r$ **F**
12. $r \vee (p \wedge \sim q)$ **F**
13. $r \rightarrow (s \vee r)$ (The truth value of the statement s is unknown.) **T**
14. $r \leftrightarrow (p \rightarrow \sim q)$ **T**
15. What are the necessary conditions for a conditional statement to be false? for a conjunction to be true? **For a conditional statement to be false, the antecedent must be true and the consequent must be false. For a conjunction to be true, both component statements must be true.**
16. Explain in your own words why, if p is a statement, the biconditional $p \leftrightarrow \sim p$ must be false.

Write a truth table for each of the following. Identify any tautologies. **In Exercises 17–18, we give the truth values found in the order in which they appear in the final column of the truth table, when the truth table is constructed in the format described in the chapter.**

17. $p \wedge (\sim p \vee q)$ **TFFF**
18. $\sim(p \wedge q) \rightarrow (\sim p \vee \sim q)$ **TTTT (tautology)**

Decide whether each statement is true *or* false.

19. All positive integers are whole numbers. **true**
20. No real numbers are integers. **false**

Write each conditional statement in the form if . . . then. **Wording may vary in the answers to Exercises 21–26.**

21. All rational numbers are real numbers. **If it is a rational number, then it is a real number.**
22. Being a rectangle is sufficient for a polygon to be a quadrilateral. **If a polygon is a rectangle, then it is a quadrilateral.**
23. Being divisible by 2 is necessary for a number to be divisible by 6. **If a number is divisible by 6, then it is divisible by 2.**
24. She cries only if she is hurt. **If she cries, then she is hurt.**

For each statement, write **(a)** *the converse,* **(b)** *the inverse, and* **(c)** *the contrapositive.*

25. If a picture paints a thousand words, the graph will help me understand it. **(a) If the graph helps me understand it, then a picture paints a thousand words. (b) If a picture doesn't paint a thousand words, then the graph won't help me understand it. (c) If the graph doesn't help me understand it, then a picture doesn't paint a thousand words.**
26. $\sim p \rightarrow (q \wedge r)$ (Use one of De Morgan's laws as necessary.) **(a) $(q \wedge r) \rightarrow \sim p$ (b) $p \rightarrow (\sim q \vee \sim r)$ (c) $(\sim q \vee \sim r) \rightarrow p$**
27. Use an Euler diagram to determine whether the following argument is *valid* or *invalid*.

All members of that music club save money.
Steve Gold is a member of that music club.
———
Steve Gold saves money. **valid**

28. Match each argument in (a)–(d) with the law that justifies its validity, or the fallacy of which it is an example.

A. Modus ponens
B. Modus tollens
C. Reasoning by transitivity
D. Disjunctive syllogism
E. Fallacy of the converse
F. Fallacy of the inverse

(a) If you like ice cream, then you'll like Blue Bell.
You don't like Blue Bell.
———
You don't like ice cream. **B**

(b) If I buckle up, I'll be safer.
I don't buckle up.
———
I'm not safer. **F**

(c) If you love me, you will let me go.
If you let me go, I'll try to forget.
———
If you love me, I'll try to forget. **C**

(d) It is March or April.
It's not March.
———
It's April. **D**

Use a truth table to determine whether each argument is valid *or* invalid.

29. If I hear that song, it reminds me of my youth. If I get sentimental, then it does not remind me of my youth. I get sentimental. Therefore, I don't hear that song. **valid**

30. $\sim p \rightarrow \sim q$
$q \rightarrow p$
———
$p \vee q$ **invalid**

CHAPTER 4 Numeration and Mathematical Systems

4.1 Exercises (Text Page 161)

Convert each Egyptian numeral to Hindu-Arabic form.

1. **2,412**

2. **13,036**

3. **3,005,231**

4. **7,630,729**

Convert each Hindu-Arabic numeral to Egyptian form.

5. 427

6. 23,145

7. 306,090

8. 8,657,000

Chapter 1 of the book of Numbers in the Bible describes a census of the draft-eligible men of Israel after Moses led them out of Egypt into the Desert of Sinai, about 1450 B.C. Write an Egyptian numeral for the number of available men from each tribe listed.

9. 46,500 from the tribe of Reuben

 CONCEPTUAL  WRITING  CHALLENGING  SCIENTIFIC CALCULATOR  GRAPHING CALCULATOR

10. 59,300 from the tribe of Simeon

11. 45,650 from the tribe of Gad

12. 74,600 from the tribe of Judah

13. 54,400 from the tribe of Issachar

14. 62,700 from the tribe of Dan

Convert each Chinese numeral to Hindu-Arabic form.

15. 二百四十六 **246**

16. 九百三十五 **935**

17. 四千九百零二 **4,902**

18. 三千零七 **3,007**

Convert each Hindu-Arabic numeral to Chinese.

19. 63 六十三

20. 960 九百六十

21. 2,416 二千四百十六

22. 7,012 

Though Chinese art forms began before written history, their highest development was achieved during four particular dynasties. Write traditional Chinese numerals for the beginning and ending dates of each dynasty listed.

23. Han (202 B.C. to A.D. 220) 二百零二 to 二百二十

24. T'ang (618 to 907) 六百十八 to 九百零七

25. Sung (960 to 1279) 九百六十 to 一千二百七十九

26. Ming (1368 to 1644) 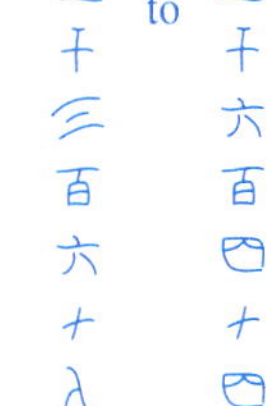 to 一千六百四十四

Work each of the following addition or subtraction problems, using regrouping as necessary. Convert each answer to Hindu-Arabic form.

27. 392

28. 216

29. 6,168

30. 53,601

31. 22

32. 113

33. 1,263

34. 7,598

Use the Egyptian algorithm to find each product.

35. 3 × 19 **57**

36. 5 × 26 **130**

37. 12 × 93 **1,116**

38. 21 × 44 **924**

▲ *Convert all numbers in the following problems to Egyptian numerals. Multiply using the Egyptian algorithm, and add using the Egyptian symbols. Give the final answer using a Hindu-Arabic numeral.*

39. King Solomon told the King of Tyre (now Lebanon) that Solomon needed the best cedar for his temple, and that he would "pay you for your men whatever sum you fix." Find the total bill to Solomon if the King of Tyre used the following numbers of men: 5,500 tree cutters at two shekels per week each, for a total of seven weeks; 4,600 sawers of wood at three shekels per week each, for a total of 32 weeks; and 900 sailors at one shekel per week each, for a total of 16 weeks. **533,000 shekels**

40. The book of Ezra in the Bible describes the return of the exiles to Jerusalem. When they rebuilt the temple, the King of Persia gave them the following items: thirty golden basins, a thousand silver basins, four hundred ten silver bowls, and thirty golden bowls. Find the total value of this treasure, if each gold basin is worth 3,000 shekels, each silver basin is worth 500 shekels, each silver bowl is worth 50 shekels, and each golden bowl is worth 400 shekels. **622,500 shekels**

▲ ◉ ✎ *Explain why each of the following steps would be an improvement in the development of numeration systems.*

41. progressing from carrying groups of pebbles to making tally marks on a stick

42. progressing from tallying to simple grouping

43. progressing from simple grouping to multiplicative grouping

44. progressing from multiplicative grouping to positional numeration

Recall that the ancient Egyptian system described in this section was simple grouping, used a base of ten, and contained seven distinct symbols. The largest number expressible in that system is 9,999,999. Identify the largest number expressible in each of the following simple grouping systems. (In Exercises 49–52, d can be any counting number.)

45. base ten, five distinct symbols **99,999**

46. base ten, ten distinct symbols **9,999,999,999**

▲ **47.** base five, five distinct symbols **3,124**

▲ **48.** base five, ten distinct symbols **9,765,624**

◉ CONCEPTUAL ✎ WRITING ▲ CHALLENGING ▦ SCIENTIFIC CALCULATOR ▦ GRAPHING CALCULATOR

49. base ten, d distinct symbols $10^d - 1$

50. base five, d distinct symbols $5^d - 1$

51. base seven, d distinct symbols $7^d - 1$

52. base b, d distinct symbols (where b is any counting number 2 or greater) $b^d - 1$

The Hindu-Arabic system is positional and uses ten as the base. Describe any advantages or disadvantages that may have resulted in each of the following cases.

53. Suppose the base had been larger, say twelve or twenty for example.

54. Suppose the base had been smaller, maybe eight or five.

4.2 Exercises (Text Page 170)

Write each number in expanded form.

1. 37 $(3 \times 10^1) + (7 \times 10^0)$

2. 814 $(8 \times 10^2) + (1 \times 10^1) + (4 \times 10^0)$

3. 2,815 $(2 \times 10^3) + (8 \times 10^2) + (1 \times 10^1) + (5 \times 10^0)$

4. 15,504 $(1 \times 10^4) + (5 \times 10^3) + (5 \times 10^2) + (0 \times 10^1) + (4 \times 10^0)$

5. three thousand, six hundred twenty-eight
$(3 \times 10^3) + (6 \times 10^2) + (2 \times 10^1) + (8 \times 10^0)$

6. fifty-three thousand, eight hundred twelve
$(5 \times 10^4) + (3 \times 10^3) + (8 \times 10^2) + (1 \times 10^1) + (2 \times 10^0)$

7. thirteen million, six hundred six thousand, ninety
$(1 \times 10^7) + (3 \times 10^6) + (6 \times 10^5) + (0 \times 10^4) + (6 \times 10^3) + (0 \times 10^2) +$
$(9 \times 10^1) + (0 \times 10^0)$

8. one hundred twelve million, fourteen thousand, one hundred twelve
$(1 \times 10^8) + (1 \times 10^7) + (2 \times 10^6) + (0 \times 10^5) + (1 \times 10^4) + (4 \times 10^3) +$
$(1 \times 10^2) + (1 \times 10^1) + (2 \times 10^0)$

Simplify each of the following expansions.

9. $(7 \times 10^1) + (3 \times 10^0)$ 73

10. $(2 \times 10^2) + (6 \times 10^1) + (0 \times 10^0)$ 260

11. $(5 \times 10^3) + (0 \times 10^2) + (7 \times 10^1) + (2 \times 10^0)$ 5,072

12. $(4 \times 10^5) + (0 \times 10^4) + (7 \times 10^3) + (7 \times 10^2) + (5 \times 10^1) + (2 \times 10^0)$ 407,752

13. $(5 \times 10^7) + (6 \times 10^5) + (2 \times 10^3) + (3 \times 10^0)$ 50,602,003

14. $(6 \times 10^8) + (5 \times 10^7) + (1 \times 10^2) + (4 \times 10^0)$ 650,000,104

In each of the following, add in expanded notation.

15. 63 + 26 89

16. 693 + 305 998

In each of the following, subtract in expanded notation.

17. 84 − 52 32

18. 673 − 412 261

Perform each addition using expanded notation.

19. 65 + 44 109

20. 536 + 279 815

21. 424 + 298 722

22. 6,755 + 4,827 11,582

Perform each subtraction using expanded notation.

23. 53 − 47 6

24. 253 − 48 205

25. 643 − 436 207

26. 826 − 345 481

Identify the number represented on each of these abaci.

27. 23

28. 256

29. 4,536

30. 63,259

Sketch an abacus to show each number.

31. 38

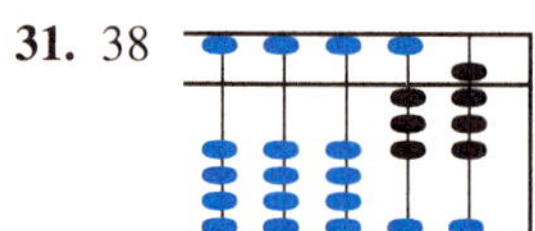

32. 183

33. 2,547

34. 70,163

Use the lattice method to find each product.

35. 63 × 28 **1,764**

36. 29 × 635 **18,415**

37. 413 × 68 **28,084**

38. 845 × 396 **334,620**

Refer to Example 10, where Napier's rods were used to find the product of 723 *and* 4,198. *Then complete Exercises 39 and 40.*

39. Find the product of 723 and 4,198 by completing the lattice process shown here. **3,035,154**

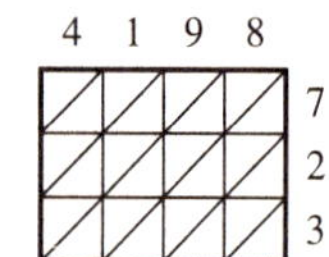

40. Explain how Napier's rods could have been used in Example 10 to set up one complete lattice product rather than adding three individual (shifted) lattice products.

Make use of Napier's rods (Figure 2) to find each product.

41. 8 × 62 **496**

42. 32 × 73 **2,336**

43. 26 × 8,354 **217,204**

44. 526 × 4,863 **2,557,938**

Use the Russian peasant method to find each product.

45. 5 × 82 **410**

46. 41 × 33 **1,353**

47. 62 × 429 **26,598**

48. 135 × 63 **8,505**

4.3 Exercises (Text Page 179)

List the first twenty counting numbers in each of the following bases.

1. seven (Only digits 0 through 6 are used in base seven.)
1, 2, 3, 4, 5, 6, 10, 11, 12, 13, 14, 15, 16, 20, 21, 22, 23, 24, 25, 26

2. eight (Only digits 0 through 7 are used.)
1, 2, 3, 4, 5, 6, 7, 10, 11, 12, 13, 14, 15, 16, 17, 20, 21, 22, 23, 24

▲ **3.** nine (Only digits 0 through 8 are used.)
1, 2, 3, 4, 5, 6, 7, 8, 10, 11, 12, 13, 14, 15, 16, 17, 18, 20, 21, 22

4. sixteen (The digits 0, 1, 2, . . . , 9, A, B, C, D, E, F are used in base sixteen.)
1, 2, 3, 4, 5, 6, 7, 8, 9, A, B, C, D, E, F, 10, 11, 12, 13, 14

For each of the following, write (in the same base) the counting numbers just before and just after the given number. (Do not convert to base ten.)

5. 14_{five} **13_{five}; 20_{five}**

6. 555_{six} **554_{six}; $1{,}000_{\text{six}}$**

7. $\text{B6F}_{\text{sixteen}}$ **$\text{B6E}_{\text{sixteen}}$; $\text{B70}_{\text{sixteen}}$**

8. $10{,}111_{\text{two}}$ **$10{,}110_{\text{two}}$; $11{,}000_{\text{two}}$**

Determine the number of distinct symbols needed in each of the following positional systems.

9. base three **3**

10. base seven **7**

11. base eleven **11**

12. base sixteen **16**

Determine, in each of the following bases, the smallest and largest four-digit numbers and their decimal equivalents.

13. three **smallest: $1{,}000_{\text{three}} = 27$; largest: $2{,}222_{\text{three}} = 80$**

14. sixteen **smallest: $1{,}000_{\text{sixteen}} = 4{,}096$; largest: $\text{FFFF}_{\text{sixteen}} = 65{,}535$**

Convert each of the following to decimal form by expanding in powers and by using the calculator shortcut.

15. 24_{five} **14**

16. 62_{seven} **44**

17. $1{,}011_{\text{two}}$ **11**

18. 35_{eight} **29**

19. $\text{3BC}_{\text{sixteen}}$ **956**

20. $34{,}432_{\text{five}}$ **2,492**

21. $2{,}366_{\text{seven}}$ **881**

22. $101{,}101{,}110_{\text{two}}$ **366**

23. $70{,}266_{\text{eight}}$ **28,854**

24. $\text{A,BCD}_{\text{sixteen}}$ **43,981**

▲ **25.** $2{,}023_{\text{four}}$ **139**

▲ **26.** $6{,}185_{\text{nine}}$ **4,532**

▲ **27.** $41{,}533_{\text{six}}$ **5,601**

▲ **28.** $88{,}703_{\text{nine}}$ **58,890**

Convert each of the following from decimal form to the given base.

29. 86 to base five **321_{five}**

30. 65 to base seven **122_{seven}**

31. 19 to base two **$10{,}011_{\text{two}}$**

32. 935 to base eight **$1{,}647_{\text{eight}}$**

33. 147 to base sixteen **93_{sixteen}**

34. 2,730 to base sixteen **$\text{AAA}_{\text{sixteen}}$**

35. 36,401 to base five **$2{,}131{,}101_{\text{five}}$**

36. 70,893 to base seven **$413{,}454_{\text{seven}}$**

37. 586 to base two **$1{,}001{,}001{,}010_{\text{two}}$**

38. 12,888 to base eight **$31{,}130_{\text{eight}}$**

39. 8,407 to base three **$102{,}112{,}101_{\text{three}}$**

40. 11,028 to base four **$2{,}230{,}110_{\text{four}}$**

41. 9,346 to base six **$111{,}134_{\text{six}}$**

42. 99,999 to base nine **$162{,}150_{\text{nine}}$**

Make the following conversions as indicated.

43. 43_{five} to base seven **32_{seven}**

44. 27_{eight} to base five **43_{five}**

45. $\text{C02}_{\text{sixteen}}$ to base seven **$11{,}651_{\text{seven}}$**

▲ **46.** $6{,}748_{\text{nine}}$ to base four **$1{,}031{,}321_{\text{four}}$**

Convert each of the following from octal form to binary form.

47. 367_{eight} **$11{,}110{,}111_{\text{two}}$**

48. $2{,}406_{\text{eight}}$ **$10{,}100{,}000{,}110_{\text{two}}$**

Convert each of the following from binary form to octal form.

49. $100{,}110{,}111_{\text{two}}$ **467_{eight}**

50. $11{,}010{,}111{,}101_{\text{two}}$ **$3{,}275_{\text{eight}}$**

Make the following conversions as indicated.

51. $\text{DC}_{\text{sixteen}}$ to binary **$11{,}011{,}100_{\text{two}}$**

52. $\text{F,}111_{\text{sixteen}}$ to binary **$1{,}111{,}000{,}100{,}010{,}001_{\text{two}}$**

53. $101{,}101_{\text{two}}$ to hexadecimal **$2\text{D}_{\text{sixteen}}$**

54. $101{,}111{,}011{,}101{,}000_{\text{two}}$ to hexadecimal **$5\text{,EE}8_{\text{sixteen}}$**

Identify the largest number from each list in Exercises 55 and 56.

55. 42_{seven}, 37_{eight}, $1\text{D}_{\text{sixteen}}$ **37_{eight}**

56. $1{,}101{,}110_{\text{two}}$, 407_{five}, $6\text{F}_{\text{sixteen}}$ **$6\text{F}_{\text{sixteen}}$**

Some people think that twelve would be a better base than ten. This is mainly because twelve has more divisors (1, 2, 3, 4, 6, 12) *than ten* (1, 2, 5, 10), *which makes fractions easier in base twelve. The base twelve system is called the* duodecimal system. *Just as in the decimal system we speak of a one, a ten, and a hundred (and so on), in the duodecimal system we say a* one, *a* dozen *(twelve), and a* gross *(twelve squared, or one hundred forty-four).*

57. Altogether, Otis Taylor's clients ordered 9 gross, 10 dozen, and 11 copies of *Math for Base Runners* during the last week of January. How many copies was that in base ten? **1,427**

One very common method of converting symbols into binary digits for computer processing is called ASCII (American Standard Code of Information Interchange). The upper case letters A *through* Z *are assigned the numbers* 65 *through* 90, *so* A *has binary code* 1000001 *and* Z *has code* 1011010. *Lowercase letters* a *through* z *have codes* 97 *through* 122 *(that is,* 1100001 *through* 1111010*). ASCII codes, as well as other numerical computer output, normally appear without commas.*

Write the binary code for each of the following letters.

58. C **1000011**

59. X **1011000**

60. k **1101011**

61. r **1110010**

Break each of the following into groups of seven digits and write as letters.

62. 1001000100010110011001010000 **HELP**

63. 10000111001000101010110000111001011 **CHUCK**

Translate each word into an ASCII string of binary digits. (Be sure to distinguish upper and lowercase letters.)

64. New **100111011001011110111**

65. Orleans **10011111110010110110011001011100001110111011110011**

66. Explain why the octal and hexadecimal systems are convenient for people who code for computers.

67. There are thirty-seven counting numbers whose base eight numerals contain two digits but whose base three numerals contain four digits. Find the smallest and largest of these numbers. **27 and 63**

68. What is the smallest counting number (expressed in base ten) that would require six digits in its base nine representation? **59,049**

 CONCEPTUAL  WRITING ▲ CHALLENGING SCIENTIFIC CALCULATOR 

69. In a pure decimal (base ten) monetary system, the first three denominations needed are pennies, dimes, and dollars. Name the first three denominations of a base five monetary system. **pennies, nickels, and quarters**

Refer to Table 9 for Exercises 70–73.

70. After observing the binary forms of the numbers 1–31, identify a common property of all Table 9 numbers in each of the following columns.
(a) Column A **The binary ones digit is 1.**
(b) Column B **The binary twos digit is 1.**
(c) Column C **The binary fours digit is 1.**
(d) Column D **The binary eights digit is 1.**
(e) Column E **The binary sixteens digit is 1.**

71. Explain how the "trick" of Table 9 works.

72. How many columns would be needed for Table 9 to include all ages up to 63? **6 columns**

73. How many columns would be needed for Table 9 to include all numbers up to 127? **7**

In our decimal system, we distinguish odd and even numbers by looking at their ones (or units) digit. If the ones digit is even (0, 2, 4, 6, *or* 8) *the number is even. If the ones digit is odd* (1, 3, 5, 7, *or* 9) *the number is odd. For Exercises 74–83, determine whether this same criterion works for numbers expressed in the given bases.*

74. two **yes** **75.** three **no** **76.** four **yes** **77.** five **no**

78. six **yes** **79.** seven **no** **80.** eight **yes** **81.** nine **no**

82. Consider all even bases. If it works for all, explain why. If not, find a criterion that does work for all even bases. **yes**

83. Consider all odd bases. If it works for all, explain why. If not, find a criterion that does work for all odd bases. **no**

84. Which of the following base five numerals represent numbers that are divisible by five?

3,204, 200, 342, 2,310, 3,041 **200, 2,310**

4.4 Exercises (Text Page 191)

Find each of the following differences on the 12-hour clock.

1. 8 − 3 **5** **2.** 4 − 9 **7** **3.** 2 − 8 **6** **4.** 0 − 3 **9**

5. Complete the 12-hour clock multiplication table at the right. You can use repeated addition and the addition table (for example, 3 × 7 = 7 + 7 + 7 = 2 + 7 = 9) or use modulo 12 multiplication techniques, as in Example 8, parts (d) and (e). **row 2: 0, 6, 10; row 3: 9, 0, 9; row 4: 0, 4, 0, 8, 0; row 5: 1, 9, 2, 7; row 6: 6, 0, 0; row 7: 4, 11, 6, 8, 3, 5; row 8: 8, 4, 0, 0; row 9: 6, 3, 9, 3, 9, 6, 3; row 10: 6, 4, 0, 10, 8, 6, 4; row 11: 10, 9, 8, 7, 6, 5, 4, 3, 2**

×	0	1	2	3	4	5	6	7	8	9	10	11
0	0	0	0	0	0	0	0	0	0	0	0	0
1	0	1	2	3	4	5	6	7	8	9	10	11
2	0	2	4	6	8	10		2	4		8	
3	0	3	6	9	0	3	6			3	6	
4	0	4	8			8		4			4	8
5	0	5	10	3	8		6	11	4			
6	0	6	0		0	6	0	6		6		6
7	0	7	2	9				1			10	
8	0	8	4	0				8	4		8	4
9	0	9			0		6		0			
10	0	10	8			2						2
11	0	11										1

By referring to your table in Exercise 5, determine which of the following properties hold for the system of 12-hour clock numbers with the operation of multiplication.

6. closure **yes**

7. commutative **yes**

8. identity **yes (1 is the identity element.)**

A 5-hour clock system utilizes the set {0, 1, 2, 3, 4}, *and relates to the clock face shown here.*

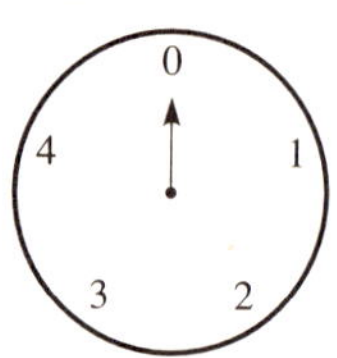

5-hour clock system

⊙ **9.** Complete this 5-hour clock addition table. **row 1: 0; row 2: 0; row 3: 0, 1, 2; row 4: 0, 1, 2**

+	**0**	**1**	**2**	**3**	**4**
0	0	1	2	3	4
1	1	2	3	4	
2	2	3	4		1
3	3	4			
4	4				3

⊙ *Which of the following properties are satisfied by the system of 5-hour clock numbers with the operation of addition?*

10. closure **yes**

11. commutative **yes**

12. identity (If so, what is the identity element?) **yes (0 is the identity element.)**

13. inverse (If so, name the inverse of each element.) **yes (0 is its own inverse, 1 and 4 are inverses of each other, and 2 and 3 are inverses of each other.)**

14. Complete this 5-hour clock multiplication table. **row 2: 1, 3; row 3: 1, 4, 2; row 4: 3, 2, 1**

×	**0**	**1**	**2**	**3**	**4**
0	0	0	0	0	0
1	0	1	2	3	4
2	0	2	4		
3	0	3			
4	0	4			

⊙ *Which of the following properties are satisfied by the system of 5-hour clock numbers with the operation of multiplication?*

15. closure **yes**

16. commutative **yes**

17. identity (If so, what is the identity element?) **yes (1 is the identity element.)**

In clock arithmetic, as in ordinary arithmetic, $a - b = d$ *is true if and only if* $b + d = a$. *Similarly,* $a \div b = q$ *if and only if* $b \times q = a$.

▲ ⊙ *Use the idea above and your 5-hour clock multiplication table of Exercise 14 to find the following quotients on a 5-hour clock.*

18. $1 \div 3$ **2**

19. $3 \div 1$ **3**

20. $2 \div 3$ **4**

21. $3 \div 2$ **4**

⊙ CONCEPTUAL  WRITING  CHALLENGING  SCIENTIFIC CALCULATOR  GRAPHING CALCULATOR

22. Is division commutative on a 5-hour clock? Explain.

23. Is there an answer for 4 ÷ 0 on a 5-hour clock? Find it or explain why not.

The military uses a 24-hour clock to avoid the problems of "A.M." *and* "P.M." *For example,* 1100 hours *is* 11 A.M., *while* 2100 hours *is* 9 P.M. *(*12 *noon* + 9 *hours). In these designations, the last two digits represent minutes, and the digits before that represent hours. Find each of the following sums in the 24-hour clock system.*

24. 1400 + 500 **1900**

25. 1300 + 1800 **0700**

26. 0750 + 1630 **0020**

27. 1545 + 0815 **0000**

28. Explain how the following three statements can *all* be true. (*Hint:* Think of clocks.)

$$1145 + 1135 = 2280$$
$$1145 + 1135 = 1120$$
$$1145 + 1135 = 2320$$

first answer: ordinary whole number arithmetic; second answer: 12-hour clock arithmetic; third answer: 24-hour clock arithmetic

Answer true *or* false *for each of the following.*

29. 5 ≡ 19 (mod 3) **false**

30. 35 ≡ 8 (mod 9) **true**

31. 5,445 ≡ 0 (mod 3) **true**

32. 7,021 ≡ 4,202 (mod 6) **false**

Do each of the following modular arithmetic problems.

33. (12 + 7) (mod 4) **3**

34. (62 + 95) (mod 9) **4**

35. (35 − 22) (mod 5) **3**

36. (82 − 45)(mod 3) **1**

37. (5 × 8) (mod 3) **1**

38. (32 × 21) (mod 8) **0**

39. [4 × (13 + 6)] (mod 11) **10**

40. [(10 + 7) × (5 + 3)] (mod 10) **6**

41. The text described how to do arithmetic modulo m when the ordinary answer comes out nonnegative. Explain what to do when the ordinary answer is negative.

Do each of the following modular arithmetic problems.

42. (3 − 27) (mod 5) **1**

43. (16 − 60) (mod 7) **5**

44. [(−8) × 11] (mod 3) **2**

45. [2 × (−23)] (mod 5) **4**

In each of Exercises 46 and 47:
(a) *Complete the given addition table.*
(b) *Decide whether the closure, commutative, identity, and inverse properties are satisfied.*
(c) *If the inverse property is satisfied, give the inverse of each number.*

46. modulo 4

+	**0**	**1**	**2**	**3**
0	0	1	2	3
1				
2				
3				

(a) row 1: 1, 2, 3, 0; row 2: 2, 3, 0, 1; row 3: 3, 0, 1, 2 (b) All four properties are satisfied. (c) 0 is its own inverse, 2 is its own inverse, and 1 and 3 are inverses of each other.

47. modulo 7

+	**0**	**1**	**2**	**3**	**4**	**5**	**6**
0	0	1	2	3	4	5	6
1	1	2	3	4	5	6	
2							
3							
4							
5							
6							

(a) row 1: 0; row 2: 2, 3, 4, 5, 6, 0, 1; row 3: 3, 4, 5, 6, 0, 1, 2; row 4: 4, 5, 6, 0, 1, 2, 3; row 5: 5, 6, 0, 1, 2, 3, 4; row 6: 6, 0, 1, 2, 3, 4, 5 (b) All four properties are satisfied. (c) 0 is its own inverse, 1 and 6 are inverses of each other, 2 and 5 are inverses of each other, and 3 and 4 are inverses of each other.

⊙ *In each of Exercises 48–51:*
(a) *Complete the given multiplication table.*
(b) *Decide whether the closure, commutative, identity, and inverse properties are satisfied.*
(c) *Give the inverse of each nonzero number that has an inverse.*

48. modulo 2

×	**0**	**1**
0	0	0
1	0	

(a) 1 (b) All properties are satisfied. (c) 1 is its own inverse.

49. modulo 3

×	**0**	**1**	**2**
0	0	0	0
1	0	1	2
2	0	2	

(a) 1 (b) All properties are satisfied. (c) 1 is its own inverse, and 2 is its own inverse.

50. modulo 4

×	**0**	**1**	**2**	**3**
0	0	0	0	0
1	0	1	2	3
2	0	2		
3	0	3		

(a) row 2: 0, 2; row 3: 2, 1 (b) no inverse property (c) 1 is its own inverse, as is 3; 2 has no inverse.

51. modulo 9

×	0	1	2	3	4	5	6	7	8
0	0	0	0	0	0	0	0	0	0
1	0	1	2	3	4	5	6	7	8
2	0	2	4	6	8			5	
3	0	3	6	0		6		3	6
4	0	4	8		7		6		5
5	0	5	1		2		3	8	
6	0	6	3	0	6	3	0	6	3
7	0	7	5			8			2
8	0	8	7			4	3		1

(a) row 2: 1, 3, 7; row 3: 3, 0; row 4: 3, 2, 1; row 5: 6, 7, 4; row 7: 3, 1, 6, 4; row 8: 6, 5, 2 (b) no inverse property (c) 1 is its own inverse, as is 8; 2 and 5 are inverses of each other, as are 4 and 7; 3 has no inverse, and 6 has no inverse.

52. Complete this statement: a modulo system satisfies the inverse property for multiplication only if the modulo number is a(n) __________ number. **prime**

Find all positive solutions for each of the following equations. Note any identities.

53. $x \equiv 3 \pmod 7$ **{3, 10, 17, 24, 31, 38, . . .}**

54. $(2 + x) \equiv 7 \pmod 3$ **{2, 5, 8, 11, 14, 17, . . .}**

55. $6x \equiv 2 \pmod 2$ **identity**

56. $(5x - 3) \equiv 7 \pmod 4$ **{2, 6, 10, 14, 18, 22, . . .}**

---

57. For many years automobile odometers showed five whole-number digits and a digit for tenths of a mile. For those odometers showing just five whole-number digits, totals are recorded in modulo what number? **100,000**

58. If a car's five-digit whole number odometer shows a reading of 29,306, *in theory* how many miles might the car have traveled? **29,306 + 100,000*k* where *k* is any nonnegative integer.**

59. Lawrence Rosenthal finds that whether he sorts his White Sox ticket stubs into piles of 10, piles of 15, or piles of 20, there are always 2 left over. What is the least number of stubs he could have (assuming he has more than 2)? **62**

60. Roxanna Parker has a collection of silver spoons from all over the world. She finds that she can arrange her spoons in sets of 7 with 6 left over, sets of 8 with 1 left over, or sets of 15 with 3 left over. If Roxanna has fewer than 200 spoons, how many are there? **153**

61. Refer to Example 9 in the text. (Recall that *next* year is a leap year.) Assuming today was Thursday, January 12, answer the following questions.

(a) How many days would the next year (starting today) contain? **365**

(b) What day of the week would occur one year from today? **Friday**

62. Assume again, as in Example 9, that *next* year is a leap year. If the next year (starting today) does *not* contain 366 days, what is the range of possible dates for today? **January 1 through February 28**

63. Robin Strang and Kristyn Wasag, flight attendants for two different airlines, are close friends and like to get together as often as possible. Robin flies a 21-day schedule (including days off), which then repeats, while Kristyn has a repeating 30-day schedule. Both of their routines include stopovers in Chicago, New Orleans, and San Francisco. The table below shows which days of each of their individual schedules they are in these cities. (Assume the first day of a cycle is day number 1.)

	Days in Chicago	**Days in New Orleans**	**Days in San Francisco**
Robin	1, 2, 8	5, 12	6, 18, 19
Kristyn	23, 29, 30	5, 6, 17	8, 10, 15, 20, 25

If today is July 1 and both are starting their schedules today (day 1), list the days during July and August that they will be able to see each other in each of the three cities.
Chicago: July 23 and 29; New Orleans: July 5 and August 16; San Francisco: August 9

The basis of the complex number system is the "imaginary" number i. The powers of i cycle through a repeating pattern of just 4 distinct values as shown here:

$$i^0 = 1, \quad i^1 = i, \quad i^2 = -1, \quad i^3 = -i, \quad i^4 = 1, \quad i^5 = i, \quad \text{and so on.}$$

Find the values of each of the following powers of i.

64. i^{16} **1**

65. i^{47} $-i$

66. i^{98} **−1**

67. i^{137} i

The following formula can be used to find the day of the week on which a given year begins. Here y represents the year (after* 1582, *when our current calendar began). First calculate*

$$a = y + [\![(y - 1)/4]\!] - [\![(y - 1)/100]\!] + [\![(y - 1)/400]\!],$$

where $[\![x]\!]$ *represents the greatest integer less than or equal to x. (For example,* $[\![9.2]\!] = 9$, and $[\![\pi]\!] = 3$.) *After finding a, find the smallest nonnegative integer b such that*

$$a \equiv b \pmod 7.$$

Then b gives the day of January 1, with b = 0 *representing Sunday, b* = 1 *Monday, and so on.*

Find the day of the week on which January 1 would occur in the following years.

68. 1812 **Wednesday**

69. 1865 **Sunday**

70. 1997 **Wednesday**

71. 1999 **Friday**

Some people believe that Friday the thirteenth is unlucky. The table below shows the months that will have a Friday the thirteenth if the first day of the year is known. A year is a leap year if it is divisible by* 4. *The only exception to this rule is that a century year* (1900, *for example) is a leap year only when it is divisible by* 400.

First Day of Year	Nonleap Year	Leap Year
Sunday	Jan., Oct.	Jan., April, July
Monday	April, July	Sept., Dec.
Tuesday	Sept., Dec.	June
Wednesday	June	March, Nov.
Thursday	Feb., March, Nov.	Feb., Aug.
Friday	August	May
Saturday	May	Oct.

Use the table to determine the months that have a Friday the thirteenth for the following years.

72. 1997 **June**

73. 1995 **Jan., Oct.**

74. 1999 **August**

75. 2200 **June**

*Given in "An Aid to the Superstitious," by G. L. Ritter, S. R. Lowry, H. B. Woodruff, and T. L. Isenhour. *The Mathematics Teacher,* May 1977, pages 456–457.

76. Modulo numbers can be used to create **modulo designs.** For example, to construct the design (11, 5) proceed as follows.

(a) Draw a circle and divide the circumference into 10 equal parts. Label the division points as 1, 2, 3, . . . , 10.

(b) Since $1 \times 5 \equiv 5 \pmod{11}$, connect 1 and 5. (We use 5 as a multiplier since we are making an (11, 5) design.)

(c) $2 \times 5 \equiv 10 \pmod{11}$ Therefore, connect 2 and ______. **10**

(d) $3 \times 5 \equiv$ ______ (mod 11) Connect 3 and ______. **4, 4**

(e) $4 \times 5 \equiv$ ______ (mod 11) Connect 4 and ______. **9, 9**

(f) $5 \times 5 \equiv$ ______ (mod 11) Connect 5 and ______. **3, 3**

(g) $6 \times 5 \equiv$ ______ (mod 11) Connect 6 and ______. **8, 8**

(h) $7 \times 5 \equiv$ ______ (mod 11) Connect 7 and ______. **2, 2**

(i) $8 \times 5 \equiv$ ______ (mod 11) Connect 8 and ______. **7, 7**

(j) $9 \times 5 \equiv$ ______ (mod 11) Connect 9 and ______. **1, 1**

(k) $10 \times 5 \equiv$ ______ (mod 11) Connect 10 and ______. **6, 6**

(l) You might want to shade some of the regions you have found to make an interesting pattern. Other modulo designs are shown at the side. For more information, see "Residue Designs," by Phil Locke in *The Mathematics Teacher,* March 1972, pages 260–263.

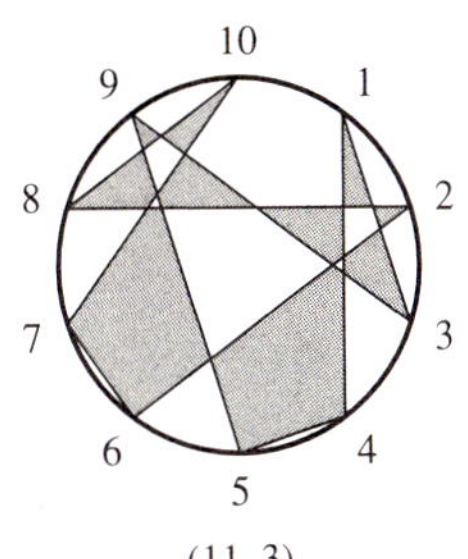

(11, 3)

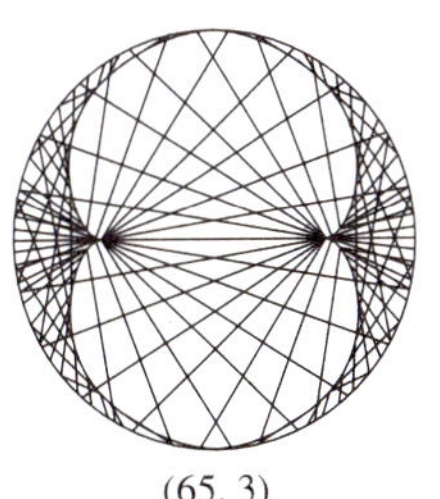
(65, 3)

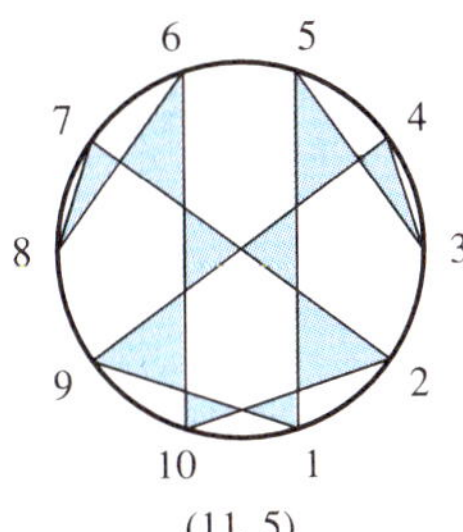

(11, 5)

Identification numbers are used in various ways for all kinds of different products. Books, for example, are assigned International Standard Book Numbers (ISBNs). Each ISBN is a ten-digit number. It includes a check digit, which is determined on the basis of modular arithmetic. The ISBN for this book is*

0-673-99893-2

The first digit, 0, *identifies the book as being published in an English-speaking country. The next digits,* 673, *identify the publisher, while* 99893 *identifies this particular book. The final digit,* 2, *is a check digit. To find this check digit, start at the left and multiply the digits of the ISBN number by* 10, 9, 8, 7, 6, 5, 4, 3, *and* 2, *respectively. Then add these products. For this book we get*

$$(10 \times 0) + (9 \times 6) + (8 \times 7) + (7 \times 3) + (6 \times 9) + (5 \times 9) + (4 \times 8) + (3 \times 9) + (2 \times 3) = 295.$$

* For an interesting general discussion, see "The Mathematics of Identification Numbers," by Joseph A. Gallian in *The College Mathematics Journal,* May 1991, page 194.

The check digit is the smallest number that must be added to this result to get a multiple of 11. *Since* 295 + 2 = 297, *a multiple of* 11, *the check digit is* 2. *(It is possible to have a check "digit" of* 10; *the letter X is used instead of* 10.*)*

When an order for this book is received, the ISBN is entered into a computer, and the check digit evaluated. If this result does not match the check digit on the order, the order will not be processed.

Which of the following ISBNs have correct check digits?

77. 0-399-13615-4 **incorrect**

78. 0-691-02356-5 **correct**

Find the appropriate check digit for each of the following ISBNs.

79. *The Beauty of Fractals,* by H. O. Peitgen and P. H. Richter, 3-540-15851- **0**

80. *Women in Science,* by Vivian Gornick, 0-671-41738- **X**

81. *Beyond Numeracy,* by John Allen Paulos, 0-394-58640- **9**

82. *Iron John,* by Robert Bly, 0-201-51720- **5**

NOTE Another application of modular arithmetic is found in the Collaborative Investigation at the end of this chapter.

4.5 Exercises (Text Page 201)

For each system in Exercises 1–10, decide which of the properties of single-operation systems are satisfied. If the identity property is satisfied, give the identity element. If the inverse property is satisfied, give the inverse of each element. If the identity property is satisfied but the inverse property is not, name the elements that have no inverses.

1. {1, 2, 3, 4}; multiplication modulo 5

×	1	2	3	4
1	1	2	3	4
2	2	4	1	3
3	3	1	4	2
4	4	3	2	1

all properties; 1 is the identity element; 1 is its own inverse, as is 4; 2 and 3 are inverses.

2. {1, 2}; multiplication modulo 3

×	1	2
1	1	2
2	2	1

all properties; 1 is the identity element; 1 is its own inverse, as is 2.

3. {1, 2, 3, 4, 5}; multiplication modulo 6

×	1	2	3	4	5
1	1	2	3	4	5
2	2	4	0	2	4
3	3	0	3	0	3
4	4	2	0	4	2
5	5	4	3	2	1

commutative, associative, and identity properties; 1 is the identity element; 2, 3, and 4 have no inverses.

4. {1, 2, 3, 4, 5, 6, 7}; multiplication modulo 8

×	1	2	3	4	5	6	7
1	1	2	3	4	5	6	7
2	2	4	6	0	2	4	6
3	3	6	1	4	7	2	5
4	4	0	4	0	4	0	4
5	5	2	7	4	1	6	3
6	6	4	2	0	6	4	2
7	7	6	5	4	3	2	1

all properties except the inverse; 1 is the identity element; 2, 4, and 6 have no inverses.

5. {1, 3, 5, 7}; multiplication modulo 8

×	**1**	**3**	**5**	**7**
1	1	3	5	7
3	3	1	7	5
5	5	7	1	3
7	7	5	3	1

all properties; 1 is the identity element; 1, 3, 5, and 7 are their own inverses.

6. {1, 3, 5, 7, 9}; multiplication modulo 10

×	**1**	**3**	**5**	**7**	**9**
1	1	3	5	7	9
3	3	9	5	1	7
5	5	5	5	5	5
7	7	1	5	9	3
9	9	7	5	3	1

all properties except the inverse; 1 is the identity element; 5 has no inverse.

7. {m, n, p}; operation J

J	m	n	p
m	n	p	n
n	p	m	n
p	n	n	m

closure and commutative properties

8. {A, B, F}; operation ☆

☆	A	B	F
A	B	F	A
B	F	A	B
F	A	B	F

all properties; F is the identity element; A and B are inverses; F is its own inverse.

9. {A, J, T, U}; operation #

#	A	J	T	U
A	A	J	T	U
J	J	T	U	A
T	T	U	A	J
U	U	A	J	T

all properties; A is the identity element; J and U are inverses; A and T are their own inverses.

10. {r, s, t, u}; operation Z

Z	r	s	t	u
r	u	t	r	s
s	t	u	s	r
t	r	s	t	u
u	s	r	u	t

all properties; t is the identity element; s and r are inverses; t and u are their own inverses.

The tables in the finite mathematical systems that we developed in this section can be obtained in a variety of ways. For example, let us begin with a square, as shown in the figure. Let the symbols a, b, c, and d be defined as shown in the figure.

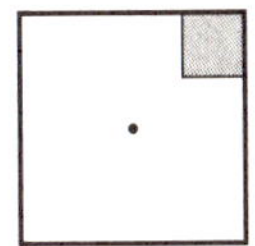

Let a represent zero rotation—leave the original square as is.

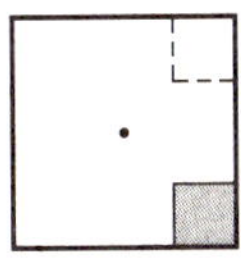

Let b represent rotation of 90° clockwise from original position.

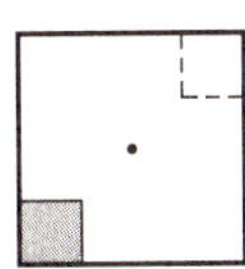

Let c represent rotation of 180° clockwise from original position.

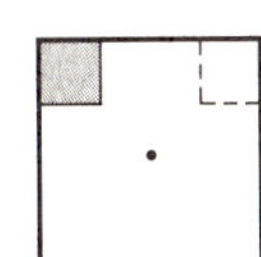

Let d represent rotation of 270° clockwise from original position.

Define an operation □ for these letters as follows. To evaluate b □ c, for example, first perform b by rotating the square 90°. (See the figure.) Then perform operation c by rotating the square an additional 180°. The net result is the same as if we had performed d only. Thus,

$$b \;\square\; c = d.$$

Start with a.

Perform b.

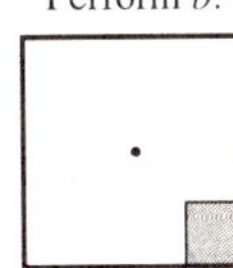

Start with b, and perform c.

Use this method to find each of the following.

11. $b \square d$ **a** **12.** $b \square b$ **c** **13.** $d \square b$ **a** **14.** $a \square b$ **b**

15. Complete the table at the right.

$\square$	a	b	c	d
a	a	b	c	d
b	b	c		a
c	c		a	
d	d	a		

row b: d; row c: d, b; row d: b, c

16. Which of the properties from this section are satisfied by this system? **all**

▲ ⊙ **17.** Define a universal set U as the set of counting numbers. Form a new set that contains all possible subsets of U. This new set of subsets together with the operation of set intersection forms a mathematical system. Which of the properties listed in this section are satisfied by this system? **associative, commutative, identity (U), closure**

▲ ⊙ **18.** Replace the word "intersection" with the word "union" in Exercise 17; then answer the same question. **associative, commutative, identity ($\emptyset$), closure**

▲ ⊙ **19.** Complete the table at the right so that the result is *not* the same as operation $\square$ of Exercise 15, but so that the five properties listed in this section still hold.
Answers may vary. One possibility is as follows.

	a	b	c	d
a				
b				
c				
d				

	a	b	c	d
a	a	b	c	d
b	b	a	d	c
c	c	d	a	b
d	d	c	b	a

Try examples to help you decide whether or not the following operations, when applied to the integers, satisfy the distributive property.

20. subtraction over multiplication **no**

21. addition over subtraction **no**

22. subtraction over addition **no**

Recall that Example 3 provided a counterexample for the general statement

$$a + (b \times c) = (a + b) \times (a + c).$$

Thus, addition is not *distributive over multiplication. Now do Exercises 23–26.*

23. Decide if the statement above is true for each of the following sets of values.
(a) $a = 2, b = -5, c = 4$ **true** **(b)** $a = -7, b = 5, c = 3$ **true**
(c) $a = -8, b = 14, c = -5$ **true** **(d)** $a = 1, b = 6, c = -6$ **true**

24. Find another set of a, b, and c values that make the statement true.
Answers will vary.

▲ ⊙ **25.** Explain what general conditions will always cause the statement above to be true.
$a + b + c = 1$ **or** $a = 0$

⊙ **26.** Explain why, regardless of the results in Exercises 23–25, addition is still *not* distributive over multiplication.

⊙ CONCEPTUAL  WRITING ▲ CHALLENGING SCIENTIFIC CALCULATOR GRAPHING CALCULATOR

27. Give the conditions under which each of the following equations would be true.
(a) $a + (b - c) = (a + b) - (a + c)$ **$a = 0$**
(b) $a - (b + c) = (a - b) + (a - c)$ **$a = 0$**

28. (a) Find values of a, b, and c such that

$$a - (b \times c) = (a - b) \times (a - c).$$

Answers will vary.

(b) Does this mean that subtraction is distributive over multiplication? Explain.

Verify for the mathematical system of Example 4 that the distributive property holds for the following cases.

29. $c \star (d \circ e) = (c \star d) \circ (c \star e)$
Each side simplifies to e.

30. $a \star (a \circ b) = (a \star a) \circ (a \star b)$
Each side simplifies to a.

31. $d \star (e \circ c) = (d \star e) \circ (d \star c)$
Each side simplifies to d.

32. $b \star (b \circ b) = (b \star b) \circ (b \star b)$
Each side simplifies to c.

These exercises are for students who have studied sets.

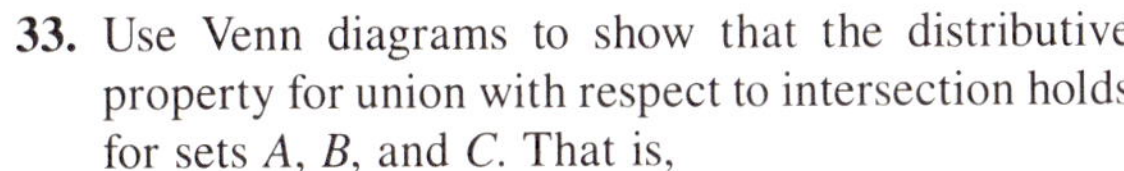

33. Use Venn diagrams to show that the distributive property for union with respect to intersection holds for sets A, B, and C. That is,

$$A \cup (B \cap C) = (A \cup B) \cap (A \cup C).$$

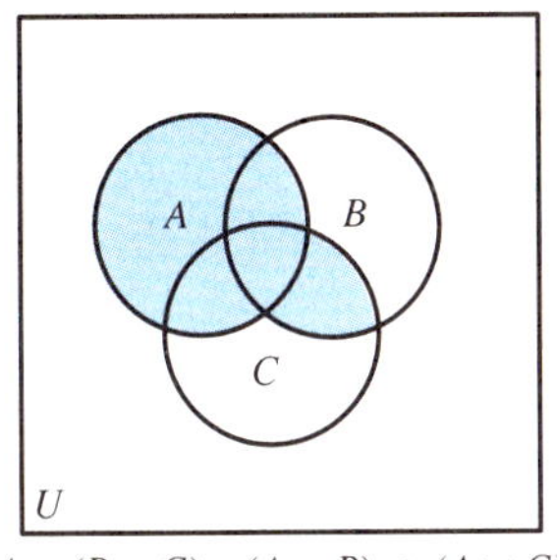

$A \cup (B \cap C) = (A \cup B) \cap (A \cup C)$

34. Use Venn diagrams to show that *another* distributive property holds for sets A, B, and C. It is the distributive property of intersection with respect to union.

$$A \cap (B \cup C) = (A \cap B) \cup (A \cap C)$$

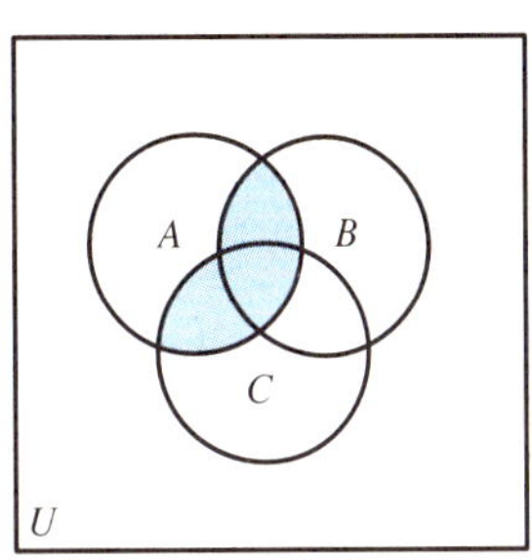

$A \cap (B \cup C) = (A \cap B) \cup (A \cap C)$

These exercises are for students who have studied logic.

35. Use truth tables to show that the following distributive property holds:

$$p \vee (q \wedge r) \equiv (p \vee q) \wedge (p \vee r).$$

TTTTTFFF for each final column

36. Use truth tables to show that *another* distributive property holds:

$$p \wedge (q \vee r) \equiv (p \wedge q) \vee (p \wedge r).$$

TTTFFFFF for each final column

4.6 Exercises (Text Page 210)

What is wrong with the way in which each of the following questions is stated?

1. Do the integers form a group?
2. Does multiplication satisfy all of the group properties?

Decide whether or not each system is a group. If not a group, identify all properties that are not satisfied. (Recall that any system failing to satisfy the identity property automatically fails to satisfy the inverse property also.) For the finite systems, it may help to construct tables. For infinite systems, try some examples to help you decide.

3. $\{0\}$; addition **yes**

4. $\{0\}$; multiplication **yes**

5. {0}; subtraction **yes**

6. {0, 1}; addition **no: closure, inverse**

7. {0, 1}; multiplication **no: inverse**

8. {−1, 1}; division **yes**

9. {−1, 0, 1}; addition **no: closure**

10. {−1, 0, 1}; multiplication **no: inverse**

11. integers; multiplication **no: inverse**

12. integers; subtraction **no: associative, identity, inverse**

13. counting numbers; addition **no: identity, inverse**

14. odd integers; multiplication **no: inverse**

15. even integers; addition **yes**

16. rational numbers; addition **yes**

17. nonzero rational numbers; multiplication **yes**

18. prime numbers; addition **no: closure, identity, inverse**

19. Explain why a *finite* group based on the operation of ordinary addition of numbers cannot contain the element 1.

20. Explain why a group based on the operation of addition of ordinary numbers *must* contain the element 0.

Exercises 21–34 apply to the system of symmetries of a square presented in the text.

Find each of the following.

21. RN S

22. PR T

23. TV N

24. VP S

Verify each of the following statements.

25. $N(TR) = (NT)R$ M

26. $V(PS) = (VP)S$ Q

27. $S(MR) = (SM)R$ M

28. $T(VN) = (TV)N$ Q

Find the inverse of each element.

29. N N

30. Q Q

31. R R

32. S S

33. T T

34. V V

Determine whether or not each of the following systems is a group.

35. {0, 1, 2}; addition modulo 3 **yes**

36. {0, 1, 2, 3}; addition modulo 4 **yes**

37. {0, 1, 2, 3, 4}; addition modulo 5 **yes**

38. {0, 1, 2, 3, 4, 5}; addition modulo 6 **yes**

39. Complete this statement: _____ modular addition systems are groups. **all**

Determine whether or not each of the following systems is a group.

40. {1, 2, 3}; multiplication modulo 4 **no**

41. {1, 2, 3, 4}; multiplication modulo 5 **yes**

42. {1, 2, 3, 4, 5}; multiplication modulo 6 **no**

43. {1, 2, 3, 4, 5, 6}; multiplication modulo 7 **yes**

44. {1, 2, 3, 4, 5, 6, 7}; multiplication modulo 8 **no**

45. {1, 2, 3, 4, 5, 6, 7, 8}; multiplication modulo 9 **no**

46. Notice that in constructing modular multiplication systems, we excluded the element 0. Complete this statement: a modular multiplication system is a group only when the modulus is _____. **prime**

A group that also satisfies the commutative property is called a commutative group (or an abelian *group, after Niels Henrik Abel).*

Determine whether each of the following groups is commutative.

47. {0, 1, 2, 3} under addition modulo 4 **yes**

48. {1, 2} under multiplication modulo 3 **yes**

49. the group of symmetries of a square **no**

50. the subgroup of Example 6 **yes**

51. the integers under addition **yes**

52. the permutation group on three symbols **no**

Give illustrations to back up your answers for Exercises 53–56.

53. Produce a mathematical system with two operations which is a group under one operation but not a group under the other operation.

54. Explain what property is gained when the system of counting numbers is extended to the system of whole numbers.

55. Explain what property is gained when the system of whole numbers is extended to the system of integers.

56. Explain what property is gained when the system of integers is extended to the rational numbers.

Chapter 4 Test (Text Page 213)

1. For the numeral 𓆼𓆼𓍢𓍢𓍢𓍢𓍢𓎆𓎆𓎆𓏺𓏺𓏺𓏺𓏺𓏺, identify the numeration system, and give the Hindu-Arabic equivalent. **ancient Egyptian; 2,536**

2. Simplify: $(8 \times 10^3) + (3 \times 10^2) + (6 \times 10^1) + (4 \times 10^0)$. **8,364**

3. Write in expanded notation: 60,923.
$(6 \times 10^4) + (0 \times 10^3) + (9 \times 10^2) + (2 \times 10^1) + (3 \times 10^0)$

Convert each of the following to base ten.

4. 424_{five} **114**

5. $100{,}110_{two}$ **38**

6. $A{,}80C_{sixteen}$ **43,020**

Convert as indicated.

7. 58 to base two $111{,}010_{two}$

8. 1,846 to base five $24{,}341_{five}$

9. $10{,}101{,}110_{two}$ to base eight 256_{eight}

10. $B52_{sixteen}$ to base two $101{,}101{,}010{,}010_{two}$

Briefly explain each of the following.

11. the advantage of multiplicative grouping over simple grouping
less repetition of symbols

12. the advantage, in a positional numeration system, of a smaller base over a larger base
fewer different symbols to learn

13. the advantage, in a positional numeration system, of a larger base over a smaller base
fewer digits in numerals

Do the following arithmetic problems on a 12-hour clock.

14. $11 + 9$ **8**

15. $6 - 9$ **9**

16. 8×9 **0**

Do the following modular arithmetic problems.

17. $(7 \times 18) \pmod 8$ **6**

18. $(52 + 39) \pmod 6$ **1**

19. $(23 \times 56) \pmod 2$ **0**

20. Construct an addition table for a modulo 6 system.

+	0	1	2	3	4	5
0	0	1	2	3	4	5
1	1	2	3	4	5	0
2	2	3	4	5	0	1
3	3	4	5	0	1	2
4	4	5	0	1	2	3
5	5	0	1	2	3	4

21. Construct a multiplication table for a modulo 6 system.

×	0	1	2	3	4	5
0	0	0	0	0	0	0
1	0	1	2	3	4	5
2	0	2	4	0	2	4
3	0	3	0	3	0	3
4	0	4	2	0	4	2
5	0	5	4	3	2	1

Find all positive solutions for the following.

22. $x \equiv 5 \pmod{12}$ **{5, 17, 29, 41, 53, 65, . . .}**

23. $5x \equiv 3 \pmod{8}$ **{7, 15, 23, 31, 39, 47, . . .}**

24. Find the smallest positive integer that satisfies both $x \equiv 3 \pmod{12}$ and $x \equiv 5 \pmod{7}$. **75**

25. Describe in general what constitutes a mathematical *group*.

A mathematical system is defined by the table here.

V	***a***	***e***	***i***	***o***	***u***
a	*o*	*e*	*u*	*a*	*i*
e	*u*	*o*	*a*	*e*	*i*
i	*e*	*u*	*o*	*i*	*a*
o	*a*	*e*	*i*	*o*	*u*
u	*i*	*a*	*e*	*u*	*o*

26. **(a)** Is there an identity element in this system? **yes** **(b)** If so, what is it? **yes;** ***O***

27. **(a)** Is closure satisfied by this system? **yes** **(b)** Explain.

28. **(a)** Is this system commutative? **no** **(b)** Explain.

29. **(a)** Is the distributive property satisfied in this system? **no** **(b)** Explain.

30. **(a)** Is the system a group? **no** **(b)** Explain.

CHAPTER 5 Number Theory

5.1 Exercises (Text Page 220)

Decide whether the following statements are true *or* false.

1. Every natural number is divisible by 1. **true**

2. No natural number is both prime and composite. **true**

3. There are no even prime numbers. **false**

4. If n is a natural number and $9 | n$, then $3 | n$. **true**

5. If n is a natural number and $5 | n$, then $10 | n$. **false**

6. 1 is the smallest prime number. **false**

7. Every natural number is both a factor and a multiple of itself. **true**

8. If 16 divides a natural number, then 2, 4, and 8 must also divide that natural number. **true**

9. The composite number 50 has exactly two prime factorizations. **false**

10. The prime number 53 has exactly two natural number factors. **true**

Find all natural number factors of each number.

11. 12 **1, 2, 3, 4, 6, 12**

12. 18 **1, 2, 3, 6, 9, 18**

13. 20 **1, 2, 4, 5, 10, 20**

14. 28 **1, 2, 4, 7, 14, 28**

15. 52 **1, 2, 4, 13, 26, 52**

16. 63 **1, 3, 7, 9, 21, 63**

17. 120 **1, 2, 3, 4, 5, 6, 8, 10, 12, 15, 20, 24, 30, 40, 60, 120**

18. 172 **1, 2, 4, 43, 86, 172**

Use divisibility tests to decide whether the given number is divisible by
(a) 2 **(b)** 3 **(c)** 4 **(d)** 5 **(e)** 6 **(f)** 8 **(g)** 9 **(h)** 10 **(i)** 12.

19. 315 **(a) no (b) yes (c) no (d) yes (e) no (f) no (g) yes (h) no (i) no**

20. 7,425 **(a) no (b) yes (c) no (d) yes (e) no (f) no (g) yes (h) no (i) no**

21. 1,092 **(a) yes (b) yes (c) yes (d) no (e) yes (f) no (g) no (h) no (i) yes**

22. 4,488 **(a) yes (b) yes (c) yes (d) no (e) yes (f) yes (g) no (h) no (i) yes**

23. 630 **(a) yes (b) yes (c) no (d) yes (e) yes (f) no (g) yes (h) yes (i) no**

24. 25,025 **(a) no (b) no (c) no (d) yes (e) no (f) no (g) no (h) no (i) no**

25. 45,815 **(a) no (b) no (c) no (d) yes (e) no (f) no (g) no (h) no (i) no**

26. 5,940 **(a) yes (b) yes (c) yes (d) yes (e) yes (f) no (g) yes (h) yes (i) yes**

27. 123,456,789 **(a) no (b) yes (c) no (d) no (e) no (f) no (g) yes (h) no (i) no**

28. 987,654,321 **(a) no (b) yes (c) no (d) no (e) no (f) no (g) yes (h) no (i) no**

29. Continue the Sieve of Eratosthenes in Table 1 from 101 to 200. (You need only check for divisibility by primes through 13.) List the primes between 100 and 200.
101, 103, 107, 109, 113, 127, 131, 137, 139, 149, 151, 157, 163, 167, 173, 179, 181, 191, 193, 197, 199

30. List two primes that are consecutive natural numbers. Can there be any others?
2, 3; no

31. Can there be three primes that are consecutive natural numbers? Explain.

32. For a natural number to be divisible by both 2 and 5, what must be true about its last digit? **It must be 0.**

33. Consider the divisibility tests for 2, 4, and 8 (all powers of 2). Use inductive reasoning to predict the divisibility test for 16. Then, use the test to show that 456,882,320 is divisible by 16. **The last four digits must form a number divisible by 16. The number 456,882,320 is divisible by 16, since 2,320 is divisible by 16: 2,320 = 16 · 145.**

Find the prime factorization of each composite number.

34. 240 **$2^4 \cdot 3 \cdot 5$** **35.** 300 **$2^2 \cdot 3 \cdot 5^2$** **36.** 360 **$2^3 \cdot 3^2 \cdot 5$** **37.** 425 **$5^2 \cdot 17$**

38. 663 **$3 \cdot 13 \cdot 17$** **39.** 885 **$3 \cdot 5 \cdot 59$** **40.** 1,280 **$2^8 \cdot 5$** **41.** 1,575 **$3^2 \cdot 5^2 \cdot 7$**

Here is a divisibility test for 7.

(a) *Double the last digit of the given number, and subtract this value from the given number with the last digit omitted.*

(b) *Repeat the process of part (a) as many times as necessary until the number obtained can easily be divided by 7.*

(c) *If the final number obtained is divisible by 7, then the given number is also divisible by 7. If the final number is not divisible by 7, then neither is the given number.*

▲ *Use this divisibility test to determine whether or not each of the following is divisible by 7.*

42. 142,891 **yes** **43.** 409,311 **yes** **44.** 458,485 **no** **45.** 287,824 **no**

Here is a divisibility test for 11.

(a) *Starting at the left of the given number, add together every other digit.*

(b) *Add together the remaining digits.*

(c) *Subtract the smaller of the two sums from the larger.*

(d) *If the final number obtained is divisible by 11, then the given number is also divisible by 11. If the final number is not divisible by 11, then neither is the given number.*

▲ *Use this divisibility test to determine whether or not each of the following is divisible by 11.*

46. 8,493,969 **yes** **47.** 847,667,942 **yes** **48.** 453,896,248 **no** **49.** 552,749,913 **no**

◎ **50.** Consider the divisibility test for the composite number 6, and make a conjecture for the divisibility test for the composite number 15. **The number must be divisible by both 3 and 5. That is, the sum of the digits must be divisible by 3, and the last digit must be 5 or 0.**

Find the smallest natural number that is divisible by all of the numbers in the group of numbers listed.

51. 2, 3, 5, 7, 8 **840**

52. 2, 3, 4, 9, 10 **180**

53. 2, 3, 4, 5, 6, 7, 8, 9 **2,520**

54. 2, 3, 4, 5, 6, 7, 8, 9, 12 **2,520**

◎ ✎ **55.** Explain why the answers in Exercises 53 and 54 must be the same.

◎ ✎ **56.** Explain why the answer in Exercise 51 would not change if 2 were omitted from the group of numbers.

▲ *Determine all possible digit replacements for x so that the first number is divisible by the second. For example, 37,58x is divisible by 2 if x = 0, 2, 4, 6, or 8.*

57. 398,87x; 2 **0, 2, 4, 6, 8** **58.** 2,45x,765; 3 **1, 4, 7** **59.** 64,537,84x; 4 **0, 4, 8**

60. 2,143,89x; 5 **0, 5** **61.** 985,23x; 6 **0, 6** **62.** 7,643,24x; 8 **0, 8**

63. 4,329,7x5; 9 **6** **64.** 23,x54,470; 10 **any digit**

◎ CONCEPTUAL ✎ WRITING ▲ CHALLENGING ▦ SCIENTIFIC CALCULATOR ▦ GRAPHING CALCULATOR

There is a method to determine the **number of divisors** *of a composite number. To do this, write the composite number in its prime factored form, using exponents. Add* 1 *to each exponent and multiply these numbers. Their product gives the number of divisors of the composite number. For example,* $24 = 2^3 \cdot 3 = 2^3 \cdot 3^1$. *Now add* 1 *to each exponent:* $3 + 1 = 4$, $1 + 1 = 2$. *Multiply* 4×2 *to get* 8. *There are* 8 *divisors of* 24. *(Since* 24 *is rather small, this can be verified easily. The divisors are* 1, 2, 3, 4, 6, 8, 12, *and* 24, *a total of eight as predicted.)*

Find the number of divisors of each composite number.

65. 36 **9**
66. 48 **10**
67. 72 **12**
68. 144 **15**
69. $2^8 \cdot 3^2$ **27**
70. $2^4 \cdot 3^4 \cdot 5^2$ **75**

Leap years occur when the year number is divisible by 4. *An exception to this occurs when the year number is divisible by* 100 *(that is, it ends in two zeros). In such a case, the number must be divisible by* 400 *in order for the year to be a leap year. Determine which of the following years are leap years.*

71. 1776 **leap year**
72. 1948 **leap year**
73. 1994 **not a leap year**
74. 1894 **not a leap year**
75. 2000 **leap year**
76. 2400 **leap year**
77. 1900 **not a leap year**
78. 1800 **not a leap year**

79. Why is the following *not* a valid divisibility test for 8? "A number is divisible by 8 if it is divisible by both 4 and 2." Support your answer with an example.

80. Choose any three consecutive natural numbers, multiply them together, and divide the product by 6. Repeat this several times, using different choices of three consecutive numbers. Make a conjecture concerning the result. **The product of any three consecutive natural numbers is divisible by 6.**

81. Explain why the product of three consecutive natural numbers must be divisible by 6.

82. Choose any 6-digit number consisting of three digits followed by the same three digits in the same order (for example, 467,467). Divide by 13. Divide by 11. Divide by 7. What do you notice? Why do you think this happens? **In each case, the number is divisible by 13, 11, and 7. This happens because such a number is divisible by 1,001, and** $1{,}001 = 13 \cdot 11 \cdot 7$.

5.2 Exercises (Text Page 229)

Based on your readings in this section, decide whether each of the following (Exercises 1–10) is true *or* false.

1. There are infinitely many prime numbers. **true**

2. The prime numbers 2 and 3 are twin primes. **false**

3. There is no perfect number between 496 and 8,128. **true**

4. Because its last digit is 4, the number 54,034 cannot possibly be perfect. **true**

5. Any prime number must be deficient. **true**

6. The equation $17 + 51 = 68$ verifies Goldbach's conjecture for the number 68. **false**

7. The number $2^5 - 1$ is an example of a Mersenne prime. **true**

8. The number 31 cannot be represented as the difference of two squares. **false**

9. The number $2^6(2^7 - 1)$ is perfect. **true**

10. A natural number greater than 1 will be one and only one of the following: perfect, deficient, or abundant. **true**

11. The proper divisors of 496 are 1, 2, 4, 8, 16, 31, 62, 124, and 248. Use this information to verify that 496 is perfect.
The sum of the proper divisors is 496: $1 + 2 + 4 + 8 + 16 + 31 + 62 + 124 + 248 = 496$.

12. The proper divisors of 8,128 are 1, 2, 4, 8, 16, 32, 64, 127, 254, 508, 1,016, 2,032, and 4,064. Use this information to verify that 8,128 is perfect.
The sum of the proper divisors is 8,128: $1 + 2 + 4 + 8 + 16 + 32 + 64 + 127 + 254 + 508 + 1{,}016 + 2{,}032 + 4{,}064 = 8{,}128$.

13. As mentioned in the text, when $2^n - 1$ is prime, $2^{n-1}(2^n - 1)$ is perfect. By letting $n = 2, 3, 5$, and 7, we obtain the first four perfect numbers. Show that $2^n - 1$ is prime for $n = 13$, and then find the decimal digit representation for the fifth perfect number. **8,191 is prime; 33,550,336**

14. In 1988, the largest known prime number was $2^{216,091} - 1$. Use the formula in Exercise 13 to write an expression for the perfect number generated by this prime number. $\mathbf{2^{216,090}(2^{216,091} - 1)}$

15. It has been proved that the reciprocals of *all* the positive divisors of a perfect number have a sum of 2. Verify this for the perfect number 6. $\mathbf{1 + 1/2 + 1/3 + 1/6 = 2}$

16. Consider the following equations.

$$6 = 1 + 2 + 3$$

$$28 = 1 + 2 + 3 + 4 + 5 + 6 + 7$$

Show that a similar equation is valid for the third perfect number, 496.
$\mathbf{496 = 1 + 2 + 3 + \cdots + 31}$

Determine whether each number is abundant *or* deficient.

17. 36 **abundant** **18.** 30 **abundant**

19. 75 **deficient** **20.** 95 **deficient**

21. There are four abundant numbers between 1 and 25. Find them. (*Hint:* They are all even, and no prime number is abundant.) **12, 18, 20, 24**

22. Explain why a prime number must be deficient.

23. The first odd abundant number is 945. Its proper divisors are 1, 3, 5, 7, 9, 15, 21, 27, 35, 45, 63, 105, 135, 189, and 315. Use this information to verify that 945 is abundant.
1 + 3 + 5 + 7 + 9 + 15 + 21 + 27 + 35 + 45 + 63 + 105 + 135 + 189 + 315 = 975, and 975 > 945, so 945 is abundant.

24. Explain in your own words the terms perfect number, abundant number, and deficient number.

25. Nicolo Paganini's numbers 1,184 and 1,210 are amicable. The proper divisors of 1,184 are 1, 2, 4, 8, 16, 32, 37, 74, 148, 296, and 592. The proper divisors of 1,210 are 1, 2, 5, 10, 11, 22, 55, 110, 121, 242, and 605. Use the definition of amicable (friendly) numbers to show that they are indeed amicable.
1 + 2 + 4 + 8 + 16 + 32 + 37 + 74 + 148 + 296 + 592 = 1,210 and 1 + 2 + 5 + 10 + 11 + 22 + 55 + 110 + 121 + 242 + 605 = 1,184.

▲ **26.** An Arabian mathematician of the ninth century stated the following.

If the three numbers

$$x = 3 \cdot 2^{n-1} - 1,$$

$$y = 3 \cdot 2^n - 1,$$

and

$$z = 9 \cdot 2^{2n-1} - 1$$

are all prime and $n \geq 2$, then $2^n xy$ and $2^n z$ are amicable numbers.

(a) Use $n = 2$, and show that the result is the pair of amicable numbers mentioned in the discussion in the text.
$\mathbf{2^2 \cdot 5 \cdot 11 = 220 \text{ and } 2^2 \cdot 71 = 284}$

(b) Use $n = 4$ to obtain another pair of amicable numbers. **17,296 and 18,416**

Write each of the following even numbers as the sum of two primes. (There may be more than one way to do this.)

27. 14 **3 + 11** **28.** 22 **11 + 11**

29. 26 **3 + 23** **30.** 32 **13 + 19**

31. Joseph Louis Lagrange (1736–1813) conjectured that every odd natural number greater than 5 can be written as a sum $a + 2b$, where a and b are both primes. Verify this for the odd natural number 11.
Let $a = 5$ and $b = 3$; $11 = 5 + 2 \cdot 3$

32. An unproved conjecture in number theory states that every natural number multiple of 6 can be written as the difference of two primes. Verify this for 6, 12, and 18.
6 = 11 − 5; 12 = 17 − 5; 18 = 23 − 5

Find one pair of twin primes between the two numbers given.

33. 65 and 80 **71 and 73**

34. 85 and 105 **101 and 103**

35. 125 and 140 **137 and 139**

While Pierre De Fermat is probably best known for his now famous "last theorem," he did provide proofs of many other theorems in number theory. Exercises 36–40 investigate some of these theorems.

36. If p is prime and the natural numbers a and p have no common factor except 1, then $a^{p-1} - 1$ is divisible by p.

(a) Verify this for $p = 5$ and $a = 3$.
$3^4 - 1 = 80$ is divisible by 5.

(b) Verify this for $p = 7$ and $a = 2$.
$2^6 - 1 = 63$ is divisible by 7.

37. Every odd prime can be expressed as the difference of two squares in one and only one way.

(a) Find this one way for the prime number 5.
$5 = 9 - 4 = 3^2 - 2^2$

(b) Find this one way for the prime number 11.
$11 = 36 - 25 = 6^2 - 5^2$

38. A prime number of the form $4k + 1$ can be represented as the sum of two squares.

(a) The prime number 5 satisfies the conditions of the theorem, with $k = 1$. Verify this theorem for 5. **$5 = 1 + 4 = 1^2 + 2^2$**

(b) Verify this theorem for 13 (here, $k = 3$).
$13 = 4 + 9 = 2^2 + 3^2$

39. There is only one solution in natural numbers for $a^2 + 2 = b^3$, and it is $a = 5$, $b = 3$. Verify this solution. **$5^2 + 2 = 27 = 3^3$**

40. There are only two solutions in integers for $a^2 + 4 = b^3$, and they are $a = 2$, $b = 2$, and $a = 11$, $b = 5$. Verify these solutions. **$2^2 + 4 = 8 = 2^3$ and $11^2 + 4 = 125 = 5^3$**

While there are infinitely many primes, no formula has ever been found that will yield only prime numbers for all natural numbers n. The formula

$$p = n^2 - n + 41$$

will give prime numbers for all natural numbers $n = 1$ *through* $n = 40$. *Verify that this is true for the following values of n.*

41. $n = 1$ **41 is prime.** **42.** $n = 5$ **61 is prime.** **43.** $n = 7$ **83 is prime.** **44.** $n = 9$ **113 is prime.**

45. Explain why the formula used in Exercises 41–44 fails when $n = 41$.

The formula

$$p = n^2 - 79n + 1{,}601$$

will give prime numbers for all values of n less than 80, *but fails with* $n = 80$, *since for* $n = 80$, $p = 1{,}681 = 41^2$. *Determine the primes generated by this formula for the following values of n.*

46. $n = 1$ **1,523** **47.** $n = 5$ **1,231** **48.** $n = 7$ **1,097** **49.** $n = 9$ **971**

50. In Euclid's proof that there is no largest prime, we formed a number M by taking the product of primes and adding 1. Observe the pattern below.

$M = 2 + 1 = 3$ (3 is prime)
$M = 2 \cdot 3 + 1 = 7$ (7 is prime)
$M = 2 \cdot 3 \cdot 5 + 1 = 31$ (31 is prime)
$M = 2 \cdot 3 \cdot 5 \cdot 7 + 1 = 211$ (211 is prime)
$M = 2 \cdot 3 \cdot 5 \cdot 7 \cdot 11 + 1 = 2{,}311$ (2,311 is prime)

It seems as though this pattern will always yield a prime number. Now evaluate

$$M = 2 \cdot 3 \cdot 5 \cdot 7 \cdot 11 \cdot 13 + 1.$$

Is M prime or composite? If composite, give its prime factorization.
composite; $M = 30{,}031 = 59 \cdot 509$

5.3 Exercises (Text Page 236)

◉ *Decide whether each of the following is* true *or* false.

1. Two even natural numbers cannot be relatively prime. **true**
2. Two different prime numbers must be relatively prime. **true**
3. If p is a prime number, then the greatest common factor of p and p^2 is p. **true**
4. If p is a prime number, then the least common multiple of p and p^2 is p^3. **false**
5. There is no prime number p such that the greatest common factor of p and 2 is 2. **false**
6. The set of all common multiples of two given natural numbers is finite. **false**
7. Two natural numbers must have at least one common factor. **true**
8. The least common multiple of two different primes is their product. **true**
9. Two composite numbers may be relatively prime. **true**
10. The set of all common factors of two given natural numbers is finite. **true**

Use the prime factors method to find the greatest common factor of each group of numbers.

11. 70 and 120 **10**
12. 180 and 300 **60**
13. 480 and 1,800 **120**
14. 168 and 504 **168**
15. 28, 35, and 56 **7**
16. 252, 308, and 504 **28**

Use the method of dividing by prime factors to find the greatest common factor of each group of numbers.

17. 60 and 84 **12**
18. 130 and 455 **65**
19. 310 and 460 **10**
20. 234 and 470 **2**
21. 12, 18, and 30 **6**
22. 450, 1,500, and 432 **6**

Use the Euclidean algorithm to find the greatest common factor of each group of numbers.

23. 36 and 60 **12**
24. 25 and 70 **5**
25. 84 and 180 **12**
26. 72 and 120 **24**
27. 210 and 560 **70**
28. 150 and 480 **30**
29. Explain in your own words how to find the greatest common factor of a group of numbers.
30. Explain in your own words how to find the least common multiple of a group of numbers.

Use the prime factors method to find the least common multiple of each group of numbers.

31. 24 and 30 **120**
32. 12 and 32 **96**
33. 56 and 96 **672**
34. 28 and 70 **140**
35. 30, 40, and 70 **840**
36. 24, 36, and 48 **144**

Use the formula given in the text and the results of Exercises 23–28 to find the least common multiple of each group of numbers.

37. 36 and 60 **180**
38. 25 and 70 **350**
39. 84 and 180 **1,260**
40. 72 and 120 **360**
41. 210 and 560 **1,680**
42. 150 and 480 **2,400**
43. ◉ If p, q, and r are different primes, and a, b, and c are natural numbers such that $a > b > c$,
   (a) what is the greatest common factor of $p^a q^c r^b$ and $p^b q^a r^c$? $p^b q^c r^c$
   (b) what is the least common multiple of $p^b q^a$, $q^b r^c$, and $p^a r^b$? $p^a q^a r^b$

◉ CONCEPTUAL WRITING ▲ CHALLENGING SCIENTIFIC CALCULATOR GRAPHING CALCULATOR

44. Find **(a)** the greatest common factor and **(b)** the least common multiple of $2^{31} \cdot 5^{17} \cdot 7^{21}$ and $2^{34} \cdot 5^{22} \cdot 7^{13}$. Leave your answers in prime factored form.
(a) $2^{31} \cdot 5^{17} \cdot 7^{13}$ (b) $2^{34} \cdot 5^{22} \cdot 7^{21}$

It is possible to extend the Euclidean algorithm in order to find the greatest common factor of more than two numbers. For example, if we wish to find the greatest common factor of 150, 210, *and* 240, *we can first use the algorithm to find the greatest common factor of two of these (say, for example,* 150 *and* 210*). Then we find the greatest common factor of that result and the third number,* 240. *The final result is the greatest common factor of the original group of numbers.*

Use the Euclidean algorithm as described above to find the greatest common factor of each group of numbers.

45. 150, 210, and 240 30

46. 12, 75, and 120 3

47. 90, 105, and 315 15

48. 48, 315, and 450 25,200

49. 144, 180, and 192 2,880

50. 180, 210, and 630 1,260

If we allow repetitions of prime factors, Venn diagrams (Chapter 2) can be used to find the greatest common factor and the least common multiple of two numbers. For example, consider $36 = 2^2 \cdot 3^2$ *and* $45 = 3^2 \cdot 5$.

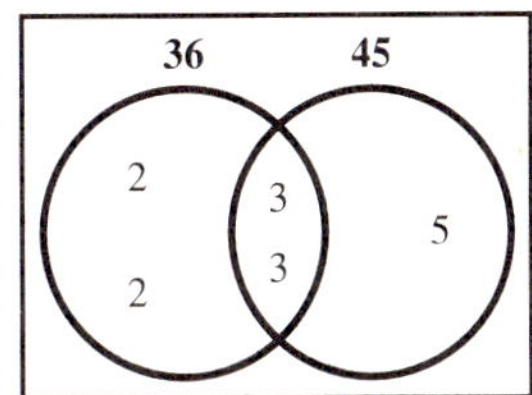

Intersection gives 3, 3.
Union gives 2, 2, 3, 3, 5.

Their greatest common factor is $3^2 = 9$, *and their least common multiple is* $2^2 \cdot 3^2 \cdot 5 = 180$.

Use this method to find **(a)** *the greatest common factor and* **(b)** *the least common multiple of the two numbers given.*

51. 12 and 18 (a) 6 (b) 36

52. 27 and 36 (a) 9 (b) 108

53. 54 and 72 (a) 18 (b) 216

54. Suppose that the least common multiple of p and q is q. What can we say about p and q? **p is a factor of q (or, q is a multiple of p).**

55. Suppose that the least common multiple of p and q is pq. What can we say about p and q? **p and q are relatively prime.**

56. Suppose that the greatest common factor of p and q is p. What can we say about p and q? **p is a factor of q (or, q is a multiple of p).**

57. Recall some of your early experiences in mathematics (for example, in the elementary grade classroom). What topic involving fractions required the use of the least common multiple? Give an example.

58. Recall some of your experiences in elementary algebra. What topics required the use of the greatest common factor? Give an example.

Refer to Examples 8 and 9 to solve the following problems.

59. Melissa Lerner and Kay LaBarre work on an assembly line, inspecting electronic calculators. Melissa inspects the electronics of every sixteenth calculator, while Kay inspects the workmanship of every thirty-sixth calculator. If they both start working at the same time, which calculator will be the first that they both inspect? **144th**

60. Daniel Lange and Donald Cole work as security guards at a publishing company. Daniel has every sixth night off, and Donald has every tenth night off. If both are off on July 1, what is the next night that they will both be off together? **July 31**

61. Diane Blake has 240 pennies and 288 nickels. She wants to place the pennies and nickels in stacks so that each stack has the same number of coins, and each stack contains only one denomination of coin. What is the largest number of coins that she can place in each stack? **48**

62. Melissa Martin and Tracy Copley are in a bicycle race, following a circular track. If they start at the same place and travel in the same direction, and Melissa completes a revolution every 40 seconds and Tracy completes a revolution every 45 seconds, how long will it take them before they reach the starting point again simultaneously? **360 seconds**

63. John Cross sold some books at $24 each, and had enough money to buy some concert tickets at $50 each. He had no money left over after buying the tickets. What is the least amount of money he could have earned from selling the books? What is the least number of books he could have sold?
$600; 25 books

64. Joan Langevin has some two-by-four pieces of lumber. Some are 60 inches long, and some are 72 inches long. She wishes to cut them so as to obtain equal length pieces. What is the longest such piece she can cut so that no lumber is left over?
12 inches

5.4 Exercises (Text Page 243)

Answer each of the following questions concerning the Fibonacci sequence or the golden ratio.

1. The sixteenth Fibonacci number is 987 and the seventeenth Fibonacci number is 1,597. What is the eighteenth Fibonacci number? **2,584**

2. Recall that F_n represents the Fibonacci number in the nth position in the sequence. What are the only two values of n such that $F_n = n$? **1 and 5**

3. $F_{23} = 28{,}657$ and $F_{25} = 75{,}025$. What is the value of F_{24}? **46,368**

4. If two successive terms of the Fibonacci sequence are both odd, is the next term even or odd? **even**

5. What is the exact value of the golden ratio? $\dfrac{1 + \sqrt{5}}{2}$

6. What is the approximate value of the golden ratio to the nearest thousandth? **1.618**

In Exercises 7–14, a pattern is established involving terms of the Fibonacci sequence. Use inductive reasoning to make a conjecture concerning the next equation in the pattern, and verify it. You may wish to refer to the first few terms of the sequence given in the text.

7.
$1 = 2 - 1$
$1 + 1 = 3 - 1$
$1 + 1 + 2 = 5 - 1$
$1 + 1 + 2 + 3 = 8 - 1$
$1 + 1 + 2 + 3 + 5 = 13 - 1$
$1 + 1 + 2 + 3 + 5 + 8 = 21 - 1$; Each expression is equal to 20.

8.
$1 = 2 - 1$
$1 + 3 = 5 - 1$
$1 + 3 + 8 = 13 - 1$
$1 + 3 + 8 + 21 = 34 - 1$
$1 + 3 + 8 + 21 + 55 = 89 - 1$
$1 + 3 + 8 + 21 + 55 + 144 = 233 - 1$; Each expression is equal to 232.

9.
$1 = 1$
$1 + 2 = 3$
$1 + 2 + 5 = 8$
$1 + 2 + 5 + 13 = 21$
$1 + 2 + 5 + 13 + 34 = 55$
$1 + 2 + 5 + 13 + 34 + 89 = 144$; Each expression is equal to 144.

10.
$1^2 + 1^2 = 2$
$1^2 + 2^2 = 5$
$2^2 + 3^2 = 13$
$3^2 + 5^2 = 34$
$5^2 + 8^2 = 89$
$8^2 + 13^2 = 233$; Each expression is equal to 233.

11.
$2^2 - 1^2 = 3$
$3^2 - 1^2 = 8$
$5^2 - 2^2 = 21$
$8^2 - 3^2 = 55$
$13^2 - 5^2 = 144$; Each expression is equal to 144.

12.
$2^3 + 1^3 - 1^3 = 8$
$3^3 + 2^3 - 1^3 = 34$
$5^3 + 3^3 - 2^3 = 144$
$8^3 + 5^3 - 3^3 = 610$
$13^3 + 8^3 - 5^3 = 2{,}584$; Each expression is equal to 2,584.

13. $1 = 1^2$
$1 - 2 = -1^2$
$1 - 2 + 5 = 2^2$
$1 - 2 + 5 - 13 = -3^2$
$1 - 2 + 5 - 13 + 34 = 5^2$
$1 - 2 + 5 - 13 + 34 - 89 = -8^2$; Each expression is equal to -64.

14. $1 - 1 = -1 + 1$
$1 - 1 + 2 = 1 + 1$
$1 - 1 + 2 - 3 = -2 + 1$
$1 - 1 + 2 - 3 + 5 = 3 + 1$
$1 - 1 + 2 - 3 + 5 - 8 = -5 + 1$
$1 - 1 + 2 - 3 + 5 - 8 + 13 = 8 + 1$; Each expression is equal to 9.

---

15. Every natural number can be expressed as a sum of Fibonacci numbers, where no number is used more than once. For example, $25 = 21 + 3 + 1$. Express each of the following in this way.
(There are other ways to do this.)
(a) 37 **$37 = 34 + 3$**
(b) 40 **$40 = 34 + 5 + 1$**
(c) 52 **$52 = 34 + 13 + 5$**

16. It has been shown that if m divides n, then F_m is a factor of F_n. Show that this is true for the following values of m and n.
(a) $m = 2, n = 6$ **2 divides 6, and $F_2 = 1$ is a factor of $F_6 = 8$.**
(b) $m = 3, n = 9$ **3 divides 9, and $F_3 = 2$ is a factor of $F_9 = 34$.**
(c) $m = 4, n = 8$ **4 divides 8, and $F_4 = 3$ is a factor of $F_8 = 21$.**

17. It has been shown that if the greatest common factor of m and n is r, then the greatest common factor of F_m and F_n is F_r. Show that this is true for the following values of m and n.
(a) $m = 10, n = 4$ **The greatest common factor of 10 and 4 is 2, and the greatest common factor of $F_{10} = 55$ and $F_4 = 3$ is $F_2 = 1$.**
(b) $m = 12, n = 6$ **The greatest common factor of 12 and 6 is 6, and the greatest common factor of $F_{12} = 144$ and $F_6 = 8$ is $F_6 = 8$.**
(c) $m = 14, n = 6$ **The greatest common factor of 14 and 6 is 2, and the greatest common factor of $F_{14} = 377$ and $F_6 = 8$ is $F_2 = 1$.**

18. For any prime number p except 2 or 5, either F_{p+1} or F_{p-1} is divisible by p. Show that this is true for the following values of p.
(a) $p = 3$ **$F_{3+1} = F_4 = 3$ is divisible by 3.**
(b) $p = 7$ **$F_{7+1} = F_8 = 21$ is divisible by 7.**
(c) $p = 11$ **$F_{11-1} = F_{10} = 55$ is divisible by 11.**

19. Earlier we saw that if a term of the Fibonacci sequence is squared and then the product of the terms on each side of the term is found, there will always be a difference of 1. Follow the steps below, choosing the seventh Fibonacci number, 13.
(a) Square 13. Multiply the terms of the sequence two positions away from 13 (i.e., 5 and 34). Subtract the smaller result from the larger, and record your answer. **$5 \cdot 34 - 13^2 = 1$**
(b) Square 13. Multiply the terms of the sequence three positions away from 13. Once again, subtract the smaller result from the larger, and record your answer. **$13^2 - 3 \cdot 55 = 4$**
(c) Repeat the process, moving four terms away from 13. **$2 \cdot 89 - 13^2 = 9$**
(d) Make a conjecture about what will happen when you repeat the process, moving five terms away. Verify your answer.
The difference will be 25, since we are obtaining the squares of the terms of the Fibonacci sequence. $13^2 - 1 \cdot 144 = 25 = 5^2$.

20. Here is a number trick that you can perform. Ask someone to pick any two numbers at random and to write them down. Ask the person to determine a third number by adding the first and second, a fourth number by adding the second and third, and so on, until ten numbers are determined. Then ask the person to add these ten numbers. You will be able to give the sum before the person even completes the list, because the sum will always be 11 times the seventh number in the list. Verify that this is true, by using x and y as the first two numbers arbitrarily chosen. (*Hint:* Remember the distributive property from algebra.)
The sum is $55x + 88y = 11(5x + 8y)$. Since $5x + 8y$ is the seventh number in the list, the process is verified.

Another Fibonacci-type *sequence that has been studied by mathematicians is the* **Lucas sequence,** *named after a French mathematician of the nineteenth century. The first ten terms of the Lucas sequence are*

1, 3, 4, 7, 11, 18, 29, 47, 76, 123.

Exercises 21–26 pertain to the Lucas sequence.

21. What is the eleventh term of the Lucas sequence?
199

22. Choose any term of the Lucas sequence and square it. Then multiply the terms on either side of the one you chose. Subtract the smaller result from the larger. Repeat this for a different term of the sequence. Do you get the same result? Make a conjecture about this pattern.
In each case, the difference is 5.

23. The first term of the Lucas sequence is 1. Add the first and third terms. Record your answer. Now add the first, third, and fifth terms and record your answer. Continue this pattern, each time adding another term that is in an *odd* position in the sequence. What do you notice about all of your results?
Each sum is 2 less than a Lucas number.

24. The second term of the Lucas sequence is 3. Add the second and fourth terms. Record your answer. Now add the second, fourth, and sixth terms and record your answer. Continue this pattern, each time adding another term that is in an *even* position of the sequence. What do you notice about all of your sums?
Each sum is 1 less than a Lucas number.

25. Many interesting patterns exist between the terms of the Fibonacci sequence and the Lucas sequence. Make a conjecture about the next equation that would appear in each of the lists and then verify it.

(a)
$1 \cdot 1 = 1$
$1 \cdot 3 = 3$
$2 \cdot 4 = 8$
$3 \cdot 7 = 21$
$5 \cdot 11 = 55$
$8 \cdot 18 = 144$; Each expression is equal to 144.

(b)
$1 + 2 = 3$
$1 + 3 = 4$
$2 + 5 = 7$
$3 + 8 = 11$
$5 + 13 = 18$
$8 + 21 = 29$; Each expression is equal to 29.

(c)
$1 + 1 = 2 \cdot 1$
$1 + 3 = 2 \cdot 2$
$2 + 4 = 2 \cdot 3$
$3 + 7 = 2 \cdot 5$
$5 + 11 = 2 \cdot 8$
$8 + 18 = 2 \cdot 13$; Each expression is equal to 26.

26. In the text we illustrate that the quotients of successive terms of the Fibonacci sequence approach the golden ratio. Make a similar observation for the terms of the Lucas sequence; that is, find the decimal approximations for the quotients

$$\frac{3}{1}, \frac{4}{3}, \frac{7}{4}, \frac{11}{7}, \frac{18}{11}, \frac{29}{18},$$

and so on, using a calculator. Then make a conjecture about what seems to be happening.
The quotients seem to approach the golden ratio. (This is indeed the case.)

Recall the **Pythagorean theorem** *from geometry: If a right triangle has legs of lengths a and b and hypotenuse of length c, then $a^2 + b^2 = c^2$. Suppose that we choose any four successive terms of the Fibonacci sequence. Multiply the first and fourth. Double the product of the second and third. Add the squares of the second and third. The three results obtained form a Pythagorean triple (three numbers that satisfy the equation $a^2 + b^2 = c^2$). Find the Pythagorean triple obtained this way using the four given successive terms of the Fibonacci sequence.*

27. 1, 1, 2, 3
3, 4, 5

28. 1, 2, 3, 5
5, 12, 13

29. 2, 3, 5, 8
16, 30, 34

30. Look at the values of the hypotenuse (c) in the answers to Exercises 27–29. What do you notice about each of them?
Each is a term of the Fibonacci sequence.

31. The following array of numbers is called **Pascal's triangle.**

```
            1
          1   1
        1   2   1
      1   3   3   1
    1   4   6   4   1
  1   5  10  10   5   1
1   6  15  20  15   6   1
```

This array is important in the study of counting techniques and probability (see later chapters) and appears in algebra in the binomial theorem. If the triangular array is written in a different form, as follows, and the sums along the diagonals as indicated by the dashed lines are found, an interesting thing occurs. What do you find when the numbers are added?

1
1 1
1 2 1
1 3 3 1
1 4 6 4 1
1 5 10 10 5 1
1 6 15 20 15 6 1

The sums are 1, 1, 2, 3, 5, 8, 13. They are terms of the Fibonacci sequence.

32. Write a paragraph explaining some of the occurrences of the Fibonacci sequence and the golden ratio in your everyday surroundings.

Exercises 33–38 require the use of a scientific calculator.

33. The positive solution of the equation $x^2 - x - 1 = 0$ is $(1 + \sqrt{5})/2$, as indicated in the text. The negative solution is $(1 - \sqrt{5})/2$. Find decimal approximations for both. What similarity do you notice between the two decimals?
$(1 + \sqrt{5})/2 \approx 1.618033989$ and $(1 - \sqrt{5})/2 \approx -.618033989$. The decimal places have the same digits.

34. In some cases, writers define the golden ratio to be the *reciprocal* of $(1 + \sqrt{5})/2$. Find a decimal approximation for the reciprocal of $(1 + \sqrt{5})/2$. What similarity do you notice between the decimals for $(1 + \sqrt{5})/2$ and its reciprocal?
$2/(1 + \sqrt{5}) \approx .618033989$. The decimal places for $(1 + \sqrt{5})/2$ and its reciprocal have the same digits.

A remarkable relationship exists between the two solutions of $x^2 - x - 1 = 0$,

$$\phi = \frac{1 + \sqrt{5}}{2} \quad \text{and} \quad \bar{\phi} = \frac{1 - \sqrt{5}}{2},$$

and the Fibonacci numbers. To find the nth Fibonacci number without using the recursion formula, evaluate

$$\frac{\phi^n - \bar{\phi}^n}{\sqrt{5}}$$

using a calculator. For example, to find the thirteenth Fibonacci number, evaluate

$$\frac{\left(\frac{1 + \sqrt{5}}{2}\right)^{13} - \left(\frac{1 - \sqrt{5}}{2}\right)^{13}}{\sqrt{5}}.$$

This form is known as the **Binet form** *of the nth Fibonacci number. Use the Binet form and a calculator to find the nth Fibonacci number for each of the following values of n.*

35. $n = 14$ **377** **36.** $n = 20$ **6,765**

37. $n = 22$ **17,711** **38.** $n = 25$ **75,025**

Extension Exercises (Text Page 248)

Given a magic square, other magic squares may be obtained by rotating the given one. For example, starting with the magic square in Figure 5, a 90° rotation in a clockwise direction gives the magic square shown at the right.

6	1	8
7	5	3
2	9	4

Start with Figure 5 and give the magic square obtained by each rotation described.

1. 180° in a clockwise direction

2	7	6
9	5	1
4	3	8

2. 90° in a counterclockwise direction

4	9	2
3	5	7
8	1	6

Start with Figure 7 and give the magic square obtained by each rotation described.

3. 90° in a clockwise direction

11	10	4	23	17
18	12	6	5	24
25	19	13	7	1
2	21	20	14	8
9	3	22	16	15

4. 180° in a clockwise direction

9	2	25	18	11
3	21	19	12	10
22	20	13	6	4
16	14	7	5	23
15	8	1	24	17

5. 90° in a counterclockwise direction

15	16	22	3	9
8	14	20	21	2
1	7	13	19	25
24	5	6	12	18
17	23	4	10	11

6. Try to construct an order 2 magic square containing the entries 1, 2, 3, 4. What happens? **Such a magic square is impossible to construct.**

Given a magic square, other magic squares may be obtained by adding or subtracting a constant value to each entry, multiplying each entry by a constant value, or dividing each entry by a nonzero constant value. In Exercises 7–10, start with the magic square whose figure number is indicated, and perform the operation described to find a new magic square. Give the new magic sum.

7. Figure 5, multiply by 3

24	9	12
3	15	27
18	21	6

Magic sum is 45.

8. Figure 5, add 7

15	10	11
8	12	16
13	14	9

Magic sum is 36.

9. Figure 7, divide by 2

$\frac{17}{2}$	12	$\frac{1}{2}$	4	$\frac{15}{2}$
$\frac{23}{2}$	$\frac{5}{2}$	$\frac{7}{2}$	7	8
2	3	$\frac{13}{2}$	10	11
5	6	$\frac{19}{2}$	$\frac{21}{2}$	$\frac{3}{2}$
$\frac{11}{2}$	9	$\frac{25}{2}$	1	$\frac{9}{2}$

Magic sum is $32\frac{1}{2}$.

10. Figure 7, subtract 10

7	14	−9	−2	5
13	−5	−3	4	6
−6	−4	3	10	12
0	2	9	11	−7
1	8	15	−8	−1

Magic sum is 15.

According to a fanciful story by Charles Trigg in Mathematics Magazine *(September 1976, page 212), the Emperor Charlemagne (742–814) ordered a five-sided fort to be built at an important point in his kingdom. As good-luck charms, he had magic squares placed on all five sides of the fort. He had one restriction for these magic squares: all the numbers in them must be prime.*

The magic squares are given in Exercises 11–15, with one missing entry. Find the missing entry in each square.

11. 479

	71	257
47	269	491
281	467	59

12. 191

389		227
107	269	431
311	347	149

13. 467

389	227	191
71	269	
347	311	149

14. 311

401	227	179
47	269	491
359		137

15. 269

401	257	149
17		521
389	281	137

16. Compare the magic sums in Exercises 11–15. Charlemagne had stipulated that each magic sum should be the year in which the fort was built. What was that year? 807

Find the missing entries in each magic square.

17.

75	68	**(a)**
(b)	72	**(c)**
71	76	**(d)**

(a) 73 (b) 70 (c) 74 (d) 69

18.

1	8	13	**(a)**
(b)	14	7	2
16	9	4	**(c)**
(d)	**(e)**	**(f)**	15

(a) 12 (b) 11 (c) 5 (d) 6 (e) 3 (f) 10

19.

3	20	**(a)**	24	11
(b)	14	1	18	10
9	21	13	**(c)**	17
16	8	25	12	**(d)**
(e)	2	**(f)**	**(g)**	**(h)**

(a) 7 (b) 22 (c) 5 (d) 4 (e) 15
(f) 19 (g) 6 (h) 23

20.

3	36	2	35	31	4
10	12	**(a)**	26	7	27
21	13	17	14	**(b)**	22
16	**(c)**	23	**(d)**	18	15
28	30	8	**(e)**	25	9
(f)	1	32	5	6	34

(a) 29 (b) 24 (c) 19 (d) 20 (e) 11 (f) 33

21. Use the "staircase method" to construct a magic square of order 7, containing the entries 1, 2, 3, . . . , 49.

30	39	48	1	10	19	28
38	47	7	9	18	27	29
46	6	8	17	26	35	37
5	14	16	25	34	36	45
13	15	24	33	42	44	4
21	23	32	41	43	3	12
22	31	40	49	2	11	20

The magic square shown in the photograph is from a woodcut by Albrecht Dürer entitled Melancholia.

The two numbers in the center of the bottom row give 1514, *the year the woodcut was created. Refer to this magic square to answer Exercises 22–30.*

16	3	2	13
5	10	11	8
9	6	7	12
4	15	14	1

Dürer's Magic Square

22. What is the magic sum? **34**

23. Verify: The sum of the entries in the four corners is equal to the magic sum. **Each sum is equal to 34.**

24. Verify: The sum of the entries in any 2 by 2 square at a corner of the given magic square is equal to the magic sum. **Each sum is equal to 34.**

25. Verify: The sum of the entries in the diagonals is equal to the sum of the entries not in the diagonals. **Each sum is equal to 68.**

26. Verify: The sum of the squares of the entries in the diagonals is equal to the sum of the squares of the entries not in the diagonals. **Each sum is equal to 748.**

27. Verify: The sum of the cubes of the entries in the diagonals is equal to the sum of the cubes of the entries not in the diagonals. **Each sum is equal to 9,248.**

28. Verify: The sum of the squares of the entries in the top two rows is equal to the sum of the squares of the entries in the bottom two rows. **Each sum is equal to 748.**

29. Verify: The sum of the squares of the entries in the first and third rows is equal to the sum of the squares of the entries in the second and fourth rows. **Each sum is equal to 748.**

30. Find another interesting property of Dürer's magic square and state it.

31. A magic square of order 4 may be constructed as follows. Lightly sketch in the diagonals of the blank magic square. Beginning at the upper left, move across each row from left to right, counting the cells

CONCEPTUAL  WRITING  CHALLENGING  SCIENTIFIC CALCULATOR  GRAPHING CALCULATOR

as you go along. If the cell is on a diagonal, count it but do not enter its number. If it is not on a diagonal, enter its number. When this is completed, reverse the procedure, beginning at the bottom right and moving across from right to left. As you count the cells, enter the number if the cell is not occupied. If it is already occupied, count it but do not enter its number. You should obtain a magic square similar to the one given for Exercises 22–30. How do they differ?

16	2	3	13
5	11	10	8
9	7	6	12
4	14	15	1

The second and third columns are interchanged.

With chosen values for a, b, and c, an order 3 magic square can be constructed by substituting these values in the generalized form shown here.

$a+b$	$a-b-c$	$a+c$
$a-b+c$	a	$a+b-c$
$a-c$	$a+b+c$	$a-b$

Use the given values of a, b, and c to construct an order 3 magic square, using this generalized form.

32. $a = 5, \quad b = 1, \quad c = -3$

6	7	2
1	5	9
8	3	4

33. $a = 16, \quad b = 2, \quad c = -6$

18	20	10
8	16	24
22	12	14

34. $a = 5, \quad b = 4, \quad c = -8$

9	9	−3
−7	5	17
13	1	1

35. It can be shown that if an order n magic square has least entry k, and its entries are consecutive counting numbers, then its magic sum is given by the formula

$$\text{MS} = \frac{n(2k + n^2 - 1)}{2}.$$

Construct an order 7 magic square whose least entry is 10 using the staircase method. What is its magic sum?

39	48	57	10	19	28	37
47	56	16	18	27	36	38
55	15	17	26	35	44	46
14	23	25	34	43	45	54
22	24	33	42	51	53	13
30	32	41	50	52	12	21
31	40	49	58	11	20	29

Magic sum is 238.

▲ **36.** Use the formula of Exercise 35 to find the missing entries in the following order 4 magic square whose least entry is 24.

(a)	38	37	27
35	**(b)**	30	32
31	33	**(c)**	28
(d)	26	25	**(e)**

(a) 24 (b) 29 (c) 34 (d) 36 (e) 39

In a 1769 letter from Benjamin Franklin to a Mr. Peter Collinson, Franklin exhibited the following magic square of order 8.

52	61	4	13	20	29	36	45
14	3	62	51	46	35	30	19
53	60	5	12	21	28	37	44
11	6	59	54	43	38	27	22
55	58	7	10	23	26	39	42
9	8	57	56	41	40	25	24
50	63	2	15	18	31	34	47
16	1	64	49	48	33	32	17

37. What is the magic sum? **260**

Verify the following properties of this magic square.

38. The first half of each row and the second half of each row are each equal to half the magic sum. **Each half is equal to 130.**

39. The four corner entries added to the four center entries is equal to the magic sum. **52 + 45 + 16 + 17 + 54 + 43 + 10 + 23 = 260**

40. The "bent diagonals" consisting of eight entries, going up four entries from left to right and down four entries from left to right, give the magic sum. (For example, starting with 16, one bent diagonal sum is 16 + 63 + 57 + 10 + 23 + 40 + 34 + 17.) **50 + 8 + 7 + 54 + 43 + 26 + 25 + 47 = 260, and so on.**

If we use a "knight's move" (up two, right one) from chess, a variation of the staircase method gives rise to the magic square shown here. (When blocked, we move to the cell just below the previous entry.)

10	18	1	14	22
11	24	7	20	3
17	5	13	21	9
23	6	19	2	15
4	12	25	8	16

▲ *Use a similar process to construct an order 5 magic square, starting with 1 in the cell described.*

41. fourth row, second column (up two, right one; when blocked, move to the cell just below the entry)

5	13	21	9	17
6	19	2	15	23
12	25	8	16	4
18	1	14	22	10
24	7	20	3	11

42. third row, third column (up one, right two; when blocked, move to the cell just to the left of the previous entry)

9	3	22	16	15
21	20	14	8	2
13	7	1	25	19
5	24	18	12	6
17	11	10	4	23

Chapter 5 Test (Text Page 252)

◎ *Decide whether each statement is* true *or* false *(Exercises 1–5).*

1. No two prime numbers differ by 1. **false**

2. There are infinitely many prime numbers. **true**

3. If a natural number is divisible by 9, then it must also be divisible by 3. **true**

4. If p and q are different primes, 1 is their greatest common factor and pq is their least common multiple. **true**

5. For all natural numbers n, 1 is a factor of n and n is a multiple of n. **true**

6. Use divisibility tests to determine whether the number

331,153,470

is divisible by each of the following.

(a) 2 **yes** **(b)** 3 **yes** **(c)** 4 **no**
(d) 5 **yes** **(e)** 6 **yes** **(f)** 8 **no**
(g) 9 **yes** **(h)** 10 **yes** **(i)** 12 **no**

◎ CONCEPTUAL  WRITING ▲ CHALLENGING SCIENTIFIC CALCULATOR  GRAPHING CALCULATOR

7. Decide whether each number is prime, composite, or neither.
   (a) 93 **composite** (b) 1 **neither**
   (c) 59 **prime**

8. Give the prime factorization of 1,440. $2^5 \cdot 3^2 \cdot 5$

9. In your own words state the fundamental theorem of arithmetic.

10. Decide whether each number is perfect, deficient, or abundant.
    (a) 17 **deficient** (b) 6 **perfect**
    (c) 24 **abundant**

11. Which of the following statements is false? **(c)**
    (a) There are no known odd perfect numbers.
    (b) Every even perfect number must end in 6 or 28.
    (c) Goldbach's Conjecture for the number 8 is verified by the equation 8 = 7 + 1.

12. Give a pair of twin primes between 40 and 50. **41 and 43**

13. Find the greatest common factor of 270 and 450. **90**

14. Find the least common multiple of 24, 36, and 60. **360**

15. Both Sherrie Firavich and Della Daniel work at a fast-food outlet. Sherrie has every sixth day off and Della has every fourth day off. If they are both off on Wednesday of this week, what will be the day of the week that they are next off together? **Monday**

16. The twenty-second Fibonacci number is 17,711 and the twenty-third Fibonacci number is 28,657. What is the twenty-fourth Fibonacci number? **46,368**

17. Make a conjecture about the next equation in the following list, and verify it.

$$8 - (1 + 1 + 2 + 3) = 1$$
$$13 - (1 + 2 + 3 + 5) = 2$$
$$21 - (2 + 3 + 5 + 8) = 3$$
$$34 - (3 + 5 + 8 + 13) = 5$$
$$55 - (5 + 8 + 13 + 21) = 8$$

89 − (8 + 13 + 21 + 34) = 13; Each expression is equal to 13.

18. (a) Give the first eight terms of a Fibonacci-type sequence with first term 1 and second term 5. **1, 5, 6, 11, 17, 28, 45, 73**
    (b) Choose any term after the first in the sequence just formed. Square it. Multiply the two terms on either side of it. Subtract the smaller result from the larger. Now repeat the process with a different term. Make a conjecture about what this process will yield for any term of the sequence. **The process will yield 19 for any term chosen.**

19. Which one of the following is the *exact* value of the golden ratio? **(a)**
    (a) $\frac{1 + \sqrt{5}}{2}$ (b) $\frac{1 - \sqrt{5}}{2}$ (c) 1.6
    (d) 1.618

20. Write a short paragraph explaining what Fermat's Last Theorem says and giving an account of the recent developments concerning it.

CHAPTER 6 The Real Number System

6.1 Exercises (Text Page 261)

Graph each set on a number line.

1. {2, 3, 4, 5}

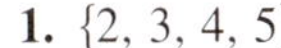

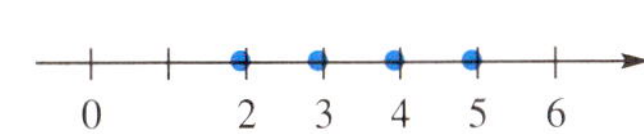

2. {0, 2, 4, 6, 8}

3. {−4, −3, −2, −1, 0, 1}

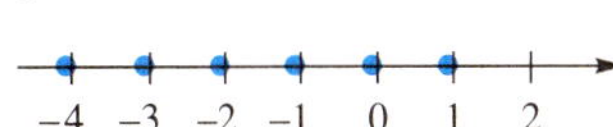

4. {−6, −5, −4, −3, −2}

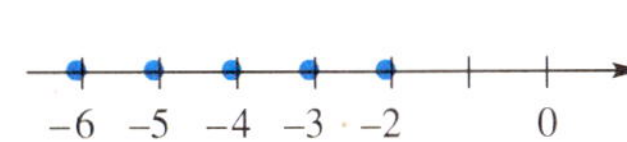

5. $\left\{-\frac{1}{2}, \frac{3}{4}, \frac{5}{3}, \frac{7}{2}\right\}$

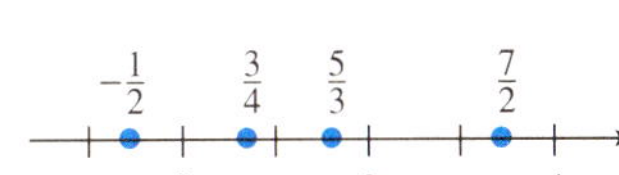

6. $\left\{-\frac{3}{5}, -\frac{1}{10}, \frac{9}{8}, \frac{12}{5}, \frac{13}{4}\right\}$

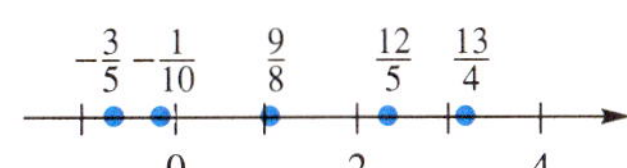

List the numbers in the given set that belong to **(a)** *the natural numbers,* **(b)** *the whole numbers,* **(c)** *the integers,* **(d)** *the rational numbers,* **(e)** *the irrational numbers, and* **(f)** *the real numbers.*

7. $\left\{-9, -\sqrt{7}, -\frac{5}{4}, -\frac{3}{5}, 0, \sqrt{5}, 3, 5.9, 7\right\}$
(a) 3, 7 (b) 0, 3, 7 (c) −9, 0, 3, 7
(d) −9, −5/4, −3/5, 0, 3, 5.9, 7 (e) $-\sqrt{7}, \sqrt{5}$
(f) All are real numbers.

8. $\left\{-5.3, -5, -\sqrt{3}, -1, -\frac{1}{9}, 0, 1.2, 1.8, 3, \sqrt{11}\right\}$
(a) 3 (b) 0, 3 (c) −5, −1, 0, 3
(d) −5.3, −5, −1, −1/9, 0, 1.2, 1.8, 3
(e) $-\sqrt{3}, \sqrt{11}$ **(f) All are real numbers.**

Decide whether each of the following statements is true *or* false.

9. $-2 < -1$ **true**
10. $-8 < -4$ **true**
11. $-3 \geq -7$ **true**
12. $-9 \geq -12$ **true**
13. $-15 \leq -20$ **false**
14. $-21 \leq -27$ **false**
15. $-8 \leq -(-4)$ **true**
16. $-9 \leq -(-6)$ **true**
17. $0 \leq -(-4)$ **true**
18. $0 \geq -(-6)$ **false**
19. $6 > -(-2)$ **true**
20. $-8 > -(-2)$ **false**

Give **(a)** *the additive inverse and* **(b)** *the absolute value of each of the following.*

21. 5 **(a) −5 (b) 5**
22. 9 **(a) −9 (b) 9**
23. −6 **(a) 6 (b) 6**
24. −8 **(a) 8 (b) 8**

---

25. A statement commonly heard is "Absolute value is always positive." Is this true? If not, explain.

26. If a is a negative number, then is $-|-a|$ positive or negative? **negative**

27. Fill in the blanks with the correct values: The opposite of −3 is ____, while the absolute value of −3 is ____. The additive inverse of −3 is ____, while the additive inverse of the absolute value of −3 is ____. **3; 3; 3; −3**

28. True or false: For all real numbers a and b,

$$|a - b| = |b - a|.$$ **true**

Specify each set by listing its elements. If there are no elements, write ∅.

29. $\{x \mid x$ is a natural number less than $7\}$
{1, 2, 3, 4, 5, 6}

30. $\{m \mid m$ is a whole number less than $9\}$
{0, 1, 2, 3, 4, 5, 6, 7, 8}

31. $\{a \mid a$ is an even integer greater than $10\}$
{12, 14, 16, 18, . . .}

32. $\{k \mid k$ is a natural number less than $1\}$ **∅**

33. $\{x \mid x$ is an irrational number that is also rational$\}$ **∅**

34. $\{r \mid r$ is a negative integer greater than $-1\}$ **∅**

35. $\{p \mid p$ is a number whose absolute value is $3\}$
{3, −3}

36. $\{w \mid w$ is a number whose absolute value is $12\}$
{12, −12}

37. $\{z \mid z$ is a whole number multiple of $5\}$
{0, 5, 10, 15, . . .}

38. $\{n \mid n$ is a natural number multiple of $3\}$
{3, 6, 9, 12, . . .}

Give three examples of numbers that satisfy the given condition. **Answers may vary in Exercises 39–44.**

39. positive real numbers but not integers
1/2, 5/8, 1 3/4

40. real numbers but not positive numbers
−1, −3/4, −5

41. real numbers but not whole numbers
−3 1/2, −2/3, 3/7

42. rational numbers but not integers
1/2, −2/3, 2/7

43. real numbers but not rational numbers
$\sqrt{5}, \sqrt{2}, -\sqrt{3}$

44. rational numbers but not negative numbers
2/3, 5/6, 0

Tell whether each statement is true *or* false.

45. Every rational number is an integer. **false**

46. Every natural number is an integer. **true**

47. Every integer is a rational number. true

48. Every whole number is a real number. true

49. Some rational numbers are irrational. false

50. Some natural numbers are whole numbers. true

51. Some rational numbers are integers. true

52. Some real numbers are integers. true

53. Every rational number is a real number. true

54. Some integers are not real numbers. false

55. Every integer is positive. false

56. Every whole number is positive. false

57. Some irrational numbers are negative. true

58. Some real numbers are not rational. true

59. Not every rational number is positive. true

60. Some whole numbers are not integers. false

Simplify each expression by removing absolute value symbols.

61. $|3|$ 3

62. $|-8|$ 8

63. $-|7|$ -7

64. $-|-9|$ -9

65. $|7-4|$ 3

66. $-|12-3|$ -9

67. $-|-(5-1)|$ -4

68. $|-(3-1)|$ 2

Decide which of the following sets the given number belongs to: **(a)** *whole numbers,* **(b)** *integers,* **(c)** *rational numbers,* **(d)** *irrational numbers,* **(e)** *real numbers.*

69. 12 (a), (b), (c), (e)

70. -4 (b), (c), (e)

71. $\frac{2}{3}$ (c), (e)

72. $\frac{4}{7}$ (c), (e)

73. 9.25 (c), (e)

74. -7.335 (c), (e)

75. 0 (a), (b), (c), (e)

76. .888 . . . (c), (e)

77. $\sqrt{2}$ (d), (e)

78. $\sqrt{-1}$ none of these sets

A ***bar graph,*** *as shown in the figure, is a convenient method for depicting data. Here we see that in 1989, $28 million was the figure for direct impact of moviemakers' spending in the state of Louisiana. In 1990, it was $23 million, and so on. Use the bar graph to answer the questions in Exercises 79–84.*

Graph, "It's Showtime: Moviemakers' spending a hit in Louisiana," from *The Times-Picayune*, August 17, 1994. Copyright © 1994 by *The Times-Picayune*. Reprinted by permission.

79. In what years was the direct impact greater than $23 million? 1989, 1992, and 1993

80. In what years was the direct impact less than $29 million? 1989, 1990, 1991, and 1992

81. In what years was the direct impact less than or equal to $29 million? 1989, 1990, 1991, and 1992

82. Explain why the answers to Exercises 80 and 81 are the same.

83. The total economic impact is usually determined by multiplying the direct impact by 3. What was the total economic impact from 1989 through 1992? $302.7 million

84. (See Exercise 83.) What was the total economic impact in 1993? $111.6 million

6.2 Exercises (Text Page 271)

Decide whether each statement is true *or* false.

1. The sum of two negative numbers must be negative. true

2. The difference between two negative numbers must be negative. false

3. The product of two negative numbers must be negative. false

4. The quotient of two negative numbers must be negative. false

5. If $a > 0$ and $b < 0$, then $\frac{a}{b} < 0$. true

6. The product of 648 and -927 is a real number. true

Perform the indicated operations, using the order of operations as necessary.

7. $-12 + (-8)$ **−20**

8. $-5 + (-2)$ **−7**

9. $12 + (-16)$ **−4**

10. $-6 + 17$ **11**

11. $-12 - (-1)$ **−11**

12. $-3 - (-8)$ **5**

13. $-5 + 11 + 3$ **9**

14. $-9 + 16 + 5$ **12**

15. $12 - (-3) - (-5)$ **20**

16. $15 - (-6) - (-8)$ **29**

17. $-9 - (-11) - (4 - 6)$ **4**

18. $-4 - (-13) + (-5 + 10)$ **14**

19. $(-12)(-2)$ **24**

20. $(-3)(-5)$ **15**

21. $9(-12)(-4)(-1)3$ **−1,296**

22. $-5(-17)(2)(-2)4$ **−1,360**

23. $\frac{-18}{-3}$ **6**

24. $\frac{-100}{-50}$ **2**

25. $\frac{36}{-6}$ **−6**

26. $\frac{52}{-13}$ **−4**

27. $\frac{0}{12}$ **0**

28. $\frac{0}{-7}$ **0**

29. $-6 + [5 - (3 + 2)]$ **−6**

30. $-8[4 + (7 - 8)]$ **−24**

31. $-8(-2) - [(4^2) + (7 - 3)]$ **−4**

32. $-7(-3) - [(2^3) - (3 - 4)]$ **12**

33. $-4 - 3(-2) + 5^2$ **27**

34. $-6 - 5(-8) + 3^2$ **43**

35. $(-6 - 3)(-2 - 3)$ **45**

36. $(-8 - 5)(-2 - 1)$ **39**

37. $\frac{(-10 + 4) \cdot (-3)}{-7 - 2}$ **−2**

38. $\frac{(-6 + 3) \cdot (-4)}{-5 - 1}$ **−2**

39. Which of the following expressions are undefined?

(a) $\frac{8}{0}$ **(b)** $\frac{9}{6 - 6}$ **(c)** $\frac{4 - 4}{5 - 5}$ **(d)** $\frac{0}{-1}$ **(a), (b), (c)**

40. If you have no money in your pocket and you divide it equally among your three siblings, how much does each get? Use this situation to explain division of zero by a nonzero number.

Identify the property illustrated by each of the following statements.

41. $6 + 9 = 9 + 6$ **commutative property of addition**

42. $8 \cdot 4 = 4 \cdot 8$ **commutative property of multiplication**

43. $7 + (2 + 5) = (7 + 2) + 5$ **associative property of addition**

44. $(3 \cdot 5) \cdot 4 = 4 \cdot (3 \cdot 5)$ **commutative property of multiplication**

45. $9 + (-9) = 0$ **inverse property of addition**

46. $12 + 0 = 12$ **identity property of addition**

47. $9 \cdot 1 = 9$ **identity property of multiplication**

48. $(-1/3) \cdot (-3) = 1$ **inverse property of multiplication**

49. $0 + 283 = 283$ **identity property of addition**

50. $6 \cdot (4 \cdot 2) = (6 \cdot 4) \cdot 2$ **associative property of multiplication**

51. $2 \cdot (4 + 3) = 2 \cdot 4 + 2 \cdot 3$ **distributive property**

52. $9 \cdot 6 + 9 \cdot 8 = 9 \cdot (6 + 8)$ **distributive property**

53. $19 + 12$ is a real number. **closure property of addition**

54. $19 \cdot 12$ is a real number. **closure property of multiplication**

55. **(a)** Evaluate $6 - 8$ and $8 - 6$. **$-2, 2$**

(b) By the results of part (a), we may conclude that subtraction is not a(n) _____ operation. **commutative**

(c) Are there *any* real numbers a and b for which $a - b = b - a$? If so, give an example. **yes; Choose $a = b$. For example, $a = b = 2$: $2 - 2 = 2 - 2$.**

56. **(a)** Evaluate $4 \div 8$ and $8 \div 4$. **1/2, 2**

(b) By the results of part (a), we may conclude that division is not a(n) _____ operation. **commutative**

(c) Are there *any* real numbers a and b for which $a \div b = b \div a$? If so, give an example. **yes; Choose $|a| = |b| \neq 0$. For example, $a = b = 2$: $2 \div 2 = 2 \div 2$.**

57. Many everyday occurrences can be thought of as operations that have opposites or inverses. For example, the inverse operation for "going to sleep" is "waking up." For each of the given activities, specify its inverse activity.

(a) cleaning up your room **messing up your room**

(b) earning money **spending money**

(c) increasing the volume on your portable radio **decreasing the volume on your portable radio**

58. Many everyday activities are commutative; that is, the order in which they occur does not affect the outcome. For example, "putting on your shirt" and "putting on your pants" are commutative operations. Decide whether the given activities are commutative.

(a) putting on your shoes; putting on your socks **no**

(b) getting dressed; taking a shower **no**

(c) combing your hair; brushing your teeth **yes**

59. The following conversation actually took place between one of the authors of this text and his son, Jack, when Jack was four years old.

Daddy: "Jack, what is 3 + 0?"
Jack: "3"
Daddy: "Jack, what is 4 + 0?"
Jack: "4 . . . and Daddy, *string* plus zero equals *string*!"

What property of addition of real numbers did Jack recognize? **identity**

60. The phrase *defective merchandise counter* is an example of a phrase that can have different meanings depending upon how the words are grouped (think of the associative properties). For example, (*defective merchandise*) *counter* is a location at which we would return an item that does not work, while *defective* (*merchandise counter*) is a broken place where items are bought and sold. For each of the following phrases, explain why the associative property does not hold.

(a) difficult test question

(b) woman fearing husband

(c) man biting dog

61. The distributive property holds for multiplication with respect to addition. Does the distributive property hold for addition with respect to multiplication? That is, is $a + (b \cdot c) = (a + b) \cdot (a + c)$ true for all values of a, b, and c? (*Hint:* Let $a = 2$, $b = 3$, and $c = 4$.) **no**

62. Suppose someone makes the following claim: The distributive property for addition with respect to multiplication (from Exercise 61) is valid, and here's why: Let $a = 2$, $b = -4$, and $c = 3$. Then $a + (b \cdot c) = 2 + (-4 \cdot 3) = 2 + (-12) = -10$, and $(a + b) \cdot (a + c) = [2 + (-4)] \cdot [2 + 3] = -2 \cdot 5 = -10$. Since both expressions are equal, the property must be valid. How would you respond to this reasoning?

63. Suppose that a student shows you the following work.

$$-3(4 - 6) = -3(4) - 3(6) = -12 - 18 = -30$$

The student has made a very common error in applying the distributive property. Write a short paragraph explaining the student's mistake, and work the problem correctly.

64. Work the following problem in two ways, first using the order of operations, and then using the distributive property: Evaluate $9(11 + 15)$. **$9(11 + 15) = 9(26) = 234$; $9(11 + 15) = 9(11) + 9(15) = 99 + 135 = 234$**

Recall from the text that an expression such as -2^4 *is evaluated as follows:*

$$-2^4 = -(2 \cdot 2 \cdot 2 \cdot 2) = -16.$$

The expression $(-2)^4$ *is evaluated as follows:*

$$(-2)^4 = (-2)(-2)(-2)(-2) = 16.$$

Each of the expressions in Exercises 65–72 is equal to either 81 *or* −81. *Decide which of these is the correct value.*

65. -3^4 **−81**

66. $-(3^4)$ **−81**

67. $(-3)^4$ **81**

68. $-(-3^4)$ **81**

69. $-(-3)^4$ **−81**

70. $[-(-3)]^4$ **81**

71. $-[-(-3)]^4$ **−81**

72. $-[-(-3^4)]$ **−81**

Solve the problem by evaluating a sum or difference of real numbers. No variables are needed.

73. In 1990, the net incomes of savings institutions (in millions of dollars) for states in the northeast United States were as follows.

State	Net Income (in millions of dollars)
Maine	0
New Hampshire	−24
Vermont	2
Massachusetts	−212
Rhode Island	−13
Connecticut	−149

What was the total of these net incomes? (*Source:* U.S. Office of Thrift Supervision) **−$396 million**

74. The 1991 state general fund balances (in millions of dollars) for states in the Pacific region of the United States were as follows.

State	General Fund Balance (in millions of dollars)
Washington	468
Oregon	380
California	−1,259
Alaska	791
Hawaii	347

What was the total of these balances? (*Source:* National Association of State Budget Officers) **$727 million**

75. The two charts show the heights of some selected mountains and the depths of some selected trenches. Use the information given to find the answers in parts (a) through (d). (*Source: The World Almanac and Book of Facts,* 1994)

Mountain	Height (in feet)
Foraker	17,400
Wilson	14,246
Pikes Peak	14,110

Trench	Depth in Feet (as a negative number)
Philippine	−32,995
Cayman	−24,721
Java	−23,376

(a) What is the difference between the height of Mt. Foraker and the depth of the Philippine Trench? **50,395 feet**

CONCEPTUAL WRITING CHALLENGING SCIENTIFIC CALCULATOR GRAPHING CALCULATOR

(b) What is the difference between the height of Pikes Peak and the depth of the Java Trench? **37,486 feet**
(c) How much deeper is the Cayman Trench than the Java Trench? **1,345 feet**
(d) How much deeper is the Philippine Trench than the Cayman Trench? **8,274 feet**

Solve each problem.

76. Shalita's checking account balance is $54.00. She then takes a gamble by writing a check for $89.00. What is her new balance? (Write the balance as a signed number.) **−$35.00**

77. The surface, or rim, of a canyon is at altitude 0. On a hike down into the canyon, a party of hikers stops for a rest at 130 meters below the surface. They then descend another 54 meters. What is their new altitude? (Write the altitude as a signed number.) **−184 meters**

78. A pilot announces to his passengers that the current altitude of their plane is 34,000 feet. Because of some unexpected turbulence, he is forced to descend 2,100 feet. What is the new altitude of the plane? (Write the altitude as a signed number.) **31,900 feet**

79. The lowest temperature ever recorded in Little Rock, Arkansas, was −5°F. The highest temperature ever recorded there was 117°F more than the lowest. What was this highest temperature? **112°F**

80. On January 23, 1943, the temperature rose 49°F in two minutes in Spearfish, South Dakota. If the starting temperature was −4°, what was the temperature two minutes later? **45°F**

81. On a series of three consecutive running plays, Herschel Walker gained 4 yards, lost 3 yards, and lost 2 yards. What positive or negative number represents his total net yardage for the series of plays? **−1 yard**

82. On three consecutive passing plays, Troy Aikman passed for a gain of 6 yards, was sacked for a loss of 12 yards, and passed for a gain of 43 yards. What positive or negative number represents the total net yardage for the plays? **37 yards**

83. The top of Mount Whitney, visible from Death Valley, has an altitude of 14,494 feet above sea level. The bottom of Death Valley is 282 feet below sea level. Using zero as sea level, find the difference between these two elevations. **14,776 feet**

84. The highest point in Louisiana is Driskill Mountain, at an altitude of 535 feet. The lowest point is at Spanish Fort, 8 feet below sea level. Using zero as sea level, find the difference between these two elevations. **543 feet**

85. A certain Greek mathematician was born in 426 B.C. His father was born 43 years earlier. In what year was his father born? **469 B.C.**

86. Kevin Carlson enjoys playing Triominoes every Thursday night. Last Thursday, on four successive turns, his scores were −19, 28, −5, and 13. What was his total score for the four turns? **17**

87. David Fleming enjoys scuba diving. He dives to 34 feet below the surface of a lake. His partner, Kim Walrath, dives to 40 feet below the surface, but then ascends 20 feet. What is the vertical distance between Kim and David? **14 feet**

88. Fontaine Evaldo, a pilot for a major airline, announced to her passengers that their plane, currently at 34,000 feet, would descend 2,500 feet to avoid turbulence, and then ascend 3,000 feet once they were out of danger from the turbulence. What would their final altitude be? **34,500 feet**

89. During the years 1980–1986, Rickey Henderson was the American League leader in stolen bases each year. The *bar graph* gives a representation of the number of steals each year. Use a signed number to represent the change in the number of steals from one year to the next. For example, from 1980 to 1981 the change was 56 − 100 = −44.
(a) from 1981 to 1982 **74**
(b) from 1982 to 1983 **−22**
(c) from 1983 to 1984 **−42**
(d) from 1984 to 1985 **14**
(e) from 1985 to 1986 **7**

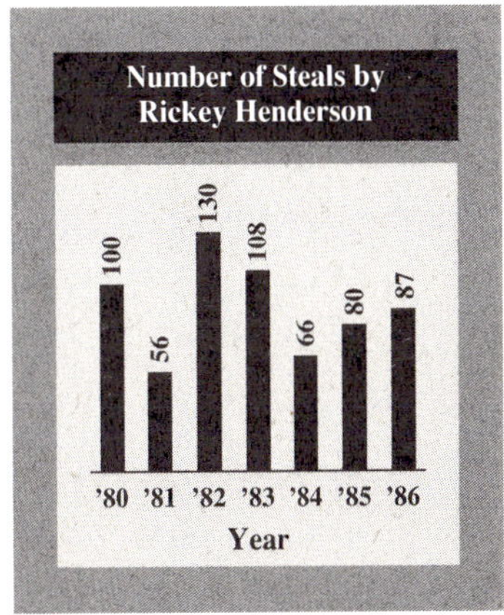

90. During the years 1985–1990, Vince Coleman was the National League leader in stolen bases each year. The *line graph* gives a representation of the number of steals each year. Use a signed number to represent the change in the number of steals from one year to the next. For example, from 1985 to 1986 the change was $107 - 110 = -3$.

(a) from 1986 to 1987 **2**

(b) from 1987 to 1988 **−28**

(c) from 1988 to 1989 **−15**

(d) from 1989 to 1990 **11**

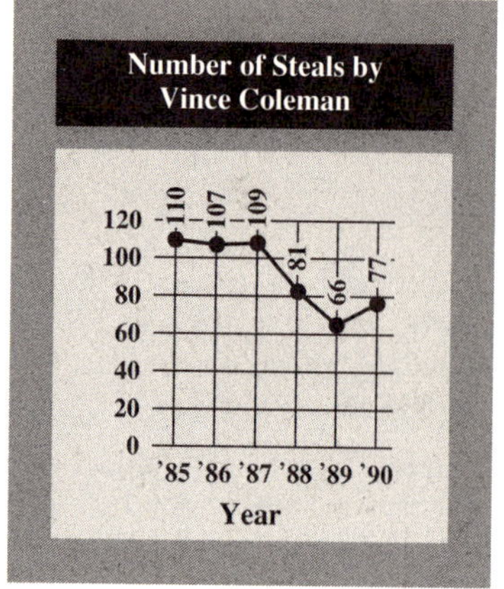

6.3 Exercises (Text Page 284)

◉ *Decide whether each statement is* true *or* false.

1. If p and q are different prime numbers, the rational number p/q is reduced to lowest terms. **true**

2. The same number may be added to both the numerator and the denominator of a fraction without changing the value of the fraction. **false**

3. A nonzero fraction and its reciprocal will always have the same sign. **true**

4. The set of integers has the property of density. **false**

Use the fundamental property of rational numbers to write each of the following in lowest terms.

5. $\frac{16}{48}$ **1/3** **6.** $\frac{21}{28}$ **3/4** **7.** $-\frac{15}{35}$ **−3/7** **8.** $-\frac{8}{48}$ **−1/6**

Use the fundamental property to write each of the following in three other ways. **Answers will vary in Exercises 9–12. We give three of infinitely many possible answers.**

9. $\frac{3}{8}$ **6/16, 9/24, 12/32** **10.** $\frac{9}{10}$ **18/20, 27/30, 36/40**

11. $-\frac{5}{7}$ **−10/14, −15/21, −20/28** **12.** $-\frac{7}{12}$ **−14/24, −21/36, −28/48**

◉ **13.** For each of the following, write a fraction in lowest terms that represents the portion of the figure that is in color.

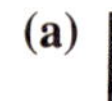

(b)
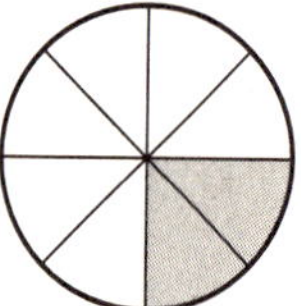
(c)
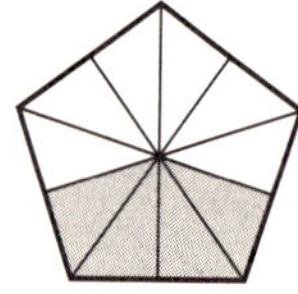
(d)
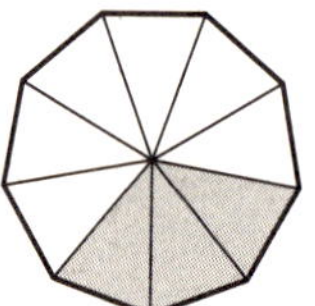

(a) 1/3 (b) 1/4 (c) 2/5 (d) 1/3

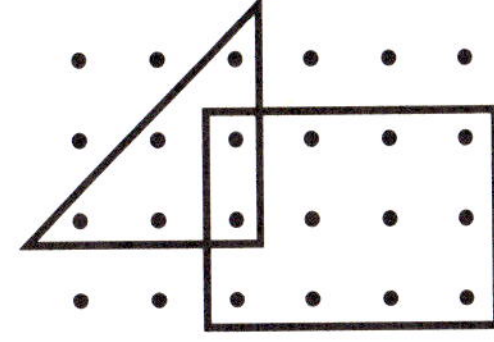

14. For each of the following, write a fraction in lowest terms that represents the region described.
    (a) the dots in the rectangle as a part of the dots in the entire figure 1/2
    (b) the dots in the triangle as a part of the dots in the entire figure 1/4
    (c) the dots in the rectangle as a part of the dots in the union of the triangle and the rectangle 3/4
    (d) the dots in the intersection of the triangle and the rectangle as a part of the dots in the union of the triangle and the rectangle 1/8

15. Refer to the figure for Exercise 14 and write a description of the region that is represented by the fraction 1/12. **the dots in the intersection of the triangle and the rectangle as a part of the dots in the entire figure**

16. In the local softball league, the first five games produced the following results: David Glenn got 8 hits in 20 at-bats, and Chalon Bridges got 12 hits in 30 at-bats. David claims that he and Chalon did equally well. Is he correct? Why or why not? **He is correct because 8/20 = 12/30 = 2/5.**

17. After ten games in the local softball league, the following batting statistics were obtained.

Player	At-bats	Hits	Home Runs
Bishop, Kelley	40	9	2
Carlton, Robert	36	12	3
Dykler, Francoise	11	5	1
Crowe, Vonalaine	16	8	0
Marshall, James	20	10	2

Answer each of the following, using estimation skills as necessary.
    (a) Which player got a hit in exactly 1/3 of his or her at-bats? **Carlton**
    (b) Which player got a hit in just less than 1/2 of his or her at-bats? **Dykler**
    (c) Which player got a home run in just less than 1/10 of his or her at-bats? **Dykler**
    (d) Which player got a hit in just less than 1/4 of his or her at-bats? **Bishop**
    (e) Which two players got hits in exactly the same fractional parts of their at-bats? What was the fractional part, reduced to lowest terms? **Crowe and Marshall; 1/2**

18. Use estimation skills to determine the best approximation for the following sum:

$$\frac{14}{26} + \frac{98}{99} + \frac{100}{51} + \frac{90}{31} + \frac{13}{27}.$$

(a) 6 (b) 7 (c) 5 (d) 8 **(b)**

Perform the indicated operations and express answers in lowest terms. Use the order of operations as necessary.

19. $\frac{3}{8} + \frac{1}{8}$ **1/2**
20. $\frac{7}{9} + \frac{1}{9}$ **8/9**
21. $\frac{5}{16} + \frac{7}{12}$ **43/48**
22. $\frac{1}{15} + \frac{7}{18}$ **41/90**
23. $\frac{2}{3} - \frac{7}{8}$ **−5/24**
24. $\frac{13}{20} - \frac{5}{12}$ **7/30**
25. $\frac{5}{8} - \frac{3}{14}$ **23/56**
26. $\frac{19}{15} - \frac{7}{12}$ **41/60**
27. $\frac{3}{4} \cdot \frac{9}{5}$ **27/20**
28. $\frac{3}{8} \cdot \frac{2}{7}$ **3/28**
29. $-\frac{2}{3} \cdot -\frac{5}{8}$ **5/12**
30. $-\frac{2}{4} \cdot \frac{3}{9}$ **−1/6**

31. $\frac{5}{12} \div \frac{15}{4}$ 1/9

32. $\frac{15}{16} \div \frac{30}{8}$ 1/4

33. $-\frac{9}{16} \div -\frac{3}{8}$ 3/2

34. $-\frac{3}{8} \div \frac{5}{4}$ −3/10

35. $\left(\frac{1}{3} \div \frac{1}{2}\right) + \frac{5}{6}$ 3/2

36. $\frac{2}{5} \div \left(-\frac{4}{5} \div \frac{3}{10}\right)$ −3/20

37. On Thursday, April 7, 1994, Hewlett-Packard stock closed at 83 3/4 dollars (per share). This was 2 1/4 dollars above the price at the start of the day. How much did one share of this stock cost at the start of the day? **81 1/2 dollars**

38. On Friday, April 8, 1994, the New York Stock Exchange reported that the 52-week high for IBM was 60 (dollars per share) while the low was 40 5/8 (dollars per share). What was the difference between these prices? **19 3/8 dollars**

39. The following chart appears on a package of Quaker Quick Grits.

Microwave		Stove Top		
Servings	**1**	**1**	**4**	**6**
Water	$\frac{3}{4}$ cup	1 cup	3 cups	4 cups
Grits	3 Tbsp	3 Tbsp	$\frac{3}{4}$ cup	1 cup
Salt (optional)	dash	dash	$\frac{1}{4}$ tsp	$\frac{1}{2}$ tsp

(a) How many cups of water would be needed for 6 microwave servings? **4 1/2 cups**

(b) How many cups of grits would be needed for 5 stove top servings? (*Hint:* 5 is halfway between 4 and 6.) **7/8 cup**

40. The pie chart shown depicts the makeup of the population of Mexico. Use the chart to answer the following.

(a) What fractional part of the population is Caucasian? **9/100**

(b) What fractional part of the population is composed of either Mestizo or Indian? **9/10**

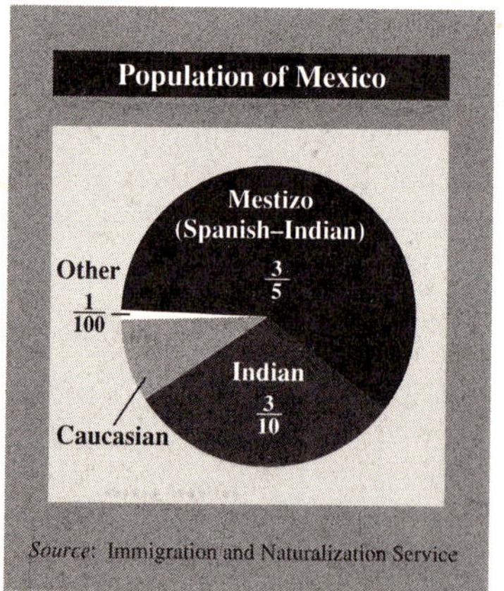

The **mixed number** 2 5/8 *represents the sum* 2 + 5/8. *We can convert* 2 5/8 *to a fraction as follows:*

$$2\frac{5}{8} = 2 + \frac{5}{8} = \frac{2}{1} + \frac{5}{8} = \frac{16}{8} + \frac{5}{8} = \frac{21}{8}.$$

The fraction 21/8 can be converted back to a mixed number by dividing 8 into 21. The quotient is 2 and the remainder is 5.

Convert each mixed number in the following exercises to a fraction, and convert each fraction to a mixed number.

41. $4\frac{1}{3}$ **13/3**

42. $3\frac{7}{8}$ **31/8**

43. $2\frac{9}{10}$ **29/10**

44. $\frac{18}{5}$ **3 3/5**

45. $\frac{27}{4}$ **6 3/4**

46. $\frac{19}{3}$ **6 1/3**

It is possible to add mixed numbers by first converting them to fractions, adding, and then converting the sum back to a mixed number. For example,

$$2\frac{1}{3} + 3\frac{1}{2} = \frac{7}{3} + \frac{7}{2} = \frac{14}{6} + \frac{21}{6} = \frac{35}{6} = 5\frac{5}{6}.$$

The other operations with mixed numbers may be performed in a similar manner.

Perform each operation and express your answer as a mixed number.

47. $3\frac{1}{4} + 2\frac{7}{8}$ **6 1/8**

48. $6\frac{1}{5} - 2\frac{7}{15}$ **3 11/15**

49. $-4\frac{7}{8} \cdot 3\frac{2}{3}$ **−17 7/8**

50. $-4\frac{1}{6} \div 1\frac{2}{3}$ **−2 1/2**

A quotient of fractions (with denominator not zero) is called a **complex fraction.** *There are two methods that are used to simplify a complex fraction.*

Method 1: *Simplify the numerator and denominator separately. Then rewrite as a division problem, and proceed as you would when dividing fractions.*

Method 2: *Multiply both the numerator and denominator by the least common denominator of all the fractions found within the complex fraction. (This is, in effect, multiplying the fraction by 1, which does not change its value.) Apply the distributive property, if necessary, and simplify.*

▲ *Use one of the methods above to simplify each of the following complex fractions.*

51. $\dfrac{\frac{1}{2} + \frac{1}{4}}{\frac{1}{2} - \frac{1}{4}}$ **3**

52. $\dfrac{\frac{2}{3} + \frac{1}{6}}{\frac{2}{3} - \frac{1}{6}}$ **5/3**

53. $\dfrac{\frac{5}{8} - \frac{1}{4}}{\frac{1}{8} + \frac{3}{4}}$ **3/7**

54. $\dfrac{\frac{3}{16} + \frac{1}{2}}{\frac{5}{16} - \frac{1}{8}}$ **11/3**

55. $\dfrac{\frac{7}{11} + \frac{3}{10}}{\frac{1}{11} - \frac{9}{10}}$ **−103/89**

56. $\dfrac{\frac{11}{15} + \frac{1}{9}}{\frac{13}{15} - \frac{2}{3}}$ **38/9**

The **continued fraction** *corresponding to the rational number p/q is an expression of the form*

$$a_1 + \cfrac{1}{a_2 + \cfrac{1}{a_3 + \cfrac{1}{a_4 + \ddots}}}$$

where each of the a's is an integer. For example, the continued fraction for 29/8 may be found using the following procedure.

$$\frac{29}{8} = 3 + \frac{5}{8} = 3 + \frac{1}{\frac{8}{5}} = 3 + \cfrac{1}{1 + \frac{3}{5}} = 3 + \cfrac{1}{1 + \cfrac{1}{\frac{5}{3}}}$$

$$= 3 + \cfrac{1}{1 + \cfrac{1}{1 + \frac{2}{3}}} = 3 + \cfrac{1}{1 + \cfrac{1}{1 + \cfrac{1}{\frac{3}{2}}}} = 3 + \cfrac{1}{1 + \cfrac{1}{1 + \cfrac{1}{1 + \frac{1}{2}}}}$$

▲ *Use this procedure to find the continued fraction representation for each of the following rational numbers.*

57. $\frac{28}{13}$ $2 + \cfrac{1}{6 + \frac{1}{2}}$

58. $\frac{73}{31}$ $2 + \cfrac{1}{2 + \cfrac{1}{1 + \cfrac{1}{4 + \frac{1}{2}}}}$

59. $\frac{52}{11}$ $4 + \cfrac{1}{1 + \cfrac{1}{2 + \cfrac{1}{1 + \frac{1}{2}}}}$

60. $\frac{29}{13}$ $2 + \cfrac{1}{4 + \frac{1}{3}}$

▲ *Write each of the following continued fractions in the form p/q, reduced to lowest terms. (Hint: Start at the bottom, and work upward.)*

61. $2 + \cfrac{1}{1 + \cfrac{1}{3 + \frac{1}{2}}}$ **25/9**

62. $4 + \cfrac{1}{2 + \cfrac{1}{1 + \frac{1}{3}}}$ **48/11**

Find the rational number halfway between the two given rational numbers.

63. $\frac{1}{2}, \frac{3}{4}$ **5/8**

64. $\frac{1}{3}, \frac{5}{12}$ **3/8**

65. $\frac{3}{5}, \frac{2}{3}$ **19/30**

66. $\frac{7}{12}, \frac{5}{8}$ **29/48**

67. $-\frac{2}{3}, -\frac{5}{6}$ **−3/4**

68. $-3, -\frac{5}{2}$ **−11/4**

In the March 1973 issue of The Mathematics Teacher *there appeared an article by Laurence Sherzer, an eighth-grade mathematics teacher, that immortalized one of his students, Robert McKay. The class was studying the density property and Sherzer was explaining how to find a rational number between two given positive rational numbers by finding the average. McKay pointed out that there was no need to go to all that trouble. To find a number (not necessarily their average) between two positive rational numbers a/b and c/d, he claimed, simply "add the tops and add the bottoms." Much to Sherzer's surprise, this method really does work. For example, to find a rational number between 1/3 and 1/4, add 1 + 1 = 2 to get the numerator and 3 + 4 = 7 to get the denominator. Therefore, by* **McKay's theorem,** *2/7 is between 1/3 and 1/4. Sherzer provided a proof of this method in the article.*

Use McKay's theorem *to find a rational number between the two given rational numbers.*

69. $\frac{5}{6}$ and $\frac{9}{13}$ **14/19**

70. $\frac{10}{11}$ and $\frac{13}{19}$ **23/30**

71. $\frac{4}{13}$ and $\frac{9}{16}$ **13/29**

72. $\frac{6}{11}$ and $\frac{8}{9}$ **14/20 or 7/10**

73. 2 and 3 **5/2**

74. 3 and 4 **7/2**

75. Apply McKay's theorem to any pair of consecutive integers, and make a conjecture about what happens in this case. **It gives the rational number halfway between the two integers.**

76. Explain in your own words how to find the rational number that is one-fourth of the way between two different rational numbers.

Convert each rational number into either a repeating or a terminating decimal. Use a calculator if your instructor so allows.

77. $\frac{3}{4}$ **.75**

78. $\frac{7}{8}$ **.875**

79. $\frac{3}{16}$ **.1875**

80. $\frac{9}{32}$ **.28125**

81. $\frac{3}{11}$ $.\overline{27}$

82. $\frac{9}{11}$ $.\overline{81}$

83. $\frac{2}{7}$ $.\overline{285714}$

84. $\frac{11}{15}$ $.7\overline{3}$

Convert each terminating decimal into a quotient of integers. Write each in lowest terms.

85. .4 **2/5**

86. .9 **9/10**

87. .85 **17/20**

88. .105 **21/200**

89. .934 **467/500**

90. .7984 **499/625**

Convert each repeating decimal into a quotient of integers. Write each in lowest terms.

91. $.\overline{8}$ **8/9**

92. $.\overline{1}$ **1/9**

93. $.\overline{54}$ **6/11**

94. $.\overline{36}$ **4/11**

95. $.4\overline{3}$ **13/30**

96. $.2\overline{6}$ **4/15**

97. $1.\overline{9}$ **2**

98. $3.0\overline{9}$ **31/10**

Use the method of Example 7 to decide whether each of the following rational numbers would yield a repeating or a terminating decimal. (Hint: Write in lowest terms before trying to decide.)

99. $\frac{8}{15}$ **repeating**

100. $\frac{8}{35}$ **repeating**

101. $\frac{13}{125}$ **terminating**

102. $\frac{3}{24}$ **terminating**

103. $\frac{22}{55}$ **terminating**

104. $\frac{24}{75}$ **terminating**

105. Follow through on each part of this exercise in order.

(a) Find the decimal for 1/3. $.\overline{3}$ **or .333. . .**

(b) Find the decimal for 2/3. $.\overline{6}$ **or .666. . .**

(c) By adding the decimal expressions obtained in parts (a) and (b), obtain a decimal expression for 1/3 + 2/3 = 3/3 = 1. $.\overline{9}$ **or .999. . .**

(d) Does your result seem bothersome? Read the margin note on terminating and repeating decimals in this section, which refers to this idea.

106. It is a fact that 1/3 = .333. . . . Multiply both sides of this equation by 3. Does your answer bother you? See the margin note on terminating and repeating decimals in this section. **3(1/3) = 3(.333. . .) gives 3/3 or 1 = .999. . . .**

6.4 Exercises (Text Page 296)

Identify each of the following as rational *or* irrational.

1. $\frac{4}{7}$ **rational**
2. $\frac{5}{8}$ **rational**
3. $\sqrt{6}$ **irrational**
4. $\sqrt{13}$ **irrational**
5. .89 **rational**
6. .76 **rational**
7. $.\overline{89}$ **rational**
8. $.\overline{76}$ **rational**
9. .878778777877778 . . . **irrational**
10. .434334333433334 . . . **irrational**
11. 3.14159 **rational**
12. $\frac{22}{7}$ **rational**
13. π **irrational**
14. 0 **rational**
15. (a) Find the following sum.

$$\begin{array}{r} .272772777277772\ldots \\ +\ .616116111611116\ldots \\ \hline \end{array}$$ **$.\overline{8}$**

(b) Based on the result of part (a), we can conclude that the sum of two _____ numbers may be a(n) _____ number. **irrational, rational**

16. (a) Find the following sum.

$$\begin{array}{r} .010110111011110\ldots \\ +\ .252552555255552\ldots \\ \hline \end{array}$$ **.262662666266662. . .**

(b) Based on the result of part (a), we can conclude that the sum of two _____ numbers may be a(n) _____ number. **irrational, irrational**

Use a calculator to find a rational decimal approximation for each of the following irrational numbers. Give as many places as your calculator shows. **The number of digits shown will vary among calculator models in Exercises 17–24.**

17. $\sqrt{39}$ **6.244997998**
18. $\sqrt{44}$ **6.633249581**
19. $\sqrt{15.1}$ **3.885871846**
20. $\sqrt{33.6}$ **5.796550698**
21. $\sqrt{884}$ **29.73213749**
22. $\sqrt{643}$ **25.35744467**
23. $\sqrt{\frac{9}{8}}$ **1.060660172**
24. $\sqrt{\frac{6}{5}}$ **1.095445115**

Use the methods of Examples 2 and 3 to simplify each of the following expressions. Then, use a calculator to approximate both the given expression and the simplified expression. (Both should be the same.)

25. $\sqrt{50}$ **$5\sqrt{2}$; 7.071067812**
26. $\sqrt{32}$ **$4\sqrt{2}$; 5.656854249**
27. $\sqrt{75}$ **$5\sqrt{3}$; 8.660254038**
28. $\sqrt{150}$ **$5\sqrt{6}$; 12.247448714**
29. $\sqrt{288}$ **$12\sqrt{2}$; 16.97056275**
30. $\sqrt{200}$ **$10\sqrt{2}$; 14.14213562**
31. $\frac{5}{\sqrt{6}}$ **$\frac{5\sqrt{6}}{6}$; 2.041241452**
32. $\frac{3}{\sqrt{2}}$ **$\frac{3\sqrt{2}}{2}$; 2.121320344**
33. $\sqrt{\frac{7}{4}}$ **$\frac{\sqrt{7}}{2}$; 1.322875656**
34. $\sqrt{\frac{8}{9}}$ **$\frac{2\sqrt{2}}{3}$; .9428090416**
35. $\sqrt{\frac{7}{3}}$ **$\frac{\sqrt{21}}{3}$; 1.527525232**
36. $\sqrt{\frac{14}{5}}$ **$\frac{\sqrt{70}}{5}$; 1.673320053**

Use the method of Example 4 to perform the indicated operations.

37. $\sqrt{17} + 2\sqrt{17}$ **$3\sqrt{17}$**

38. $3\sqrt{19} + \sqrt{19}$ **$4\sqrt{19}$**

39. $5\sqrt{7} - \sqrt{7}$ **$4\sqrt{7}$**

40. $3\sqrt{27} - \sqrt{27}$ **$6\sqrt{3}$**

41. $3\sqrt{18} + \sqrt{2}$ **$10\sqrt{2}$**

42. $2\sqrt{48} - \sqrt{3}$ **$7\sqrt{3}$**

43. $-\sqrt{12} + \sqrt{75}$ **$3\sqrt{3}$**

44. $2\sqrt{27} - \sqrt{300}$ **$-4\sqrt{3}$**

Solve the problems in Exercises 45–50. Use a calculator as necessary.

45. *Heron's formula* gives a method of finding the area of a triangle if the lengths of its sides are known. Suppose that a, b, and c are the lengths of the sides. Let s denote one-half of the perimeter of the triangle (called the *semiperimeter*); that is,

$$s = \frac{1}{2}(a + b + c).$$

Then the area of the triangle is

$$A = \sqrt{s(s - a)(s - b)(s - c)}.$$

Find the area of the Bermuda Triangle, if the "sides" of this triangle measure approximately 850 miles, 925 miles, and 1,300 miles. Round your answer to the nearest thousand. **392,000 square miles**

46. According to an article in *The World Scanner Report* (August 1991), the distance, D, in miles, to the horizon from an observer's point of view over water or "flat" earth is given by

$$D = \sqrt{2H},$$

where H is the height of the point of view, in feet. If a person whose eyes are 6 feet above ground level is standing at the top of a hill 28 feet above the "flat" earth, approximately how far to the horizon will she be able to see? Round your answer to the nearest tenth of a mile. **8.2 miles**

47. The illumination I, in footcandles, produced by a light source is related to the distance d, in feet, from the light source by the equation

$$d = \sqrt{\frac{k}{I}},$$

where k is a constant. If $k = 640$, how far from the light source will the illumination be 2 footcandles? Give the exact value, and then round to the nearest tenth of a foot. **$8\sqrt{5}$ ft; 17.9 ft**

48. The length of the diagonal of a box is given by

$$D = \sqrt{L^2 + W^2 + H^2},$$

where L, W, and H are the length, the width, and the height of the box. Find the length of the diagonal, D, of a box that is 4 feet long, 3 feet high, and 2 feet wide. Give the exact value, then round to the nearest tenth of a foot. **$\sqrt{29}$ ft; 5.4 ft**

49. The period of a pendulum in seconds depends on its length, L, in feet, and is given by the formula

$$P = 2\pi\sqrt{\frac{L}{32}}.$$

If a pendulum is 5.1 feet long, what is its period? Use 3.14 for π. **2.5 seconds**

50. The length, L, of an animal, in centimeters, is related to its surface area, S, by the formula

$$L = \sqrt{\frac{S}{a}},$$

where a is a constant that depends on the type of animal. If an animal has $a = .51$ and has a surface area of 1,379 square centimeters, what is the length of the animal? **52 cm**

51. Find the first eight digits in the decimal for 355/113. Compare the result to the decimal for π given in the margin note. What do you find? **The result is 3.1415929, which agrees with the first seven digits in the decimal form of π.**

52. Using a calculator with a square root key, divide 2,143 by 22, and then press the square root key twice. Compare your result to the decimal for π given in the margin note. **The result agrees with the first nine digits.**

53. A **mnemonic** device is a scheme whereby one is able to recall facts by memorizing something completely unrelated to the facts. One way of learning the first few digits of the decimal for π is to memorize a sentence (or several sentences) and count the letters in each word of the sentence. For example, "See, I know a digit," will give the first 5 digits of π: "See" has 3 letters, "I" has 1 letter, "know" has 4 letters, "a" has 1 letter, and "digit" has 5 letters. So the first five digits are 3.1415.

Verify that the following mnemonic devices work. Use the decimal for π given in the margin note.

(a) "May I have a large container of coffee?" **3.1415926**

(b) "See, I have a rhyme assisting my feeble brain, its tasks ofttimes resisting." **3.141592653589**

(c) "How I want a drink, alcoholic of course, after the heavy lectures involving quantum mechanics." **3.14159265358979**

54. Make up your own mnemonic device to obtain the first eight digits of π. **One possibility is: "Hey, I want a Merry Christmas in Mobile."**

55. You may have seen the statements "use 22/7 for π" and "use 3.14 for π." Since 22/7 is the quotient of two integers, and 3.14 is a terminating decimal, do these statements suggest that π is rational? **They are only rational *approximations* of π.**

56. In algebra the expression $a^{1/2}$ is defined to be $\sqrt{a}$ for nonnegative values of a. Use a calculator with an exponential key to evaluate each of the following, and compare with the value obtained with the square root key. (Both should be the same.)

(a) $2^{1/2}$ **1.414213562**

(b) $7^{1/2}$ **2.645751311**

(c) $13.2^{1/2}$ **3.633180425**

(d) $25^{1/2}$ **5**

57. The method for simplifying square root radicals, as explained in Examples 2 and 3, can be generalized to cube roots, using the perfect cubes $8 = 2^3$, $27 = 3^3$, $64 = 4^3$, $125 = 5^3$, $216 = 6^3$, and so on. Simplify each of the following cube roots.

(a) $\sqrt[3]{16}$ **$2\sqrt[3]{2}$** (b) $\sqrt[3]{54}$ **$3\sqrt[3]{2}$**

(c) $\sqrt[3]{24}$ **$2\sqrt[3]{3}$** (d) $\sqrt[3]{250}$ **$5\sqrt[3]{2}$**

58. Based on Exercises 56 and 57, answer the following.

(a) How do you think that the expression $a^{1/3}$ is defined as a radical? **$\sqrt[3]{a}$**

(b) Use a calculator with a cube root key to approximate $\sqrt[3]{16}$. **2.5198421**

(c) Use a calculator with an exponential key to approximate $16^{1/3}$. **2.5198421**

(d) Compare your results in parts (b) and (c). What do you find? **They are the same.**

(e) Compare the calculator approximation for the answer in Exercise 57(a) to the results in parts (b) and (c). What do you find? **They are the same.**

59. The threshold weight, T, for a person is the weight above which the risk of death increases greatly. One researcher found that the threshold weight in pounds for men aged 40–49 is related to height, h, in inches by the equation

$$h = 12.3\sqrt[3]{T}.$$

What height corresponds to a threshold of 180 pounds for a 43-year-old man? Round your answer to the nearest tenth of a foot, and then to the nearest inch. **5.8 ft; 69 in.**

60. Use a calculator with an exponential key to find values for the following: $(1.1)^{10}$, $(1.01)^{100}$, $(1.001)^{1,000}$, $(1.0001)^{10,000}$, and $(1.00001)^{100,000}$. Compare your results to the approximation given for the irrational number e in the margin note in this section. What do you find? **The values seem to approach the value of e.**

6.5 Exercises (Text Page 307)

Decide whether each of the following is true *or* false.

1. 50% of a quantity is the same as 1/2 of the quantity. **true**
2. 200% of 8 is 16. **true**
3. When 435.67 is rounded to the nearest ten, the answer is 435.7. **false**
4. When 668.342 is rounded to the nearest hundredth, the answer is 668.34. **true**
5. A football team that wins 10 games and loses 6 games has a winning percentage of 60%. **false**
6. To find 25% of a quantity, we may simply divide the quantity by 4. **true**
7. If 60% is a passing grade and a test has 40 items, then answering more than 22 items correctly will assure you of a passing grade. **false**

  WRITING   SCIENTIFIC CALCULATOR  GRAPHING CALCULATOR

8. To find 40% of a quantity, we may find 10% of that quantity and multiply our answer by 4. **true**

9. 15 is less than 30% of 45. **false**

10. If an item usually costs $50.00 and it is discounted 10%, the sale price is $5.00. **false**

Work each of the following using either a calculator or paper-and-pencil methods, as directed by your instructor.

11. 8.53 + 2.785 **11.315**

12. 9.358 + 7.2137 **16.5717**

13. 8.74 − 12.955 **−4.215**

14. 2.41 − 3.997 **−1.587**

15. 25.7 × .032 **.8224**

16. 45.1 × 8.344 **376.3144**

17. 1,019.825 ÷ 21.47 **47.5**

18. −262.563 ÷ 125.03 **−2.1**

19. $\frac{118.5}{1.45 + 2.3}$ **31.6**

20. 2.45(1.2 + 3.4 − 5.6) **−2.45**

Solve each of the following problems. Use a calculator as necessary.

The table shows the annual percent change in the Consumer Price Index for the years 1986 and 1987. Use the information provided to answer Exercises 21–24.

	Percent Change	
Category	**1986**	**1987**
Food	3.2	4.1
Shelter	5.5	4.7
Rent, residential	5.8	4.1
Fuel and other utilities	−2.3	−1.1
Apparel and upkeep	.9	4.4
Private transportation	−4.7	−3.0
New cars	4.2	3.6
Gasoline	−21.9	−4.0
Public transportation	5.9	3.5
Medical care	7.5	6.6
Entertainment	3.4	3.3
Commodities	−.9	3.2

Source: Bureau of Labor Statistics, U.S. Dept. of Labor

21. What category of what year represents the greatest drop? **gasoline in 1986**

22. Which percent change is represented by a number with a larger absolute value: 1986 fuel and other utilities or 1987 gasoline? **1987 gasoline**

23. True or false? The absolute value of the change in apparel and upkeep in 1986 is less than the absolute value of the change in commodities during the same year. **false**

24. True or false? The absolute value of the change in private transportation in 1986 was less than the absolute value of the change in private transportation in 1987. **false**

---

25. The bank balance of Tammy's Hobby Shop was $1,856.12 on March 1. During March, Tammy deposited $1,742.18 received from the sale of goods, $9,271.94 paid by customers on their accounts, and a $28.37 tax refund. She paid out $7,195.14 for merchandise, $511.09 for salaries, and $1,291.03 for other expenses.

(a) How much did Tammy deposit during March? **$11,042.49**

(b) How much did she pay out? **$8,997.26**

(c) What was her bank balance at the end of March? **$3,901.35**

26. On a recent trip to Target, David Horwitz bought three curtain rods at $4.57 apiece, five picture frames at $2.99 each, and twelve packs of gum at $.39 per pack. If 6% sales tax was added to his purchase, what was his total bill? **$35.34**

Exercises 27–30 are based on formulas found in Auto Math Handbook: Mathematical Calculations, Theory, and Formulas for Automotive Enthusiasts, *by John Lawlor (1991, HP Books).*

27. The Blood Alcohol Concentration (BAC) of a person who has been drinking is given by the formula

$$\text{BAC} = \frac{(\text{ounces} \times \text{percent alcohol} \times .075)}{\text{body weight in lb}} - (\text{hours of drinking} \times .015).$$

Suppose a policeman stops a 190-pound man who, in two hours, has ingested four 12-ounce beers with each beer having a 3.2 percent alcohol content. The formula would then read

$$\text{BAC} = \frac{[(4 \times 12) \times 3.2 \times .075]}{190} - (2 \times .015).$$

(a) Find this BAC. **.031**

(b) Find the BAC for a 135-pound woman who, in three hours, has drunk three 12-ounce beers with each beer having a 4.0 percent alcohol content. **.035**

28. The approximate rate of an automobile in miles per hour (MPH) can be found in terms of the engine's revolutions per minute (rpm), the tire diameter in inches, and the overall gear ratio by the formula

$$\text{MPH} = \frac{\text{rpm} \times \text{tire diameter}}{\text{gear ratio} \times 336}.$$

If a certain automobile has an rpm of 5,600, a tire diameter of 26 inches, and a gear ratio of 3.12, what is its approximate rate (MPH)? **139 mph**

29. Horsepower can be found from indicated mean effective pressure (mep) in pounds per square inch, engine displacement in cubic inches, and revolutions per minute (rpm) using the formula

$$\text{Horsepower} = \frac{\text{mep} \times \text{displacement} \times \text{rpm}}{792{,}000}.$$

Suppose that an engine has displacement of 302 cubic inches, and indicated mep of 195 pounds per square inch at 4,000 rpm. What is its approximate horsepower? **297**

30. To determine the torque at a given value of rpm, the formula below applies:

$$\text{Torque} = \frac{5{,}252 \times \text{horsepower}}{\text{rpm}}.$$

If the horsepower of a certain vehicle is 400 at 4,500 rpm, what is the approximate torque? **467**

Round each of the following numbers to the nearest **(a)** *tenth;* **(b)** *hundredth. Always round from the original number.*

31. 78.414 **(a) 78.4** **(b) 78.41**

32. 3,689.537 **(a) 3,689.5** **(b) 3,689.54**

33. .0837 **(a) .1** **(b) .08**

34. .0658 **(a) .1** **(b) .07**

35. 12.68925 **(a) 12.7** **(b) 12.69**

36. 43.99613 **(a) 44.0** **(b) 44.00**

Convert each decimal to a percent.

37. .42 **42%**

38. .87 **87%**

39. .365 **36.5%**

40. .792 **79.2%**

41. .008 **.8%**

42. .0093 **.93%**

43. 2.1 **210%**

44. 8.9 **890%**

Convert each fraction to a percent.

45. $\frac{1}{5}$ **20%**

46. $\frac{2}{5}$ **40%**

47. $\frac{1}{100}$ **1%**

48. $\frac{1}{50}$ **2%**

49. $\frac{3}{8}$ **37 1/2%**

50. $\frac{5}{6}$ **83 1/3%**

51. $\frac{3}{2}$ **150%**

52. $\frac{7}{4}$ **175%**

53. Explain the difference between 1/2 of a quantity and 1/2% of the quantity.

CONCEPTUAL  WRITING CHALLENGING SCIENTIFIC CALCULATOR GRAPHING CALCULATOR

54. In the left column of the chart below there are some common percents, found in many everyday situations. In the right column are fractional equivalents of these percents. Match the fractions in the right column with their equivalent percents in the left column.

(a) 25% E   A. $\frac{1}{3}$

(b) 10% D   B. $\frac{1}{50}$

(c) 2% B   C. $\frac{3}{4}$

(d) 20% F   D. $\frac{1}{10}$

(e) 75% C   E. $\frac{1}{4}$

(f) $33\frac{1}{3}\%$ A   F. $\frac{1}{5}$

55. Fill in each blank with the appropriate numerical response.
(a) 5% means _____ in every 100. 5
(b) 25% means 6 in every _____ . 24
(c) 200% means _____ for every 4. 8
(d) .5% means _____ in every 100. .5 or 1/2
(e) _____% means 12 for every 2. 600

56. The following Venn diagram shows the number of elements in the four regions formed.

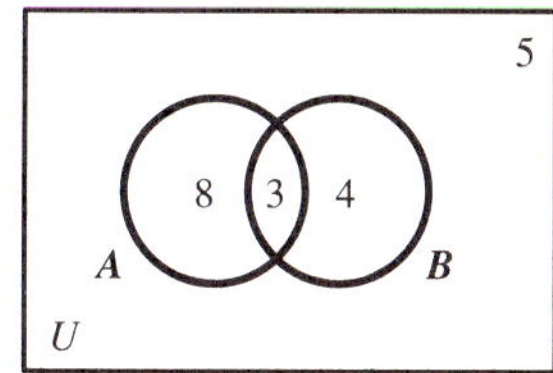

(a) What percent of the elements in the universe are in $A \cap B$? 15%
(b) What percent of the elements in the universe are in A but not in B? 40%
(c) What percent of the elements in $A \cup B$ are in $A \cap B$? 20%
(d) What percent of the elements in the universe are in neither A nor B? 25%

57. Suppose that an item regularly costs $50.00 and it is discounted 10%. If it is then marked up 10%, is the resulting price $50.00? If not, what is it?
No, it is $49.50.

58. At the start of play on July 24, 1995, the standings of the teams in the Central Division of the American League were as follows:

	Wins	Losses
Cleveland	56	22
Milwaukee	40	39
Kansas City	37	40
Chicago	33	45
Minnesota	28	51

"Winning percentage" is commonly expressed as a decimal rounded to the nearest thousandth. To find the winning percentage of a team, divide the number of wins by the total number of games played. Find the winning percentage of each of the teams in the division.
In order, they are: .718, .506, .481, .423, .354.

59. Refer to the figures in Exercise 13 of Section 3 of this chapter, and express each of the fractional parts represented by the shaded areas as percents.
(a) 33 1/3% (b) 25% (c) 40% (d) 33 1/3%

60. If there is a 40% chance of rain tomorrow, what *fraction* in lowest terms represents the chance that it will not rain? 3/5

Work each of the following problems involving percent.

61. What is 26% of 480? 124.8

62. Find 38% of 12. 4.56

63. Find 10.5% of 28. 2.94

64. What is 48.6% of 19? 9.234

65. What percent of 30 is 45? 150%

66. What percent of 48 is 20? 41 2/3% or $41.\overline{6}\%$

67. 25% of what number is 150? 600

68. 12% of what number is 3,600? 30,000

69. .392 is what percent of 28? 1.4%

70. 78.84 is what percent of 292? 27%

Use mental techniques to answer the questions in Exercises 71–74. Try to avoid using paper and pencil or a calculator.

71. Johnny Cross's allowance was raised from $4.00 per week to $5.00 per week. What was the percent of the increase?
(a) 25% (b) 20% (c) 50%
(d) 30% (a)

72. Jennifer Twomey bought a boat five years ago for $5,000 and sold it this year for $2,000. What percent of her original purchase did she lose on the sale?
(a) 40% **(b)** 50% **(c)** 20%
(d) 60% (d)

73. The 1990 U.S. census showed that the population of Alabama was 4,040,587, with 25.3% representing African-Americans. What is the best estimate of the African-American population of Alabama?
(a) 500,000 **(b)** 750,000
(c) 1,000,000 **(d)** 1,500,000 (c)

74. The 1990 U.S. census showed that the population of New Mexico was 1,515,069, with 38.2% being Hispanic. What is the best estimate of the Hispanic population of New Mexico?
(a) 600,000 **(b)** 60,000 **(c)** 750,000
(d) 38,000 (a)

The total revenue of all securities firms in the United States between 1987 and 1990 is illustrated in the accompanying bar graph. Use the graph to answer the questions in Exercises 75 and 76.

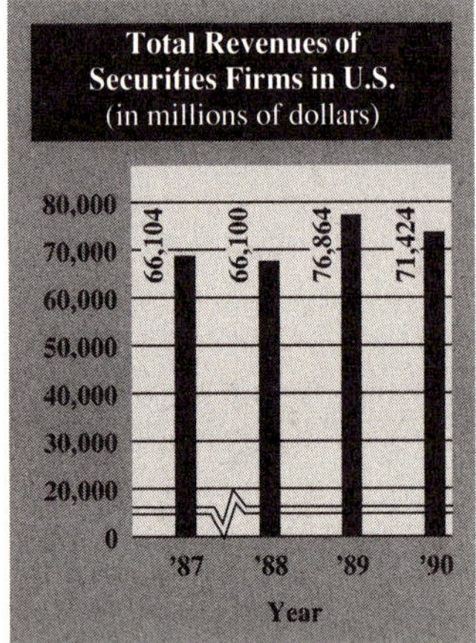

75. What was the percent increase from 1988 to 1989? 16.3%

76. What was the percent decrease from 1989 to 1990? 7.1%

Work each of the following problems. Round all money amounts to the nearest dollar.

77. According to a Knight-Ridder Newspapers report, as of May 31, 1992, the nation's "consumer-debt burden" was 16.4%. This means that the average American had consumer debts, such as credit card bills, auto loans, and so on, totaling 16.4% of his or her take-home pay. Suppose that George Duda has a take-home pay of $3,250 per month. What is 16.4% of his monthly take-home pay? $533

78. In 1992 General Motors announced that it would raise prices on its 1993 vehicles by an average of 1.6%. If a certain vehicle had a 1992 price of $10,526 and this price was raised 1.6%, what would the 1993 price be? $10,694

79. The 1916 dime minted in Denver is quite rare. The 1979 edition of *A Guide Book of United States Coins* listed its value in Extremely Fine condition as $625.00. The 1995 value had increased to $2,400.00. What was the percent increase in the value of this coin? 284%

80. In 1963, the value of a 1903 Morgan dollar minted in New Orleans in Uncirculated condition was $1,500. Due to a discovery of a large hoard of these dollars late that year, the value plummeted. Its value as listed in the 1995 edition of *A Guide Book of United States Coins* was $175. What percent of its 1963 value was its 1995 value? 11 2/3%

81. The manufacturer's suggested retail price of a Mazda MX 3 was $14,295. A dealer advertised a $2,300 discount. What percent discount was this? 16.1%

82. According to a report by Freeport-McMoRan Copper Co., Inc., its Grasberg prospect in Indonesia has a large copper find. It has at least 50 million metric tons of ore at an average grade of 1.4% copper and 1.3 grams of gold per ton. Assuming that there actually are 50 million metric tons of ore,
(a) find the number of metric tons of pure copper; .7 million metric tons
(b) find the number of grams of gold. 65 million grams

It is customary in our society to "tip" waiters and waitresses when eating in restaurants. The usual rate of tipping is 15%. A quick way of figuring a tip that will give a close approximation of 15% is as follows:

1. *Round off the bill to the nearest dollar.*
2. *Find 10% of this amount by moving the decimal point one place to the left.*
3. *Take half of the amount obtained in Step 2 and add it to the result of Step 2.*

This will give you approximately 15% of the bill. The amount obtained in Step 3 is 5%, and 10% + 5% = 15%.

Use the method above to find an approximation of 15% of each of the following restaurant bills.

83. \$29.57 **\$4.50** **84.** \$38.32 **\$5.70**

85. \$5.15 **\$.75** **86.** \$7.89 **\$1.20**

87. Example 2(a) shows a paper-and-pencil method of multiplying 4.613×2.52. The following discussion gives a mathematical justification of this method. Fill in the blanks with the appropriate responses.

$$4.613 = 4\frac{613}{1{,}000} = \frac{4{,}613}{1{,}000} = \frac{4{,}613}{10^3}$$

$$2.52 = 2\frac{52}{100} = \frac{252}{100} = \frac{252}{10^2}$$

$$4.613 \times 2.52 = \frac{4{,}613}{10^3} \cdot \frac{252}{10^2} \qquad (*)$$

(a) In algebra, we learn that multiplying powers of the same number is accomplished by *adding* the exponents. Thus,

$10^3 \cdot 10^2 = 10^{\_\_+\_\_} = 10^{\_\_}$. **3; 2; 5**

(b) The product in the line indicated by (*) is obtained by multiplying the fractions.

$$\frac{4{,}613 \cdot 252}{10^5} = \frac{1{,}162{,}476}{10^5} = 11.62476$$

The _____ places to the right of the decimal point in the product are the result of division by $10^{\_\_}$. **5; 5**

▲ **88.** Develop an argument justifying the paper-and-pencil method of dividing decimal numbers, as shown in Example 2(b).

89. A television reporter once asked a professional wrist-wrestler what percent of his sport was physical and what percent was mental. The athlete responded "I would say it's 50% physical and 90% mental." Comment on this response.

90. We often hear the claim "(S)he gave 110%." Comment on this claim. Do you think that this is actually possible?

Extension Exercises (Text Page 314)

Use the method of Examples 1–3 to write each as a real number or a product of a real number and i.

1. $\sqrt{-144}$ **$12i$** **2.** $\sqrt{-196}$ **$14i$** **3.** $-\sqrt{-225}$ **$-15i$** **4.** $-\sqrt{-400}$ **$-20i$**

5. $\sqrt{-3}$ **$i\sqrt{3}$** **6.** $\sqrt{-19}$ **$i\sqrt{19}$** **7.** $\sqrt{-75}$ **$5i\sqrt{3}$** **8.** $\sqrt{-125}$ **$5i\sqrt{5}$**

9. $\sqrt{-5} \cdot \sqrt{-5}$ **-5** **10.** $\sqrt{-3} \cdot \sqrt{-3}$ **-3** **11.** $\sqrt{-9} \cdot \sqrt{-36}$ **-18**

12. $\sqrt{-4} \cdot \sqrt{-81}$ **-18** **13.** $\sqrt{-16} \cdot \sqrt{-100}$ **-40** **14.** $\sqrt{-81} \cdot \sqrt{-121}$ **-99**

15. $\dfrac{\sqrt{-200}}{\sqrt{-100}}$ **$\sqrt{2}$** **16.** $\dfrac{\sqrt{-50}}{\sqrt{-2}}$ **5** **17.** $\dfrac{\sqrt{-54}}{\sqrt{6}}$ **$3i$**

18. $\dfrac{\sqrt{-90}}{\sqrt{10}}$ **$3i$** **19.** $\dfrac{\sqrt{-288}}{\sqrt{-8}}$ **6** **20.** $\dfrac{\sqrt{-48} \cdot \sqrt{-3}}{\sqrt{-2}}$ **$6i\sqrt{2}$**

21. Why is it incorrect to use the product rule for radicals to multiply $\sqrt{-3} \cdot \sqrt{-12}$?

22. In your own words describe the relationship between complex numbers and real numbers.

Use the method of Example 4 to find each power of i.

23. i^8 **1** **24.** i^{16} **1** **25.** i^{42} **-1** **26.** i^{86} **-1**

27. i^{47} **$-i$** **28.** i^{63} **$-i$** **29.** i^{101} **i** **30.** i^{141} **i**

31. Explain the difference between $\sqrt{-1}$ and $-\sqrt{1}$. Which one of these is defined as i?

32. Is it possible to give an example of a real number that is not a complex number? Why or why not?

Chapter 6 Test (Text Page 315)

1. List the numbers in the set $\{-8, -\sqrt{6}, -4/3, -.6, 0, \sqrt{2}, 3.9, 10\}$ that are **(a)** natural numbers, **(b)** whole numbers, **(c)** integers, **(d)** rational numbers, **(e)** irrational numbers, **(f)** real numbers. **(a) 10 (b) 0, 10 (c) −8, 0, 10 (d) −8, −4/3, −.6, 0, 3.9, 10 (e) $-\sqrt{6}, \sqrt{2}$ (f) All are real numbers.**

2. Explain what is wrong with the following statement: "The absolute value of a number is always positive."

3. Specify the set $\{x \mid x \text{ is a positive integer less than } 4\}$ by listing its elements. **{1, 2, 3}**

4. Give three examples of a number that is a positive rational number, but not an integer. **Answers may vary. Three examples are 1/2, 3/4, 5/6.**

5. True or false: The absolute value of −5 is −(−5). **true**

6. Perform the indicated operations, using the order of operations as necessary.

(a) $5^2 - 3(2 + 6)$ **1**

(b) $(-3)(-2) - [5 + (8 - 10)]$ **3**

(c) $\dfrac{(-8 + 3) - (5 + 10)}{7 - 9}$ **10**

7. The U.S. official reserve assets for the years 1990–1992 are shown in the accompanying bar graph. What is the sum of the assets for these years? (*Source:* Bureau of Economic Analysis) **$7,506 million**

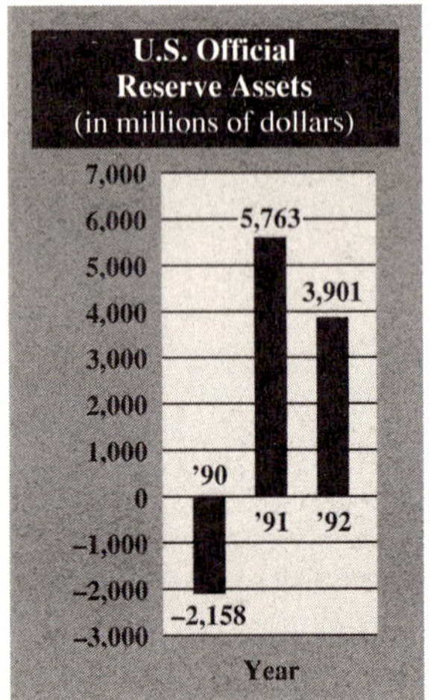

8. Match each of the statements on the left with the property that justifies it.

(a) $7 \cdot (8 \cdot 5) = (7 \cdot 8) \cdot 5$ **E**
(b) $3x + 3y = 3(x + y)$ **A**
(c) $8 \cdot 1 = 1 \cdot 8 = 8$ **B**
(d) $7 + (6 + 9) = (6 + 9) + 7$ **D**
(e) $9 + (-9) = -9 + 9 = 0$ **F**
(f) $5 \cdot 8$ is a real number. **C**

A. Distributive property
B. Identity property
C. Closure property
D. Commutative property
E. Associative property
F. Inverse property

9. The highest temperature ever recorded in Wyoming was 114°F, while the lowest was −63°F. What is the difference between the highest and lowest temperatures? **177°F**

10. The funds available from the National Endowment for the Arts for the years 1986 through 1990 are shown in the chart. Determine the change from one year to the next by subtraction for each year indicated with a blank. For example, from 1986 to 1987, the change was $170.9 - 167.1 = 3.8$ million dollars. (*Source:* U.S. National Endowment for the Arts) **$.2 million, −$4.4 million, $4.1 million**

Year	Funds Available (in millions of dollars)	Change
1986	167.1	
1987	170.9	3.8
1988	171.1	______
1989	166.7	______
1990	170.8	______

CONCEPTUAL    

11. The five starters on the local high school basketball team had the following shooting statistics after the first three games.

Player	Field Goal Attempts	Field Goals Made
Camp, Jim	40	13
Cooper, Daniel	10	4
Cornett, Bill	20	8
Hickman, Chuck	6	4
Levinson, Harold	7	2

Answer each of the following, using estimation skills as necessary.

(a) Which player made more than half of his attempts? **Hickman**

(b) Which players made just less than 1/3 of their attempts? **Camp and Levinson**

(c) Which player made exactly 2/3 of his attempts? **Hickman**

(d) Which two players made the same fractional parts of their attempts? What was the fractional part, reduced to lowest terms? **Cooper and Cornett; 2/5**

Perform each operation. Reduce your answer to lowest terms.

12. $\frac{3}{16} + \frac{1}{2}$ **11/16**

13. $\frac{9}{20} - \frac{3}{32}$ **57/160**

14. $\frac{3}{8} \cdot \left(-\frac{16}{15}\right)$ **−2/5**

15. $\frac{7}{9} \div \frac{14}{27}$ **3/2**

16. Based on the pie chart shown, what fractional part of total legal gaming revenue in 1992 came from sources other than lotteries, casinos, and pari-mutuel wagering? (*Source:* Gaming and Wagering Business) $\frac{4}{25}$

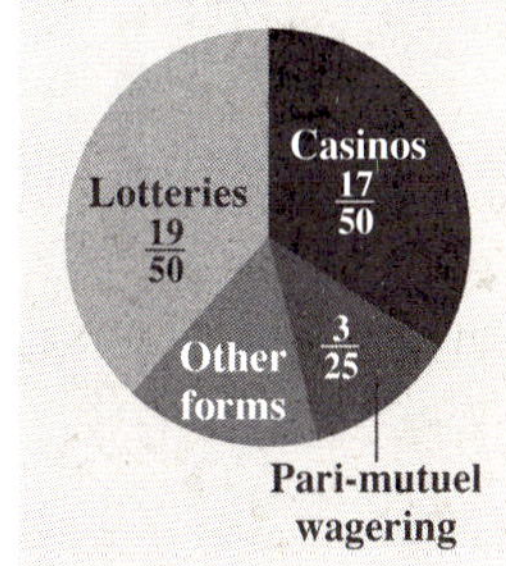

Convert each rational number into a repeating or terminating decimal. Use a calculator if your instructor so allows.

17. $\frac{9}{20}$ **.45**

18. $\frac{5}{12}$ $.41\overline{6}$

Convert each decimal into a quotient of integers, reduced to lowest terms.

19. .72 **18/25**

20. $.\overline{58}$ **58/99**

21. Identify each number as rational or irrational.

(a) $\sqrt{10}$ **irrational** **(b)** $\sqrt{16}$ **rational** **(c)** .01 **rational**

(d) $.\overline{01}$ **rational** **(e)** .0101101110 . . . **irrational**

For each of the following **(a)** *use a calculator to find a decimal approximation and* **(b)** *simplify the radical according to the guidelines in this chapter.*

22. $\sqrt{150}$ **(a) 12.247448714 (b)** $5\sqrt{6}$

23. $\frac{13}{\sqrt{7}}$ **(a) 4.913538149 (b)** $\frac{13\sqrt{7}}{7}$

24. $2\sqrt{32} - 5\sqrt{128}$ **(a) −45.254834 (b)** $-32\sqrt{2}$

25. A student using her powerful new calculator states that the *exact* value of $\sqrt{65}$ is 8.062257748. Is she correct? If not, explain.

Work each of the following using either a calculator or paper-and-pencil methods, as directed by your instructor.

26. 4.6 + 9.21 13.81

27. 12 − 3.725 − 8.59 −.315

28. 86(.45) 38.7

29. 236.439 ÷ (−9.73) −24.3

30. Round 9.0449 to the following place values: **(a)** hundredths **(b)** thousandths. (a) 9.04 (b) 9.045

31. Find 18.5% of 90. 16.65

32. What number is 145% of 70? 101.5

33. Use mental techniques to answer the following: In 1995, James Ertl sold $150,000 worth of books. In 1996, he sold $450,000 worth of books. His 1996 sales were _____ of his 1995 sales.

(a) 200% **(b)** 33 1/3% **(c)** 300% **(d)** 30% (c)

34. The bar graph shows the number of units of elementary/high school texts, in millions, sold during the years 1987 through 1990. What was the percent increase from 1988 to 1989? 5.45%

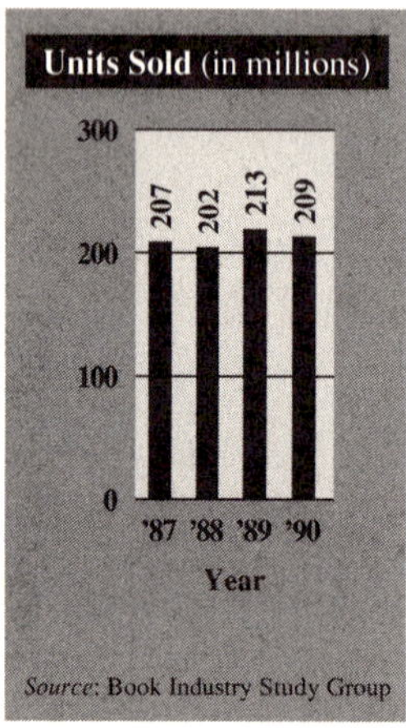

35. Consider the figure.

(a) What percent of the total number of shapes are circles? 26 2/3%

(b) What percent of the total number of shapes are not stars? 66 2/3%

CHAPTER 7 The Basic Concepts of Algebra

7.1 Exercises (Text Page 326)

Decide whether the given number is a solution of the equation.

1. $-6x = -24;$ 4 yes

2. $8r = 56;$ 7 yes

3. $5x + 2 = 3;$ $\frac{2}{5}$ no

4. $6y - 4 = 4;$ $\frac{1}{2}$ no

5. $9x + 2x = 6x;$ 0 yes

6. $-2p + 10p = 7p;$ 0 yes

7. Which one of the following equations is not a linear equation in x?

(a) $3x + x - 1 = 0$ (b) $8 = x^2$ (c) $6x + 2 = 9$ (d) $\frac{1}{2}x - \frac{1}{4}x = 0$ (b)

8. If two equations are equivalent, they have the same ____________ _____.
solution set

Solve the equation.

9. $7k + 8 = 1$ $\{-1\}$

10. $5m - 4 = 21$ $\{5\}$

11. $8 - 8x = -16$ $\{3\}$

12. $9 - 2r = 15$ $\{-3\}$

13. $7y - 5y + 15 = y + 8$ $\{-7\}$

14. $2x + 4 - x = 4x - 5$ $\{3\}$

15. $12w + 15w - 9 + 5 = -3w + 5 - 9$ $\{0\}$

16. $-4t + 5t - 8 + 4 = 6t - 4$ $\{0\}$

17. $2(x + 3) = -4(x + 1)$ $\left\{-\frac{5}{3}\right\}$

18. $4(y - 9) = 8(y + 3)$ $\{-15\}$

19. $3(2w + 1) - 2(w - 2) = 5$ $\left\{-\frac{1}{2}\right\}$

20. $4(x - 2) + 2(x + 3) = 6$ $\left\{\frac{4}{3}\right\}$

21. $2x + 3(x - 4) = 2(x - 3)$ $\{2\}$

22. $6y - 3(5y + 2) = 4(1 - y)$ $\{-2\}$

23. $6p - 4(3 - 2p) = 5(p - 4) - 10$ $\{-2\}$

24. $-2k - 3(4 - 2k) - 2(k - 3) + 2$ $\{4\}$

25. $-[2z - (5z + 2)] = 2 + (2z + 7)$ $\{7\}$

26. $-[6x - (4x + 8)] = 9 + (6x + 3)$ $\left\{-\frac{1}{2}\right\}$

27. $-(9 - 3a) - (4 + 2a) - 3 = -(2 - 5a) + (-a) + 1$ $\{-5\}$

28. $-(-2 + 4x) - (3 - 4x) + 5 = -(-3 + 6x) + x + 1$ $\{0\}$

29. Explain in your own words the steps used to solve a linear equation.

30. In order to solve the linear equation $\frac{8y}{3} - \frac{2y}{4} = 13$, we may multiply both sides by the least common denominator of all the fractions in the equation. What is this least common denominator? 12

31. In order to solve the linear equation $.05y + .12(y + 5{,}000) = 940$, we may multiply both sides by a power of 10 so that all coefficients are integers. What is the smallest power of 10 that will accomplish this goal? 2 (that is, 10^2)

32. Suppose that in solving the equation

$$\frac{1}{3}y + \frac{1}{2}y = \frac{1}{6}y,$$

you begin by multiplying both sides by 12, rather than the *least* common denominator, 6. Should you get the correct solution anyway? yes

Solve the equation.

33. $\frac{3x}{4} + \frac{5x}{2} = 13$ {4}

34. $\frac{8y}{3} - \frac{2y}{4} = -13$ {−6}

35. $\frac{x - 8}{5} + \frac{8}{5} = -\frac{x}{3}$ {0}

36. $\frac{2r - 3}{7} + \frac{3}{7} = -\frac{r}{3}$ {0}

37. $\frac{4t + 1}{3} = \frac{t + 5}{6} + \frac{t - 3}{6}$ {0}

38. $\frac{2x + 5}{5} = \frac{3x + 1}{2} + \frac{-x + 7}{2}$ {−5}

39. $.05y + .12(y + 5{,}000) = 940$ {2,000}

40. $.09k + .13(k + 300) = 61$ {100}

41. $.02(50) + .08r = .04(50 + r)$ {25}

42. $.20(14{,}000) + .14t = .18(14{,}000 + t)$ {7,000}

43. $.05x + .10(200 - x) = .45x$ {40}

44. $.08x + .12(260 - x) = .48x$ {60}

◉ **45.** The equation $x^2 + 2 = x^2 + 2$ is called a(n) ____________, because its solution set is {all real numbers}. The equation $x + 1 = x + 2$ is called a(n) ____________, because it has no solutions. **identity; contradiction**

◉ **46.** Which one of the following is a conditional equation?
(a) $2x + 1 = 3$ **(b)** $x = 3x - 2x$ **(c)** $2(x + 2) = 2x + 2$
(d) $5x - 3 = 4x + x - 5 + 2$ (a)

Decide whether the equation is conditional, *an* identity, *or a* contradiction. *Give the solution set.*

47. $-2p + 5p - 9 = 3(p - 4) - 5$ **contradiction; ∅**

48. $-6k + 2k - 11 = -2(2k - 3) + 4$ **contradiction; ∅**

49. $-11m + 4(m - 3) + 6m = 4m - 12$ **conditional; {0}**

50. $3p - 5(p + 4) + 9 = -11 + 15p$ **conditional; {0}**

51. $7[2 - (3 + 4r)] - 2r = -9 + 2(1 - 15r)$ **identity; {all real numbers}**

52. $4[6 - (1 + 2m)] + 10m = 2(10 - 3m) + 8m$ **identity; {all real numbers}**

Solve the problem.

53. The mathematical model $y = 420x + 720$ approximates the worldwide credit card fraud losses between the years 1989 and 1993, where $x = 0$ corresponds to 1989, $x = 1$ corresponds to 1990, and so on, and y is in millions of dollars. Based on this model, what would be the approximate amount of credit card fraud losses in 1994? In what year would losses reach 3,660 million dollars (that is, $3,660,000,000)?
2,820 million dollars; 1996 (when $x = 7$)

54. According to research done by the Beverage Marketing Corporation, ready-to-drink iced tea sales have boomed during the past few years. The model $y = 310x + 260$ approximates the revenue generated, where $x = 0$ corresponds to 1991 and y is in millions of dollars. Based on this model, what would be the revenue generated in 1992? In what year would revenue reach 2,430 million dollars?
570 million dollars; 1998 (when $x = 7$)

55. The accompanying bar graph gives a pictorial representation of bank credit card charges in each year from 1989 to 1992. Use the graph to estimate the charges in each of the following years:
(a) 1989 (b) 1990
(c) 1991 (d) 1992.
(a) 230 billion dollars (b) 275 billion dollars
(c) 300 billion dollars (d) 335 billion dollars

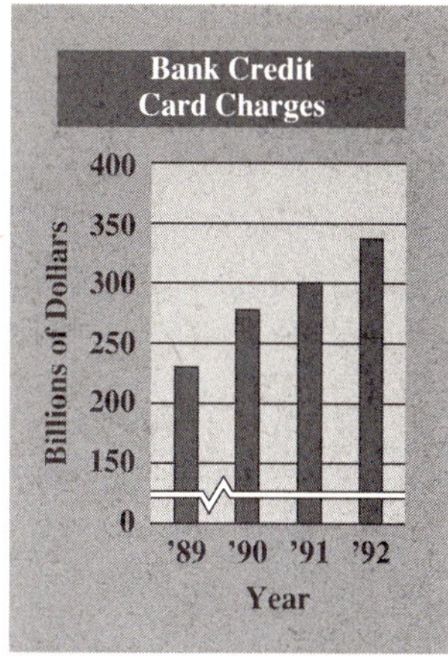

56. Based on information from RAM Research, the amount of bank credit card charges in each year between 1989 and 1992 can be approximated by the linear model $y = 37x + 230$, where $x = 0$ corresponds to 1989, and y is in billions of dollars.
(a) According to this model, what would be the charges in 1993? 378 billion dollars
(b) In 1993, the actual charges were approximately 422 billion dollars. How might you explain the discrepancy between the predicted amount from part (a) and the actual amount charged?

57. Suppose the formula $A = 2HW + 2LW + 2LH$ is "solved for L" as follows:

$$A = 2HW + 2LW + 2LH$$
$$A - 2LW - 2HW = 2LH$$
$$\frac{A - 2LW - 2HW}{2H} = L.$$

While there are no algebraic errors here, what is wrong with the final line, if we are interested in solving for L? L should not appear on both sides of the equation.

58. When a formula is solved for a particular variable, several different equivalent forms may be possible. If we solve $A = \frac{1}{2}bh$ for h, one possible correct answer is

$$h = \frac{2A}{b}.$$

Which one of the following is *not* equivalent to this?

(a) $h = 2\left(\frac{A}{b}\right)$ (b) $h = 2A\left(\frac{1}{b}\right)$ (c) $h = \frac{A}{\frac{1}{2}b}$ (d) $h = \frac{\frac{1}{2}A}{b}$ (d)

Solve the formula for the specified variable.

59. $d = rt$; for r (distance) $r = \frac{d}{t}$

60. $I = prt$; for r (simple interest) $r = \frac{I}{pt}$

61. $A = bh$; for b (area of a parallelogram) $b = \frac{A}{h}$

62. $P = 2L + 2W$; for L (perimeter of a rectangle) $L = \dfrac{P - 2W}{2}$

63. $P = a + b + c$; for a (perimeter of a triangle) $a = P - b - c$

64. $V = LWH$; for W (volume of a rectangular solid) $W = \dfrac{V}{LH}$

65. $A = \dfrac{1}{2}bh$; for h (area of a triangle) $h = \dfrac{2A}{b}$

66. $C = 2\pi r$; for r (circumference of a circle) $r = \dfrac{C}{2\pi}$

67. $S = 2\pi rh + 2\pi r^2$; for h (surface area of a right circular cylinder)
$h = \dfrac{S - 2\pi r^2}{2\pi r}$ or $h = \dfrac{S}{2\pi r} - r$

68. $A = \dfrac{1}{2}(B + b)h$; for B (area of a trapezoid)
$B = \dfrac{2A}{h} - b$ or $B = \dfrac{2A - bh}{h}$

69. $\text{C} = \dfrac{5}{9}(\text{F} - 32)$; for F (Fahrenheit to Celsius) $\text{F} = \dfrac{9}{5}\text{C} + 32$

70. $\text{F} = \dfrac{9}{5}\text{C} + 32$; for C (Celsius to Fahrenheit) $\text{C} = \dfrac{5}{9}(\text{F} - 32)$

71. $A = 2HW + 2LW + 2LH$; for H (surface area of a rectangular solid)
$H = \dfrac{A - 2LW}{2W + 2L}$

72. $V = \dfrac{1}{3}Bh$; for h (volume of a right pyramid) $h = \dfrac{3V}{B}$

7.2 Exercises (Text Page 338)

1. In your own words, list the six steps suggested in the box marked "Solving an Applied Problem" in this section.

2. If x represents the number, express the following using algebraic symbols:
(a) 9 less than a number
(b) 9 is less than a number.
Which one of these is an *expression*?
(a) $x - 9$; **(b)** $9 < x$; **(a) is an expression.**

Translate the verbal phrase into a mathematical expression. Use x to represent the unknown number.

3. a number decreased by 13 $x - 13$

4. a number decreased by 12 $x - 12$

5. 7 increased by a number $7 + x$

6. 12 more than a number $x + 12$

7. the product of 8 and 12 more than a number $8(x + 12)$

8. the product of 9 less than a number and 6 more than the number $(x - 9)(x + 6)$

9. the quotient of a number and 6 $\dfrac{x}{6}$

10. the quotient of 6 and a nonzero number $\dfrac{6}{x}$ $(x \neq 0)$

CONCEPTUAL

WRITING

CHALLENGING
SCIENTIFIC CALCULATOR

GRAPHING CALCULATOR

11. the ratio of a number and 12 $\frac{x}{12}$

12. the ratio of 12 and a nonzero number $\frac{12}{x}$ $(x \neq 0)$

13. $\frac{6}{7}$ of a number $\frac{6}{7}x$

14. 19% of a number $.19x$ or $\frac{19}{100}x$

15. Which one of the following is *not* a valid translation of "30% of a number"?

(a) $.30x$ (b) $.3x$ (c) $\frac{3x}{10}$ (d) $30x$

(d)

16. Explain the difference between the answers to Exercises 11 and 12.

Use the methods of Examples 1 and 2 or your own method to solve the problem.

17. The U.S. Senate has 100 members. In 1995, there were 6 fewer Democrats than Republicans, with no other parties represented. How many members of each party were there in the Senate?
47 Democrats, 53 Republicans

18. The total number of Democrats and Republicans in the U.S. House of Representatives in 1995 was 434. There were 26 more Republicans than Democrats. How many members of each party were there in the House of Representatives?
204 Democrats, 230 Republicans

19. In his coaching career with the Boston Celtics, Red Auerbach had 558 more wins than losses. His total number of games coached was 1,516. How many wins did Auerbach have? 1,037

20. Is the first Super Bowl, played in 1966, Green Bay and Kansas City scored a total of 45 points. Green Bay won by 25 points. What was the score of the first Super Bowl? Green Bay 35, Kansas City 10

21. On Professor Brandsma's algebra test, the highest grade was 34 points higher than the lowest grade. The sum of the two grades was 160. What were the highest and lowest grades?
highest grade: 97; lowest grade: 63

22. In one day, Gwen Boyle received 13 packages. Federal Express delivered three times as many as Airborne Express, while Airborne Express delivered two more than United Parcel Service. How many packages did each service deliver to Gwen?
Federal Express: 9; Airborne Express: 3; United Parcel Service: 1

23. In her job at the post office, Janie Quintana works a $6\frac{1}{2}$-hour day. She sorts mail, sells stamps, and does supervisory work. On one day, she sold stamps twice as long as she sorted mail, and sold stamps 1 hour longer than the time she spent doing supervisory work. How many hours did she spend at each task?
sorting mail: $1\frac{1}{2}$ hours; selling stamps: 3 hours; supervising: 2 hours

24. Venus is 31.2 million miles farther from the sun than Mercury, while Earth is 25.7 million miles farther from the sun than Venus. If the total of the distances for these three planets from the sun is 196.1 million miles, how far away from the sun is Mercury? (All distances given here are *mean (average)* distances.)
36 million miles

25. It is believed that Saturn has 5 more satellites (moons) than the known number of satellites for Jupiter, and 20 more satellites than the known number for Mars. If the total of these numbers is 41, how many satellites does Mars have? 2

26. On a recent model of the Eagle Premier, the suggested list price for the antilock brake system is $\frac{10}{3}$ the suggested list price of power door locks. Together, these two options cost $1,040. What is the suggested list price for each of these options?
antilock brakes: $800; power door locks: $240

27. Labrador retrievers and Rottweilers rank as the top two dog breeds using figures provided by the American Kennel Club. In a recent year, there were 25,434 more Labrador retrievers registered than Rottweilers, and together there were 216,324 dogs of these two breeds registered. How many of each breed were registered?
Labrador retrievers: 120,879; Rottweilers: 95,445

28. According to figures provided by the Air Transport Association of America, the Boeing B747–400 and the McDonnell Douglas L1011–100/200 are among the air carriers with maximum passenger seating. The Boeing seats 110 more passengers than the McDonnell Douglas, and together the two models seat 696 passengers. What is the seating capacity of each model?
Boeing: 403; McDonnell Douglas: 293

Refer to the accompanying graph to help you answer the questions in Exercises 29 and 30.

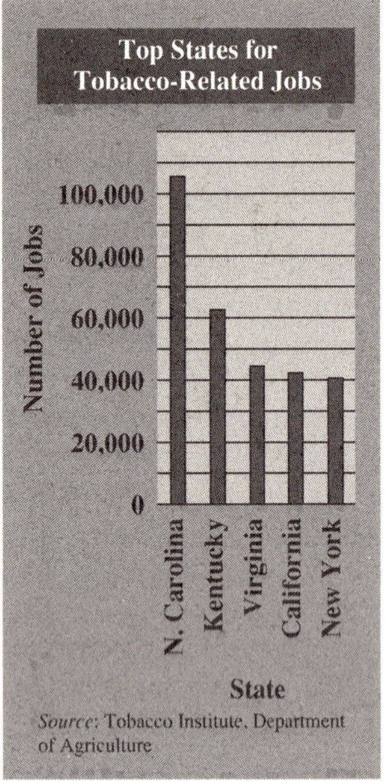

29. According to figures provided by the Tobacco Institute, Department of Agriculture, the top two states in the tobacco industry together employ 167,281 people. The leading state employs 17,663 fewer than twice as many as the second leading state. What are these states and how many people does each employ?
The leading state, North Carolina, employs 105,633 people. Second is Kentucky with 61,648 people.

30. New York employs 1,446 fewer people than California, and California employs 2,511 fewer people than Virginia. The three states together employ 127,143 people. How many people work in the tobacco industry in each of these states?
New York: 40,580; California: 42,026; Virginia: 44,537

Use the method of Example 3 or your own method to solve the problem.

31. How much pure acid is in 250 milliliters of a 14% acid solution? **35 milliliters**

32. How much pure alcohol is in 150 liters of a 30% alcohol solution? **45 liters**

33. If \$10,000 is invested for one year at 3.5% simple interest, how much interest is earned? **\$350**

34. If \$25,000 is invested at 3% simple interest for 2 years, how much interest is earned? **\$1,500**

35. What is the monetary amount of 283 nickels? **\$14.15**

36. What is the monetary amount of 35 half-dollars? **\$17.50**

Use the method of Example 4 or your own method to solve the problem.

37. How many gallons of 50% antifreeze must be mixed with 80 gallons of 20% antifreeze to get a mixture that is 40% antifreeze? **160 gallons**

38. How many liters of 25% acid solution must be added to 80 liters of 40% solution to get a solution that is 30% acid? **160 liters**

39. How many gallons of milk that is 2% butterfat must be mixed with milk that is 3.5% butterfat to get 10 gallons of milk that is 3% butterfat? (*Hint:* Let x represent the number of gallons that are 2% butterfat. Then $10 - x$ represents the number of gallons that are 3.5% butterfat.)
$3\frac{1}{3}$ gallons

40. A pharmacist has 20 liters of a 10% drug solution. How many liters of 5% solution must be added to get a mixture that is 8%?
$13\frac{1}{3}$ liters

41. How many liters of a 60% acid solution must be mixed with a 75% acid solution to get 20 liters of a 72% solution? **4 liters**

42. How many gallons of a 12% indicator solution must be mixed with a 20% indicator solution to get 10 gallons of a 14% solution?
$7\frac{1}{2}$ gallons

43. Minoxidil is a drug that has recently proven to be effective in treating male pattern baldness. A pharmacist wishes to mix a solution that is 2% minoxidil. She has on hand 50 milliliters of a 1% solution, and she wishes to add some 4% solution to it to obtain the desired 2% solution. How much 4% solution should she add? **25 milliliters**

44. Water must be added to 20 milliliters of a 4% minoxidil solution to dilute it to a 2% solution. How many milliliters of water should be used? **20 milliliters**

◉ *Use the concepts of this section to solve the problem.*

45. Suppose that a chemist is mixing two acid solutions, one of 20% concentration and the other of 30% concentration. Which one of the following concentrations could *not* be obtained?
(a) 22% **(b)** 24%
(c) 28% **(d)** 32%
(d)

46. Suppose that pure alcohol is added to a 24% alcohol mixture. Which one of the following concentrations could *not* be obtained?
(a) 22% **(b)** 26%
(c) 28% **(d)** 30%
(a)

Use the method of Example 5 or your own method to solve the problem.

47. Li Nguyen invested some money at 3% and $4,000 less than that amount at 5%. The two investments produced a total of $200 interest in one year. How much was invested at each rate?
$5,000 at 3%; $1,000 at 5%

48. LaShondra Williams inherited some money from her uncle. She deposited part of the money in a savings account paying 2%, and $3,000 more than that amount in a different account paying 3%. Her annual interest income was $690. How much did she deposit at each rate?
$12,000 at 2%; $15,000 at 3%

49. Fran Liberto sold a painting that she found at a garage sale to an art dealer. With the $12,000 she got, she invested part at 4% in a certificate of deposit and the rest at 5% in a municipal bond. Her total annual income from the two investments was $515. How much did she invest at each rate? (*Hint:* Let x represent the amount invested at 4%. Then $12{,}000 - x$ represents the amount invested at 5%.)
$8,500 at 4%; $3,500 at 5%

50. Two investments produce an annual interest income of $114. The total amount of money invested is $4,000, and the two interest rates paid are 2.5% and 3.5%. How much is invested at each rate?
$2,600 at 2.5%; $1,400 at 3.5%

51. With income earned by selling the rights to his life story, an actor invests some of the money at 3% and $30,000 more than twice as much at 4%. The total annual interest earned from the investments is $5,600. How much is invested at each rate?
$40,000 at 3%; $110,000 at 4%

52. An artist invests her earnings in two ways. Some goes into a tax-free bond paying 6%, and $6,000 more than three times as much goes into mutual funds paying 5%. Her total annual interest income from the investments is $825. How much does she invest at each rate?
$2,500 at 6%; $13,500 at 5%

The problems in Exercises 53–56 involve percent. Use your own method to solve the problem.

53. The pie chart shows the approximate percents for different sizes of Health Maintenance Organizations (HMOs) for the year 1991. During that year there were approximately 560 such organizations. Find the approximate number for each size group.

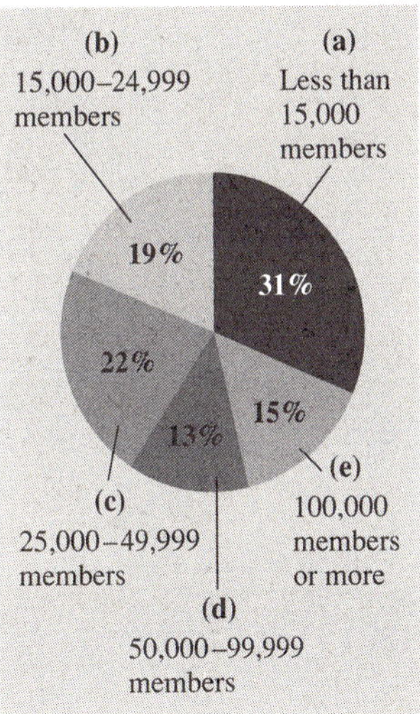

(a) 174 (b) 106 (c) 123 (d) 73 (e) 84

54. At the start of play on September 18, 1995, the standings of the Central Division of the American League were as shown. "Winning percentage" is commonly expressed as a decimal rounded to the nearest thousandth. To find the winning percentage of a team, divide the number of wins by the total

number of games played. Find the winning percentage of each of the following teams.
(a) Cleveland (b) Chicago (c) Milwaukee

	Won	Lost
Cleveland	91	41
Kansas City	67	63
Milwaukee	62	69
Chicago	60	70
Minnesota	48	81

(a) .689 (b) .462 (c) .473

55. The pie chart shows the breakdown, by approximate percents, of age groups buying aerobic shoes in 1990. According to figures of the National Sporting Goods Association, $582,000,000 was spent on aerobic shoes that year. How much was spent by each age group?
(a) younger than 14 (b) 18 to 24
(c) 45 to 64

Aerobic Shoes
Breakdown of shoe sales by age group

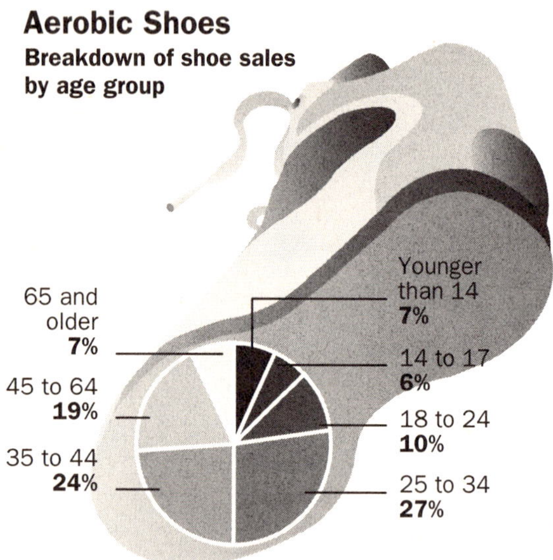

(a) $40,740,000 (b) $58,200,000
(c) $110,580,000

56. The pie chart shows the breakdown for the year 1990 regarding the sales and purchases of exercise equipment. If $2,295,000,000 was spent that year, determine the amount spent by each age group.
(a) 24 to 34 (b) 35 to 44 (c) 65 and older

Pumping Sales
Sales of exercise equipment by age group

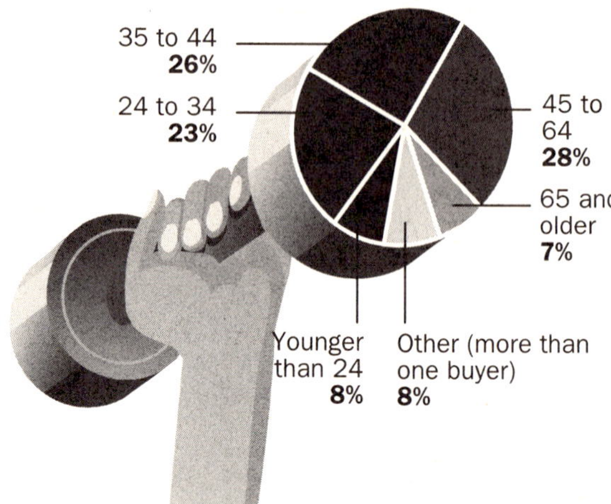

(a) $527,850,000 (b) $596,700,000
(c) $160,650,000

Use the method of Example 6 or your own method to solve the problem.

57. Leslie Cobar's piggy bank has 36 coins. Some are quarters and the rest are half-dollars. If the total value of the coins is $14.75, how many of each denomination does Leslie have?
13 quarters, 23 half-dollars

58. Liz Harold has a jar in her office that contains 47 coins. Some are pennies and the rest are dimes. If the total value of the coins is $2.18, how many of each denomination does she have?
28 pennies, 19 dimes

59. Sam Abo-zahrah has a box of coins that he uses when playing poker with his friends. The box currently contains 44 coins, consisting of pennies, dimes, and quarters. The number of pennies is equal to the number of dimes, and the total value is $4.37. How many of each denomination of coin does he have in the box?
17 pennies, 17 dimes, 10 quarters

60. Roma Sherry found some coins while looking under her sofa pillows. There were equal numbers of nickels and quarters, and twice as many half-dollars as quarters. If she found $2.60 in all, how many of each denomination of coin did she find?
2 nickels, 2 quarters, 4 half-dollars

 CONCEPTUAL  WRITING  CHALLENGING  SCIENTIFIC CALCULATOR  GRAPHING CALCULATOR

61. In the nineteenth century, the United States minted two-cent and three-cent pieces. Toni Spahn has three times as many three-cent pieces as two-cent pieces, and the face value of these coins is $1.21. How many of each denomination does she have?
11 two-cent pieces, 33 three-cent pieces

62. Kathy Jordan collects U.S. gold coins. She has a collection of 80 coins. Some are $10 coins and the rest are $20 coins. If the face value of the coins is $1,060, how many of each denomination does she have?
54 $10 coins, 26 $20 coins

63. The school production of *Our Town* was a big success. For opening night, 410 tickets were sold. Students paid $1.50 each, while nonstudents paid $3.50 each. If a total of $825 was collected, how many students and how many nonstudents attended?
305 students, 105 nonstudents

64. A total of 550 people attended a Frankie Valli concert. Floor tickets cost $20 each, while balcony tickets cost $14 each. If a total of $10,400 was collected, how many of each type of ticket were sold?
450 floor tickets, 100 balcony tickets

65. In your own words, explain how the problems in Exercises 57–64 are similar to the investment problem explained in Example 5.

66. In the nineteenth century, the United States minted half-cent coins. If an applied problem involved half-cent coins, what decimal number would represent this denomination? **.005**

In Exercises 67–70, find the time based on the information provided. Use a calculator and round your answer to the nearest thousandth.

	Event and Year	Participant	Distance	Rate
67.	Indianapolis 500, 1992	Al Unser, Jr. (Galmer-Chevrolet)	500 miles	134.479 mph
68.	Daytona 500, 1992	Davey Allison (Ford)	500 miles	160.256 mph
69.	Indianapolis 500, 1980	Johnny Rutherford (Hy-Gain McLaren/ Goodyear)	255 miles (rain-shortened)	148.725 mph
70.	Indianapolis 500, 1975	Bobby Unser (Jorgensen Eagle)	435 miles (rain-shortened)	149.213 mph

67. 3.718 hours 68. 3.120 hours 69. 1.715 hours 70. 2.915 hours

In Exercises 71–74, find the rate based on the information provided. Use a calculator and round your answer to the nearest hundredth.

	Event and Year	Participant	Distance	Time
71.	Summer Olympics 400-meter Hurdles, Women, 1992	Sally Gunnell (Great Britain)	400 meters	53.23 seconds
72.	Summer Olympics, 100-meter Dash, Women, 1992	Gail Devers (United States)	100 meters	10.82 seconds
73.	Summer Olympics, 100-meter Dash, Men, 1988	Carl Lewis (United States)	100 meters	9.92 seconds
74.	Winter Olympics, 500-meter Speed Skating, Women, 1992	Bonnie Blair (United States)	500 meters	40.33 seconds

71. 7.51 meters per second 72. 9.24 meters per second 73. 10.08 meters per second 74. 12.40 meters per second

75. A driver averaged 53 miles per hour and took 10 hours to travel from Memphis to Chicago. What is the distance between Memphis and Chicago? **530 miles**

76. A small plane traveled from Warsaw to Rome, averaging 164 miles per hour. The trip took two hours. What is the distance from Warsaw to Rome? **328 miles**

77. Suppose that an automobile averages 45 miles per hour, and travels for 30 minutes. Is the distance traveled $45 \times 30 = 1{,}350$ miles? If not, explain why not, and give the correct distance.

78. Which of the following choices is the best *estimate* for the average speed of a trip of 405 miles that lasted 8.2 hours?
(a) 50 miles per hour **(b)** 30 miles per hour
(c) 60 miles per hour **(d)** 40 miles per hour **(a)**

Use the method of Example 8 or your own method to solve the problem.

79. St. Louis and Portland are 2,060 miles apart. A small plane leaves Portland, traveling toward St. Louis at an average speed of 90 miles per hour. Another plane leaves St. Louis at the same time, traveling toward Portland, averaging 116 miles per hour. How long will it take them to meet? **10 hours**

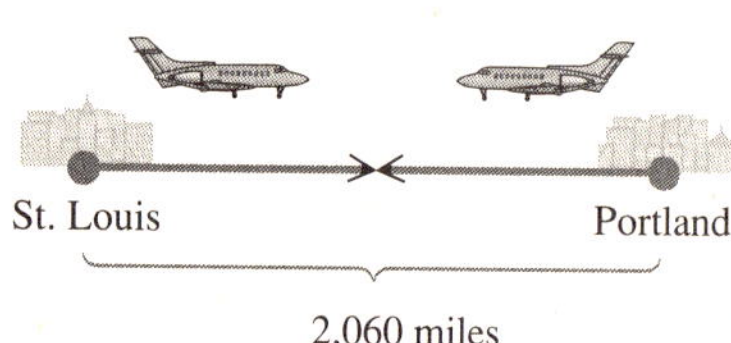

80. Atlanta and Cincinnati are 440 miles apart. John leaves Cincinnati, driving toward Atlanta at an average speed of 60 miles per hour. Pat leaves Atlanta at the same time, driving toward Cincinnati in her antique auto, averaging 28 miles per hour. How long will it take them to meet? **5 hours**

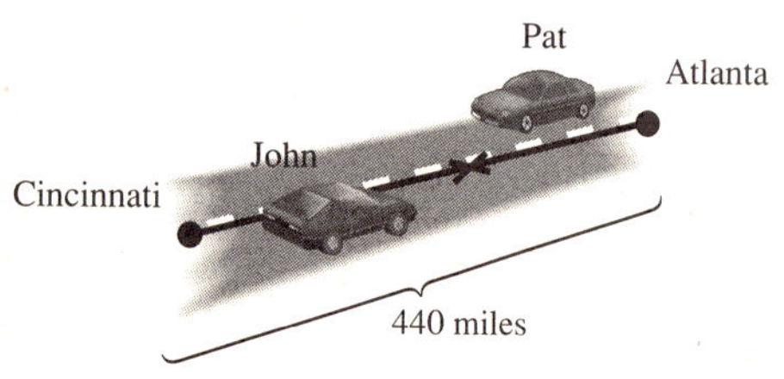

81. A train leaves Little Rock, Arkansas, and travels north at 85 kilometers per hour. Another train leaves at the same time and travels south at 95 kilometers per hour. How long will it take before they are 315 kilometers apart?

$1\frac{3}{4}$ hours

82. Two steamers leave a port on a river at the same time, traveling in opposite directions. Each is traveling 22 miles per hour. How long will it take for them to be 110 miles apart?

$2\frac{1}{2}$ hours

83. Joey and Mark commute to work, traveling in opposite directions. Joey leaves the house at 8:00 A.M. and averages 35 miles per hour. Mark leaves at 8:15 A.M. and averages 40 miles per hour. At what time will they be 140 miles apart? **10:00 A.M.**

84. Jeff leaves his house on his bicycle at 8:30 A.M. and averages 5 miles per hour. His wife, Joan, leaves at 9:00 A.M., following the same path and averaging 8 miles per hour. At what time will Joan catch up with Jeff? **9:50 A.M.**

85. When Tri drives his car to work, the trip takes 30 minutes. When he rides the bus, it takes 45 minutes. The average speed of the bus is 12 miles per hour less than his speed when driving. Find the distance he travels to work. **18 miles**

86. Latoya can get to school in 15 minutes if she rides her bike. It takes her 45 minutes if she walks. Her speed when walking is 10 miles per hour slower than her speed when riding. What is her speed when she rides? **15 miles per hour**

7.3 Exercises (Text Page 353)

1. Which one of the following ratios is not the same as 3 to 4?
   (a) .75 **(b)** 6 to 8 **(c)** 4 to 3 **(d)** 30 to 40 (c)

2. Give three ratios that are equivalent to 3 to 1.
   Answers will vary. Three examples are 6 to 2, 9 to 3, and 300 to 100.

Determine the ratio and write it in lowest terms.

3. 40 miles to 30 miles $\frac{4}{3}$

4. 60 feet to 70 feet $\frac{6}{7}$

5. 120 people to 90 people $\frac{4}{3}$

6. 72 dollars to 220 dollars $\frac{18}{55}$

7. 20 yards to 8 feet $\frac{15}{2}$

8. 30 inches to 8 feet $\frac{5}{16}$

9. 24 minutes to 2 hours $\frac{1}{5}$

10. 16 minutes to 1 hour $\frac{4}{15}$

11. 8 days to 40 hours $\frac{24}{5}$

12. 50 hours to 5 days $\frac{5}{12}$

13. Explain the distinction between *ratio* and *proportion*.

14. Suppose that someone told you to use cross products in order to multiply fractions. How would you explain to the person what is wrong with his or her thinking?

Decide whether the proportion is true *or* false.

15. $\frac{5}{35} = \frac{8}{56}$ true

16. $\frac{4}{12} = \frac{7}{21}$ true

17. $\frac{120}{82} = \frac{7}{10}$ false

18. $\frac{27}{160} = \frac{18}{110}$ false

19. $\frac{\frac{1}{2}}{5} = \frac{1}{10}$ true

20. $\frac{\frac{1}{3}}{6} = \frac{1}{18}$ true

Solve the equation.

21. $\frac{k}{4} = \frac{175}{20}$ $\{35\}$

22. $\frac{49}{56} = \frac{z}{8}$ $\{7\}$

23. $\frac{x}{6} = \frac{18}{4}$ $\{27\}$

24. $\frac{z}{80} = \frac{20}{100}$ $\{16\}$

25. $\frac{3y - 2}{5} = \frac{6y - 5}{11}$ $\{-1\}$

26. $\frac{2p + 7}{3} = \frac{p - 1}{4}$ $\left\{-\frac{31}{5}\right\}$

27. $\frac{2r + 8}{4} = \frac{3r - 9}{3}$ $\{10\}$

28. $\frac{5k + 1}{6} = \frac{3k - 2}{3}$ $\{5\}$

Solve the problem by setting up and solving a proportion.

29. According to the Home and Garden Bulletin No. 72, four spears of asparagus contain 15 calories. How many spears of asparagus would contain 50 calories?
    $13\frac{1}{3}$ spears

30. According to the source indicated in Exercise 29, three ounces of bluefish baked with butter or margarine provide 22 grams of protein. How many ounces would provide 242 grams of protein?
    33 ounces

31. A chain saw requires a mixture of 2-cycle engine oil and gasoline. According to the directions on a bottle of Oregon 2-cycle Engine Oil, for a 50 to 1 ratio requirement, 2.5 fluid ounces of oil are required for 1 gallon of gasoline. If the tank of the chain saw holds 2.75 gallons, how many fluid ounces of oil are required? **6.875 ounces**

32. The directions on the bottle mentioned in Exercise 31 indicate that if the ratio requirement is 24 to 1, 5.5 ounces of oil are required for 1 gallon of gasoline. If gasoline is to be mixed with 22 ounces of oil, how much gasoline is to be used?
4 gallons

33. In 1992, the average exchange rate between U.S. dollars and United Kingdom pounds was 1 pound to \$1.7655. Margaret went to London and exchanged her U.S. currency for U.K. pounds, and received 400 pounds. How much in U.S. money did Margaret exchange? **\$706.20**

34. If 3 U.S. dollars can be exchanged for 4.2186 Swiss francs, how many Swiss francs can be obtained for \$49.20? **69.19 francs**

35. If 6 gallons of premium unleaded gasoline cost \$9.36, how much would it cost to completely fill a 15-gallon tank? **\$23.40**

36. If sales tax on a \$16.00 compact disc is \$1.32, how much would the sales tax be on a \$120.00 compact disc player? **\$9.90**

37. The distance between Kansas City, Missouri, and Denver is 600 miles. On a certain wall map, this is represented by a length of 2.4 feet. On the map, how many feet would there be between Memphis and Philadelphia, two cities that are actually 1,000 miles apart? **4 feet**

38. The distance between Singapore and Tokyo is 3,300 miles. On a certain wall map, this distance is represented by 11 inches. The actual distance between Mexico City and Cairo is 7,700 miles. How far apart are they on the same map?
$25\frac{2}{3}$ inches

39. A recipe for green salad for 70 people calls for 18 heads of lettuce. How many heads of lettuce would be needed if 175 people were to be served?
45 heads

40. A recipe for oatmeal macaroons calls for 1 2/3 cups of flour to make four dozen cookies. How many cups of flour would be needed for six dozen cookies?
$2\frac{1}{2}$ cups

41. A piece of property assessed at \$42,000 requires an annual property tax of \$273. How much property tax would be charged for a similar piece of property assessed at \$52,000? **\$338**

42. If 4 pounds of fertilizer will cover 50 square feet of garden, how many pounds would be needed for 225 square feet? **18 pounds**

43. Biologists tagged 250 fish in Willow Lake on October 5. On a later date they found 7 tagged fish in a sample of 350. Estimate the total number of fish in Willow Lake to the nearest hundred. **12,500 fish**

44. On May 13 researchers at Argyle Lake tagged 420 fish. When they returned a few weeks later, their sample of 500 fish contained 9 that were tagged. Give an approximation of the fish population in Argyle Lake to the nearest hundred. **23,300 fish**

45. The pie chart indicates the number of people who responded in various ways to a survey. Suppose that 1,680 people participated. How many would we expect to answer in each way if 4,200 people were surveyed?

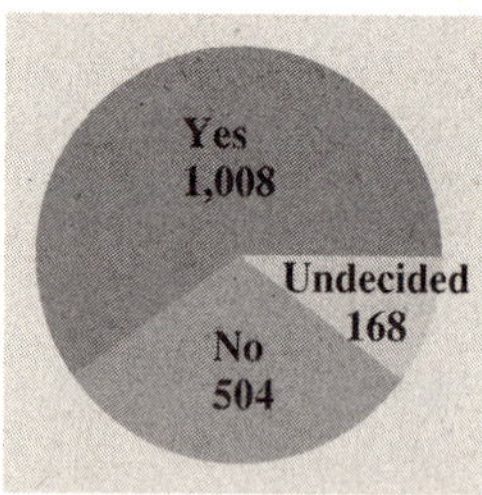

yes: 2,520; no: 1,260; undecided: 420

46. According to McDonald's Corporation, Canada has 40,113 people for every McDonald's restaurant. The 1992 estimated population of Canada was 27,351,000. According to these figures, about how many McDonald's restaurants would we have expected Canada to have in 1992?
682 restaurants

CONCEPTUAL  WRITING ▲ CHALLENGING SCIENTIFIC CALCULATOR GRAPHING CALCULATOR

A supermarket was surveyed to find the prices charged for items in various sizes. Find the best buy (based on price per unit) for the particular item.

47. Trash bags
20-count: $3.09
30-count: $4.59 **30-count size**

48. Black pepper
1-ounce size: $ .99
2-ounce size: $1.65
4-ounce size: $4.39 **2-ounce size**

49. Breakfast cereal
15-ounce size: $2.99
25-ounce size: $4.49
31-ounce size: $5.49 **31-ounce size**

50. Cocoa mix
8-ounce size: $1.39
16-ounce size: $2.19
32-ounce size: $2.99 **32-ounce size**

51. Tomato ketchup
14-ounce size: $ .89
32-ounce size: $1.19
64-ounce size: $2.95 **32-ounce size**

52. Cut green beans
8-ounce size: $ .45
16-ounce size: $ .49
50-ounce size: $1.59 **16-ounce size**

Two triangles are said to be ***similar*** *if they have the same shape (but not necessarily the same size). Similar triangles have sides that are proportional. For example, the figure shows two similar triangles. Notice that the ratios of the corresponding sides are all equal to* $\frac{3}{2}$:

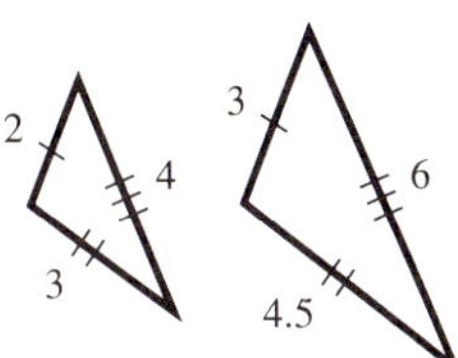

$$\frac{3}{2} = \frac{3}{2} \qquad \frac{4.5}{3} = \frac{3}{2} \qquad \frac{6}{4} = \frac{3}{2}.$$

If we know that two triangles are similar, we can set up a proportion to solve for the length of an unknown side. Use a proportion to find the length x, given that the pair of triangles are similar.

53. 3, 5, x

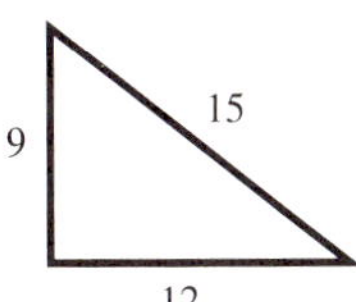

4

54. 3, 3, 2; 12, 12, x

8

55.

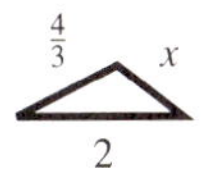

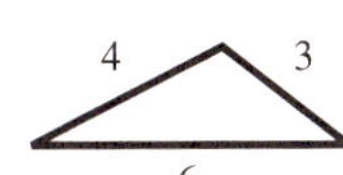

1

56.

2, 2, x

2

For the problems in Exercises 57 and 58, **(a)** *draw a sketch consisting of two right triangles, depicting the situation described, and* **(b)** *solve the problem.*

57. An enlarged version of the chair used by George Washington at the Constitutional Convention casts a shadow 18 feet long at the same time a vertical pole 12 feet high casts a shadow 4 feet long. How tall is the chair?

(a) 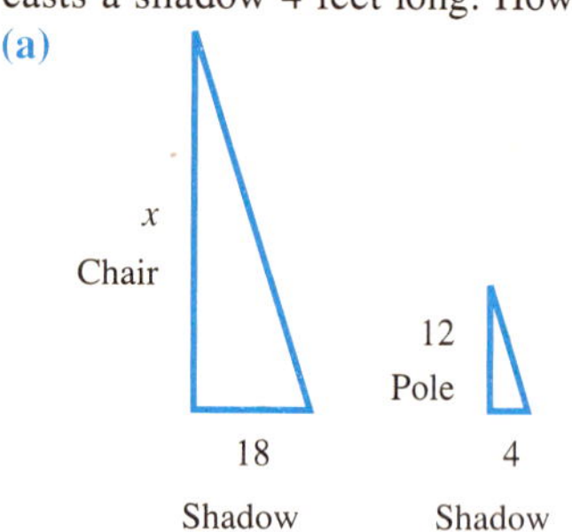

(b) 54 feet

58. One of the tallest candles ever constructed was exhibited at the 1897 Stockholm Exhibition. If it cast a shadow 5 feet long at the same time a vertical pole 32 feet high cast a shadow 2 feet long, how tall was the candle?

(a) 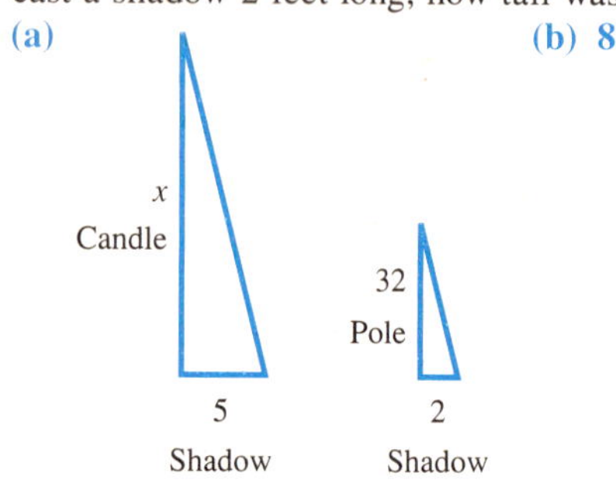

(b) 80 feet

The Consumer Price Index, issued by the U.S. Bureau of Labor Statistics, provides a means of determining the purchasing power of the U.S. dollar from one year to the next. Using the period from 1982 to 1984 as a measure of 100.0, *the Consumer Price Index for electricity from 1975 to 1991 is shown here.*

Year	Consumer Price Index	Year	Consumer Price Index
1975	50.0	1983	98.9
1976	53.1	1984	105.3
1977	56.6	1985	108.9
1978	60.9	1986	110.4
1979	65.6	1987	110.0
1980	75.8	1988	111.5
1981	87.2	1989	114.7
1982	95.8	1990	117.4
		1991	121.8

To use the Consumer Price Index, we can set up a proportion as follows:

$$\frac{\text{price of electricity in year } a}{\text{Consumer Price Index in year } a} = \frac{\text{price of electricity in year } b}{\text{Consumer Price Index in year } b}.$$

For example, if an electricity bill in 1985 was \$150, *we could predict the amount of the bill in 1989 by using the following proportion:*

$$\begin{matrix}\text{1985 price} \rightarrow \\ \text{1985 index} \rightarrow\end{matrix} \quad \frac{150}{108.9} = \frac{x}{114.7}. \quad \begin{matrix}\leftarrow \text{1989 price} \\ \leftarrow \text{1989 index}\end{matrix}$$

By using cross products and a calculator, we can determine that the 1989 price would be approximately \$158.

Use the Consumer Price Index figures given to find the amount that would be charged for the use of the same amount of electricity that cost \$225 *in 1980. Give your answer to the nearest dollar.*

59. in 1983 \$294 **60.** in 1986 \$328 **61.** in 1988 \$331 **62.** in 1991 \$362

63. The Consumer Price Index figures for shelter for the years A and B are 90.5 and 146.3. If shelter for a particular family cost \$3,000 in year A, what would be the comparable cost in year B? Give your answer to the nearest dollar. \$4,850

64. Due to a volatile fuel oil market in the early 1980s, the price of fuel decreased during the first three quarters of the decade. The Consumer Price Index figures for 1982 and 1986 were 105.0 and 74.1. If it cost you \$21.50 to fill your tank with fuel oil in 1982, how much would it have cost to fill the same tank in 1986? Give your answer to the nearest ccnt. \$15.17

Solve the problem involving variation.

65. If x varies directly as y, and $x = 27$ when $y = 6$, find x when $y = 2$. 9

66. If z varies directly as x, and $z = 30$ when $x = 8$, find z when $x = 4$. 15

67. If m varies directly as p^2, and $m = 20$ when $p = 2$, find m when p is 5. 125

68. If a varies directly as b^2, and $a = 48$ when $b = 4$, find a when $b = 7$. 147

69. If p varies inversely as q^2, and $p = 4$ when $q = \frac{1}{2}$, find p when $q = \frac{3}{2}$. $\frac{4}{9}$

70. If z varies inversely as x^2, and $z = 9$ when $x = \frac{2}{3}$, find z when $x = \frac{5}{4}$. $\frac{64}{25}$

71. Assume that the constant of variation, k, is positive.

(a) If y varies directly as x, then as y increases, x $\underset{\text{(decreases/increases)}}{\underline{\quad ? \quad}}$.

(b) If y varies inversely as x, then as y increases, x $\underset{\text{(decreases/increases)}}{\underline{\quad ? \quad}}$.

(a) increases (b) decreases

72. (a) The more gasoline you pump, the more you will have to pay. Is this an example of direct or inverse variation?

(b) The longer the term of your subscription to *Monitoring Times,* the less you will have to pay per year. Is this an example of direct or inverse variation?

(a) direct (b) inverse

Solve the problem.

73. The interest on an investment varies directly as the rate of interest. If the interest is \$48 when the interest rate is 5%, find the interest when the rate is 4.2%. \$40.32

74. For a given base, the area of a triangle varies directly as its height. Find the area of a triangle with a height of 6 inches, if the area is 10 square inches when the height is 4 inches. 15 square inches

75. Over a specified distance, speed varies inversely with time. If a car goes a certain distance in one-half hour at 30 miles per hour, what speed is needed to go the same distance in three-fourths of an hour?
20 miles per hour

76. For a constant area, the length of a rectangle varies inversely as the width. The length of a rectangle is 27 feet when the width is 10 feet. Find the length of a rectangle with the same area if the width is 18 feet.
15 feet

77. The weight of an object on the moon varies directly as the weight of the object on Earth. According to *The Guiness Book of World Records,* "Shad," a goat owned by a couple in California, is the largest known goat, weighing 352 pounds. Shad would weigh about 59 pounds on the moon. A bull moose weighing 1,800 pounds was shot in Canada and is the largest confirmed moose. How much would the moose have weighed on the moon?
about 302 pounds

78. According to *The Guiness Book of World Records,* the longest recorded voyage in a paddle boat is 2,226 miles in 103 days by the foot power of two boaters down the Mississippi River. Assuming a constant rate, how far would they have gone if they had traveled 120 days? (Distance varies directly as time.) **about 2,593 miles**

79. The pressure exerted by a certain liquid at a given point varies directly as the depth of the point beneath the surface of the liquid. The pressure at a depth of 10 feet is 50 pounds per square inch. What is the pressure at a depth of 20 feet?
100 pounds per square inch

80. If the volume is constant, the pressure of gas in a container varies directly as the temperature. If the pressure is 5 pounds per square inch at a temperature of 200 Kelvin, what is the pressure at a temperature of 300 Kelvin?
$7\frac{1}{2}$ **pounds per square inch**

81. If the temperature is constant, the pressure of a gas in a container varies inversely as the volume of the container. If the pressure is 10 pounds per square foot in a container with 3 cubic feet, what is the pressure in a container with 1.5 cubic feet?
20 pounds per square foot

82. The force required to compress a spring varies directly as the change in the length of the spring. If a force of 12 pounds is required to compress a certain spring 3 inches, how much force is required to compress the spring 5 inches? **20 pounds**

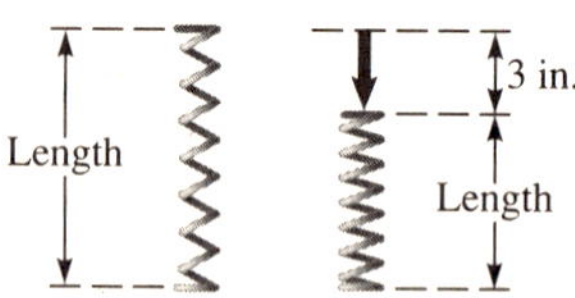

83. For a body falling freely from rest (disregarding air resistance), the distance the body falls varies directly as the square of the time. If an object is dropped from the top of a tower 400 feet high and hits the ground in 5 seconds, how far did it fall in the first 3 seconds? **144 feet**

84. The illumination produced by a light source varies inversely as the square of the distance from the source. If the illumination produced 4 feet from a light source is 75 foot-candles, find the illumination produced 9 feet from the same source.
$\frac{400}{27}$ **or** $14\frac{22}{27}$ **foot-candles**

7.4 Exercises (Text Page 364)

Write each of the following sets using interval notation.

1. $\{x \mid x < 3\}$ $(-\infty, 3)$ **2.** $\{x \mid x < -2\}$ $(-\infty, -2)$ **3.** $\{y \mid y \le 8\}$ $(-\infty, 8]$

4. $\{t \mid t \le 1\}$ $(-\infty, 1]$ **5.** $\{r \mid r > 8\}$ $(8, \infty)$ **6.** $\{s \mid s > 4\}$ $(4, \infty)$

7. $\{w \mid w \ge 2\}$ $[2, \infty)$ **8.** $\{m \mid m \ge 0\}$ $[0, \infty)$ **9.** $\{x \mid -3 \le x \le 4\}$ $[-3, 4]$

10. $\{x \mid -2 < x \le 0\}$ $(-2, 0]$ **11.** $\{x \mid 2 < x \le 9\}$ $(2, 9]$ **12.** $\{t \mid -1 < t < 1\}$ $(-1, 1)$

Using the variable x, write each of the following intervals as an inequality with set-builder notation.

13. $(-\infty, 4)$ $\{x \mid x < 4\}$ **14.** $(-\infty, 0)$ $\{x \mid x < 0\}$ **15.** $[1.5, \infty)$ $\{x \mid x \ge 1.5\}$

16. $[3.2, \infty)$ $\{x \mid x \geq 3.2\}$

17. $[-3, 10)$ $\{x \mid -3 \leq x < 10\}$

18. $(-2, 9]$ $\{x \mid -2 < x \leq 9\}$

19. $\{x \mid -4 \leq x \leq -2\}$

20. $\{x \mid x < -1\}$

21. Match each set given in interval notation with its description.

(a)	$(0, \infty)$	**A.**	positive real numbers
(b)	$[0, \infty)$	**B.**	negative real numbers
(c)	$(-\infty, 0]$	**C.**	nonpositive real numbers
(d)	$(-\infty, 0)$	**D.**	nonnegative real numbers

(a) A (b) D (c) C (d) B

22. Explain how to determine whether a parenthesis or a square bracket is used when graphing the solution set of a linear inequality.

Solve each inequality. Give the solution set in both interval and graph forms.

23. $4x > 8$ $(2, \infty)$

24. $6y > 18$ $(3, \infty)$

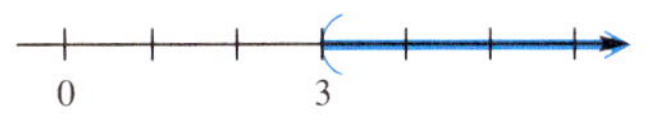

25. $2m \leq -6$ $(-\infty, -3]$

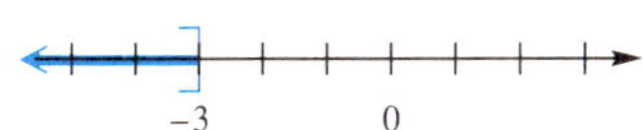

26. $5k \leq -10$ $(-\infty, -2]$

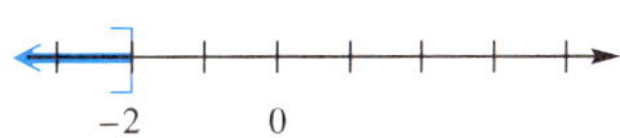

27. $3r + 1 \geq 16$ $[5, \infty)$

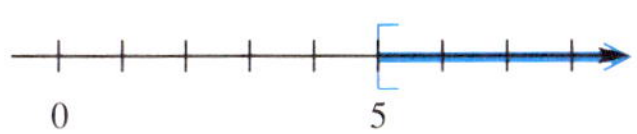

28. $2m - 5 \geq 15$ $[10, \infty)$

29. $-r \leq -7$ $[7, \infty)$

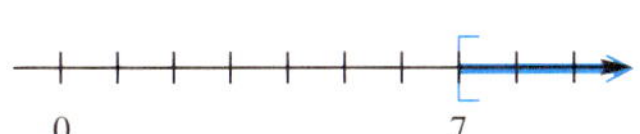

30. $-m > -12$ $(-\infty, 12)$

31. $-4x + 3 < 15$ $(-3, \infty)$

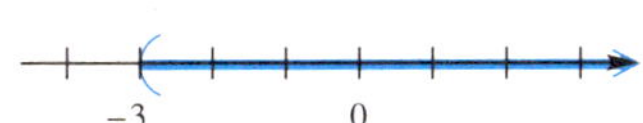

32. $-6p - 2 \geq 16$ $(-\infty, -3]$

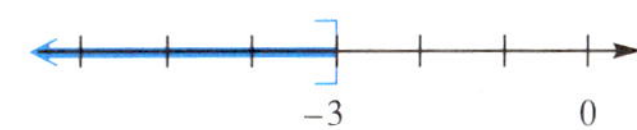

33. $-3(z - 6) > 2z - 5$ $(-\infty, 23/5)$

34. $-2(y + 4) \leq 6y + 8$ $[-2, \infty)$

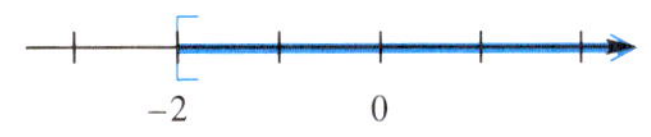

35. $-2(m - 4) \leq -3(m + 1)$ $(-\infty, -11]$

36. $-(9 + k) - 5 + 4k \geq 1$ $[5, \infty)$

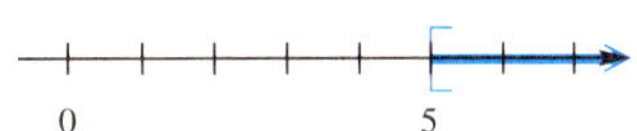

37. $-3 < x - 5 < 6$ $(2, 11)$

38. $-6 < x + 1 < 8$ $(-7, 7)$

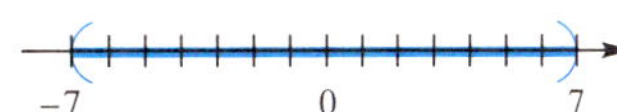

39. $-19 \leq 3x - 5 \leq -9$ $[-14/3, -4/3]$

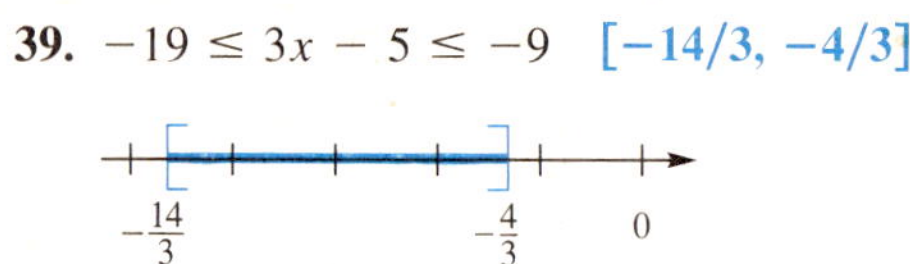

40. $-16 < 3t + 2 < -11$ $(-6, -13/3)$

-6 $\quad -\frac{13}{3}$ $\quad 0$

41. $-4 \leq \dfrac{2x - 5}{6} \leq 5$ $[-19/2, 35/2]$

$-\frac{19}{2}$ $\quad 0$ $\quad 5$ $\quad \frac{35}{2}$

42. $-8 \leq \dfrac{3m + 1}{4} \leq 3$ $[-11, 11/3]$

-11 $\quad -4$ $\quad 0$ $\quad \frac{11}{3}$

Answer the questions in Exercises 43–46 based on the given graph.

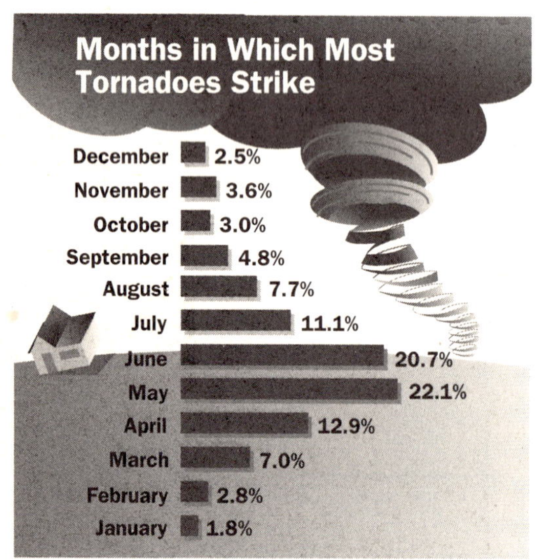

Source: The USA Today Weather Book

43. In which months did the percent of tornadoes exceed 7.7%? **April, May, June, July**

44. In which months was the percent of tornadoes at least 12.9%? **April, May, June**

45. The data used to determine the graph was based on the number of tornadoes sighted in the United States during the last twenty years. A total of 17,252 tornadoes were reported. In which months were fewer than 1,500 reported?
January, February, March, August, September, October, November, December

46. How many more tornadoes occurred during March than October? (Use the total given in Exercise 45.)
690

Solve the problem.

47. According to information provided by the Taxi and Limousine Commission, in January 1990, taxicabs in New York charged \$1.50 for the first 1/5 mile and \$.25 for each additional 1/5 mile. Amos has only \$3.75 in his pocket. What is the maximum distance he can travel (not including a tip for the cabbie)?
2 miles

48. In July 1979, New York taxicab fares were \$.90 for the first 1/7 mile and \$.10 for each additional 1/7 mile. Based on the information given in Exercise 47 and the answer you found, how much farther could Amos have traveled in 1979 than in 1990?
approximately 2 miles farther

49. Clare Lynch earned scores of 90 and 82 on her first two tests in English Literature. What score must she make on her third test to keep an average of 84 or greater? **at least 80**

50. Jay O'Callaghan scored 92 and 96 on his first two tests in Methods in Teaching Mathematics. What score must he make on his third test to keep an average of 90 or greater? **at least 82**

51. A couple wishes to rent a car for one day while on vacation. Ford Automobile Rental wants \$15.00 per day and 14¢ per mile, while Chevrolet-For-A-Day wants \$14.00 per day and 16¢ per mile. After how many miles would the price to rent the Chevrolet exceed the price to rent a Ford? **50 miles**

52. Jane and Terry Brandsma went to Long Island for a week. They needed to rent a car, so they checked out two rental firms. Avis wanted \$28 per day, with no mileage fee. Downtown Toyota wanted \$108 per week and 14¢ per mile. How many miles would they have to drive before the Avis price is less than the Toyota price? 628.6 miles

A product will produce a profit only when the revenue R from selling the product exceeds the cost C of producing it. In Exercises 53 and 54 find the smallest whole number of units x that must be sold for the business to show a profit for the item described.

53. Peripheral Visions, Inc. finds that the cost to produce x studio quality videotapes is $C = 20x + 100$, while the revenue produced from them is $R = 24x$ (C and R in dollars). 26 tapes

54. Speedy Delivery finds that the cost to make x deliveries is $C = 3x + 2{,}300$, while the revenue produced from them is $R = 5.50x$ (C and R in dollars). 921 deliveries

55. Write an explanation of what is wrong with the following argument.

Let a and b be numbers, with $a > b$. Certainly, $2 > 1$. Multiply both sides of the inequality by $b - a$.

$$2(b - a) > 1(b - a)$$
$$2b - 2a > b - a$$
$$2b - b > 2a - a$$
$$b > a$$

But the final inequality is impossible, since we know that $a > b$ from the given information.

56. Assume that $0 < a < b$, and go through the following steps.

$$a < b$$
$$ab < b^2$$
$$ab - b^2 < 0$$
$$b(a - b) < 0$$

Divide both sides by $a - b$ to get $b < 0$. This implies that b is negative. We originally assumed that b is positive. What is wrong with this argument?

7.5 Exercises (Text Page 375)

Evaluate each exponential expression.

1. 5^4 625 **2.** 10^3 1,000 **3.** $\left(\frac{5}{3}\right)^2$ $\frac{25}{9}$ **4.** $\left(\frac{3}{5}\right)^3$ $\frac{27}{125}$ **5.** $(-2)^5$ -32

6. $(-5)^4$ 625 **7.** -2^3 -8 **8.** -3^2 -9 **9.** $-(-3)^4$ -81 **10.** $-(-5)^2$ -25

11. Do $(-a)^n$ and $-a^n$ mean the same thing? Explain.

12. In *some* cases, $-a^n$ and $(-a)^n$ do give the same result. Using $a = 2$, and $n = 2, 3, 4$, and 5, draw a conclusion as to when they are equal and when they are opposites.

13. Which one of the following is equal to 1 ($a \neq 0$)?
(a) $3a^0$ **(b)** $-3a^0$ **(c)** $(3a)^0$ **(d)** $3(-a)^0$ (c)

14. Which one of the following represents a negative number?
(a) $(-3)^{-2}$ **(b)** $(-1{,}000)^0$ **(c)** $(-4)^0 - (-3)^0$ **(d)** $(-5)^{-3}$ (d)

Evaluate each exponential expression.

15. 7^{-2} $\frac{1}{49}$ **16.** 4^{-1} $\frac{1}{4}$ **17.** -7^{-2} $-\frac{1}{49}$ **18.** -4^{-1} $-\frac{1}{4}$ **19.** $\frac{2}{(-4)^{-3}}$ -128

20. $\frac{2^{-3}}{3^{-2}}$ $\frac{9}{8}$ **21.** $\frac{5^{-1}}{4^{-2}}$ $\frac{16}{5}$ **22.** $\left(\frac{1}{2}\right)^{-3}$ 8 **23.** $\left(\frac{1}{5}\right)^{-3}$ 125 **24.** $\left(\frac{2}{3}\right)^{-2}$ $\frac{9}{4}$

25. $\left(\frac{4}{5}\right)^{-2}$ $\frac{25}{16}$ **26.** $3^{-1} + 2^{-1}$ $\frac{5}{6}$ **27.** $4^{-1} + 5^{-1}$ $\frac{9}{20}$ **28.** 8^0 1 **29.** 12^0 1

30. $(-23)^0$ 1 **31.** $(-4)^0$ 1 **32.** -2^0 -1 **33.** $3^0 - 4^0$ 0 **34.** $-8^0 - 7^0$ -2

35. In order to raise a fraction to a negative power, we may change the fraction to its _____ and change the exponent to the _____ _____ of the original exponent.
reciprocal; additive inverse

36. Explain in your own words how we raise a power to a power.

37. Which one of the following is correct?

(a) $-\frac{3}{4} = \left(\frac{3}{4}\right)^{-1}$ **(b)** $\frac{3^{-1}}{4^{-1}} = \left(\frac{4}{3}\right)^{-1}$ **(c)** $\frac{3^{-1}}{4} = \frac{3}{4^{-1}}$ **(d)** $\frac{3^{-1}}{4^{-1}} = \left(\frac{3}{4}\right)^{-1}$ (d)

38. Which one of the following is incorrect?

(a) $(3r)^{-2} = 3^{-2}r^{-2}$ **(b)** $3r^{-2} = (3r)^{-2}$ **(c)** $(3r)^{-2} = \frac{1}{(3r)^2}$ **(d)** $(3r)^{-2} = \frac{r^{-2}}{9}$ (b)

Use the product, quotient, and power rules to simplify each expression. Write answers with only positive exponents. Assume that all variables represent nonzero real numbers.

39. $x^{12} \cdot x^4$ x^{16}

40. $\frac{x^{12}}{x^4}$ x^8

41. $\frac{5^{17}}{5^{16}}$ 5

42. $\frac{3^{12}}{3^{13}}$ $\frac{1}{3}$

43. $\frac{3^{-5}}{3^{-2}}$ $\frac{1}{27}$

44. $\frac{2^{-4}}{2^{-3}}$ $\frac{1}{2}$

45. $\frac{9^{-1}}{9}$ $\frac{1}{81}$

46. $\frac{12}{12^{-1}}$ 144

47. t^5t^{-12} $\frac{1}{t^7}$

48. p^5p^{-6} $\frac{1}{p}$

49. $(3x)^2$ $9x^2$

50. $(-2x^{-2})^2$ $\frac{4}{x^4}$

51. $a^{-3}a^2a^{-4}$ $\frac{1}{a^5}$

52. $k^{-5}k^{-3}k^4$ $\frac{1}{k^4}$

53. $\frac{x^7}{x^{-4}}$ x^{11}

54. $\frac{p^{-3}}{p^5}$ $\frac{1}{p^8}$

55. $\frac{r^3r^{-4}}{r^{-2}r^{-5}}$ r^6

56. $\frac{z^{-4}z^{-2}}{z^3z^{-1}}$ $\frac{1}{z^8}$

57. $7k^2(-2k)(4k^{-5})$ $-\frac{56}{k^2}$

58. $3a^2(-5a^{-6})(-2a)$ $\frac{30}{a^3}$

59. $(z^3)^{-2}z^2$ $\frac{1}{z^4}$

60. $(p^{-1})^3p^{-4}$ $\frac{1}{p^7}$

61. $-3r^{-1}(r^{-3})^2$ $-\frac{3}{r^7}$

62. $2(y^{-3})^4(y^6)$ $\frac{2}{y^6}$

63. $(3a^{-2})^3(a^3)^{-4}$ $\frac{27}{a^{18}}$

64. $(m^5)^{-2}(3m^{-2})^3$ $\frac{27}{m^{16}}$

65. $(x^{-5}y^2)^{-1}$ $\frac{x^5}{y^2}$

66. $(a^{-3}b^{-5})^2$ $\frac{1}{a^6b^{10}}$

67. Which one of the following does *not* represent the reciprocal of x ($x \neq 0$)?

(a) x^{-1} **(b)** $\frac{1}{x}$ **(c)** $\left(\frac{1}{x^{-1}}\right)^{-1}$ **(d)** $-x$ (d)

68. Which one of the following is *not* in scientific notation?

(a) 6.02×10^{23} **(b)** 14×10^{-6} **(c)** 1.4×10^{-5} **(d)** 3.8×10^3 (b)

Write each number in scientific notation.

69. 230 2.3×10^2

70. 46,500 4.65×10^4

71. .02 2×10^{-2}

72. .0051 5.1×10^{-3}

Write each number without scientific notation.

73. 6.5×10^3 6,500

74. 2.317×10^5 231,700

75. 1.52×10^{-2} .0152

76. 1.63×10^{-4} .000163

Use scientific notation to perform each of the following computations. Leave the answers in scientific notation.

77. $\dfrac{.002 \times 3{,}900}{.000013}$ 6×10^5

78. $\dfrac{.009 \times 600}{.02}$ 2.7×10^2

79. $\dfrac{.0004 \times 56{,}000}{.000112}$ 2×10^5

80. $\dfrac{.018 \times 20{,}000}{300 \times .0004}$ 3×10^3

81. $\dfrac{840{,}000 \times .03}{.00021 \times 600}$ 2×10^5

82. $\dfrac{28 \times .0045}{140 \times 1{,}500}$ 6×10^{-7}

If the number in the statement is written in scientific notation, write it without exponents. if it is written without exponents, write it in scientific notation.

83. The number of possible hands in contract bridge is about 6.35×10^{11}. 635,000,000,000

84. If there are 40 numbers to choose from in a lottery, and a player must choose 6 different ones, the player has about 3.84×10^6 ways to make a choice. 3,840,000

85. In a recent year, ESPN's regular season baseball telecasts averaged slightly fewer than 1,150,000 viewers. 1.15×10^6

86. In a recent year, ESPN's estimated losses were at least 36,000,000 dollars. 3.6×10^7

87. The body of a 150-pound person contains about 2.3×10^{-4} pound of copper and about 6×10^{-3} pound of iron. 1.5×10^2; .00023; .006

88. The mean distance from Venus to the sun is about 6.7×10^6 miles. 6,700,000

The quote is taken from the source cited. Write the number given in scientific notation in the quote without exponents.

89. The muon, a close relative of the electron produced by the bombardment of cosmic rays against the upper atmosphere, has a half-life of 2 millionths of a second (2×10^{-6} s). (Excerpt from *Conceptual Physics*, 6th Edition by Paul G. Hewitt. Copyright © by Paul G. Hewitt. Published by HarperCollins College Publishers.) .000002

90. There are 13 red balls and 39 black balls in a box. Mix them up and draw 13 out one at a time without returning any ball . . . the probability that the 13 drawings each will produce a red ball is . . . 1.6×10^{-12}. (Warren Weaver, *Lady Luck*, New York: Doubleday & Company, Inc., 1963, pp. 298–299.) .0000000000016

The quote is taken from the source cited. Write the given number(s) in scientific notation.

91. An electron and a positron attract each other in two ways: the electromagnetic attraction of their opposite electric charges, and the gravitational attraction of their two masses. The electromagnetic attraction is 4,200,000,000,000,000,000,000,000,000,000,000,000,000,000,000 times as strong as the gravitational. (Isaac Asimov, *Isaac Asimov's Book of Facts*, New York: Bell Publishing Company, 1981, p. 106.) 4.2×10^{42}

92. How is it that the average CEO in Japan receives an income of $300,000, while the average CEO in the United States earns $2.8 million? (Andrew Zimbalist, *Baseball and Billions*. New York: BasicBooks, 1992, p. 78.) 3×10^5; 2.8×10^6

Solve the problem.

93. The distance to Earth from the planet Pluto is 4.58×10^9 kilometers. In April 1983, Pioneer 10 transmitted radio signals from Pluto to Earth at the speed of light, 3.00×10^5 kilometers per second. How long (in seconds) did it take for the signals to reach Earth? about 15,300 seconds

94. In Exercise 93, how many hours did it take for the signals to reach Earth? about 4.24 hours

95. A *light-year* is the distance that light travels in one year. Find the number of miles in a light-year if light travels 1.86×10^5 miles per second. 5.87×10^{12} miles

96. The planet Mercury has a mean distance from the sun of 3.6×10^7 miles, while the mean distance of Venus from the sun is 6.7×10^7 miles. How long would it take a spacecraft traveling at 1.55×10^3 miles per hour to travel the distance represented by the difference of these two planets' mean distances from the sun? 2×10^4 hours

97. When the distance between the centers of the moon and the earth is 4.60×10^8 meters, an object on the line joining the centers of the moon and the earth exerts the same gravitational force on each when it is 4.14×10^8 meters from the center of the earth. How far is the object from the center of the moon at that point? **4.6×10^7 meters**

98. Assume that the volume of the earth is 5×10^{14} cubic meters and that the volume of a bacterium is 2.5×10^{-16} cubic meter. If the earth could be packed full of bacteria, how many would it contain? **2×10^{30} bacteria**

99. The graph shows the number of engineers employed in the United States during the years 1984–1993. (*Source:* Engineering Workforce Commission)

(a) Use the graph to approximate the number of engineers employed during 1992 and write the number in scientific notation.

(b) Suppose the average annual salary for an engineer in 1992 was $\$5.5 \times 10^4$. What would be the total of all salaries of engineers in that year? (This is hypothetical and not based on actual data.)

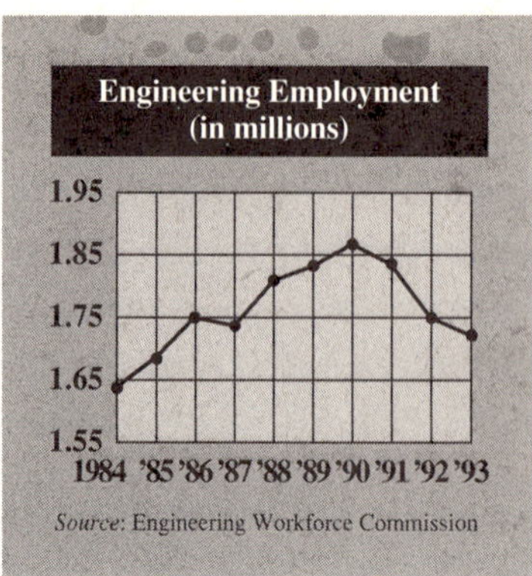

(a) 1.75×10^6 (b) about $\$9.63 \times 10^{10}$

100. In 1992, the state of Texas had about 1.3×10^6 farms with an average of 7.1×10^2 acres per farm. What was the total number of acres devoted to farmland in Texas that year? (*Source:* National Agricultural Statistics Service, U.S. Dept. of Agriculture). **about 9.2×10^8 acres**

7.6 Exercises (Text Page 385)

Find each of the following sums and differences.

1. $(3x^2 - 4x + 5) + (-2x^2 + 3x - 2)$ $x^2 - x + 3$

2. $(4m^3 - 3m^2 + 5) + (-3m^3 - m^2 + 5)$ $m^3 - 4m^2 + 10$

3. $(12y^2 - 8y + 6) - (3y^2 - 4y + 2)$ $9y^2 - 4y + 4$

4. $(8p^2 - 5p) - (3p^2 - 2p + 4)$ $5p^2 - 3p - 4$

5. $(6m^4 - 3m^2 + m) - (2m^3 + 5m^2 + 4m) + (m^2 - m)$ $6m^4 - 2m^3 - 7m^2 - 4m$

6. $-(8x^3 + x - 3) + (2x^3 + x^2) - (4x^2 + 3x - 1)$ $-6x^3 - 3x^2 - 4x + 4$

7. $5(2x^2 - 3x + 7) - 2(6x^2 - x + 12)$ $-2x^2 - 13x + 11$

8. $8x^2y - 3xy^2 + 2x^2y - 9xy^2$ $10x^2y - 12xy^2$

Find each of the following products.

9. $(x + 3)(x - 8)$ $x^2 - 5x - 24$

10. $(y - 3)(y - 9)$ $y^2 - 12y + 27$

11. $(4r - 1)(7r + 2)$ $28r^2 + r - 2$

12. $(5m - 6)(3m + 4)$ $15m^2 + 2m - 24$

13. $4x^2(3x^3 + 2x^2 - 5x + 1)$ $12x^5 + 8x^4 - 20x^3 + 4x^2$

14. $2b^3(b^2 - 4b + 3)$ $2b^5 - 8b^4 + 6b^3$

15. $(2m + 3)(2m - 3)$ $4m^2 - 9$

16. $(8s - 3t)(8s + 3t)$ $64s^2 - 9t^2$

17. $(4m + 2n)^2$ $16m^2 + 16mn + 4n^2$

18. $(a - 6b)^2$ $a^2 - 12ab + 36b^2$

19. $(5r + 3t^2)^2$ $25r^2 + 30rt^2 + 9t^4$

20. $(2z^4 - 3y)^2$ $4z^8 - 12z^4y + 9y^2$

21. $(2z - 1)(-z^2 + 3z - 4)$ $-2z^3 + 7z^2 - 11z + 4$

22. $(k + 2)(12k^3 - 3k^2 + k + 1)$ $12k^4 + 21k^3 - 5k^2 + 3k + 2$

23. $(m - n + k)(m + 2n - 3k)$ $m^2 + mn - 2n^2 - 2km + 5kn - 3k^2$

24. $(r - 3s + t)(2r - s + t)$ $2r^2 - 7rs + 3s^2 + 3rt - 4st + t^2$

25. $(a - b + 2c)^2$ $a^2 - 2ab + b^2 + 4ac - 4bc + 4c^2$

26. $(k - y + 3m)^2$ $k^2 - 2ky + y^2 + 6km - 6ym + 9m^2$

27. Which one of the following is a trinomial in descending powers, having degree 6?
(a) $5x^6 - 4x^5 + 12$ **(b)** $6x^5 - x^6 + 4$ **(c)** $2x + 4x^2 - x^6$
(d) $4x^6 - 6x^4 + 9x + 1$ (a)

28. Give an example of a polynomial of four terms in the variable x, having degree 5, written in descending powers, lacking a fourth degree term.
One example is $12x^5 - 6x^3 + 2x^2 - 1$.

29. The exponent in the expression 6^3 is 3. Explain why the degree of 6^3 is not 3. What is its degree?

30. Explain in your own words how to square a binomial.

Factor the greatest common factor from each polynomial.

31. $8m^4 + 6m^3 - 12m^2$ $2m^2(4m^2 + 3m - 6)$

32. $2p^5 - 10p^4 + 16p^3$ $2p^3(p^2 - 5p + 8)$

33. $4k^2m^3 + 8k^4m^3 - 12k^2m^4$ $4k^2m^3(1 + 2k^2 - 3m)$

34. $28r^4s^2 + 7r^3s - 35r^4s^3$ $7r^3s(4rs + 1 - 5rs^2)$

35. $2(a + b) + 4m(a + b)$ $2(a + b)(1 + 2m)$

36. $4(y - 2)^2 + 3(y - 2)$ $(y - 2)(4y - 5)$

37. $2(m - 1) - 3(m - 1)^2 + 2(m - 1)^3$ $(m - 1)(2m^2 - 7m + 7)$

38. $5(a + 3)^3 - 2(a + 3) + (a + 3)^2$ $(a + 3)(5a^2 + 31a + 46)$

Factor each of the following polynomials by grouping.

39. $6st + 9t - 10s - 15$ $(2s + 3)(3t - 5)$

40. $10ab - 6b + 35a - 21$ $(5a - 3)(2b + 7)$

41. $rt^3 + rs^2 - pt^3 - ps^2$ $(t^3 + s^2)(r - p)$

42. $2m^4 + 6 - am^4 - 3a$ $(m^4 + 3)(2 - a)$

43. $16a^2 + 10ab - 24ab - 15b^2$ $(8a + 5b)(2a - 3b)$

44. $15 - 5m^2 - 3r^2 + m^2r^2$ $(3 - m^2)(5 - r^2)$

45. $20z^2 - 8zx - 45zx + 18x^2$ $(5z - 2x)(4z - 9x)$

46. Consider the polynomial $1 - a + ab - b$. One acceptable factored form is $(1 - a)(1 - b)$. However there are other acceptable factored forms. Which one is *not* a factored form of this polynomial?
(a) $(a - 1)(b - 1)$ **(b)** $(-a + 1)(-b + 1)$ **(c)** $(-1 + a)(-1 + b)$
(d) $(1 - a)(b + 1)$ (d)

Factor each trinomial.

47. $x^2 - 2x - 15$ $(x - 5)(x + 3)$

48. $r^2 + 8r + 12$ $(r + 6)(r + 2)$

49. $y^2 + 2y - 35$ $(y + 7)(y - 5)$

50. $x^2 - 7x + 6$ $(x - 6)(x - 1)$

51. $6a^2 - 48a - 120$ $6(a - 10)(a + 2)$

52. $8h^2 - 24h - 320$ $8(h - 8)(h + 5)$

53. $3m^3 + 12m^2 + 9m$ $3m(m + 1)(m + 3)$

54. $9y^4 - 54y^3 + 45y^2$ $9y^2(y - 5)(y - 1)$

55. $6k^2 + 5kp - 6p^2$ $(3k - 2p)(2k + 3p)$

56. $14m^2 + 11mr - 15r^2$ $(7m - 5r)(2m + 3r)$

57. $5a^2 - 7ab - 6b^2$ $(5a + 3b)(a - 2b)$

58. $12s^2 + 11st - 5t^2$ $(4s + 5t)(3s - t)$

59. $9x^2 - 6x^3 + x^4$ $x^2(3 - x)^2$

60. $30a^2 + am - m^2$ $(5a + m)(6a - m)$

61. $24a^4 + 10a^3b - 4a^2b^2$ $2a^2(4a - b)(3a + 2b)$

62. $18x^5 + 15x^4z - 75x^3z^2$ $3x^3(3x - 5z)(2x + 5z)$

63. When a student was given the polynomial $4x^2 + 2x - 20$ to factor completely on a test, she lost some credit by giving the answer $(4x + 10)(x - 2)$. She complained to her teacher that the product $(4x + 10)(x - 2)$ is indeed $4x^2 + 2x - 20$. Do you think that the teacher was justified in not giving her full credit? Explain.

64. Write an explanation as to why most people would find it more difficult to factor $36x^2 - 44x - 15$ than $37x^2 - 183x - 10$.

Factor each, using the method for factoring a perfect square trinomial. It may be necessary to factor out a common factor first.

65. $9m^2 - 12m + 4$ $(3m - 2)^2$

66. $16p^2 - 40p + 25$ $(4p - 5)^2$

67. $32a^2 - 48ab + 18b^2$ $2(4a - 3b)^2$

68. $20p^2 - 100pq + 125q^2$ $5(2p - 5q)^2$

69. $4x^2y^2 + 28xy + 49$ $(2xy + 7)^2$

70. $9m^2n^2 - 12mn + 4$ $(3mn - 2)^2$

Factor each difference of two squares.

71. $x^2 - 36$ $(x + 6)(x - 6)$

72. $t^2 - 64$ $(t + 8)(t - 8)$

73. $y^2 - w^2$ $(y + w)(y - w)$

74. $25 - w^2$ $(5 + w)(5 - w)$

75. $9a^2 - 16$ $(3a + 4)(3a - 4)$

76. $16q^2 - 25$ $(4q + 5)(4q - 5)$

77. $25s^4 - 9t^2$ $(5s^2 + 3t)(5s^2 - 3t)$

78. $36z^2 - 81y^4$ $9(2z + 3y^2)(2z - 3y^2)$

79. $p^4 - 625$ $(p^2 + 25)(p + 5)(p - 5)$

80. $m^4 - 81$ $(m^2 + 9)(m + 3)(m - 3)$

Factor each sum or difference of cubes.

81. $8 - a^3$ $(2 - a)(4 + 2a + a^2)$

82. $r^3 + 27$ $(r + 3)(r^2 - 3r + 9)$

83. $125x^3 - 27$ $(5x - 3)(25x^2 + 15x + 9)$

84. $8m^3 - 27n^3$ $(2m - 3n)(4m^2 + 6mn + 9n^2)$

85. $27y^9 + 125z^6$ $(3y^3 + 5z^2)(9y^6 - 15y^3z^2 + 25z^4)$

86. $27z^3 + 729y^3$ $27(z + 3y)(z^2 - 3zy + 9y^2)$

Each of the following may be factored using one of the methods described in this section. Decide on the method, and then factor the polynomial completely.

87. $x^2 + xy - 5x - 5y$ $(x + y)(x - 5)$

88. $8r^2 - 10rs - 3s^2$ $(4r + s)(2r - 3s)$

89. $p^4(m - 2n) + q(m - 2n)$ $(m - 2n)(p^4 + q)$

90. $36a^2 + 60a + 25$ $(6a + 5)^2$

91. $4z^2 + 28z + 49$ $(2z + 7)^2$

92. $6p^4 + 7p^2 - 3$ $(3p^2 - 1)(2p^2 + 3)$

93. $1{,}000x^3 + 343y^3$
$(10x + 7y)(100x^2 - 70xy + 49y^2)$

94. $b^2 + 8b + 16 - a^2$
$(b + 4 + a)(b + 4 - a)$

95. $125m^6 - 216$ $(5m^2 - 6)(25m^4 + 30m^2 + 36)$

96. $q^2 + 6q + 9 - p^2$ $(q + 3 + p)(q + 3 - p)$

97. $12m^2 + 16mn - 35n^2$ $(6m - 7n)(2m + 5n)$

98. $216p^3 + 125q^3$ $(6p + 5q)(36p^2 - 30pq + 25q^2)$

99. The sum of two squares usually cannot be factored. For example, $x^2 + y^2$ is prime. Notice that

$$x^2 + y^2 \neq (x + y)(x + y).$$

By choosing $x = 4$ and $y = 2$, show that the above inequality is true.
$4^2 + 2^2 = 20$ and $(4 + 2)(4 + 2) = 36$; $20 \neq 36$

100. The binomial $9x^2 + 36$ is a sum of two squares. Can it be factored? If so, factor it.
yes; $9(x^2 + 4)$

CONCEPTUAL  WRITING ▲ CHALLENGING SCIENTIFIC CALCULATOR GRAPHING CALCULATOR

7.7 Exercises (Text Page 393)

Solve each of the following equations by the zero-factor property.

1. $(x + 3)(x - 9) = 0$ $\{-3, 9\}$

2. $(m + 6)(m + 4) = 0$ $\{-6, -4\}$

3. $(2t - 7)(5t + 1) = 0$ $\left\{\frac{7}{2}, -\frac{1}{5}\right\}$

4. $(7y - 3)(6y + 4) = 0$ $\left\{\frac{3}{7}, -\frac{2}{3}\right\}$

5. $x^2 - x - 12 = 0$ $\{-3, 4\}$

6. $m^2 + 4m - 5 = 0$ $\{-5, 1\}$

7. $y^2 + 9y + 14 = 0$ $\{-7, -2\}$

8. $15r^2 + 7r = 2$ $\left\{-\frac{2}{3}, \frac{1}{5}\right\}$

9. $12x^2 + 4x = 1$ $\left\{-\frac{1}{2}, \frac{1}{6}\right\}$

10. $x(x + 3) = 4$ $\{-4, 1\}$

11. $(x + 4)(x - 6) = -16$ $\{-2, 4\}$

12. $(w - 1)(3w + 2) = 4w$ $\left\{-\frac{1}{3}, 2\right\}$

13. $(r - 5)(r - 3) = 3r(r - 3)$ $\left\{-\frac{5}{2}, 3\right\}$

14. In trying to solve $(x + 4)(x - 1) = 1$, a student reasons that since $1 \cdot 1 = 1$, the equation is solved by solving

$$x + 4 = 1 \quad \text{or} \quad x - 1 = 1.$$

Explain the error in this reasoning. What is the correct way to solve this equation?

Solve each of the following by using the square root property. Give only real number solutions.

15. $x^2 = 64$ $\{\pm 8\}$

16. $w^2 = 16$ $\{\pm 4\}$

17. $t^2 = 7$ $\{\pm\sqrt{7}\}$

18. $p^2 = 13$ $\{\pm\sqrt{13}\}$

19. $x^2 = 24$ $\{\pm 2\sqrt{6}\}$

20. $x^2 = 48$ $\{\pm 4\sqrt{3}\}$

21. $r^2 = -5$ $\emptyset$

22. $y^2 = -10$ $\emptyset$

23. $(x - 4)^2 = 3$ $\{4 \pm \sqrt{3}\}$

24. $(x + 3)^2 = 11$ $\{-3 \pm \sqrt{11}\}$

25. $(2x - 5)^2 = 13$ $\left\{\frac{5 \pm \sqrt{13}}{2}\right\}$

26. $(4x + 1)^2 = 19$ $\left\{\frac{-1 \pm \sqrt{19}}{4}\right\}$

Solve each of the following by the quadratic formula. Give only real number solutions.

27. $4x^2 - 8x + 1 = 0$ $\left\{\frac{2 \pm \sqrt{3}}{2}\right\}$

28. $m^2 + 2m - 5 = 0$ $\{-1 \pm \sqrt{6}\}$

29. $2y^2 = 2y + 1$ $\left\{\frac{1 \pm \sqrt{3}}{2}\right\}$

30. $9r^2 + 6r = 1$ $\left\{\frac{-1 \pm \sqrt{2}}{3}\right\}$

31. $q^2 - 1 = q$ $\left\{\frac{1 \pm \sqrt{5}}{2}\right\}$

32. $2p^2 - 4p = 5$ $\left\{\frac{2 \pm \sqrt{14}}{2}\right\}$

33. $4k(k + 1) = 1$ $\left\{\frac{-1 \pm \sqrt{2}}{2}\right\}$

34. $4r(r - 1) = 19$ $\left\{\frac{1 \pm 2\sqrt{5}}{2}\right\}$

35. $(g + 2)(g - 3) = 1$ $\left\{\frac{1 \pm \sqrt{29}}{2}\right\}$

36. $(y - 5)(y + 2) = 6$
$\left\{\frac{3 \pm \sqrt{73}}{2}\right\}$

37. $m^2 - 6m = -14$
$\emptyset$

38. $y^2 = 2y - 2$
$\emptyset$

39. Can the quadratic formula be used to solve the equation $2x^2 - 5 = 0$? Explain, and solve it if the answer is yes.

40. Can the quadratic formula be used to solve the equation $4y^2 + 3y = 0$? Explain, and solve it if the answer is yes.

41. Explain why the quadratic formula cannot be used to solve the equation $2x^3 + 3x - 4 = 0$.

42. A student gave the quadratic formula incorrectly as follows: $x = -b \pm \frac{\sqrt{b^2 - 4ac}}{2a}$.
What is wrong with this?

The expression $b^2 - 4ac$, the radicand in the quadratic formula, is called the **discriminant** *of the quadratic equation $ax^2 + bx + c = 0$, $a \neq 0$. By evaluating it we can determine, without actually solving the equation, the number and nature of the solutions of the equation. Suppose that a, b, and c are integers. Then the following chart shows how the discriminant can be used to analyze the solutions.*

Discriminant	Solutions
Positive, and the square of an integer	Two different rational solutions
Positive, but not the square of an integer	Two different irrational solutions
Zero	One rational solution (a double solution)
Negative	No real solutions

In Exercises 43–48, evaluate the discriminant, and then determine whether the equation has **(a)** *two different rational solutions,* **(b)** *two different irrational solutions,* **(c)** *one rational solution (a double solution), or* **(d)** *no real solutions.*

43. $x^2 + 6x + 9 = 0$
0; (c)

44. $4x^2 + 20x + 25 = 0$
0; (c)

45. $6m^2 + 7m - 3 = 0$
121; (a)

46. $2x^2 + x - 3 = 0$
25; (a)

47. $9x^2 - 30x + 15 = 0$
360; (b)

48. $2x^2 - x + 1 = 0$
−7; (d)

Solve the problem by using a quadratic equation. Use a calculator as necessary, and round the answer to the nearest tenth.

49. The Mart Hotel in Dallas, Texas, is 400 feet high. Suppose that a ball is projected upward from the top of the Mart, and its position s in feet above the ground is given by the equation $s = -16t^2 + 45t + 400$, where t is the number of seconds elapsed. How long will it take for the ball to reach a height of 200 feet above the ground? 5.2 seconds

50. The Toronto Dominion Center in Winnipeg, Manitoba, is 407 feet high. Suppose that a ball is projected upward from the top of the Center, and its position s in feet above the ground is given by the equation $s = -16t^2 + 75t + 407$, where t is the number of seconds elapsed. How long will it take for the ball to reach a height of 450 feet above the ground? .7 second and 4.0 seconds

51. Refer to the equations in Exercises 49 and 50. Suppose that the first sentence in each problem did not give the height of the building. How could you use the equation to determine the height of the building? Find s when $t = 0$.

52. In Exercises 49 and 50, one problem has only one solution while the other has two solutions. Explain why this is so.

CONCEPTUAL    SCIENTIFIC CALCULATOR GRAPHING CALCULATOR

53. A search light moves horizontally back and forth along a wall with the distance of the light from a starting point at t minutes given by $s = 100t^2 - 300t$. How long will it take before the light returns to the starting point? 3 minutes

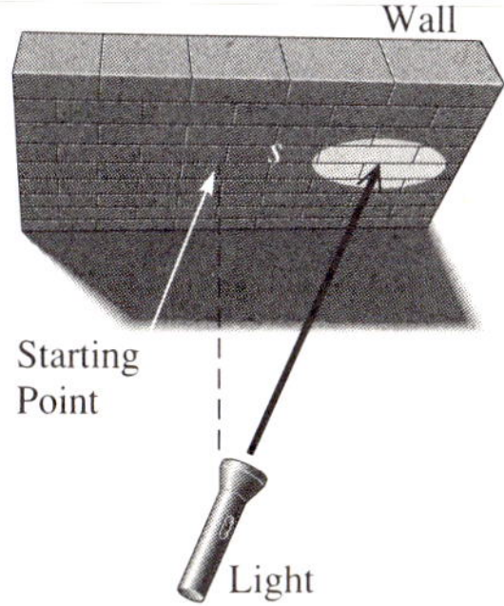

54. An object is projected directly upward from the ground. After t seconds its distance in feet above the ground is $s = 144t - 16t^2$.

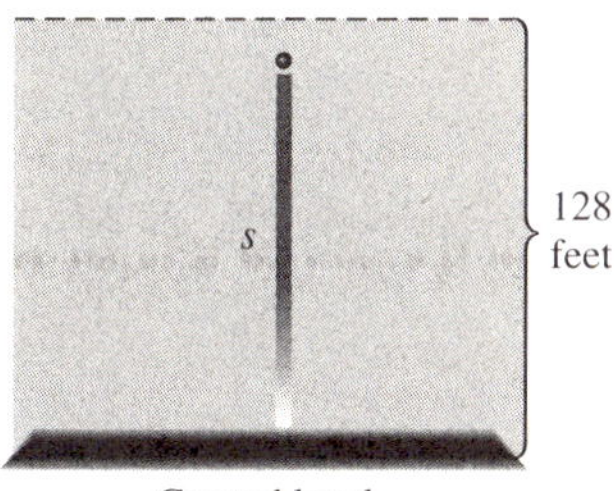

(a) After how many seconds will the object be 128 feet above the ground? (*Hint:* Look for a common factor before solving the equation.)
(b) When does the object strike the ground?
(a) 1 second and 8 seconds (b) 9 seconds after it is projected

55. The formula $D = 100t - 13t^2$ gives the distance in feet a car going approximately 68 miles per hour (initially) will skid in t seconds. Find the time it would take for the car to skid 190 feet. (*Hint:* Your answer must be less than the time it takes the car to stop, which is 3.8 seconds.) 3.4 seconds

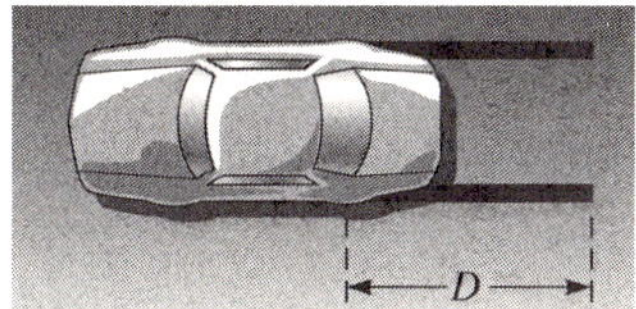

56. The formula in Exercise 55 becomes $D = 73t - 13t^2$ for a car going 50 miles per hour initially. Find the time for such a car to skid 100 feet. (*Hint:* The car will stop at about 2.8 seconds.) 2.4 seconds

57. Find the lengths of the sides of the triangle.

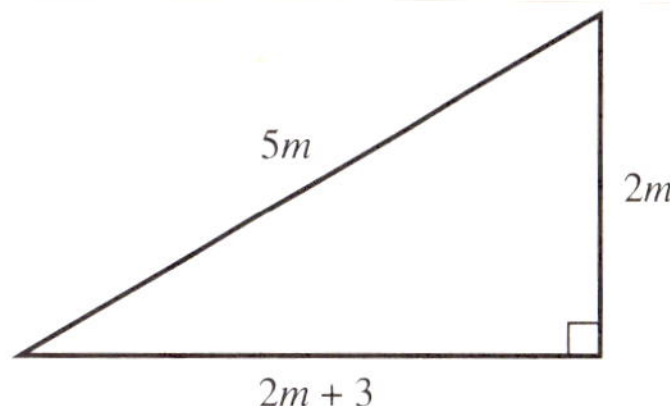

2.3, 5.3, 5.8

58. Find the lengths of the sides of the triangle.

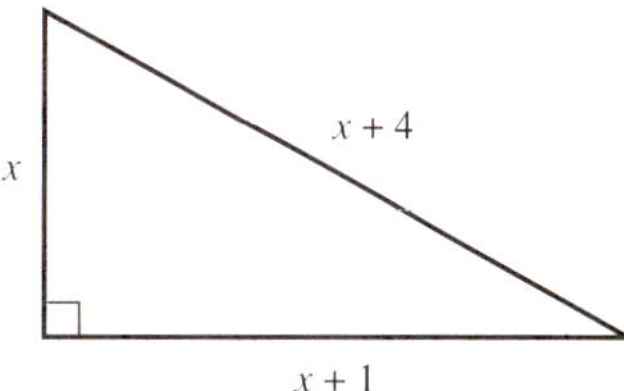

7.9, 8.9, 11.9

59. Refer to Exercise 49. Suppose that a wire is attached to the top of the Mart, and pulled tight. It is attached to the ground 100 feet from the base of the building, as shown in the figure. How long is the wire?

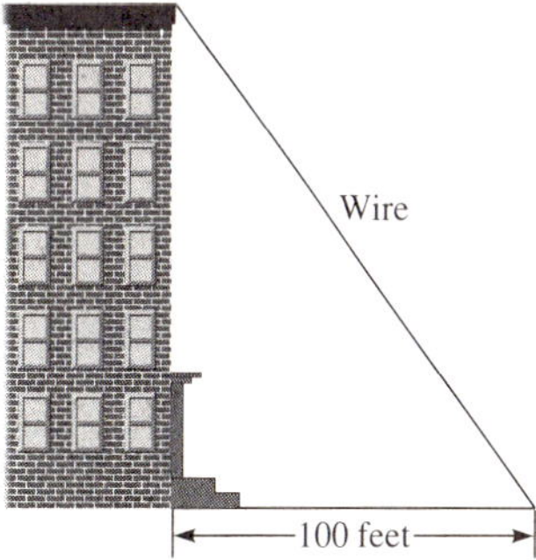

412.3 feet

60. Refer to Exercise 50. Suppose that a wire is attached to the top of the Center, and pulled tight. The length of the wire is twice the distance between the base of the Center and the point on the ground where the wire is attached. How long is the wire?
470.0 feet

61. Two ships leave port at the same time, one heading due south and the other heading due east. Several hours later, they are 170 miles apart. If the ship traveling south has traveled 70 miles farther than the other, how many miles has each traveled?

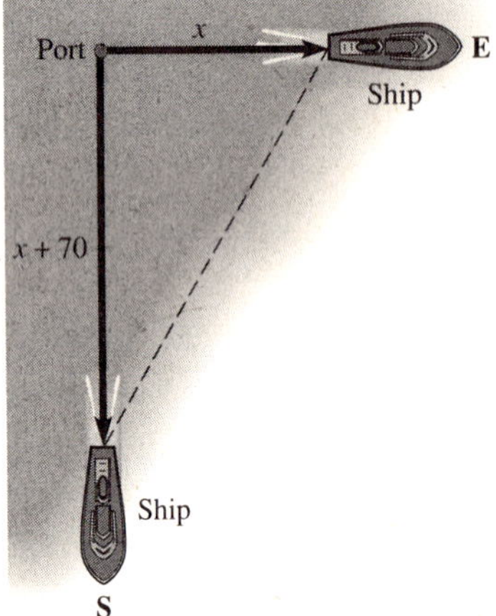

eastbound ship: 80 miles; southbound ship: 150 miles

62. Joyce Smith is flying a kite that is 30 feet farther above her hand than its horizontal distance from her. The string from her hand to the kite is 150 feet long. How high is the kite? 120 feet

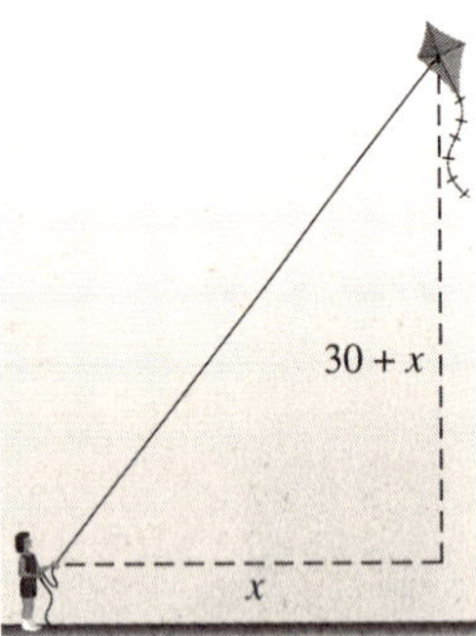

63. A toy manufacturer needs a piece of plastic in the shape of a right triangle with the longer leg 2 centimeters more than twice as long as the shorter leg, and the hypotenuse 1 centimeter more than the longer leg. How long should the three sides of the triangular piece be? 5, 12, and 13 centimeters

64. Adair Milmoe, a developer, owns a piece of land enclosed on three sides by streets, giving it the shape of a right triangle. The hypotenuse is 8 meters longer than the longer leg, and the shorter leg is 9 meters shorter than the hypotenuse. Find the lengths of the three sides of the property. 20, 21, and 29 meters

65. Two pieces of a large wooden puzzle fit together to form a rectangle with a length 1 centimeter less than twice the width. The diagonal, where the two pieces meet, is 2.5 centimeters in length. Find the length and width of the rectangle. length: 2 centimeters; width: 1.5 centimeters

66. A 13-foot ladder is leaning against a house. The distance from the bottom of the ladder to the house is 7 feet less than the distance from the top of the ladder to the ground. How far is the bottom of the ladder from the house? 5 feet

67. Catarina and José want to buy a rug for a room that is 15 by 20 feet. They want to leave an even strip of flooring uncovered around the edges of the room. How wide a strip will they have if they buy a rug with an area of 234 square feet? 1 foot

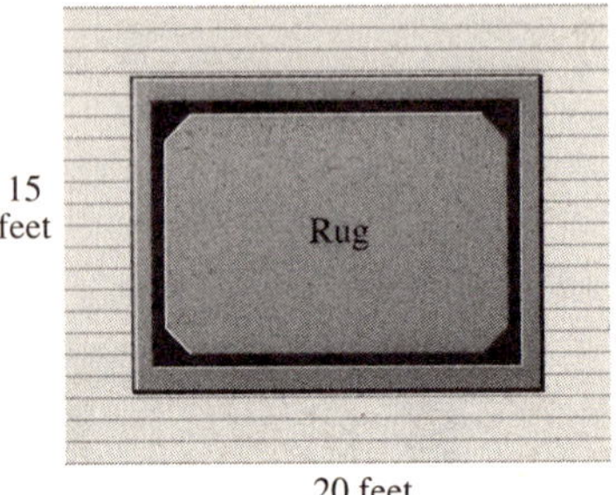

68. A club swimming pool is 30 feet wide and 40 feet long. The club members want an exposed aggregate border in a strip of uniform width around the pool. They have enough material for 296 square feet. How wide can the strip be? 2 feet

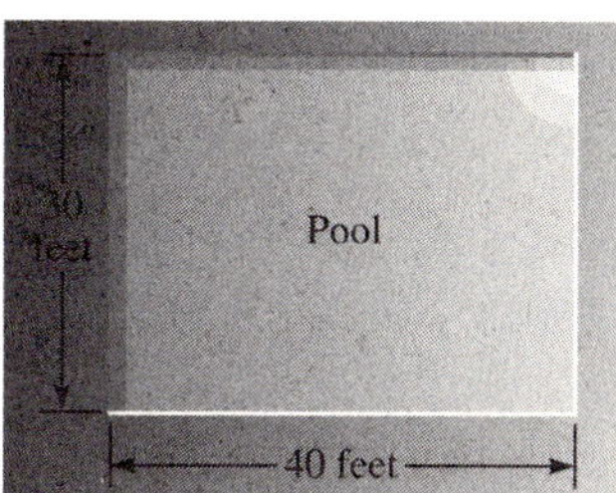

69. Arif's backyard is 20 by 30 meters. He wants to put a flower garden in the middle of the backyard, leaving a strip of grass of uniform width around the flower garden. To be happy, Arif must have 184 square meters of grass. Under these conditions what will the length and width of the garden be?
length: 26 meters; width: 16 meters

70. A rectangular piece of sheet metal has a length that is 4 inches less than twice the width. A square piece 2 inches on a side is cut from each corner. The sides are then turned up to form an uncovered box of volume 256 cubic inches. Find the length and width of the original piece of metal.

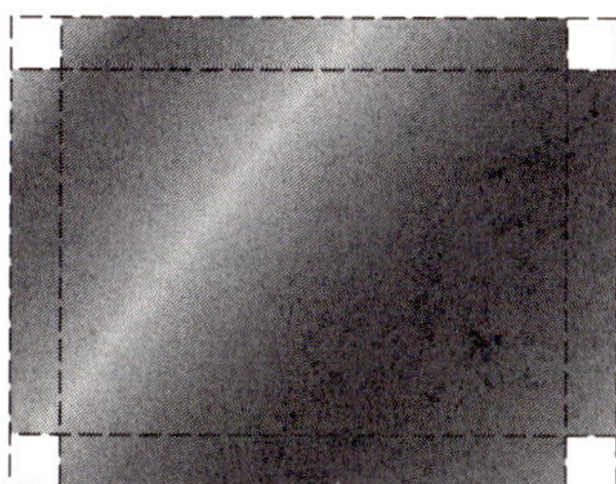

length: 20 inches; width: 12 inches

71. If a square piece of cardboard has 3-inch squares cut from its corners and then has the flaps folded up to form an open-top box, the volume of the box is given by the formula $V = 3(x - 6)^2$, where x is the length of each side of the original piece of cardboard in inches. What original length would yield a box with a volume of 432 cubic inches? 18 inches

72. The adjusted poverty threshold for a single person from the year 1984 to the year 1990 is approximated by the quadratic model $T = 18.7x^2 + 105.3x + 4814.1$, where $x = 0$ corresponds to 1984, and T is in dollars. In what year during this period was the threshold T approximately \$5,300? (*Source:* Congressional Budget Office) 1987

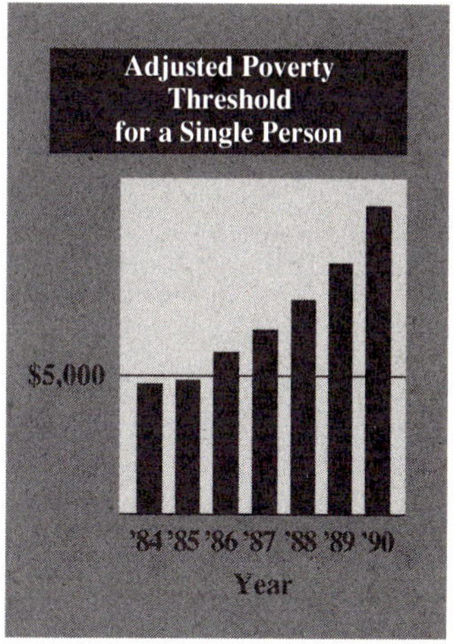

Recall from the exercises in Section 3 of this chapter that the corresponding sides of similar triangles are proportional. Use this fact to find the lengths of the indicated sides of the pair of similar triangles. Check all possible solutions in both triangles. Sides of a triangle cannot be negative.

73. side AC 5 or 14

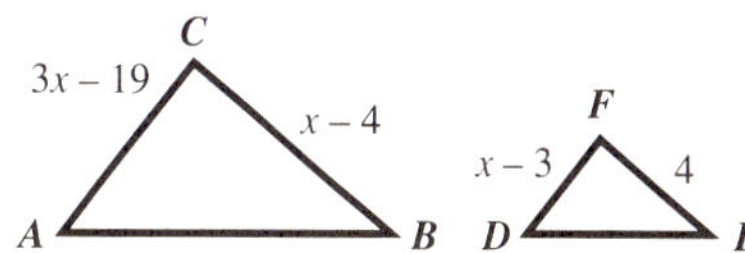

74. side RQ 4

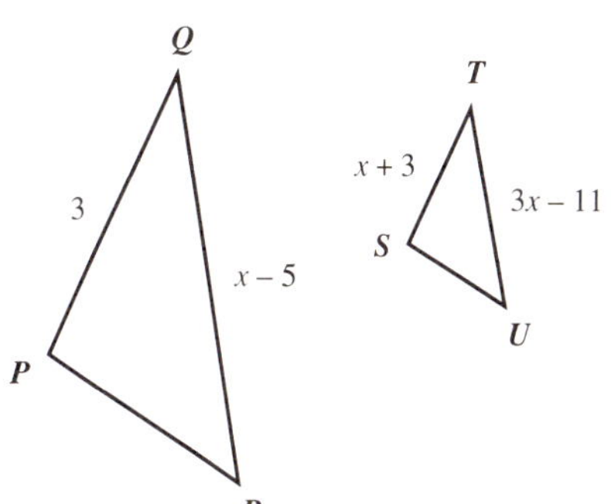

Chapter 7 Test (Text Page 398)

Solve each equation.

1. $5x - 3 + 2x = 3(x - 2) + 11$ {2}

2. $\frac{2p - 1}{3} + \frac{p + 1}{4} = \frac{43}{12}$ {4}

3. Decide whether the equation

$$3x - (2 - x) + 4x = 7x - 2 - (-x)$$

is conditional, an identity, or a contradiction. Solve and give its solution set.
identity; {all real numbers}

4. Solve for v: $S = vt - 16t^2$. $v = \dfrac{S + 16t^2}{t}$

Solve each of the following applications.

5. The three largest islands in the Hawaiian island chain are Hawaii (the Big Island), Maui, and Kauai. Together, their areas total 5,300 square miles. The island of Hawaii is 3,293 square miles larger than the island of Maui, and Maui is 177 square miles larger than Kauai. What is the area of each island?
Hawaii: 4,021 square miles; Maui: 728 square miles; Kauai: 551 square miles

6. How many liters of a 20% solution of a chemical should Michelle Jennings mix with 10 liters of a 50% solution to obtain a mixture that is 40% chemical? **5 liters**

7. A passenger train and a freight train leave a town at the same time and travel in opposite directions. Their speeds are 60 mph and 75 mph, respectively. How long will it take for them to be 297 miles apart? **2.2 hours**

8. Which is the better buy for processed cheese slices: 8 slices for \$2.19 or 12 slices for \$3.30? **8 slices for \$2.19**

9. The distance between Milwaukee and Boston is 1,050 miles. On a certain map this distance is represented by 21 inches. On the same map Seattle and Cincinnati are 46 inches apart. What is the actual distance between Seattle and Cincinnati?
2,300 miles

10. The current in a simple electrical circuit is inversely proportional to the resistance. If the current is 80 amps when the resistance is 30 ohms, find the current when the resistance is 12 ohms. **200 amps**

Solve the inequality. Give the solution set in both interval and graph forms.

11. $-4x + 2(x - 3) \geq 4x - (3 + 5x) - 7$

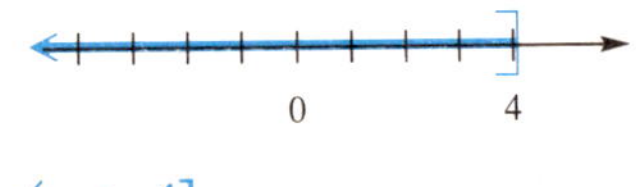

$(-\infty, 4]$

12. $-10 < 3k - 4 \leq 14$

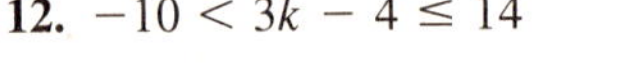

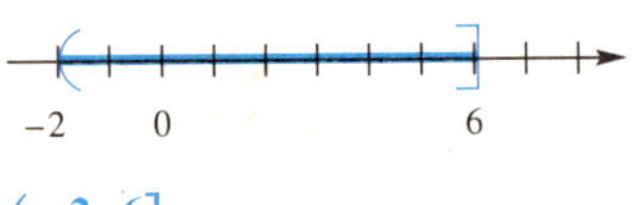

$(-2, 6]$

13. Which one of the following inequalities is equivalent to $x < -3$?
(a) $-3x < 9$ **(b)** $-3x > -9$ **(c)** $-3x > 9$ **(d)** $-3x < -9$ **(c)**

14. Edison Diest has grades of 83, 76, and 79 on his first three tests in Math 1031 (Survey of Mathematics). If he wants an average of at least 80 after his fourth test, what are the possible scores he can make on his fourth test? **at least 82**

Evaluate each exponential expression.

15. $\left(\dfrac{4}{3}\right)^2$ $\dfrac{16}{9}$

16. $-(-2)^6$ -64

17. $\left(\dfrac{3}{4}\right)^{-3}$ $\dfrac{64}{27}$

18. $-5^0 + (-5)^0$ 0

CONCEPTUAL    GRAPHING CALCULATOR

Use the properties of exponents to simplify each expression. Write answers with positive exponents only. Assume that all variables represent nonzero real numbers.

19. $9(4p^3)(6p^{-7})$ $\frac{216}{p^4}$

20. $\frac{m^{-2}(m^3)^{-3}}{m^{-4}m^7}$ $\frac{1}{m^{14}}$

21. Use scientific notation to evaluate $\frac{(2{,}500{,}000)(.00003)}{(.05)(5{,}000{,}000)}$. Leave the answer in scientific notation. 3×10^{-4}

22. In the food chain that links the largest sea creature, the whale, to the smallest, the diatom, 4×10^{14} diatoms sustain a medium-sized whale for only a few hours. Write this number without using exponents. **400,000,000,000,000**

23. Based on the accompanying graph, give the approximate federal government expenditure on crime-fighting in 1991 using scientific notation. $\$1.2 \times 10^{10}$

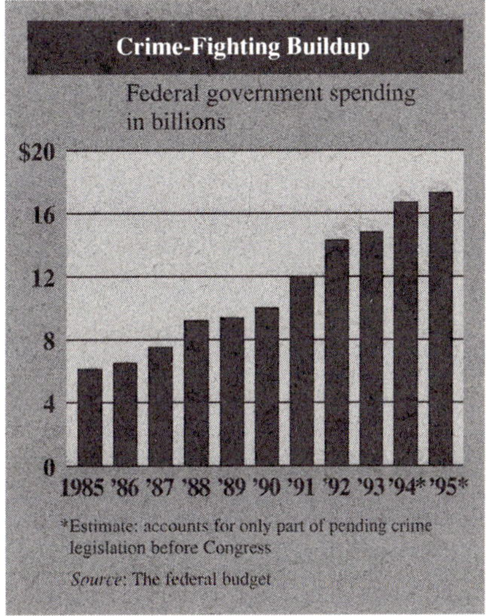

Perform the indicated operations.

24. $(3k^3 - 5k^2 + 8k - 2) - (3k^3 - 9k^2 + 2k - 12)$ $4k^2 + 6k + 10$

25. $(5x + 2)(3x - 4)$ $15x^2 - 14x - 8$

26. $(4x^2 - 3)(4x^2 + 3)$ $16x^4 - 9$

27. $(x + 4)(3x^2 + 8x - 9)$ $3x^3 + 20x^2 + 23x - 36$

28. Give an example of a polynomial in the variable t, such that it is fifth degree, in descending powers of the variable, with exactly six terms, and having a negative coefficient for its quadratic term. One example is $t^5 + 2t^4 + 3t^3 - 4t^2 + 5t + 6$.

Factor each polynomial completely.

29. $2p^2 - 5pq + 3q^2$ $(2p - 3q)(p - q)$

30. $100x^2 - 49y^2$ $(10x + 7y)(10x - 7y)$

31. $27y^3 - 125x^3$ $(3y - 5x)(9y^2 + 15yx + 25x^2)$

32. $4x + 4y - mx - my$ $(4 - m)(x + y)$

Solve each quadratic equation.

33. $6x^2 + 7x - 3 = 0$ $\left\{-\frac{3}{2}, \frac{1}{3}\right\}$

34. $x^2 - x = 7$ $\left\{\frac{1 \pm \sqrt{29}}{2}\right\}$

Solve the problem. Use a calculator and round the answer to the nearest hundredth.

35. The equation $s = 16t^2 + 15t$ gives the distance s in feet an object thrown off a building has descended in t seconds. Find the time t when the object has descended 25 feet. **.87 second**

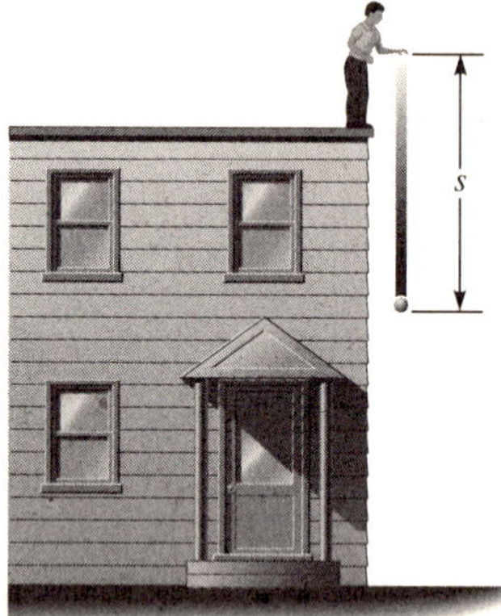

CHAPTER 8 Functions, Graphs, and Systems of Equations and Inequalities

8.1 Exercises (Text Page 406)

Fill in the blank with the correct response.

1. The point with coordinates (0, 0) is called the ______ of a rectangular coordinate system. **origin**
2. For any value of x, the point $(x, 0)$ lies on the ______ -axis. x
3. For any value of y, the point $(0, y)$ lies on the ______ -axis. y
4. The circle $x^2 + y^2 = 9$ has the point ______ as its center. **(0, 0)**
5. The point (______, 0) is the center of the circle $(x - 2)^2 + y^2 = 16$. **2**
6. What is the radius of the circle $(x + 1)^2 + (y - 12)^2 = 36$? **6**

Name the quadrant, if any, in which the point is located.

7. (a) (1, 6)
(b) $(-4, -2)$
(c) $(-3, 6)$
(d) $(7, -5)$
(e) $(-3, 0)$
(a) I (b) III (c) II (d) IV (e) none

8. (a) $(-2, -10)$
(b) (4, 8)
(c) $(-9, 12)$
(d) $(3, -9)$
(e) $(0, -8)$
(a) III (b) I (c) II (d) IV (e) none

9. Use the given information to determine the possible quadrants in which the point (x, y) must lie.

(a) $xy > 0$ (b) $xy < 0$ (c) $\frac{x}{y} < 0$ (d) $\frac{x}{y} > 0$

(a) I or III (b) II or IV (c) II or IV (d) I or III

10. What must be true about the coordinates of any point that lies along an axis?
One of the coordinates must be zero.

Locate the following points on the rectangular coordinate system.

11. (2, 3) **12.** (−1, 2)

13. (−3, −2) **14.** (1, −4)

15. (0, 5) **16.** (−2, −4)

17. (−2, 4) **18.** (3, 0)

19. (−2, 0) **20.** (3, −3)

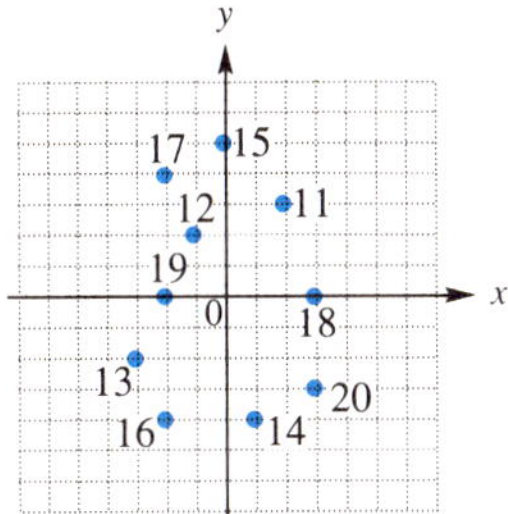

Find **(a)** *the distance between the pair of points, and* **(b)** *the coordinates of the midpoint of the segment having the points as endpoints.*

21. (3, 4) and (−2, 1)
(a) $\sqrt{34}$ (b) $\left(\frac{1}{2}, \frac{5}{2}\right)$

22. (−2, 1) and (3, −2)
(a) $\sqrt{34}$ (b) $\left(\frac{1}{2}, -\frac{1}{2}\right)$

23. (−2, 4) and (3, −2)
(a) $\sqrt{61}$ (b) $\left(\frac{1}{2}, 1\right)$

24. (1, −5) and (6, 3)
(a) $\sqrt{89}$ (b) $\left(\frac{7}{2}, -1\right)$

25. (−3, 7) and (2, −4)
(a) $\sqrt{146}$ (b) $\left(-\frac{1}{2}, \frac{3}{2}\right)$

26. (0, 5) and (−3, 12)
(a) $\sqrt{58}$ (b) $\left(-\frac{3}{2}, \frac{17}{2}\right)$

Match the equation with the correct graph.

27. $(x - 3)^2 + (y - 2)^2 = 25$
B

28. $(x - 3)^2 + (y + 2)^2 = 25$
C

29. $(x + 3)^2 + (y - 2)^2 = 25$
D

30. $(x + 3)^2 + (y + 2)^2 = 25$
A

A.

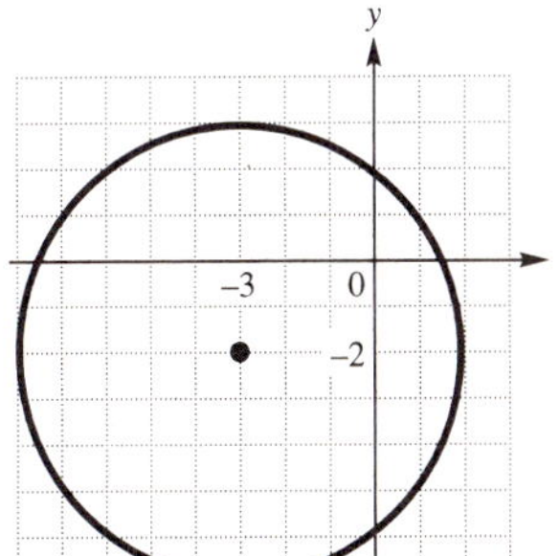

B.

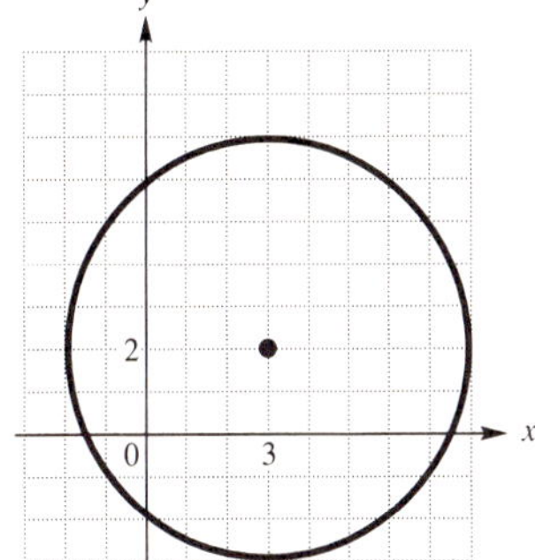

C.

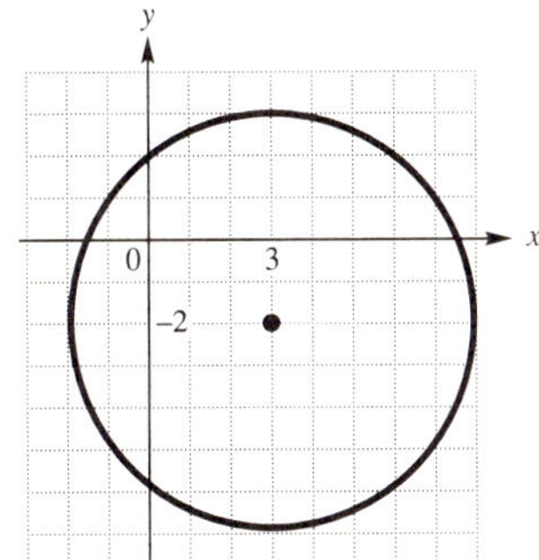

D.

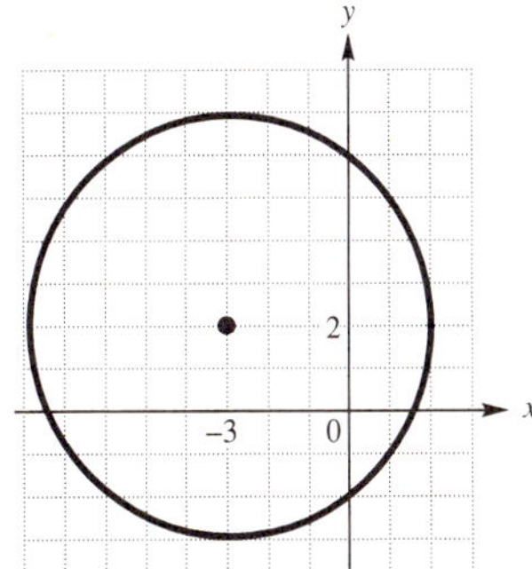

Write an equation of the circle with the given center and radius.

31. (0, 0); $r = 6$
$x^2 + y^2 = 36$

32. (0, 0); $r = 5$
$x^2 + y^2 = 25$

33. (−1, 3); $r = 4$
$(x + 1)^2 + (y - 3)^2 = 16$

34. (2, −2); $r = 3$
$(x - 2)^2 + (y + 2)^2 = 9$

35. (0, 4); $r = \sqrt{3}$
$x^2 + (y - 4)^2 = 3$

36. (−2, 0); $r = \sqrt{5}$
$(x + 2)^2 + y^2 = 5$

37. Suppose that a circle has an equation of the form $x^2 + y^2 = r^2$, $r > 0$. What is the center of the circle? What is the radius of the circle? **center: (0, 0); radius: r**

38. (a) How many points are there on the graph of $(x - 4)^2 + (y - 1)^2 = 0$? Explain your answer. **There is only one point, (4, 1). The only way for the sum of two squares to equal zero is for both of the squares to equal zero. This is satisfied when $x = 4$ and $y = 1$.**

(b) How many points are there on the graph of $(x - 4)^2 + (y - 1)^2 = -1$? Explain your answer. **There are no points because the sum of two squares cannot be negative.**

Find the center and the radius of the circle. (Hint: In Exercises 43 and 44 divide both sides by the greatest common factor.)

39. $x^2 + y^2 + 4x + 6y + 9 = 0$
center: (−2, 3); radius: 2

40. $x^2 + y^2 - 8x - 12y + 3 = 0$
center: (4, 6); radius: 7

41. $x^2 + y^2 + 10x - 14y - 7 = 0$
center: (−5, 7); radius: 9

42. $x^2 + y^2 - 2x + 4y - 4 = 0$
center: (1, −2); radius: 3

43. $3x^2 + 3y^2 - 12x - 24y + 12 = 0$
center: (2, 4); radius: 4

44. $2x^2 + 2y^2 + 20x + 16y + 10 = 0$
center: (−5, −4) radius: 6

Graph the circle.

45. $x^2 + y^2 = 36$

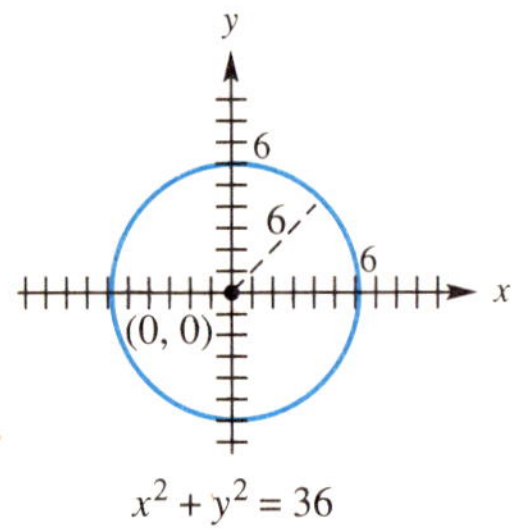

46. $x^2 + y^2 = 81$

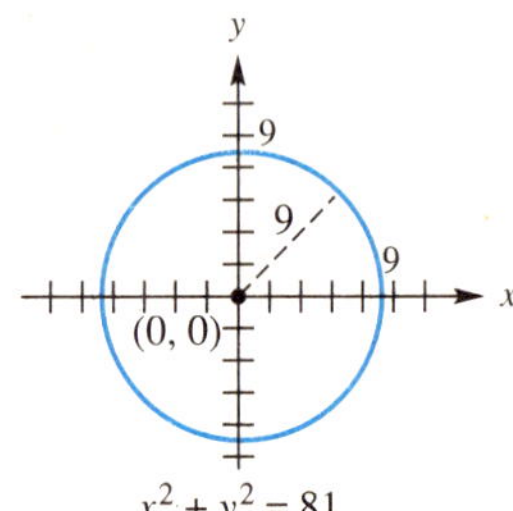

47. $(x - 2)^2 + y^2 = 36$

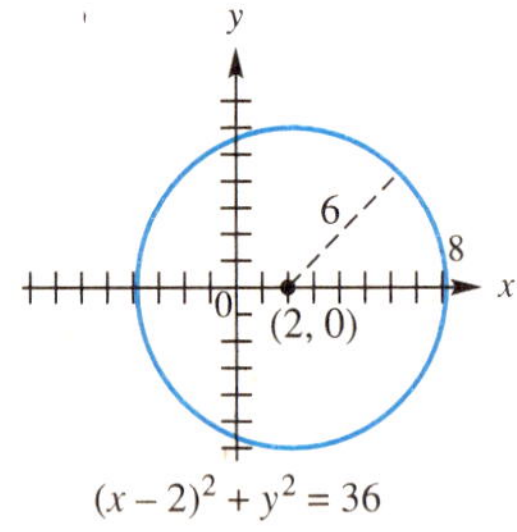

48. $x^2 + (y + 3)^2 = 49$

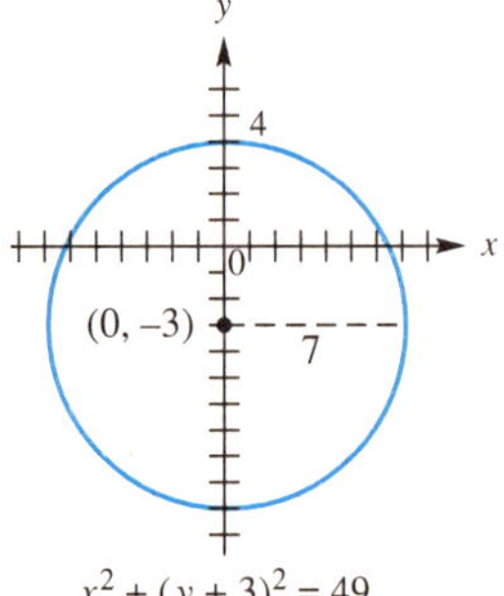

49. $(x + 2)^2 + (y - 5)^2 = 16$

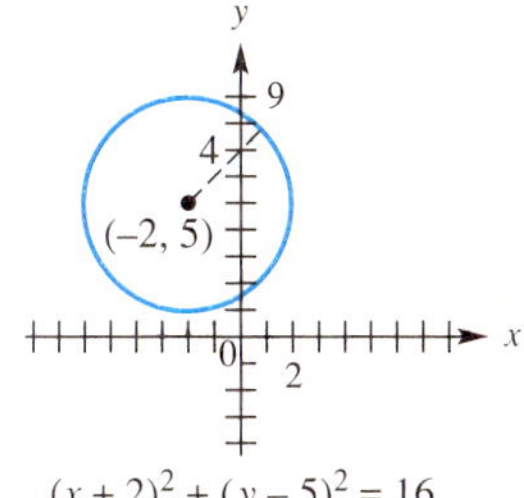

50. $(x - 4)^2 + (y - 3)^2 = 25$

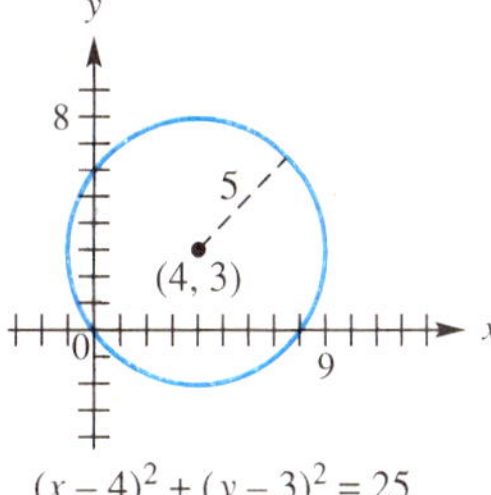

51. $(x + 3)^2 + (y + 2)^2 = 36$

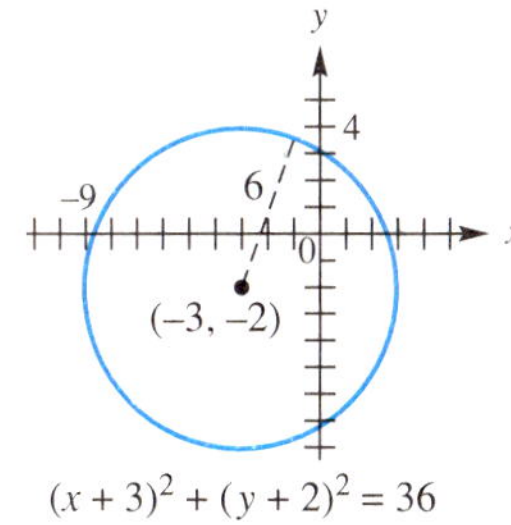

52. $(x - 5)^2 + (y + 4)^2 = 49$

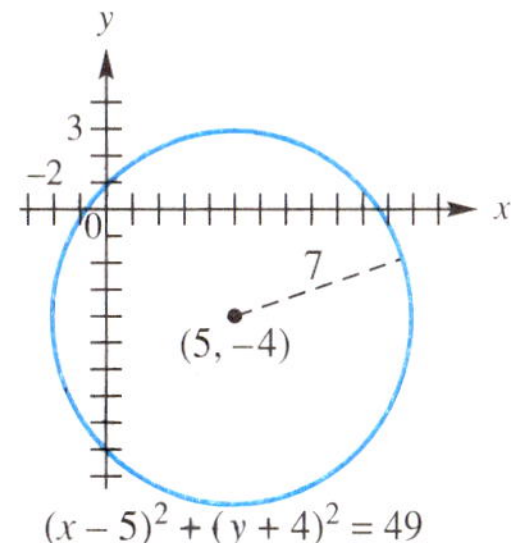

---

53. An alternate form of the distance formula is

$$d = \sqrt{(x_1 - x_2)^2 + (y_1 - y_2)^2}.$$

Compare this to the form given in this section, and explain why the two forms are equivalent.

54. A student was asked to find the distance between the points (5, 8) and (2, 14), and wrote the following:

$$d = \sqrt{(5 - 8)^2 + (2 - 14)^2}.$$

Explain why this is incorrect.

55. A circle can be drawn on a piece of posterboard by fastening one end of a string, pulling the string taut with a pencil, and tracing a curve as shown in the figure. Explain why this method works.

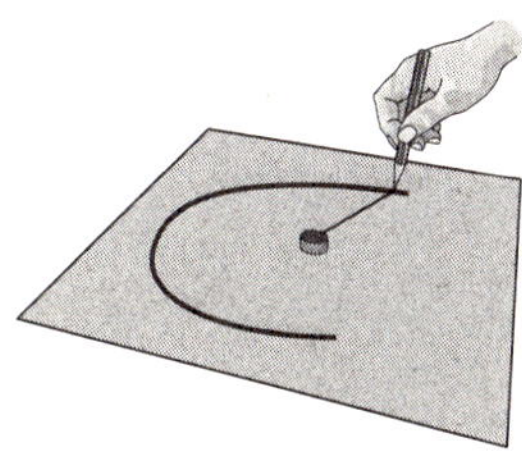

56. This figure shows how the crawfish race is held at the Crawfish Festival in Breaux Bridge, Louisiana. Explain why a circular "racetrack" is appropriate for such a race.

57. Show algebraically that if three receiving stations at (1, 4), (−6, 0), and (5, −2) record distances to an earthquake epicenter of 4 units, 5 units, and 10 units, respectively, the epicenter would lie at (−3, 4). **(−3, 4) is a solution for each of the following equations: $(x - 1)^2 + (y - 4)^2 = 16$; $(x + 6)^2 + y^2 = 25$; $(x - 5)^2 + (y + 2)^2 = 100$**

58. Three receiving stations record the presence of an earthquake. The location of the receiving center and the distance to the epicenter are contained in the following three equations: $(x - 2)^2 + (y - 1)^2 = 25$, $(x + 2)^2 + (y - 2)^2 = 16$, and $(x - 1)^2 + (y + 2)^2 = 9$. Graph the circles and determine the location of the earthquake epicenter.

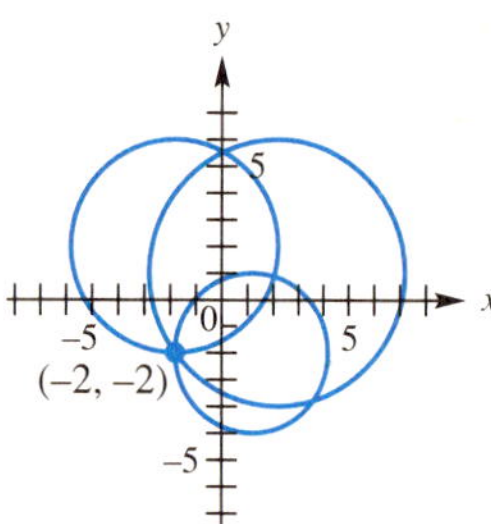

59. Without actually graphing, state whether or not the graphs of $x^2 + y^2 = 4$ and $x^2 + y^2 = 25$ will intersect. Explain your answer.

60. Can a circle have its center at (2, 4) and be tangent to both axes? (*Tangent to* means touching in one point.) Explain.

61. Suppose that the endpoints of a line segment have coordinates (x_1, y_1) and (x_2, y_2).

(a) Show that the distance between (x_1, y_1) and $\left(\frac{x_1 + x_2}{2}, \frac{y_1 + y_2}{2}\right)$ is the same as the distance between (x_2, y_2) and $\left(\frac{x_1 + x_2}{2}, \frac{y_1 + y_2}{2}\right)$.

(b) Show that the sum of the distances between (x_1, y_1) and $\left(\frac{x_1 + x_2}{2}, \frac{y_1 + y_2}{2}\right)$, and (x_2, y_2) and $\left(\frac{x_1 + x_2}{2}, \frac{y_1 + y_2}{2}\right)$ is equal to the distance between (x_1, y_1) and (x_2, y_2).

(c) From the results of parts (a) and (b), what conclusion can be made?

62. If the coordinates of one endpoint of a line segment are $(3, -8)$ and the coordinates of the midpoint of the segment are (6, 5), what are the coordinates of the other endpoint? **(9, 18)**

63. Which one of the following has a circle as its graph?
(a) $x^2 - y^2 = 9$ **(b)** $x^2 = 9 - y^2$
(c) $y^2 - x^2 = 9$ **(d)** $-x^2 - y^2 = 9$ **(b)**

64. For the three choices that are not circles in Exercise 63, explain why their equations are not those of circles.

8.2 Exercises (Text Page 415)

Complete the given ordered pairs for the equation. Then graph the equation.

1. $2x + y = 5$; (0,), (, 0), (1,), (, 1)
(0, 5), $\left(\frac{5}{2}, 0\right)$, (1, 3), (2, 1)

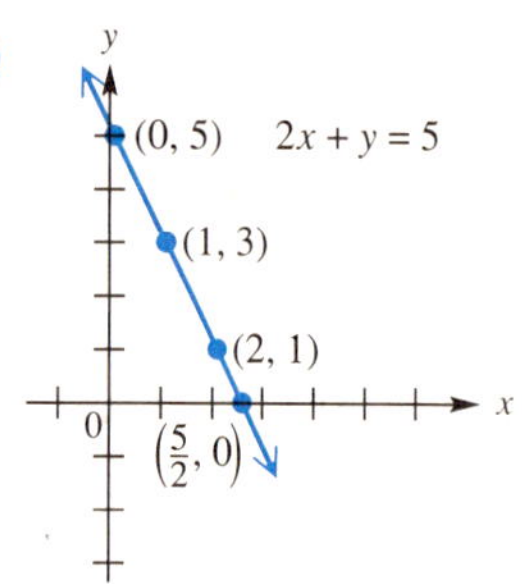

2. $3x - 4y = 24$; (0,), (, 0), (6,), (, −3)
(0, −6), (8, 0), $\left(6, -\frac{3}{2}\right)$, (4, −3)

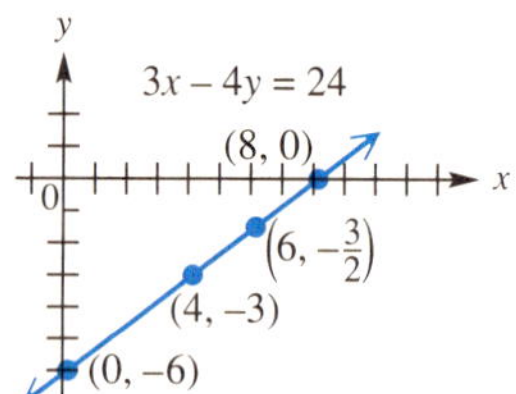

3. $x - y = 4$; (0,), (, 0), (2,), (, −1)
(0, −4), (4, 0), (2, −2), (3, −1)

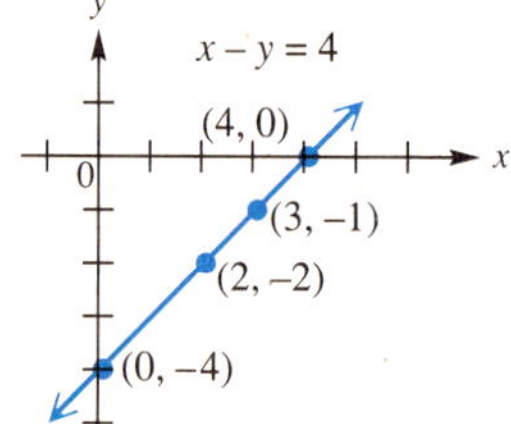

4. $x + 3y = 12$; (0,), (, 0), (3,), (, 6)
(0, 4), (12, 0), (3, 3), (−6, 6)

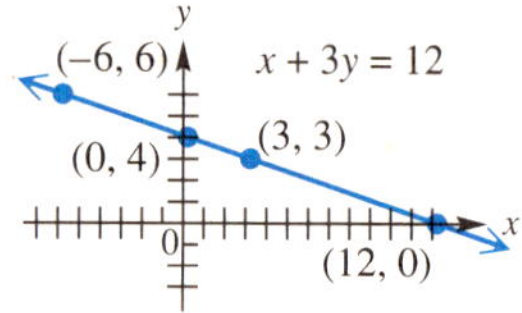

CONCEPTUAL WRITING CHALLENGING SCIENTIFIC CALCULATOR GRAPHING CALCULATOR

5. $4x + 5y = 20$; $(0, \quad)$, $(\quad, 0)$, $(3, \quad)$, $(\quad, 2)$

$(0, 4)$, $(5, 0)$, $\left(3, \frac{8}{5}\right)$, $\left(\frac{5}{2}, 2\right)$

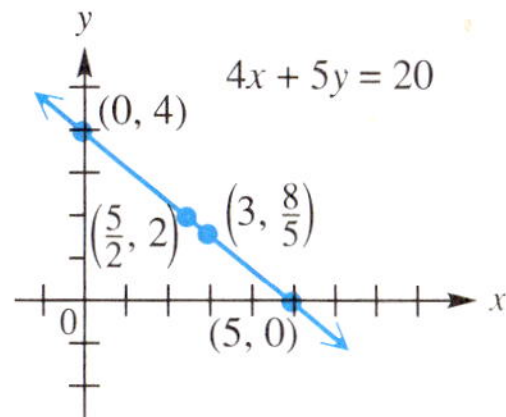

6. $2x - 5y = 12$; $(0, \quad)$, $(\quad, 0)$, $(\quad, -2)$, $(-2, \quad)$

$\left(0, -\frac{12}{5}\right)$, $(6, 0)$, $(1, -2)$, $\left(-2, -\frac{16}{5}\right)$

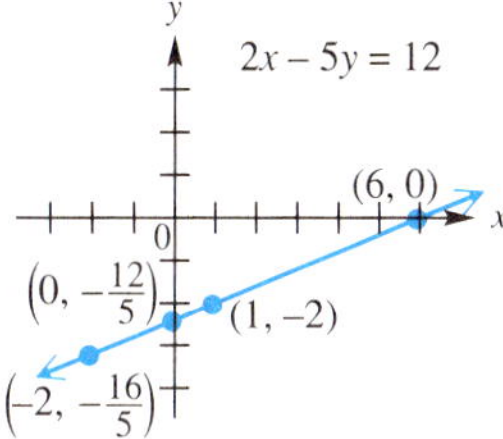

7. $3x + 2y = 8$

x	y
0	4
$\frac{8}{3}$	0
2	1
4	−2

8. $5x + y = 12$

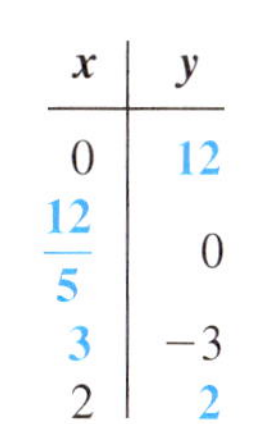

x	y
0	12
$\frac{12}{5}$	0
3	−3
2	2

9. Explain how to find the x-intercept of a linear equation in two variables.

10. Explain how to find the y-intercept of a linear equation in two variables.

11. Which one of the following has as its graph a horizontal line?
(a) $2y = 6$ **(b)** $2x = 6$ **(c)** $x - 4 = 0$ **(d)** $x + y = 0$ (a)

12. What is the minimum number of points that must be determined in order to graph a linear equation in two variables? 2

For each equation, give the x-intercept and the y-intercept. Then graph the equation.

13. $3x + 2y = 12$
$(4, 0)$; $(0, 6)$

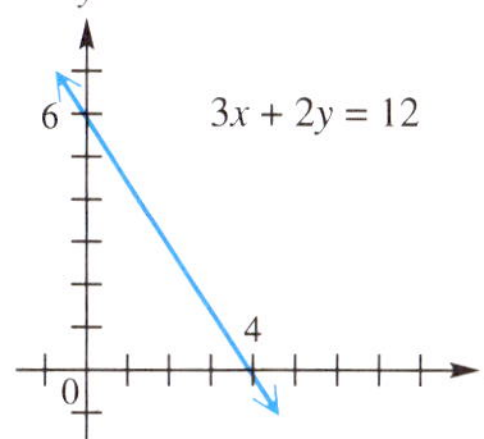

14. $2x + 5y = 10$
$(5, 0)$; $(0, 2)$

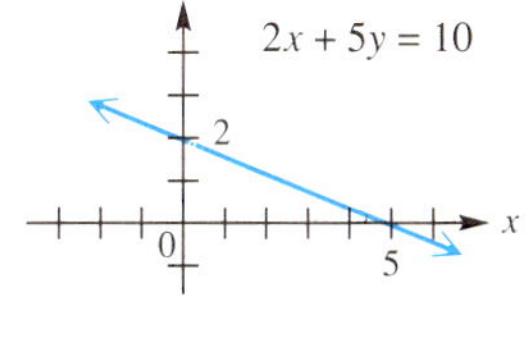

15. $5x + 6y = 10$
$(2, 0)$; $\left(0, \frac{5}{3}\right)$

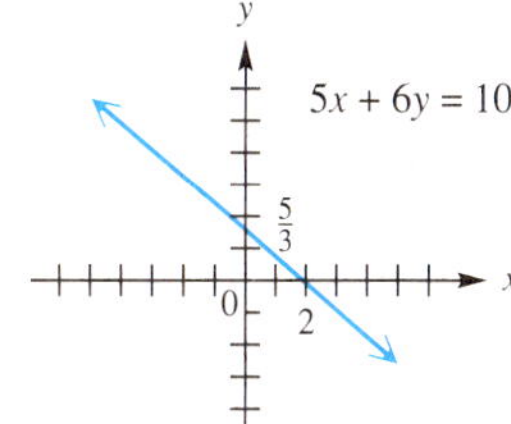

16. $3y + x = 6$

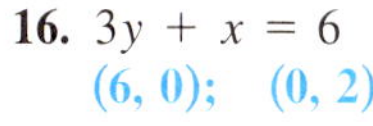
$(6, 0)$; $(0, 2)$

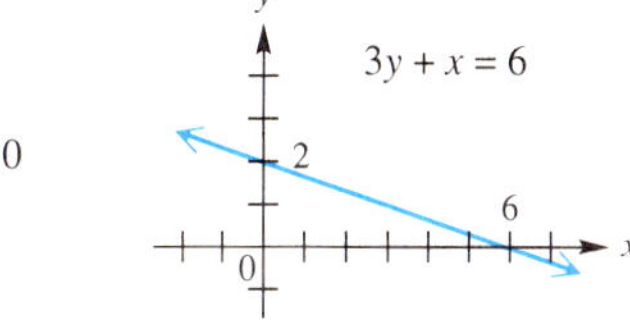

17. $2x - y = 5$

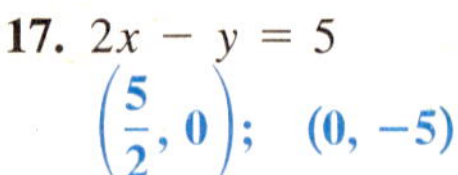

$\left(\frac{5}{2}, 0\right)$; $(0, -5)$

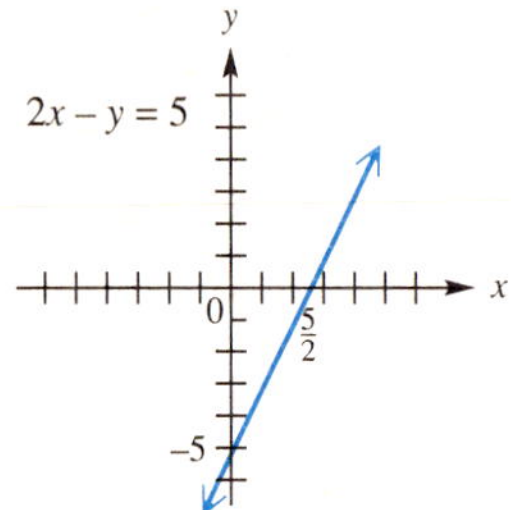

18. $3x - 2y = 4$

$\left(\frac{4}{3}, 0\right)$; $(0, -2)$

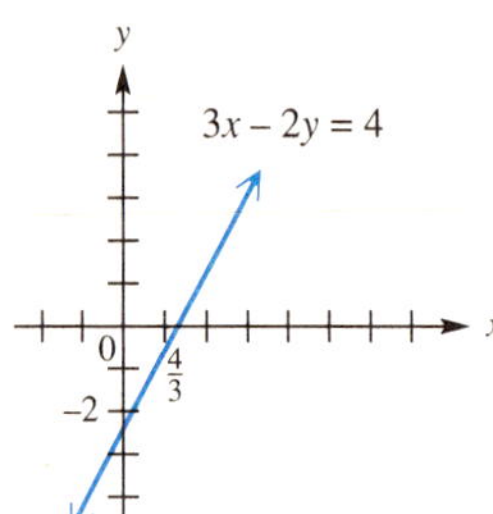

19. $x - 3y = 2$

$(2, 0)$; $\left(0, -\frac{2}{3}\right)$

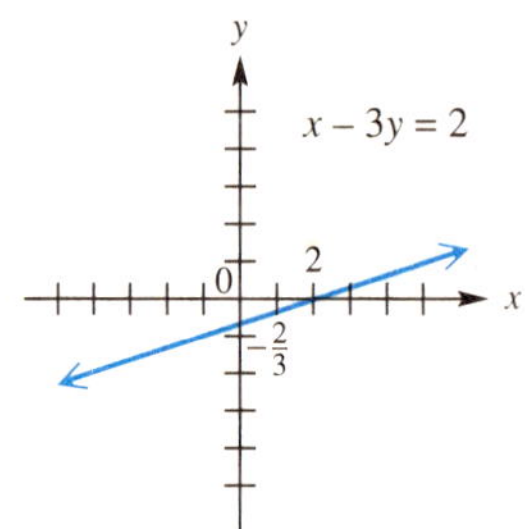

20. $y - 4x = 3$

$\left(-\frac{3}{4}, 0\right)$; $(0, 3)$

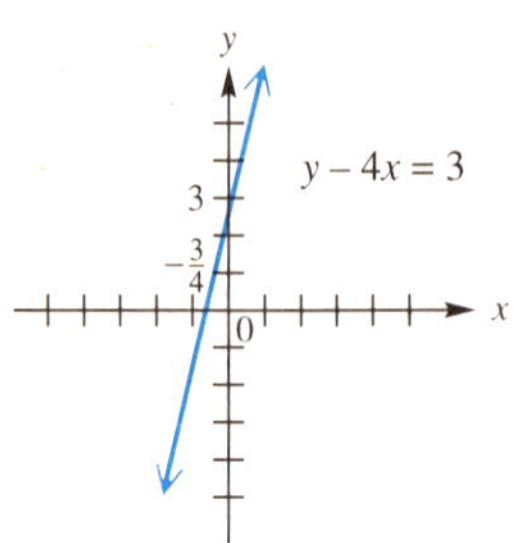

21. $y + x = 0$

$(0, 0)$; $(0, 0)$

22. $2x - y = 0$

$(0, 0)$; $(0, 0)$

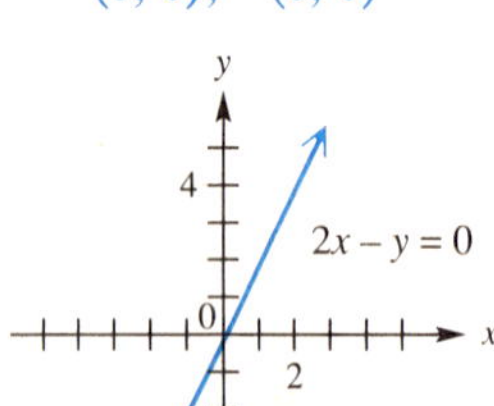

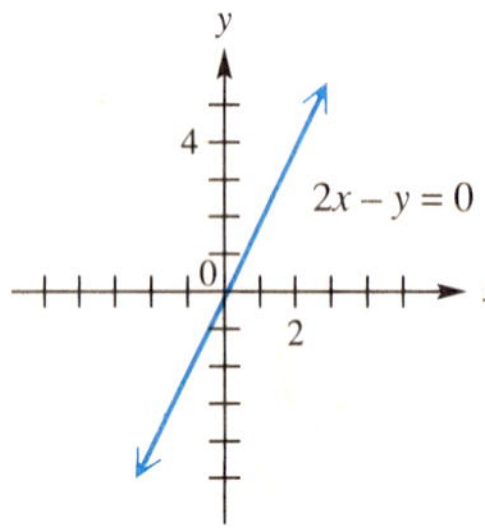

23. $3x = y$

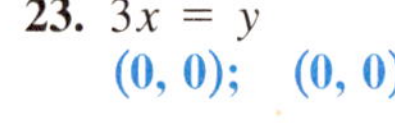

$(0, 0)$; $(0, 0)$

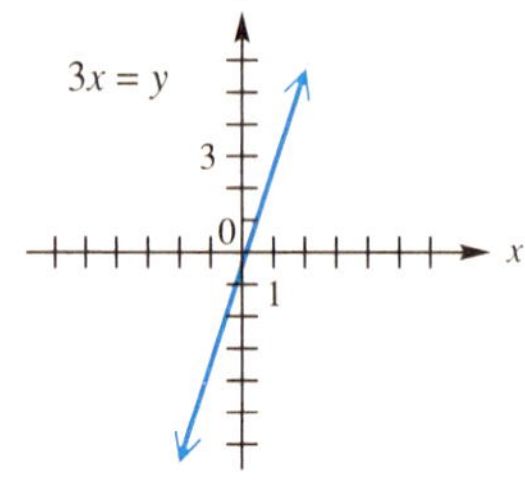

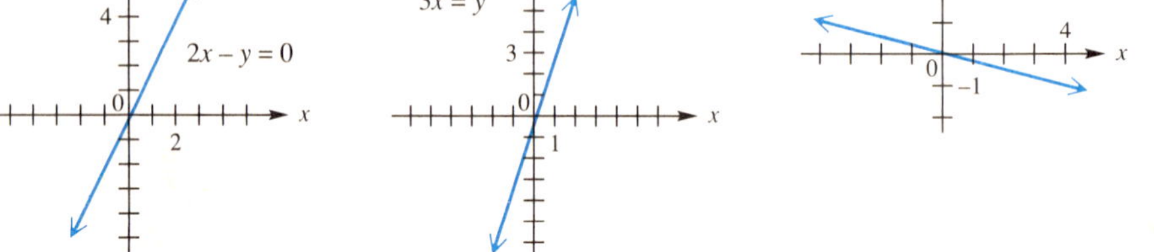

24. $x = -4y$

$(0, 0)$; $(0, 0)$

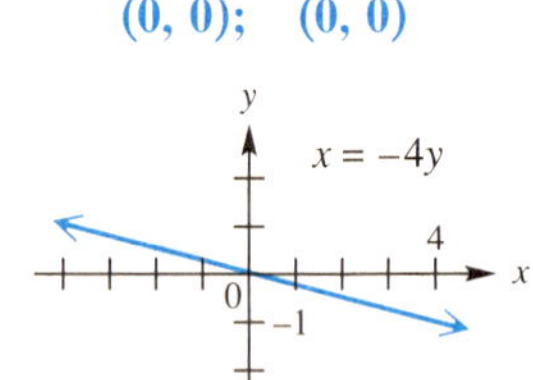

25. $x = 2$

$(2, 0)$; none

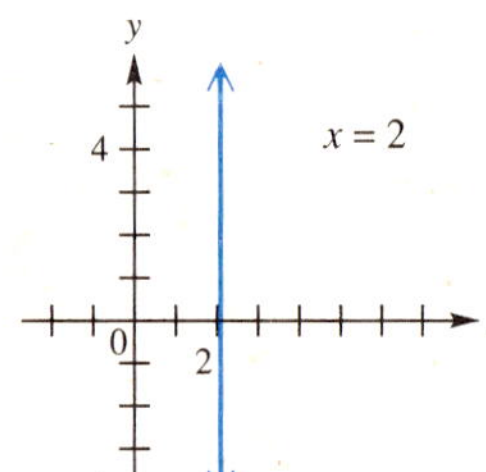

26. $y = -3$

none; $(0, -3)$

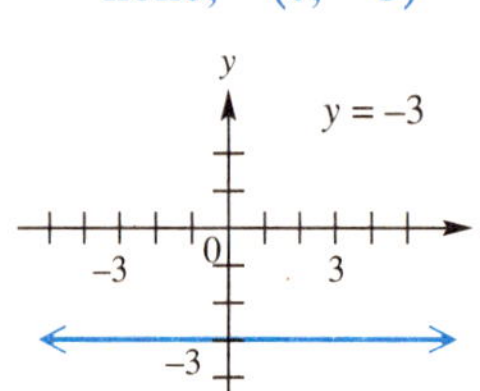

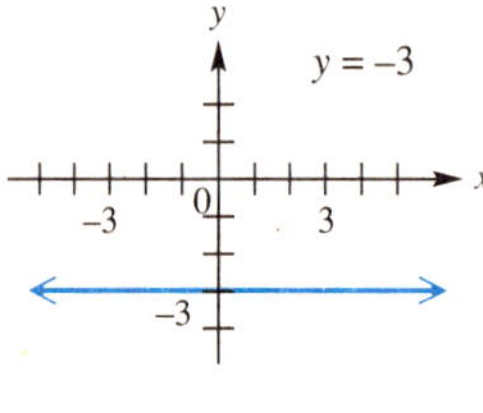

27. $y = 4$

none; $(0, 4)$

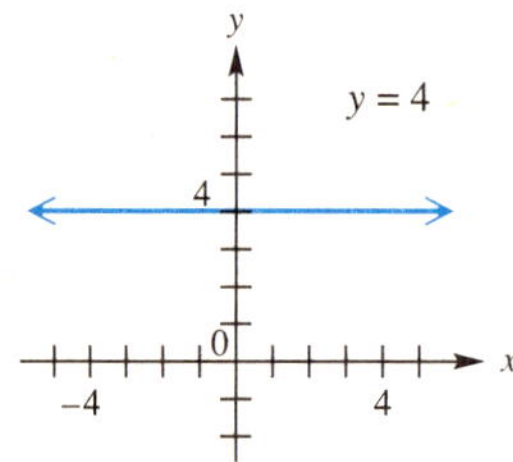

28. $x = -2$

$(-2, 0)$; none

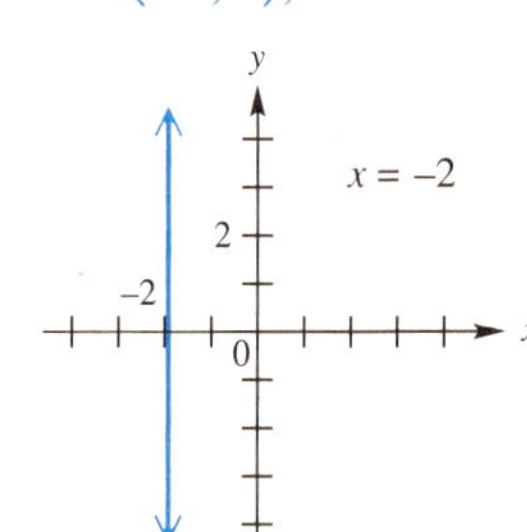

In Exercises 29–36, match the equation with the figure that most closely resembles its graph.

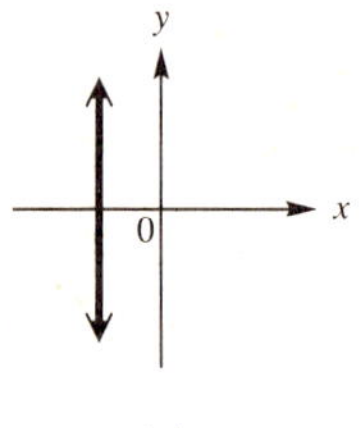

(a)

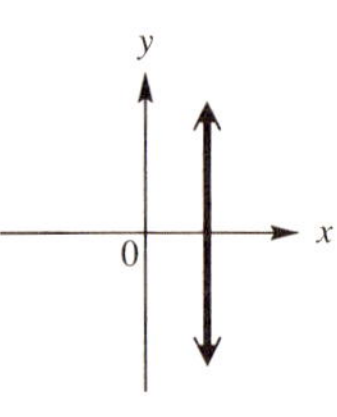

(b)

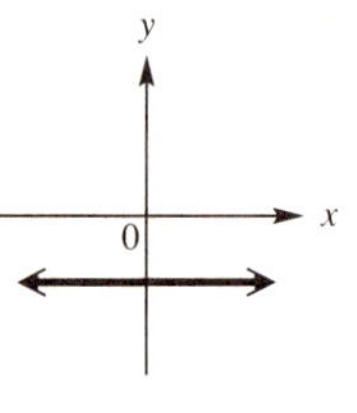

(c)

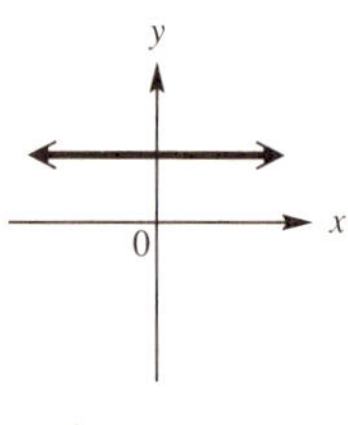

(d)

29. $y + 2 = 0$ **(c)**

30. $y + 4 = 0$ **(c)**

31. $x + 3 = 0$ **(a)**

32. $x + 7 = 0$ **(a)**

33. $y - 2 = 0$ **(d)**

34. $y - 4 = 0$ **(d)**

35. $x - 3 = 0$ **(b)**

36. $x - 7 = 0$ **(b)**

CONCEPTUAL WRITING CHALLENGING SCIENTIFIC CALCULATOR GRAPHING CALCULATOR

37. If a walkway rises 2 feet for every 10 feet on the horizontal, which of the following express its slope (or grade)? (There are several correct choices.)

(a) .2 (b) $\frac{2}{10}$ (c) $\frac{1}{5}$ (d) 20%

(e) 5 (f) $\frac{20}{100}$ (g) 500% (h) $\frac{10}{2}$

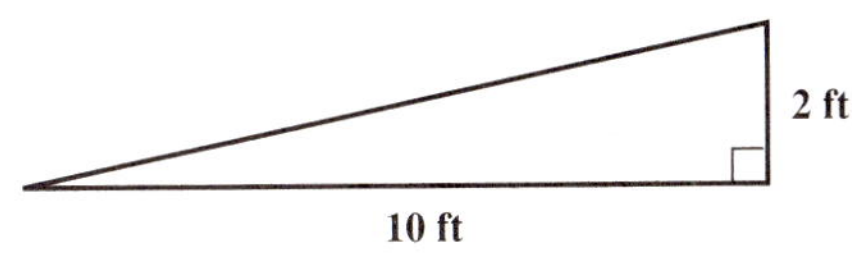

38. If the pitch of a roof is $\frac{1}{4}$, how many feet in the horizontal direction corresponds to a rise of 3 feet? 12 feet

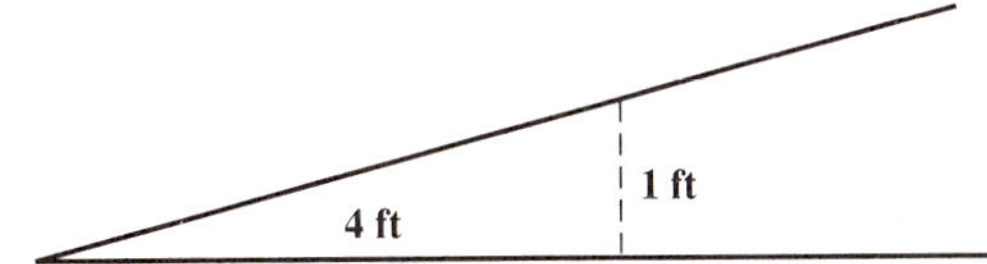

Find the slope of the line through the pair of points by using the slope formula.

39. $(-2, -3)$ and $(-1, 5)$ 8

40. $(-4, 3)$ and $(-3, 4)$ 1

41. $(-4, 1)$ and $(2, 6)$ $\frac{5}{6}$

42. $(-3, -3)$ and $(5, 6)$ $\frac{9}{8}$

43. $(2, 4)$ and $(-4, 4)$ 0

44. $(-6, 3)$ and $(2, 3)$ 0

Tell whether the slope of the given line is positive, negative, zero, *or* undefined.

45.

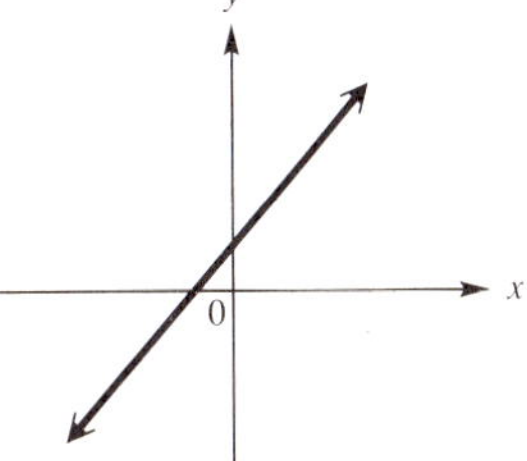

positive

46.

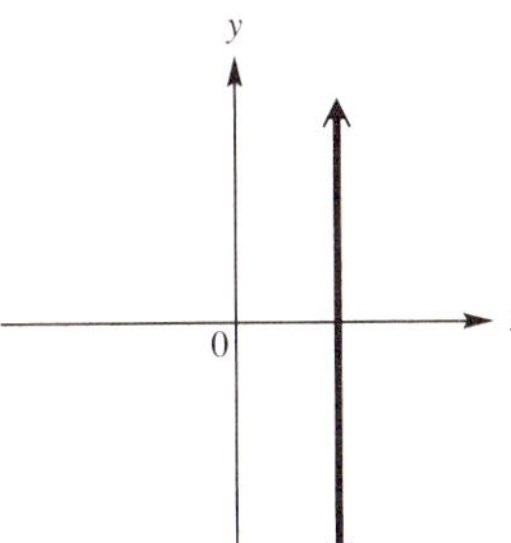

47.

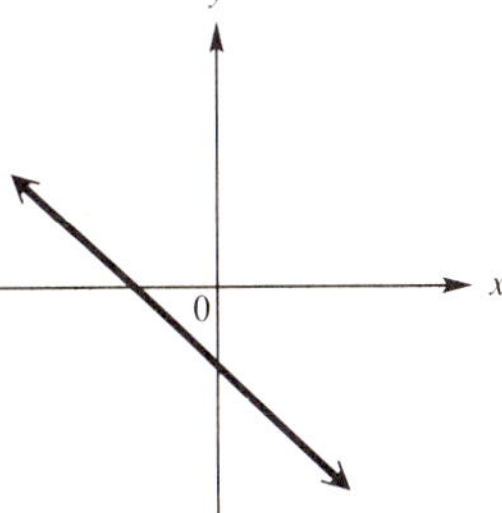

negative

48.

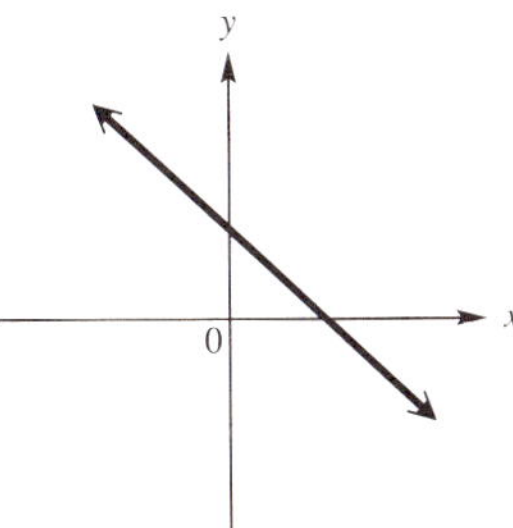

negative

49.

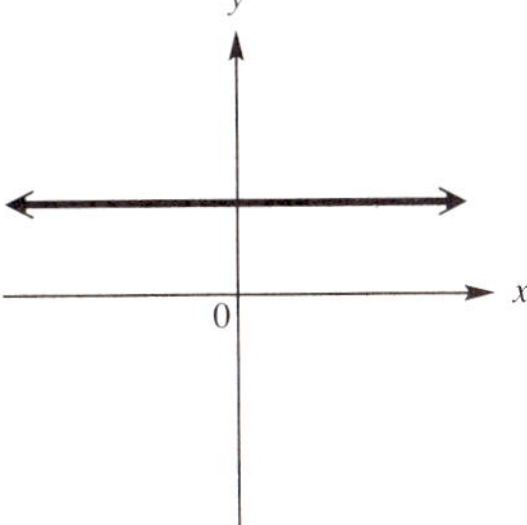

zero

50.

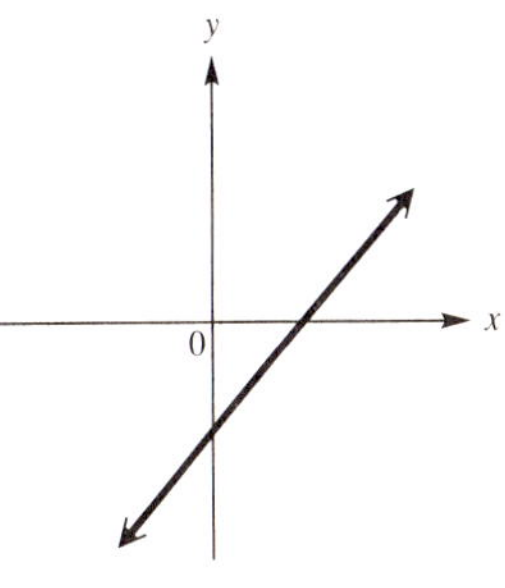

positive

Use the method of Example 5 to graph each of the following lines.

51. $m = \frac{1}{2}$, through $(-3, 2)$

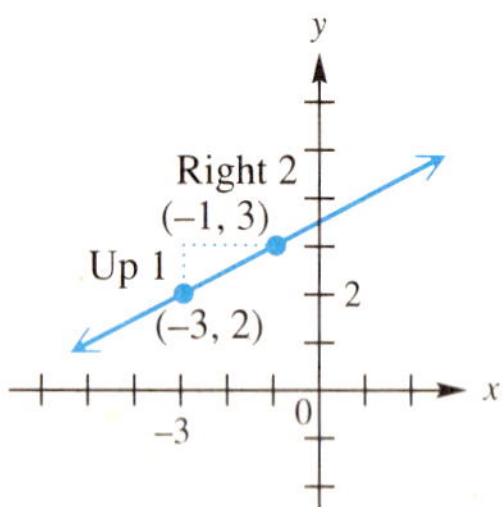

52. $m = \frac{2}{3}$, through $(0, 1)$

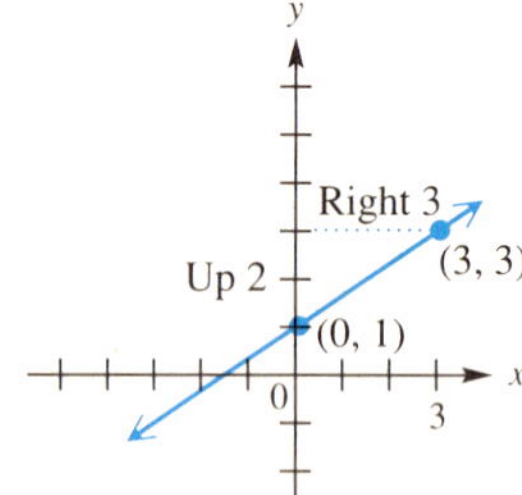

53. $m = -\frac{5}{4}$, through $(-2, -1)$

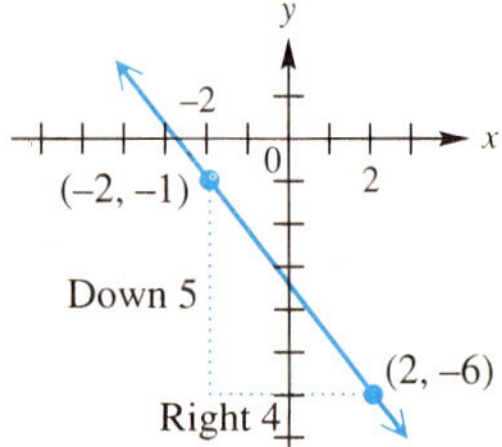

54. $m = -\frac{3}{2}$, through $(-1, -2)$

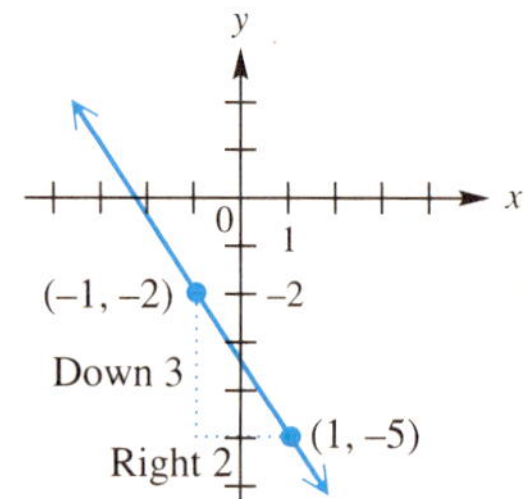

55. $m = -2$, through $(-1, -4)$

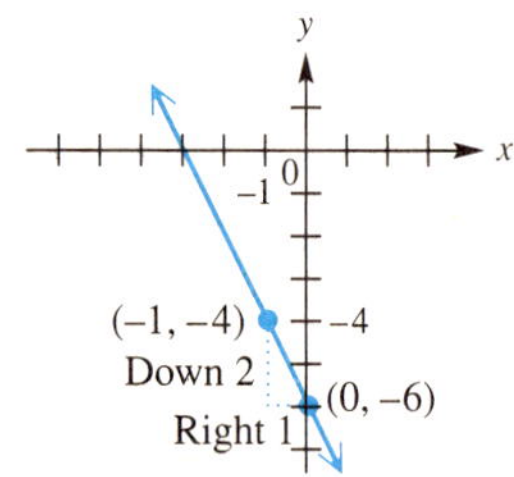

56. $m = 3$, through $(1, 2)$

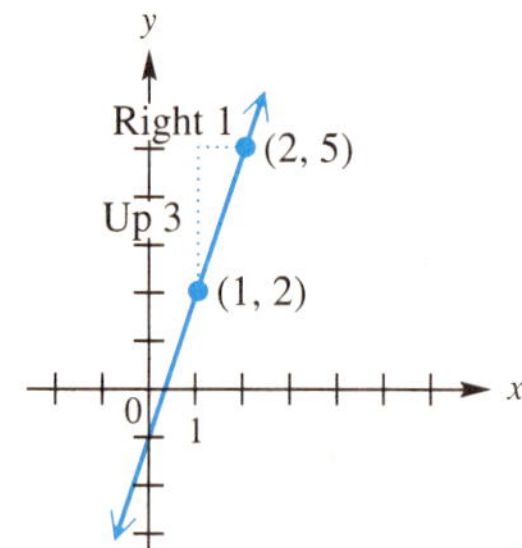

57. $m = 0$, through $(2, -5)$

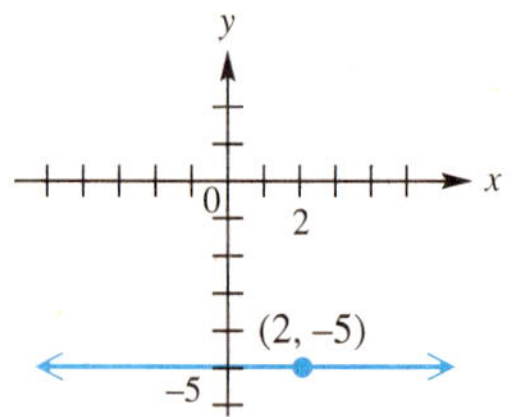

58. undefined slope, through $(-3, 1)$

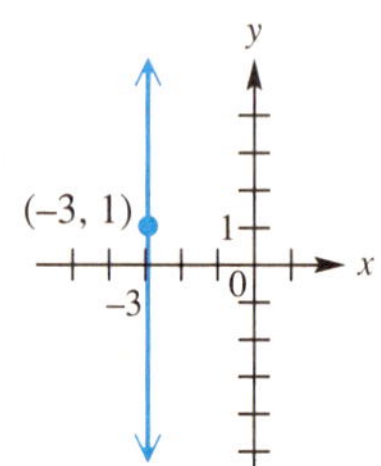

Determine whether the lines described are parallel, perpendicular, *or* neither parallel nor perpendicular.

59. L_1 through (4, 6) and (−8, 7), and L_2 through (7, 4) and (−5, 5) **parallel**

60. L_1 through (9, 15) and (−7, 12), and L_2 through (−4, 8) and (−20, 5) **parallel**

61. L_1 through (2, 0) and (5, 4), and L_2 through (6, 1) and (2, 4) **perpendicular**

62. L_1 through (0, −7) and (2, 3), and L_2 through (0, −3) and (1, −2)
neither parallel nor perpendicular

63. L_1 through (0, 1) and (2, −3), and L_2 through (10, 8) and (5, 3)
neither parallel nor perpendicular

64. L_1 through (1, 2) and (−7, −2), and L_2 through (1, −1) and (5, −9) **perpendicular**

Use the concept of slope or average rate of change to solve the problem.

65. The upper deck at the new Comiskey Park in Chicago has produced, among other complaints, displeasure with its steepness. It's been compared to a ski jump. It is 160 feet from home plate to the front of the upper deck and 250 feet from home plate to the back. The top of the upper deck is 63 feet above the bottom. What is its slope? $\frac{7}{10}$

66. When designing the new arena in Boston to replace the old Boston Garden, architects were careful to design the ramps leading up to the entrances so that circus elephants would be able to march up the ramps. The maximum grade (or slope) that an elephant will walk on is 13%. Suppose that such a ramp was constructed with a horizontal run of 150 feet. What would be the maximum vertical rise the architects could use? **19.5 feet**

67. The graph shows how average monthly rates for cable television increased from 1980 to 1992. The graph can be approximated by a straight line.

(a) Use the information provided for 1980 and 1992 to determine the average rate of change in price per year.

(b) How does a *positive* rate of change affect the consumer in a situation such as the one illustrated by this graph?

(a) $.92 (b) It means an *increase* in price.

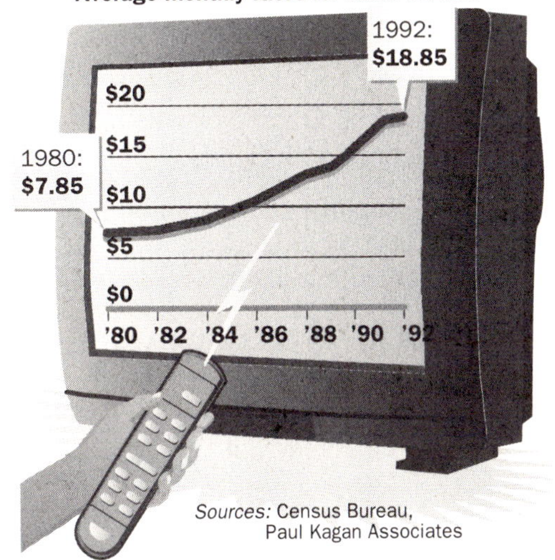

68. Assuming a linear relationship, what is the average rate of change for cable industry revenues over the period from 1990 to 1992? **$1.6 billion**

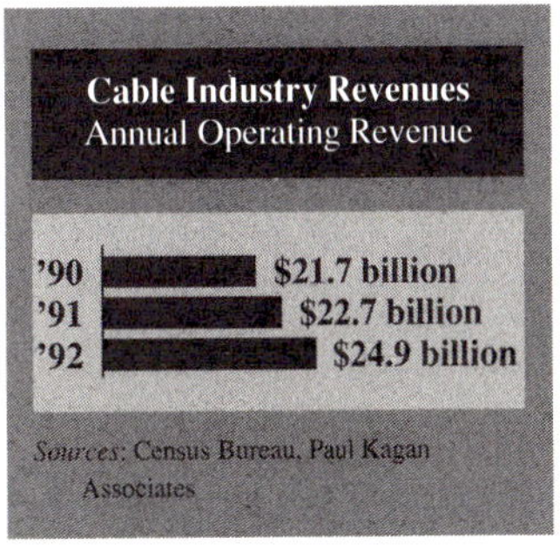

69. The 1993 Annual Report of AT&T cites the following figures concerning the global growth of international telephone calls: From 47.5 billion minutes in 1993, traffic is expected to rise to 60 billion minutes in 1995. Assuming a linear relationship, what is the average rate of change for this time period? (*Source:* TeleGeography, 1993, Washington, D.C.)
6.25 billion minutes per year

70. The market for international phone calls during the ten-year period from 1986 to 1995 is depicted in the accompanying bar graph. The tops of the bars approximate a straight line. Assuming that the traffic volume at the beginning of this period was 18 billion minutes and at the end was 60 billion minutes, what was the average rate of change for the ten-year period? 4.2 billion minutes per year

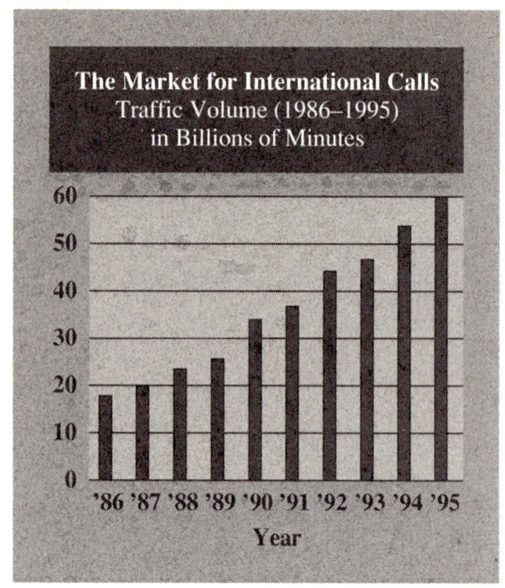

8.3 Exercises (Text Page 423)

In Exercises 1–8, choose the one of the four graphs given here which most closely resembles the graph of the given equation. Each equation is given in the form $y = mx + b$, so consider the signs of m and b in making your choice.

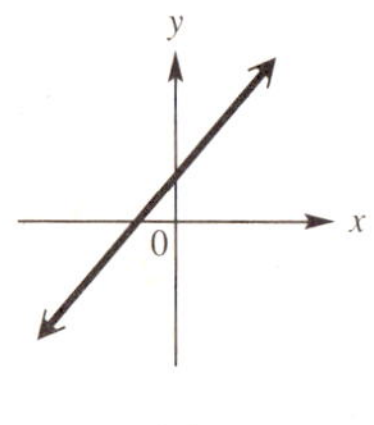

(a)

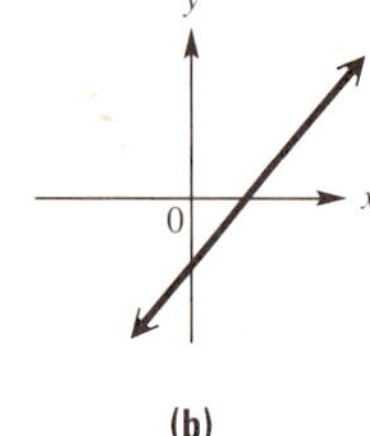

(b)

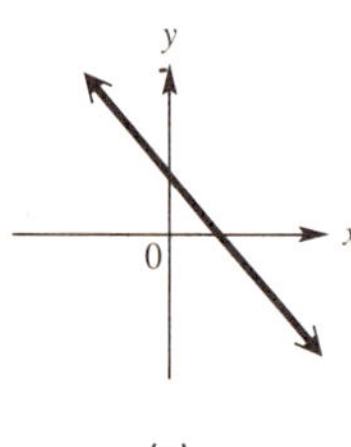

(c)

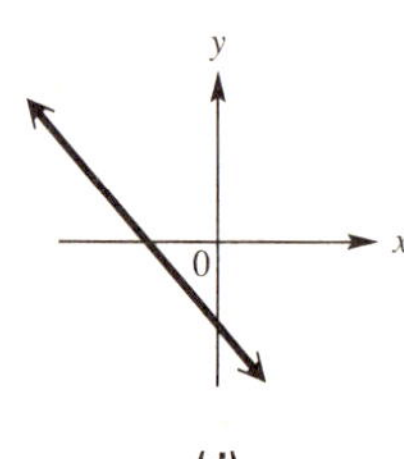

(d)

1. $y = 3x + 6$ (a) **2.** $y = 4x + 5$ (a) **3.** $y = -3x + 6$ (c) **4.** $y = -4x + 5$ (c)

5. $y = 3x - 6$ (b) **6.** $y = 4x - 5$ (b) **7.** $y = -3x - 6$ (d) **8.** $y = -4x - 5$ (d)

Write the equation of the line satisfying the given conditions.

9. through $(-2, 4)$; $m = -\frac{3}{4}$ $3x + 4y = 10$

10. through $(-1, 6)$; $m = -\frac{5}{6}$ $5x + 6y = 31$

11. through $(5, 8)$; $m = -2$ $2x + y = 18$

12. through $(12, 10)$; $m = 1$ $x - y = 2$

13. through $(-5, 4)$; $m = \frac{1}{2}$ $x - 2y = -13$

14. through $(7, -2)$; $m = \frac{1}{4}$ $x - 4y = 15$

15. through $(-4, 12)$; horizontal $y = 12$

16. through $(1, 5)$; horizontal $y = 5$

17. x-intercept $(3, 0)$; $m = 4$ $4x - y = 12$

18. x-intercept $(-2, 0)$; $m = -5$ $5x + y = -10$

19. Explain why the point-slope form of an equation cannot be used to find the equation of a vertical line.

20. Which one of the following equations is in standard form, according to the definition of standard form given in this text?
(a) $3x + 2y - 6 = 0$ **(b)** $y = 5x - 12$ **(c)** $2y = 3x + 4$ **(d)** $6x - 5y = 12$ (d)

Write an equation in the form $x = k$ or $y = k$ for some constant k for the line described.

21. through $(9, 10)$; undefined slope $x = 9$

22. through $(-2, 8)$; undefined slope $x = -2$

23. through $(.5, .2)$; vertical $x = .5$

24. through $\left(\frac{5}{8}, \frac{2}{9}\right)$; vertical $x = \frac{5}{8}$

25. through $(-7, 8)$; horizontal $y = 8$

26. through $(\sqrt{2}, \sqrt{7})$; horizontal $y = \sqrt{7}$

Write the standard form of the equation of the line passing through the two points.

27. $(3, 4)$ and $(5, 8)$ $2x - y = 2$

28. $(5, -2)$ and $(-3, 14)$ $2x + y = 8$

29. $(6, 1)$ and $(-2, 5)$ $x + 2y = 8$

30. $(-2, 5)$ and $(-8, 1)$ $2x - 3y = -19$

31. $(2, 5)$ and $(1, 5)$ $y = 5$

32. $(-2, 2)$ and $(4, 2)$ $y = 2$

33. $(7, 6)$ and $(7, -8)$ $x = 7$

34. $(13, 5)$ and $(13, -1)$ $x = 13$

Find the equation of the line satisfying the given conditions. Write it in slope-intercept form.

35. $m = 5$; $b = 15$ $y = 5x + 15$

36. $m = -2$; $b = 12$ $y = -2x + 12$

37. $m = -\frac{2}{3}$; $b = \frac{4}{5}$ $y = -\frac{2}{3}x + \frac{4}{5}$

38. $m = -\frac{5}{8}$; $b = -\frac{1}{3}$ $y = -\frac{5}{8}x - \frac{1}{3}$

39. slope $\frac{2}{5}$; y-intercept $(0, 5)$ $y = \frac{2}{5}x + 5$

40. slope $-\frac{3}{4}$; y-intercept $(0, 7)$ $y = -\frac{3}{4}x + 7$

For the given equation **(a)** *write in slope-intercept form,* **(b)** *give the slope of the line, and* **(c)** *give the y-intercept.*

41. $x + y = 12$
(a) $y = -x + 12$ (b) -1 (c) $(0, 12)$

42. $x - y = 14$
(a) $y = x - 14$ (b) 1 (c) $(0, -14)$

43. $5x + 2y = 20$
(a) $y = -\frac{5}{2}x + 10$ (b) $-\frac{5}{2}$ (c) $(0, 10)$

44. $6x + 5y = 40$
(a) $y = -\frac{6}{5}x + 8$ (b) $-\frac{6}{5}$ (c) $(0, 8)$

45. $2x - 3y = 10$
(a) $y = \frac{2}{3}x - \frac{10}{3}$ (b) $\frac{2}{3}$ (c) $\left(0, -\frac{10}{3}\right)$

46. $4x - 3y = 10$
(a) $y = \frac{4}{3}x - \frac{10}{3}$ (b) $\frac{4}{3}$ (c) $\left(0, -\frac{10}{3}\right)$

Use the method of Example 4 to graph the line whose equation is given in slope-intercept form.

47. $y = \frac{1}{3}x + 2$

48. $y = \frac{1}{4}x + 3$

49. $y = -\frac{3}{2}x - 4$

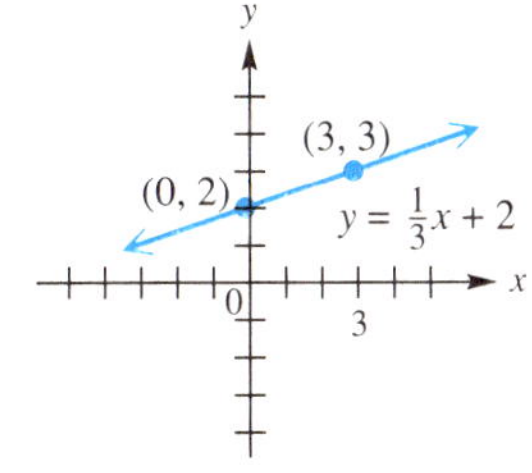

y
(4, 4)
(0, 3)
$y = \frac{1}{4}x + 3$
0
4
x

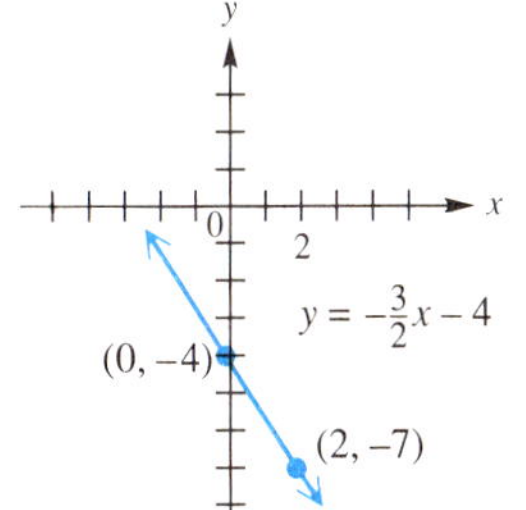

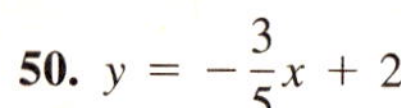

50. $y = -\frac{3}{5}x + 2$

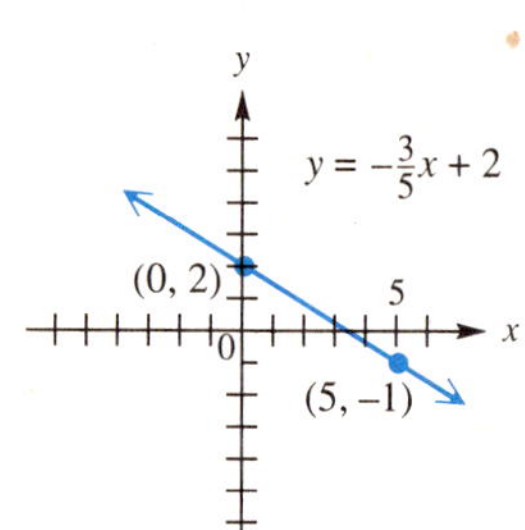

51. $y = -4x - 7$

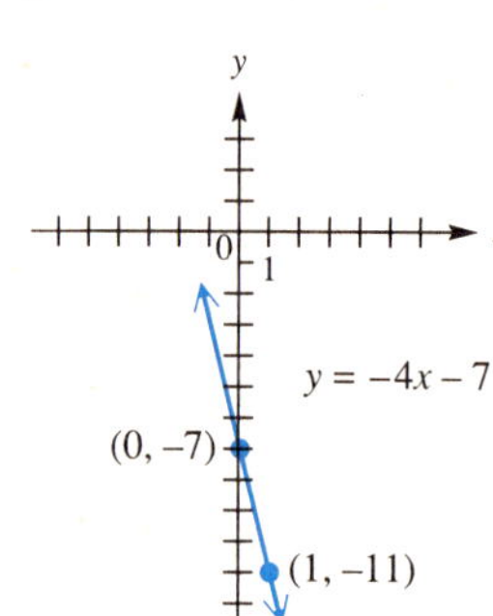

52. $y = -3x + 8$

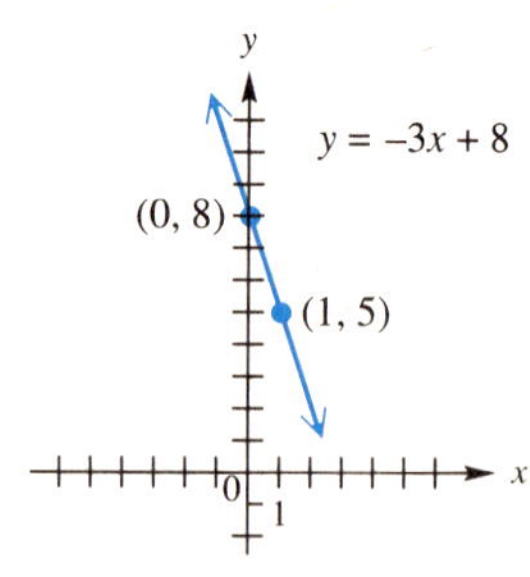

Write an equation in standard form of the line satisfying the given conditions.

53. through (7, 2); parallel to $3x - y = 8$ $\mathbf{3x - y = 19}$

54. through (4, 1); parallel to $2x + 5y = 10$ $\mathbf{2x + 5y = 13}$

55. through $(-2, -2)$; parallel to $-x + 2y = 10$ $\mathbf{x - 2y = 2}$

56. through $(-1, 3)$; parallel to $-x + 3y = 12$ $\mathbf{x - 3y = -10}$

57. through (8, 5); perpendicular to $2x - y = 7$ $\mathbf{x + 2y = 18}$

58. through $(2, -7)$; perpendicular to $5x + 2y = 18$ $\mathbf{2x - 5y = 39}$

59. through $(-2, 7)$; perpendicular to $x = 9$ $\mathbf{y = 7}$

60. through (8, 4); perpendicular to $x = -3$ $\mathbf{y = 4}$

Many real-world situations can be modeled approximately by straight-line graphs. One way to find the equation of such a line is to use two typical data points from the information provided, and then apply the point-slope form of the equation of a line. Because of the usefulness of the slope-intercept form, such equations are often given in the form $y = mx + b$. *Assume that the situation described can be modeled by a straight-line graph, and use the information to find the* $y = mx + b$ *form of the equation of the line.*

61. Let $x = 0$ represent the year 1985. In 1985, U.S. sales of Volkswagen vehicles were about 220,000. In 1993, sales had declined to 43,000. Let y represent sales in thousands. (Thus, the data points for this model are (0, 220) and (8, 43).) (*Source:* Autodata) $\mathbf{y = -22.125x + 220}$

62. Let $x = 0$ represent the year 1989. In 1989, the number of auto accidents resulting in "catastrophic claims"—those amounting to \$100,000 or more—led to an average expected payment of \$575,000. In 1991, the average expected payment had risen to \$892,000. Let y represent the average expected payment in thousands of dollars. (*Source:* Insurance Research Council) $\mathbf{y = 158.5x + 575}$

63. The number of post offices in the United States has been declining in recent years. Use the information given on the bar graph for the years 1984 and 1990, letting $x = 0$ represent the year 1984 and letting y represent the number of post offices. (*Source:* U.S. Postal Service, Annual Report of the Postmaster General)

$y = -\frac{791}{6}x + 29{,}750$

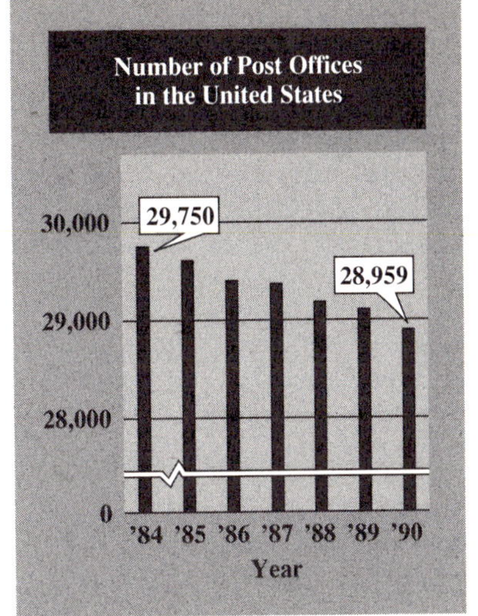

64. The number of motor vehicle deaths in the United States declined from 1988 to 1990 as seen in the accompanying bar graph. Use the information given, with $x = 0$ representing 1988 and y representing the number of deaths. (*Source:* National Safety Council)

$y = -1{,}545.5x + 49{,}391$

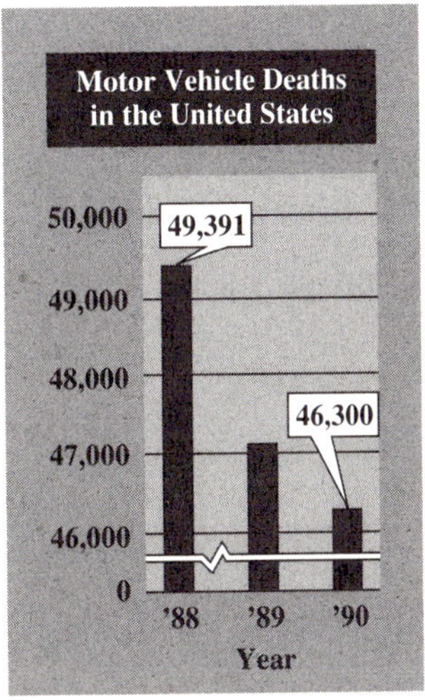

8.4 Exercises (Text Page 433)

1. Explain the meaning of the term.
(a) relation **(b)** domain of a relation **(c)** range of a relation **(d)** function

2. Give an example of a relation that is not a function having domain $\{-3, 2, 6\}$ and range $\{4, 6\}$. (There are many possible correct answers.)
One example is $\{(-3, 4), (2, 4), (2, 6), (6, 4)\}$.

Decide whether the relation is a function and give the domain and the range of the relation. Use the vertical line test in Exercises 9–12.

3. $\{(5, 1), (3, 2), (4, 9), (7, 3)\}$
function; domain: $\{5, 3, 4, 7\}$; range: $\{1, 2, 9, 3\}$

4. $\{(8, 0), (5, 4), (9, 3), (3, 9)\}$
function; domain: $\{8, 5, 9, 3\}$; range: $\{0, 4, 3, 9\}$

5. $\{(2, 4), (0, 2), (2, 6)\}$
not a function; domain: $\{2, 0\}$; range: $\{4, 2, 6\}$

6. $\{(9, -2), (-3, 5), (9, 1)\}$
not a function; domain: $\{9, -3\}$; range: $\{-2, 5, 1\}$

7.

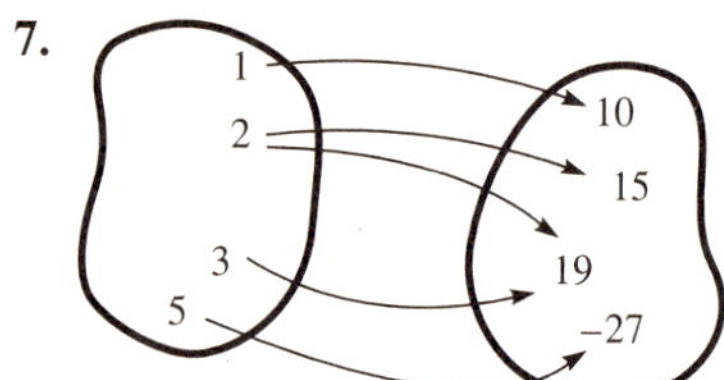

not a function; domain: $\{1, 2, 3, 5\}$; range: $\{10, 15, 19, -27\}$

8.

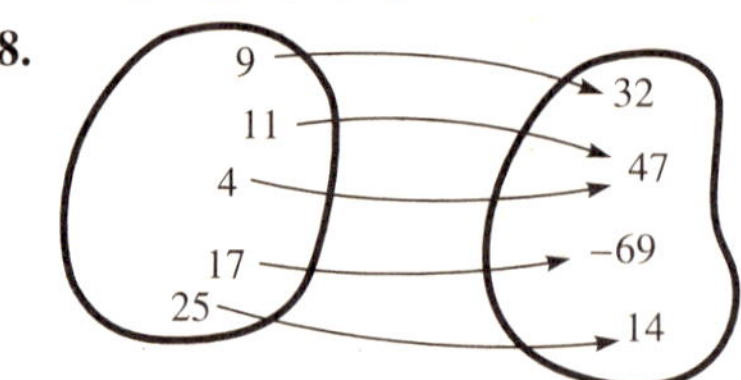

function; domain: $\{9, 11, 4, 17, 25\}$; range: $\{32, 47, -69, 14\}$

9.

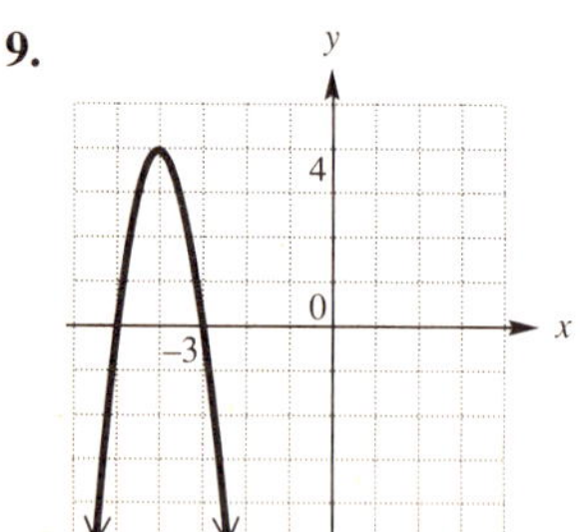
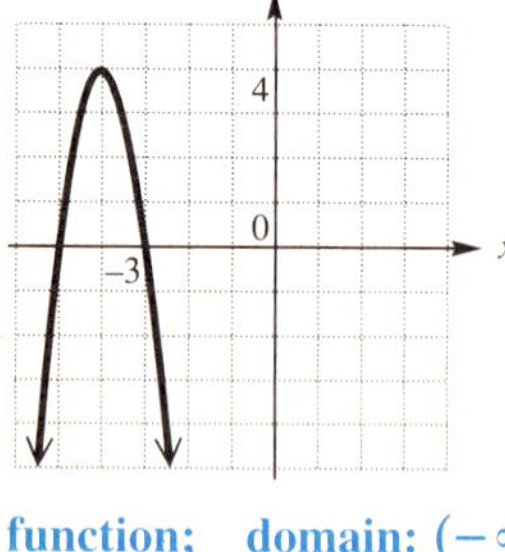

function; domain: $(-\infty, \infty)$; range: $(-\infty, 4]$

10.

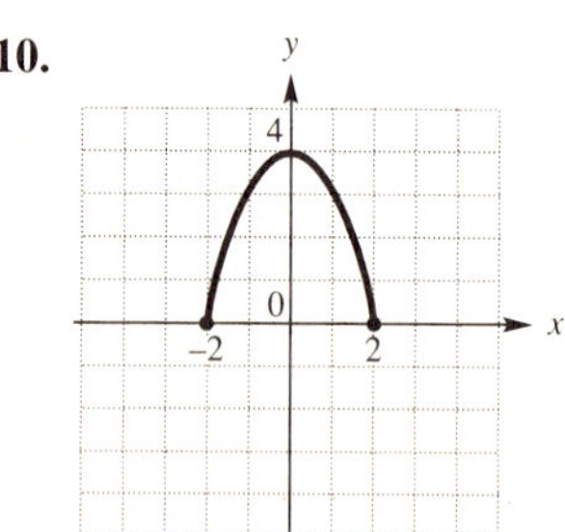

function; domain: $[-2, 2]$; range: $[0, 4]$

11.

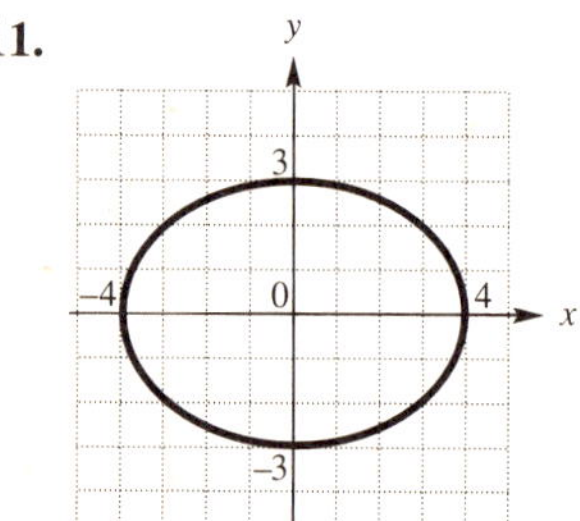

not a function; domain: $[-4, 4]$; range: $[-3, 3]$

12.

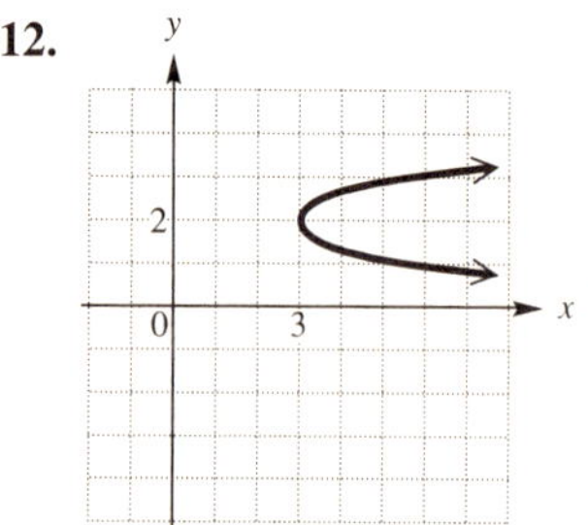

not a function; domain: $[3, \infty)$; range: $(-\infty, \infty)$

Decide whether the given equation defines y as a function of x. Give the domain. Identify any linear functions.

13. $y = x^2$
function; domain: $(-\infty, \infty)$

14. $y = x^3$
function; domain: $(-\infty, \infty)$

15. $x = y^2$
not a function; domain: $[0, \infty)$

16. $x = y^4$
not a function; domain: $[0, \infty)$

17. $x + y < 4$
not a function; domain: $(-\infty, \infty)$

18. $x - y < 3$
not a function; domain: $(-\infty, \infty)$

19. $y = \sqrt{x}$
function; domain: $[0, \infty)$

20. $y = -\sqrt{x}$
function; domain: $[0, \infty)$

21. $xy = 1$
function; domain: $(-\infty, 0) \cup (0, \infty)$

22. $xy = -3$
function; domain: $(-\infty, 0) \cup (0, \infty)$

23. $y = 2x - 6$
function (also a linear function); domain: $(-\infty, \infty)$

24. $y = -6x + 8$
function (also a linear function); domain: $(-\infty, \infty)$

 CONCEPTUAL  WRITING  CHALLENGING   SCIENTIFIC CALCULATOR  GRAPHING CALCULATOR

25. $y = \sqrt{4x + 2}$ **function; domain:** $\left[-\frac{1}{2}, \infty\right)$

26. $y = \sqrt{9 - 2x}$ **function; domain:** $\left(-\infty, \frac{9}{2}\right]$

27. $y = \dfrac{2}{x^2 - 9}$ **function; domain:** $(-\infty, -3) \cup (-3, 3) \cup (3, \infty)$

28. $y = \dfrac{-7}{x^2 - 16}$ **function; domain:** $(-\infty, -4) \cup (-4, 4) \cup (4, \infty)$

Let $f(x) = 3 + 2x$ and $g(x) = x^2 - 2$. *Find the function value.*

29. $f(1)$ **5**

30. $f(4)$ **11**

31. $g(2)$ **2**

32. $g(0)$ **−2**

33. $g(-1)$ **−1**

34. $g(-3)$ **7**

35. $f(-8)$ **−13**

36. $f(-5)$ **−7**

Sketch the graph of each linear function. Give the domain and range.

37. $f(x) = -2x + 5$

$f(x) = -2x + 5$

domain and range: $(-\infty, \infty)$

38. $g(x) = 4x - 1$

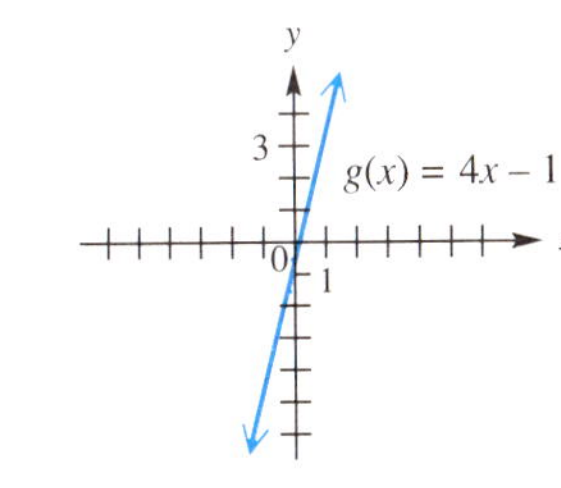

domain and range: $(-\infty, \infty)$

39. $h(x) = \frac{1}{2}x + 2$

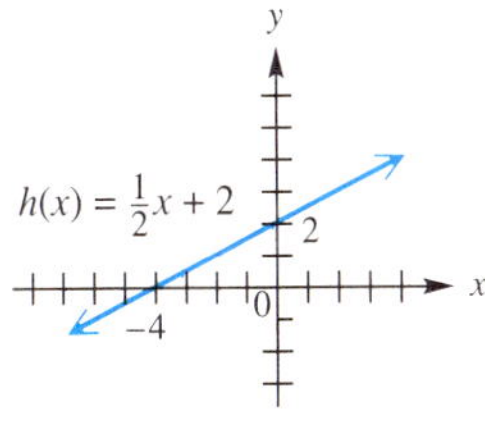

domain and range: $(-\infty, \infty)$

40. $F(x) = -\frac{1}{4}x + 1$

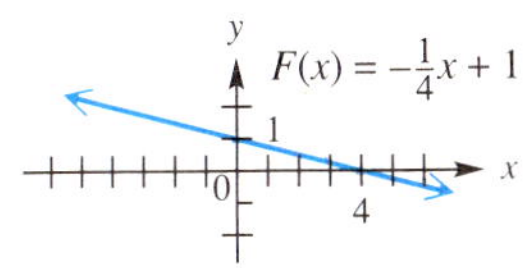

domain and range: $(-\infty, \infty)$

41. $G(x) = 2x$

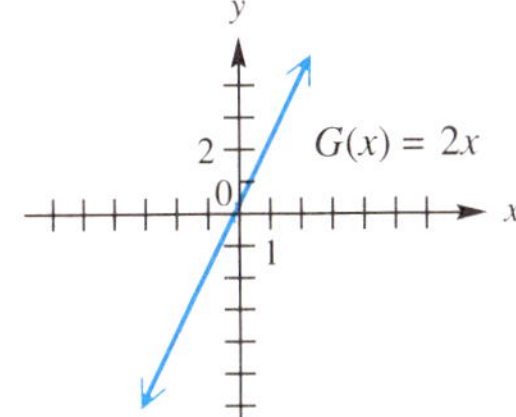

domain and range: $(-\infty, \infty)$

42. $H(x) = -3x$

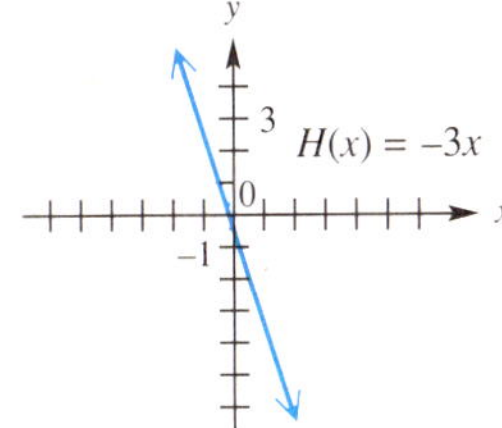

domain and range: $(-\infty, \infty)$

43. $f(x) = 5$

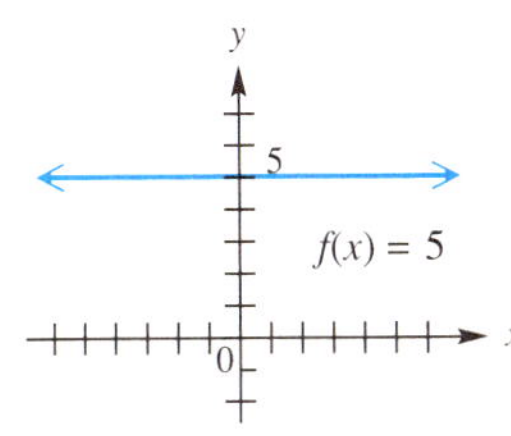

domain: $(-\infty, \infty)$;
range: $\{5\}$

44. $g(x) = -4$

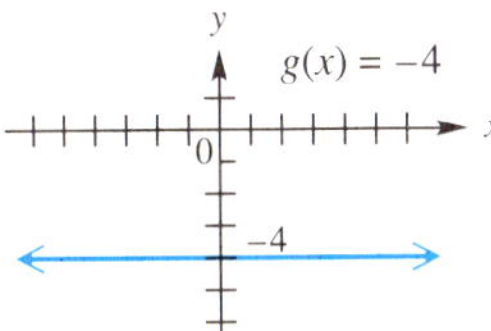

domain: $(-\infty, \infty)$;
range: $\{-4\}$

45. Fill in the blanks with the correct responses.

The equation $2x + y = 4$ has a straight _____ as its graph. One point that lies on the line is (3, _____). If we solve the equation for y and use function notation, we have a linear function $f(x) =$ _____________. For this function, $f(3) =$ _____, meaning that the point (_____, _____) lies on the graph of the function.
line; -2; $-2x + 4$; -2; 3; -2

46. Which one of the following defines a linear function?

(a) $y = \dfrac{x - 5}{4}$ **(b)** $y = \sqrt[3]{x}$

(c) $y = x^2$ **(d)** $y = x^{1/2}$ **(a)**

47. Which one of the functions in Exercise 46 has domain $[0, \infty)$? **(d)**

48. Which one of the functions in Exercise 46 does not have negative numbers in its domain? **(d)**

49. (a) Suppose that a taxicab driver charges \$.50 per mile. Fill in the chart with the correct response for the price $R(x)$ she charges for a trip of x miles.

x	$R(x)$
0	
1	
2	
3	

\$0; \$.50; \$1.00; \$1.50

(b) The linear function that gives a rule for the amount charged is $R(x) =$ _____. **\$.50x**

(c) Graph this function for x, where x is an element of $\{0, 1, 2, 3\}$.

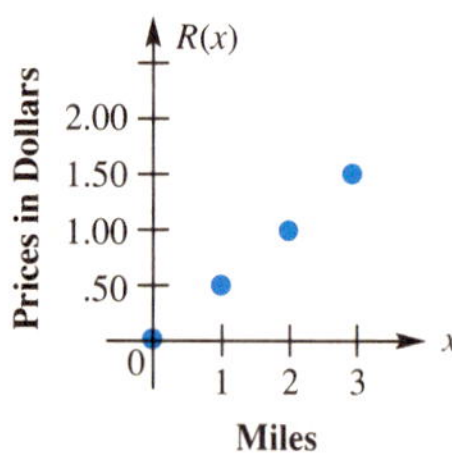

50. Suppose that a package weighing x pounds costs $C(x)$ dollars to mail to a given location, where

$$C(x) = 2.75x.$$

(a) What is the value of $C(3)$? **8.25 (dollars)**

(b) In your own words, describe what 3 and the value $C(3)$ mean in part (a), using the terminology *domain* and *range*.

51. The graphing calculator screen shows the graph of a linear function $y = f(x)$, along with the display of coordinates of a point on the graph. Use function notation to write what the display indicates.

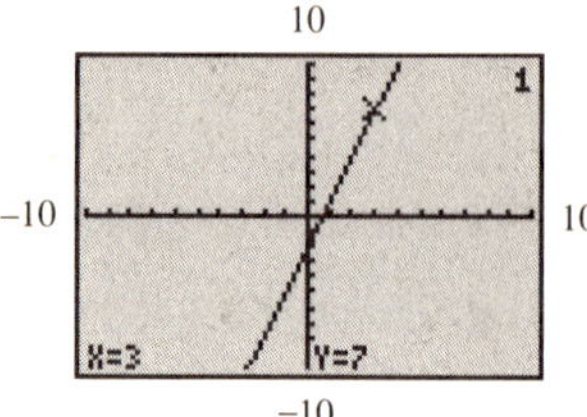

$f(3) = 7$

52. The table was generated by a graphing calculator for a linear function $y_1 = f(x)$. Use the table to answer.
(a) What is $f(2)$?
(b) If $f(x) = -3.7$, what is the value of x?
(c) What is the slope of the line?
(d) What is the y-intercept of the line?
(e) Find the expression for $f(x)$.

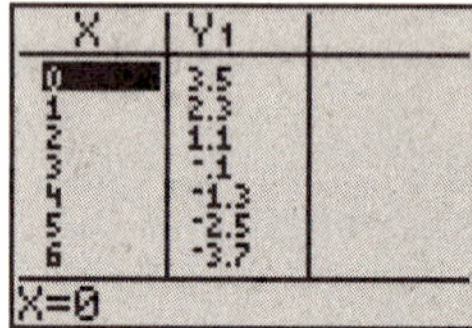

X	Y1
0	3.5
1	2.3
2	1.1
3	-.1
4	-1.3
5	-2.5
6	-3.7

X=0

(a) $f(2) = 1.1$ (b) $x = 6$ (c) -1.2 (d) 3.5
(e) $f(x) = -1.2x + 3.5$

53. The two screens show the graph of the same linear function $y = f(x)$. Find the expression for $f(x)$.

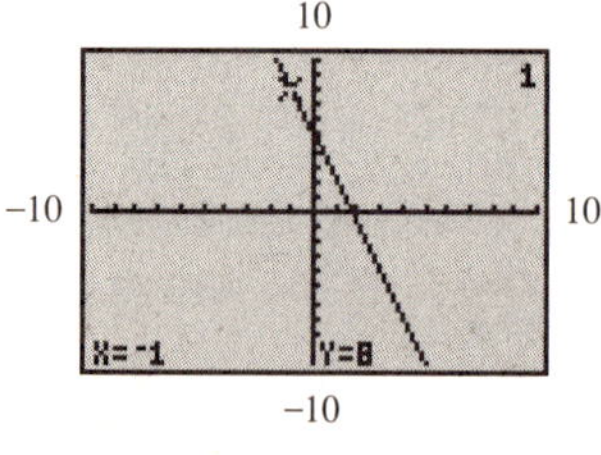

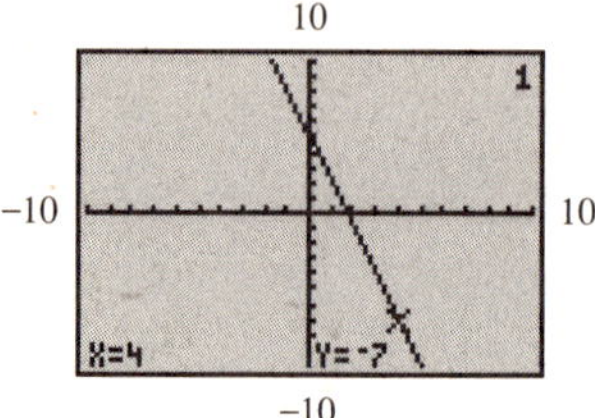

$f(x) = -3x + 5$

 CONCEPTUAL  WRITING  CHALLENGING  SCIENTIFIC CALCULATOR  GRAPHING CALCULATOR

54. As determined in the previous section, the formula for converting Celsius to Fahrenheit is $F = 1.8C + 32$. If we graph $y = f(x) = 1.8x + 32$ on a graphing calculator screen, we obtain the accompanying picture. (We used the interval $[-50, 50]$ on the x-axis and $[-75, 50]$ on the y-axis.) The point $(-40, -40)$ lies on the graph, as indicated by the display. Interpret the meaning of this in the context of this exercise.

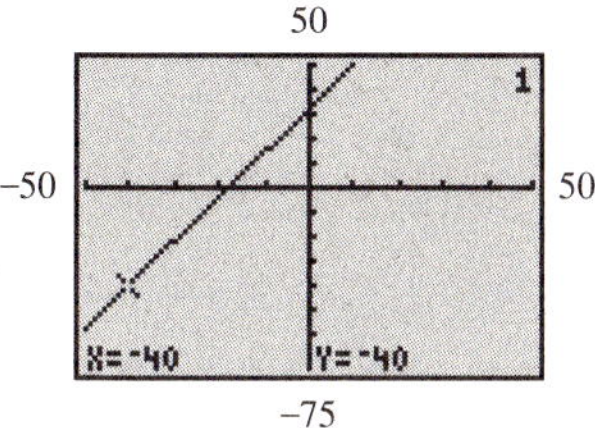

When the Celsius temperature is $-40°$, the Fahrenheit temperature is also $-40°$.

55. The linear function $f(x) = -123x + 29{,}685$ provides a model for the number of post offices in the United States from 1984 to 1990, where $x = 0$ corresponds to 1984, $x = 1$ corresponds to 1985, and so on. Use this model to give the approximate number of post offices during the given year. (*Source:* U.S. Postal Service, Annual Report of the Postmaster General)

(a) 1985 **(b)** 1987 **(c)** 1990

(d) The graphing calculator screen shows a portion of the graph of $y = f(x)$ with the coordinates of a point on the graph displayed at the bottom of the screen. Interpret the meaning of the display in the context of this exercise.

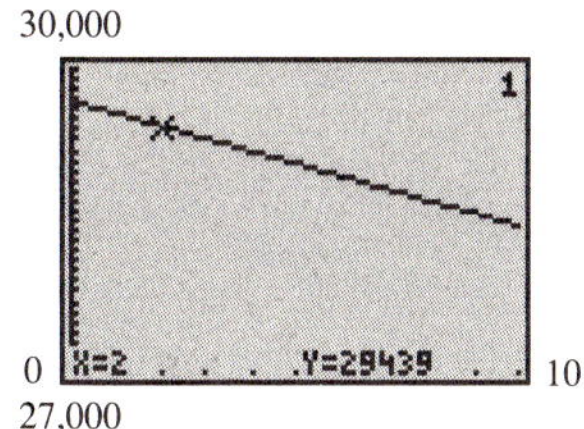

(a) 29,562 (b) 29,316 (c) 28,947
(d) In 1986 (when $x = 2$), the number of post offices was approximately 29,439.

56. The linear function $f(x) = 1{,}650x + 3{,}817$ provides a model for the United States defense budget for the decade of the 1980s, where $x = 0$ corresponds to 1980, $x = 1$ corresponds to 1981, and so on, with $f(x)$ representing the budget in millions of dollars. Use this model to approximate the defense budget during the given year. (*Source:* U.S. Office of Management and Budget)

(a) 1983 **(b)** 1985 **(c)** 1988

(d) The graphing calculator screen shows a portion of the graph of $y = f(x)$ with the coordinates of a point on the graph displayed at the bottom of the screen. Interpret the meaning of the display in the context of this exercise.

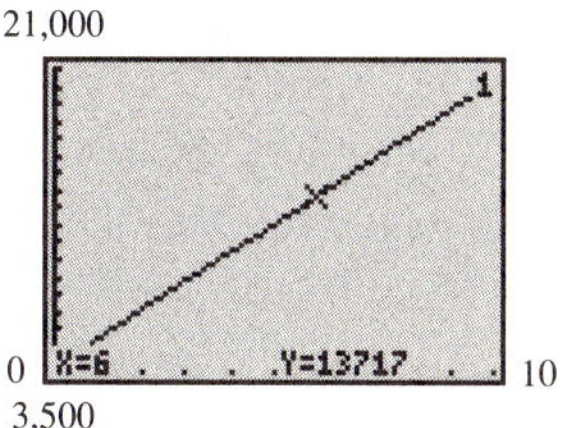

(a) \$8,767 million (b) \$12,067 million
(c) \$17,017 million
(d) In 1986 (when $x = 6$), the defense budget was approximately \$13,717 million.

In each of the following, **(a)** *express the cost C as a function of x, where x represents the quantity of items as given;* **(b)** *express the revenue R as a function of x;* **(c)** *determine the value of x for which revenue equals cost;* **(d)** *graph $y = C(x)$ and $y = R(x)$ on the same axes, and interpret the graph.*

57. Christy Lawrence stuffs envelopes for extra income during her spare time. Her initial cost to obtain the necessary information for the job was \$200.00. Each envelope costs \$.02 and she gets paid \$.04 per envelope stuffed. Let x represent the number of envelopes stuffed.

(a) $C(x) = .02x + 200$ (b) $R(x) = .04x$
(c) 10,000 (d)

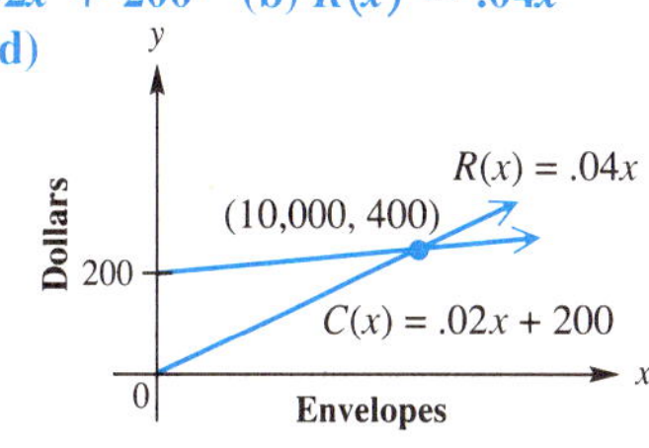

For $x < 10{,}000$, a loss
For $x > 10{,}000$, a profit

58. Evan Turner runs a copying service in his home. He paid \$3,500 for the copier and a lifetime service contract. Each sheet of paper he uses costs \$.01, and

he gets paid \$.05 per copy he makes. Let x represent the number of copies he makes.
(a) $C(x) = .01x + 3{,}500$ (b) $R(x) = .05x$
(c) 87,500 (d)

For $x < 87{,}500$, a loss
For $x > 87{,}500$, a profit

59. Ray Kelley operates a delivery service in a southern city. His start-up costs amounted to \$2,300. He estimates that it costs him (in terms of gasoline, wear and tear on his car, etc.) \$3.00 per delivery. He charges \$5.50 per delivery. Let x represent the number of deliveries he makes.
(a) $C(x) = 3.00x + 2{,}300$ (b) $R(x) = 5.50x$
(c) 920 (d)

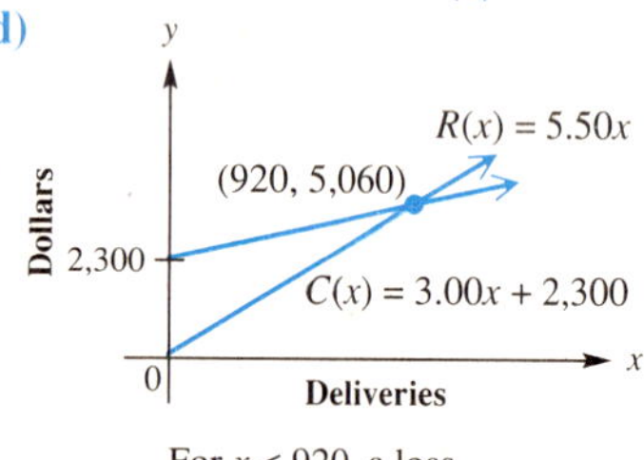

For $x < 920$, a loss
For $x > 920$, a profit

60. Julie Campbell bakes cakes and sells them at county fairs. Her initial cost for the Washington Parish fair in 1996 was \$40.00. She figures that each cake costs \$2.50 to make, and she charges \$6.50 per cake. Let x represent the number of cakes sold. (Assume that there were no cakes left over.)
(a) $C(x) = 2.50x + 40$ (b) $R(x) = 6.50x$
(c) 10 (d)

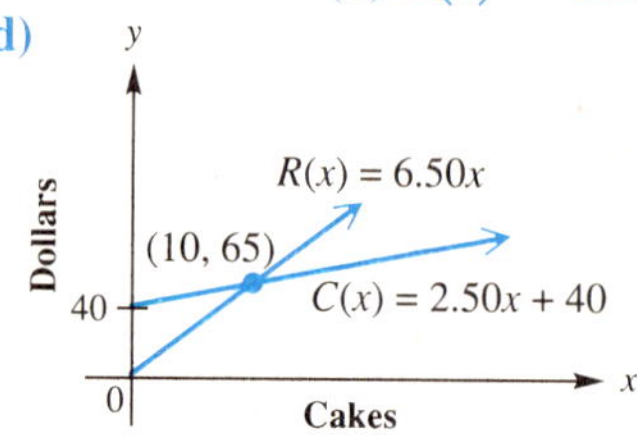

For $x < 10$, a loss
For $x > 10$, a profit

8.5 Exercises (Text Page 443)

1. Explain in your own words the meaning of the term.
(a) vertex of a parabola **(b)** axis of a parabola

2. Explain why the axis of the graph of a quadratic function cannot be a horizontal line.

Identify the vertex of the graph of the quadratic function.

3. $f(x) = -3x^2$
(0, 0)

4. $f(x) = -.5x^2$
(0, 0)

5. $f(x) = x^2 + 4$
(0, 4)

6. $f(x) = x^2 - 4$
(0, −4)

7. $f(x) = (x - 1)^2$
(1, 0)

8. $f(x) = (x + 3)^2$
(−3, 0)

9. $f(x) = (x + 3)^2 - 4$
(−3, −4)

10. $f(x) = (x - 5)^2 - 8$
(5, −8)

For the quadratic function, tell whether the graph opens upward or downward, and tell whether the graph is wider, narrower, or the same as the graph of $f(x) = x^2$.

11. $f(x) = -2x^2$
downward; narrower

12. $f(x) = -3x^2 + 1$
downward; narrower

13. $f(x) = .5x^2$
upward; wider

14. $f(x) = \frac{2}{3}x^2 - 4$
upward; wider

CONCEPTUAL  WRITING  CHALLENGING  SCIENTIFIC CALCULATOR GRAPHING CALCULATOR

15. Describe how the graph of each parabola in Exercises 9 and 10 is shifted compared to the graph of $y = x^2$.

16. Describe how the sign and the absolute value of a in $f(x) = a(x - h)^2 + k$ affects the graph of the function when compared to the graph of $y = x^2$.

17. For $f(x) = a(x - h)^2 + k$, in what quadrant is the vertex if:
(a) $h > 0, k > 0$; **(b)** $h > 0, k < 0$; **(c)** $h < 0, k > 0$; **(d)** $h < 0, k < 0$?
(a) I (b) IV (c) II (d) III

18. **(a)** What is the value of h if the graph of $f(x) = a(x - h)^2 + k$ has vertex on the y-axis?
(b) What is the value of k if the graph of $f(x) = a(x - h)^2 + k$ has vertex on the x-axis?
(a) 0 (b) 0

Match the graph of each equation with the graph at the right that it most closely resembles.

19. $f(x) = x^2 + 2$ **H**

20. $g(x) = x^2 - 5$ **G**

21. $h(x) = -x^2 + 4$ **B**

22. $k(x) = -x^2 - 4$ **C**

23. $F(x) = (x - 1)^2$ **D**

24. $G(x) = (x + 1)^2$ **A**

25. $H(x) = (x - 1)^2 + 1$ **F**

26. $K(x) = (x + 1)^2 + 1$ **E**

A.

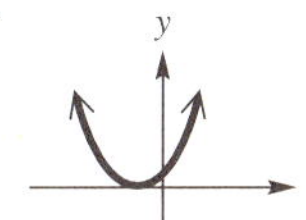

B.

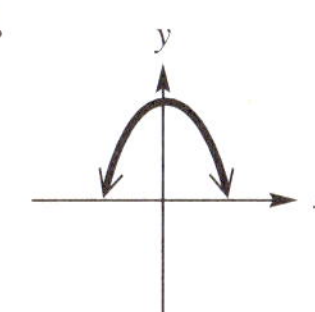

C.

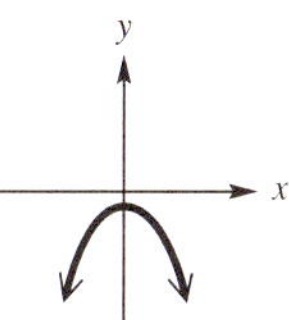

D.

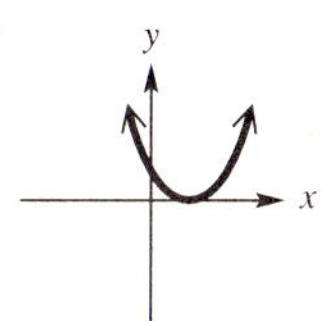

E. **F.**

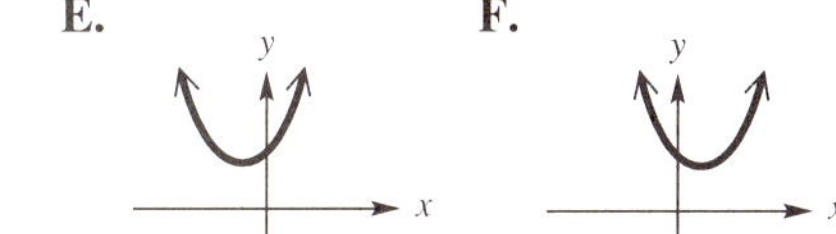

G.

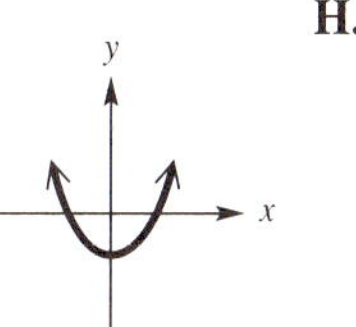

H.

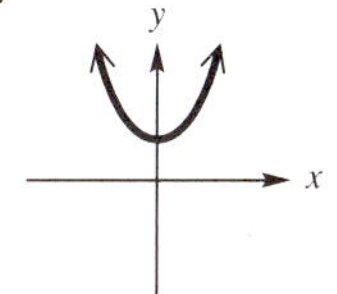

Sketch the graph of each quadratic function using the methods described in this section. Indicate two points on each graph.

27. $f(x) = 3x^2$

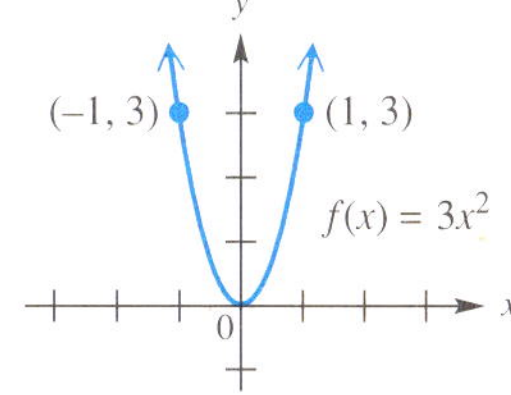

28. $f(x) = -2x^2$

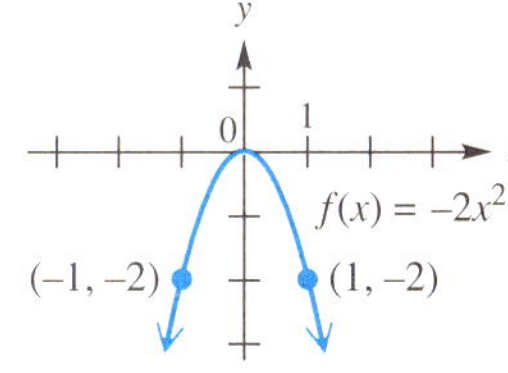

29. $f(x) = -\frac{1}{4}x^2$

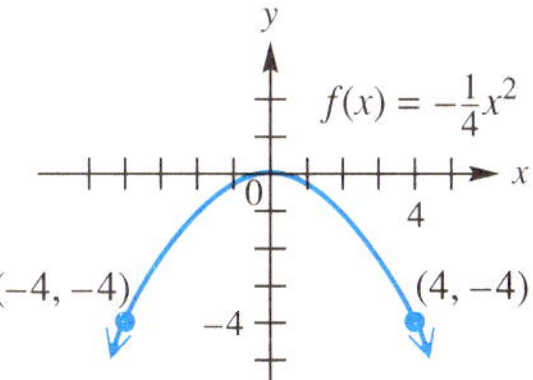

30. $f(x) = \frac{1}{3}x^2$

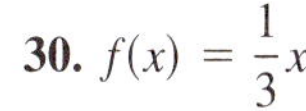

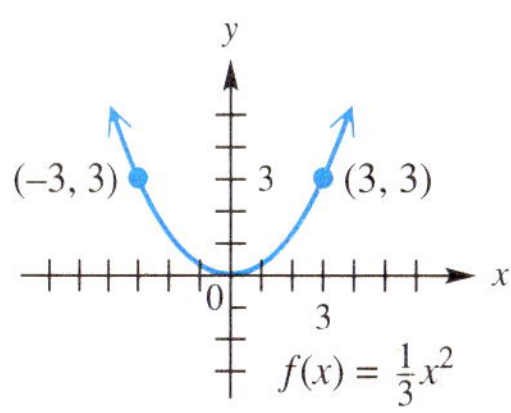

31. $f(x) = x^2 - 1$

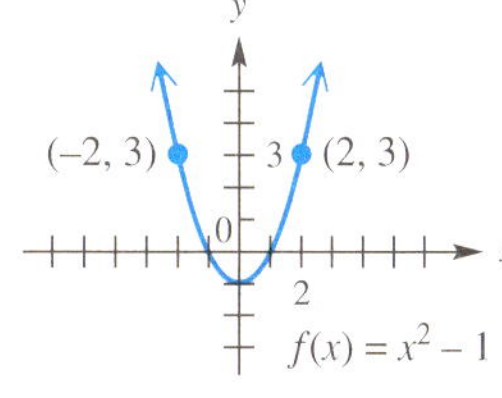

32. $f(x) = x^2 + 3$

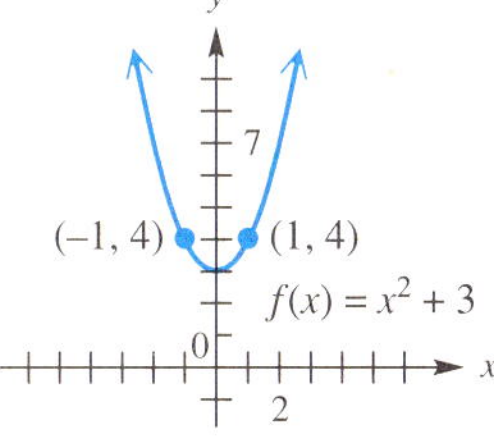

33. $f(x) = -x^2 + 2$

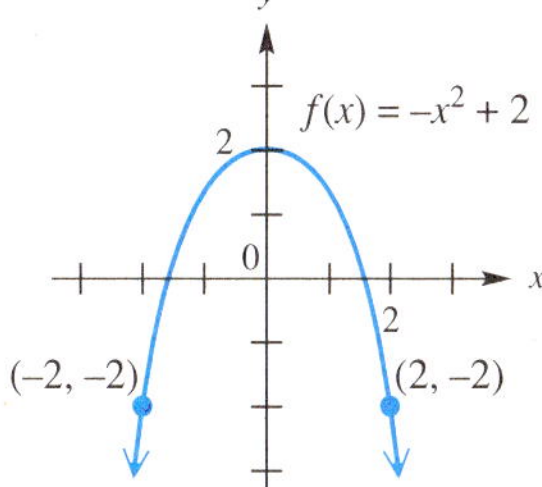

34. $f(x) = -x^2 - 4$

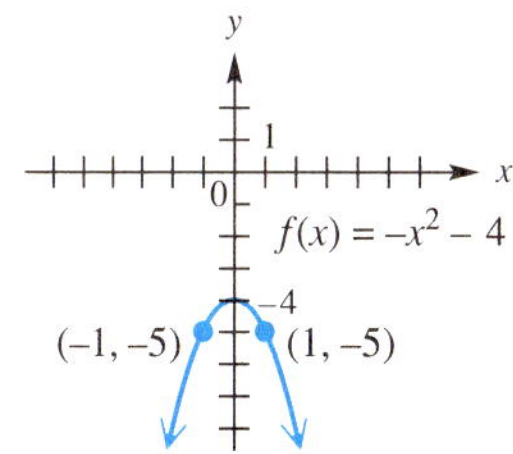

35. $f(x) = 2x^2 - 2$

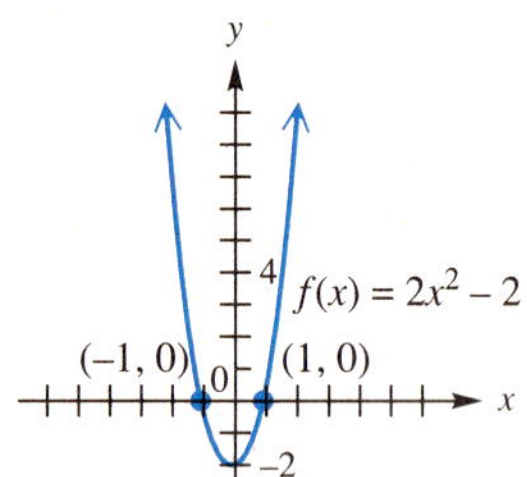

36. $f(x) = -3x^2 + 1$

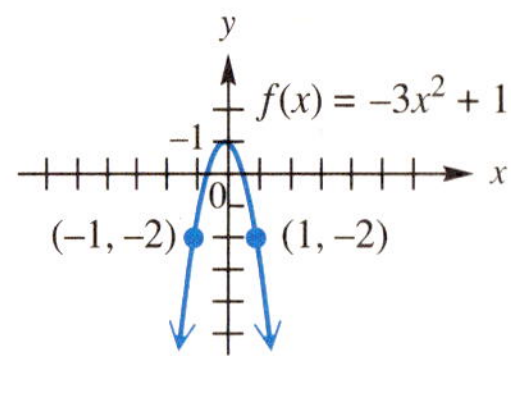

37. $f(x) = (x - 4)^2$

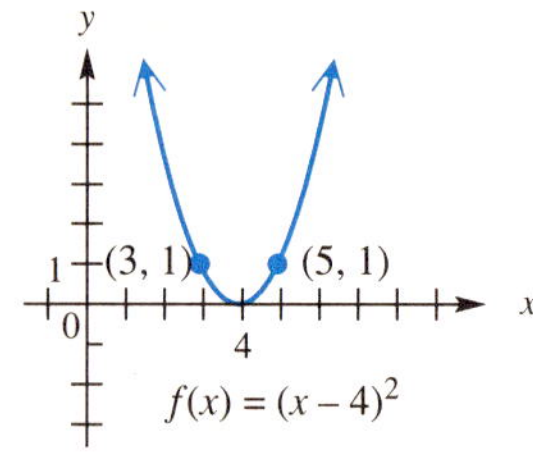

38. $f(x) = (x - 3)^2$

39. $f(x) = 3(x + 1)^2$

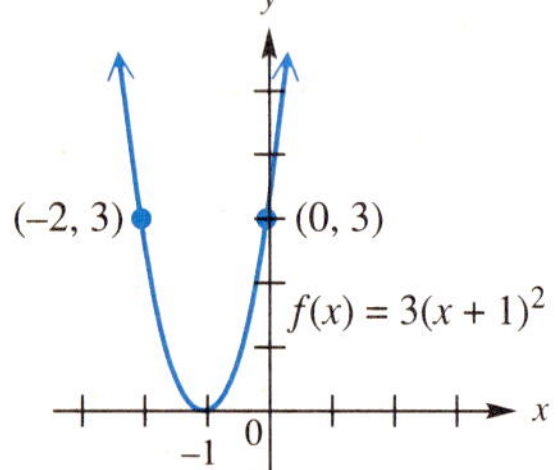

40. $f(x) = -2(x + 1)^2$

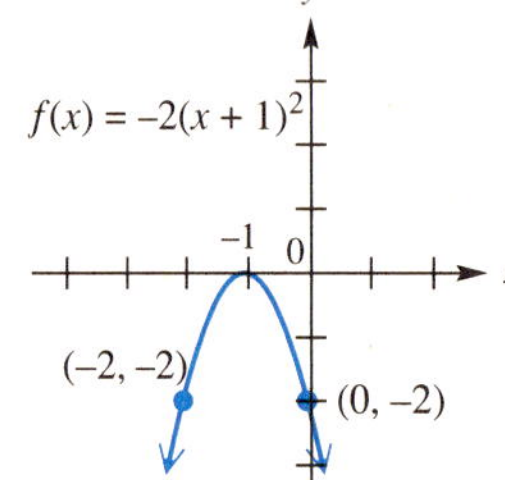

41. $f(x) = (x + 1)^2 - 2$

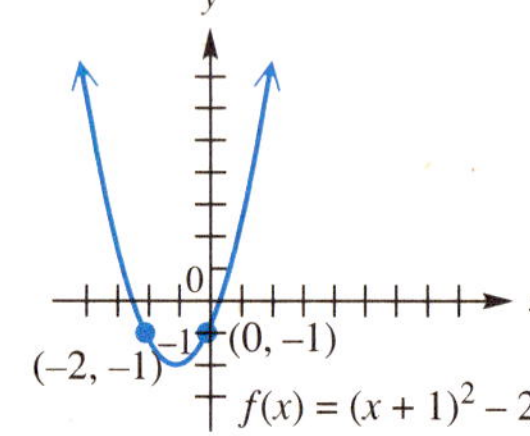

42. $f(x) = (x - 2)^2 + 3$

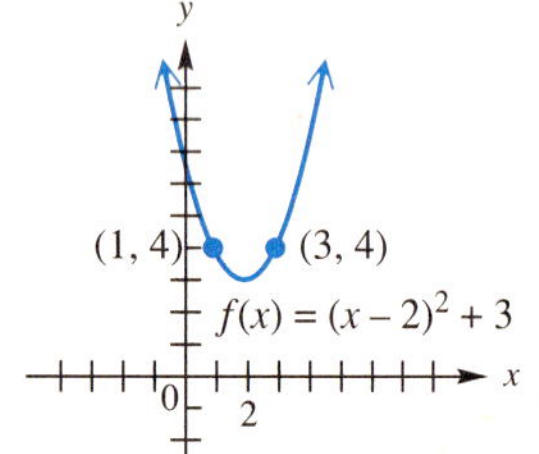

Sketch the graph of each quadratic function. Indicate the coordinates of the vertex of the graph.

43. $f(x) = x^2 + 8x + 14$

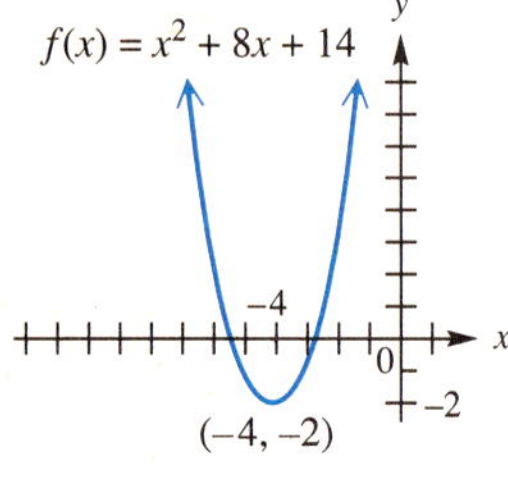

44. $f(x) = x^2 + 10x + 23$

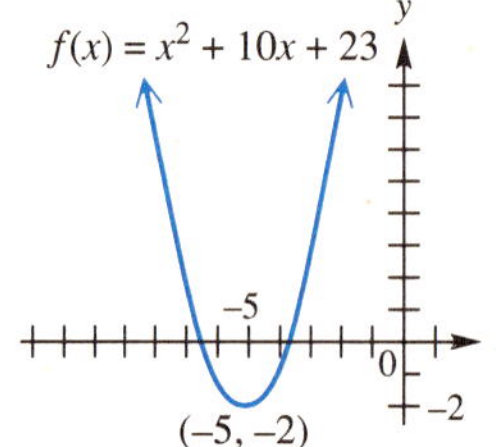

45. $f(x) = x^2 + 2x - 4$

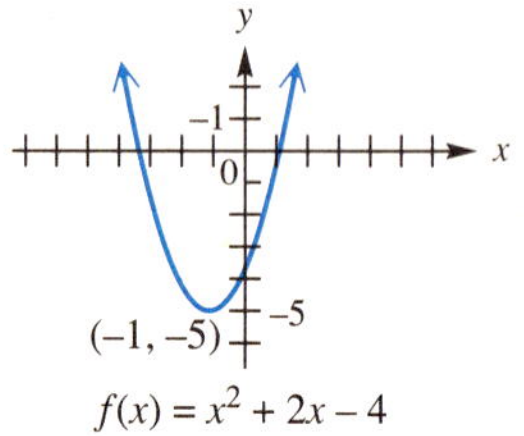

46. $f(x) = 3x^2 - 9x + 8$

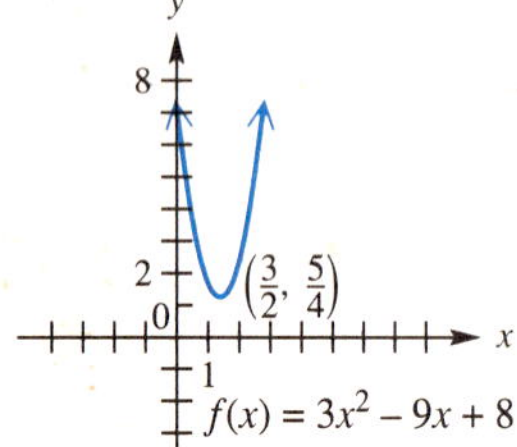

47. $f(x) = -2x^2 + 4x + 5$

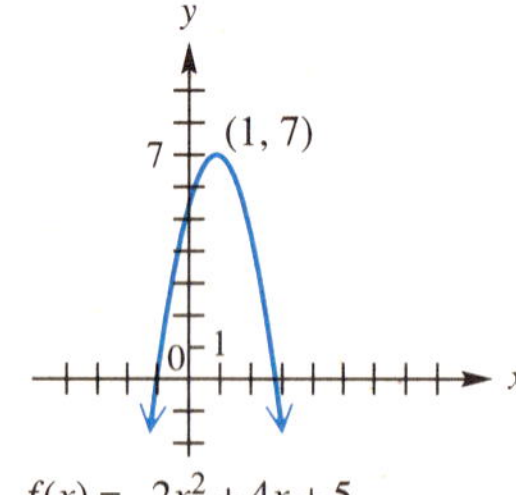

48. $f(x) = -5x^2 - 10x + 2$

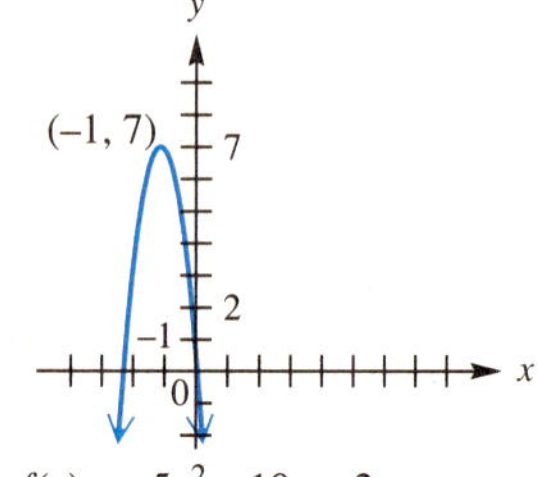

The accompanying bar graph shows the annual average number of nonfarm payroll jobs in California for the years 1988 through 1992. If the tops of the bars were joined by a smooth curve, the curve would resemble the graph of a quadratic function (that is, a parabola). Using a technique from statistics it can be determined that this function can be described approximately as

$$f(x) = -.10x^2 + .42x + 11.90,$$

where $x = 0$ corresponds to 1988, $x = 1$ corresponds to 1989, and so on, and $f(x)$ represents the number of payroll jobs in millions. (Source: California Employment Development Department)

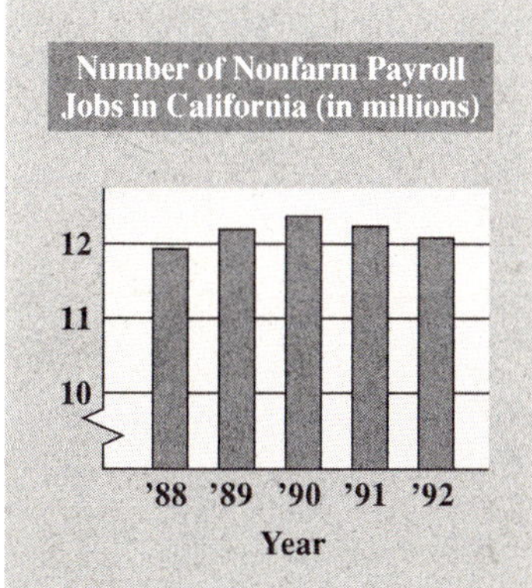

49. Explain why the coefficient of x^2 in the function is negative, based on the graph formed by joining the tops of the bars.
The parabola opens downward, indicating $a < 0$.

50. Determine the coordinates of the vertex of the graph using algebraic methods. (2.1, 12.34)

51. How does the x-coordinate of the vertex of the parabola indicate that during the time period under consideration, the maximum number of payroll jobs was in 1990?
Since $x = 2.1 \approx 2$, the year corresponding to 2, which is 1990, is the year of maximum number of jobs.

52. What does the y-coordinate of the vertex of the parabola indicate?
It indicates the maximum number of jobs, which was approximately 12.34 million.

Solve the problem.

53. Keisha Hughes has 100 meters of fencing material to enclose a rectangular exercise run for her dog. What width will give the enclosure the maximum area?
25 meters

54. Morgan's Department Store wants to construct a rectangular parking lot on land bordered on one side by a highway. It has 280 feet of fencing that is to be used to fence off the other three sides. What should be the dimensions of the lot if the enclosed area is to be a maximum? What is the maximum area?
140 feet by 70 feet; 9,800 square feet

55. The world consumption of oil in millions of barrels daily is given by

$$f(x) = -.1x^2 + 2x + 58,$$

where x is the number of years since 1985 (1985 corresponds to 0). According to this equation, when will oil consumption reach a maximum? What will the maximum consumption be? Discuss how realistically this equation describes oil consumption.
in 1995; 68 million barrels daily

56. Recent annual inflation rates (in percent) in Mexico are given by the function

$$f(x) = 4x^2 - 48x + 154,$$

where x represents the number of years since 1987. In what year does this equation indicate that the inflation rate will be a minimum? What is the minimum rate? Discuss how realistic this equation may or may not be. in 1993; 10%

57. If an object is thrown upward with an initial velocity of 32 feet per second, then its height after t seconds is given by

$$h(t) = 32t - 16t^2.$$

Find the maximum height attained by the object and the number of seconds it takes to hit the ground.
16 feet; 2 seconds

58. A projectile is fired straight upward so that its distance (in feet) above the ground t seconds after firing is

$$s(t) = -16t^2 + 400t.$$

Find the maximum height it reaches and the number of seconds it takes to reach that height.
2,500 feet; 12.5 seconds

59. If air resistance is neglected, a projectile shot straight upward with an initial velocity of 40 meters per second will be at a height s in meters given by the function

$$s(t) = -4.9t^2 + 40t,$$

where t is the number of seconds elapsed after projection. After how many seconds will it reach its maximum height, and what is this maximum height? Round your answers to the nearest tenth.
4.1 seconds; 81.6 meters

60. For a trip to a resort, a charter bus company charges a fare of \$48 per person, plus \$2 per person for each unsold seat on the bus. If the bus has 42 seats and x represents the number of unsold seats, obtain the following:

(a) a function that defines the total revenue, R, from the trip (*Hint:* Multiply the total number riding, $42 - x$, by the price per ticket, $48 + 2x$); **$R(x) = (42 - x)(48 + 2x) = -2x^2 + 36x + 2{,}016$**

(b) the graph of the function from part (a);

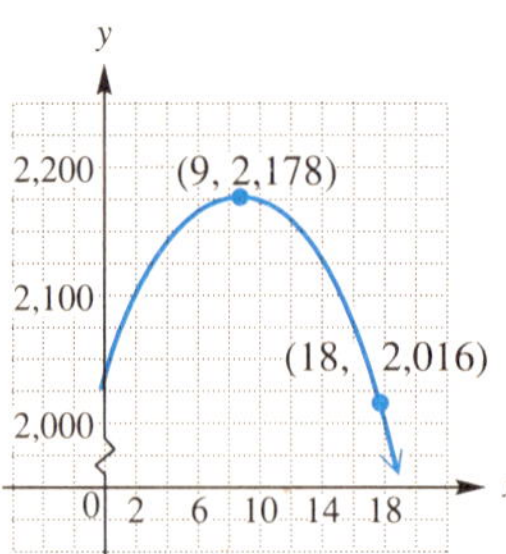

(c) the number of unsold seats that produces the maximum revenue; **9**

(d) the maximum revenue. **\$2,178**

8.6 Exercises (Text Page 455)

Fill in each blank with the correct response.

1. For an exponential function $f(x) = a^x$, if $a > 1$, the graph ________ (rises/falls) from left to right. If $0 < a < 1$, the graph ________ (rises/falls) from left to right. **rises; falls**

2. The y-intercept of the graph of $y = a^x$ is ____. **(0, 1)**

3. The graph of the exponential function $f(x) = a^x$ ________ (does/does not) have an x-intercept. **does not**

4. The point (2, ____) is on the graph of $f(x) = 3^{4x-3}$. **243**

5. For a logarithmic function $g(x) = \log_a x$, if $a > 1$, the graph ________ (rises/falls) from left to right. If $0 < a < 1$, the graph ________ (rises/falls) from left to right. **rises; falls**

6. The x-intercept of the graph of $y = \log_a x$ is ____. **(1, 0)**

7. The graph of the exponential function $g(x) = \log_a x$ ________ (does/does not) have a y-intercept. **does not**

8. The point (98, ____) lies on the graph of $g(x) = \log_{10}(x + 2)$. **2**

Use a calculator to find an approximation for each of the following numbers. Give as many digits as the calculator displays.

9. $9^{3/7}$ **2.56425419972**

10. $14^{2/7}$ **2.12551979078**

11. $(.83)^{-1.2}$ **1.25056505582**

12. $(.97)^{3.4}$ **.901620746784**

13. $(\sqrt{6})^{\sqrt{5}}$ **7.41309466896**

14. $(\sqrt{7})^{\sqrt{3}}$ **5.39357064307**

15. $\left(\frac{1}{3}\right)^{9.8}$ **.0000210965628481**

16. $\left(\frac{2}{5}\right)^{8.1}$ **.000597978996117**

Sketch the graph of each of the following functions.

17. $f(x) = 3^x$

18. $f(x) = 5^x$

19. $f(x) = \left(\frac{1}{4}\right)^x$

20. $f(x) = \left(\frac{1}{3}\right)^x$

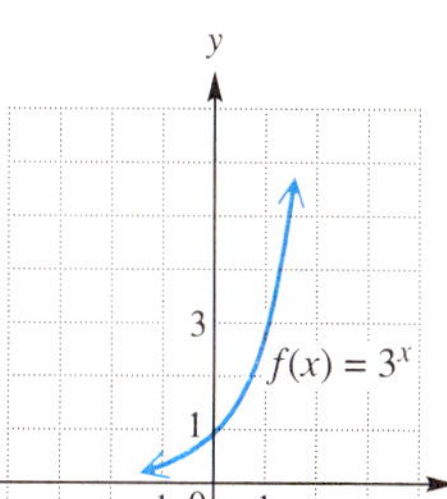

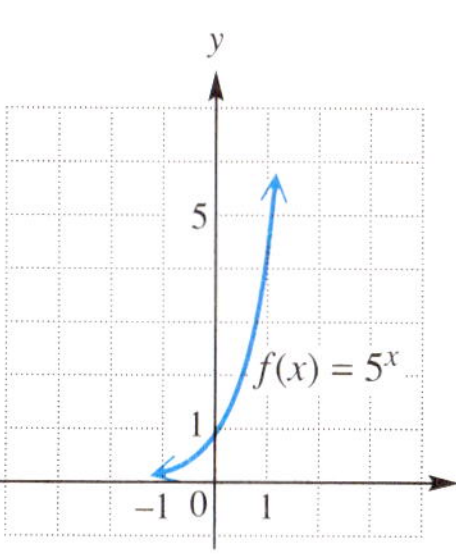

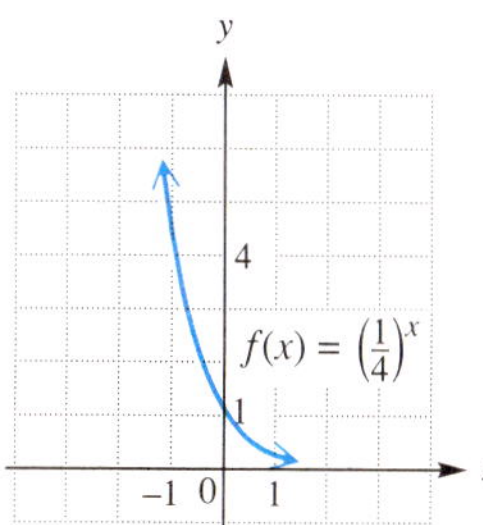

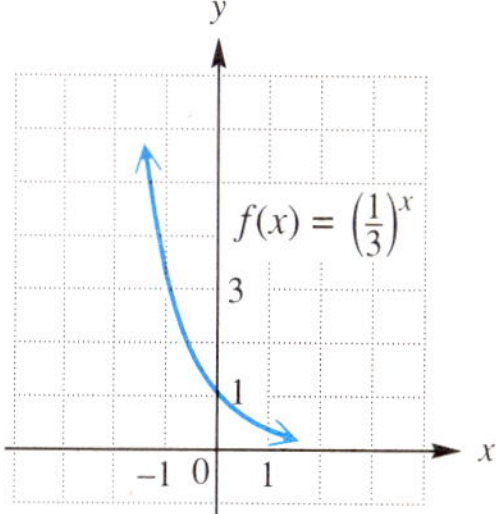

Use a calculator to find each of the following. Give as many digits as the calculator displays.

21. e^3
20.0855369232

22. e^4
54.5981500331

23. e^{-4}
.018315638889

24. e^{-3}
.049787068368

In Exercises 25–28, rewrite the exponential equation as a logarithmic equation. In Exercises 29–32, rewrite the logarithmic equation as an exponential equation.

25. $4^2 = 16$
$\mathbf{2 = \log_4 16}$

26. $5^3 = 125$
$\mathbf{3 = \log_5 125}$

27. $\left(\frac{2}{3}\right)^{-3} = \frac{27}{8}$
$\mathbf{-3 = \log_{2/3}\left(\frac{27}{8}\right)}$

28. $\left(\frac{1}{10}\right)^{-4} = 10{,}000$
$\mathbf{-4 = \log_{1/10} 10{,}000}$

29. $5 = \log_2 32$
$\mathbf{2^5 = 32}$

30. $3 = \log_4 64$
$\mathbf{4^3 = 64}$

31. $1 = \log_3 3$
$\mathbf{3^1 = 3}$

32. $0 = \log_{12} 1$
$\mathbf{12^0 = 1}$

Use a calculator to find each of the following. Give as many digits as the calculator displays.

33. ln 4
1.38629436112

34. ln 6
1.79175946923

35. ln .35
−1.0498221245

36. ln 2.45
.896088024557

Sketch the graph of each of the following functions. (Hint: Use the graphs of the exponential functions in Exercises 17–20 to help.)

37. $g(x) = \log_3 x$

38. $g(x) = \log_5 x$

39. $g(x) = \log_{1/4} x$

40. $g(x) = \log_{1/3} x$

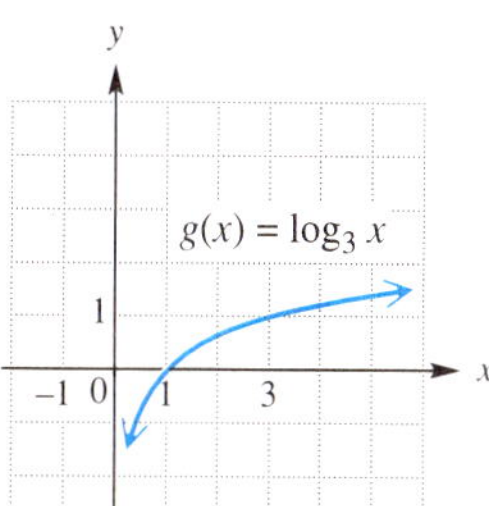

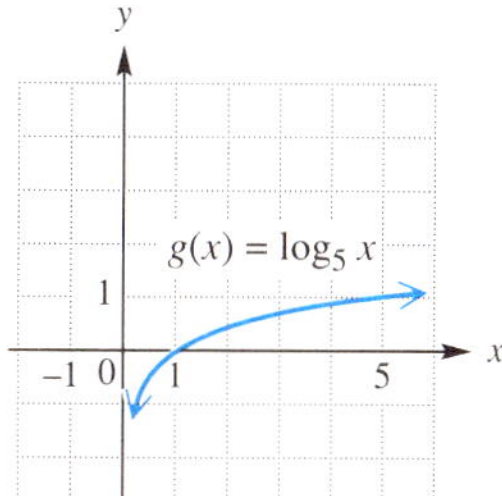

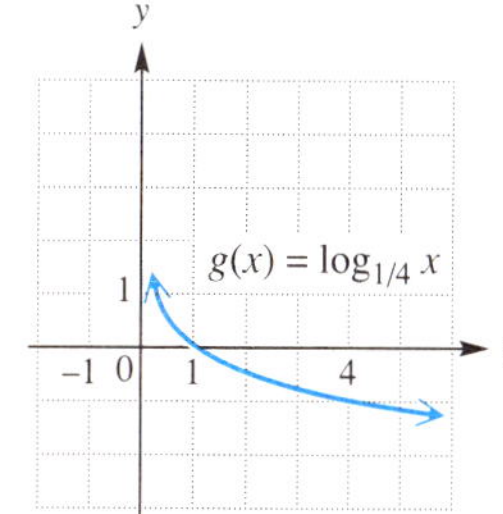

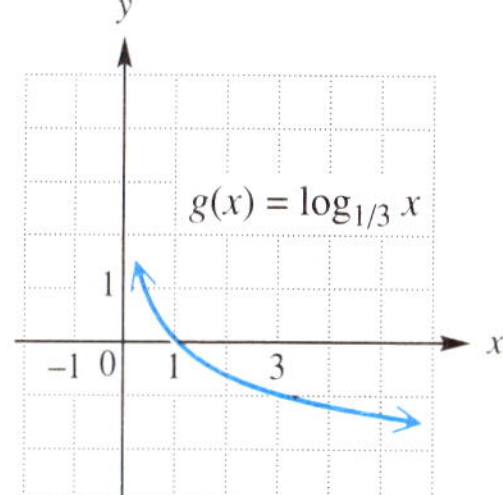

Determine the amount of money that will be accumulated in an account that pays compound interest, given the initial principal in each of the following.

41. \$4,292 at 6% compounded annually for 10 years **\$7,686.32**

42. \$8,906.54 at 5% compounded semiannually for 9 years **\$13,891.16**

43. \$56,780 at 5.3% compounded quarterly for 23 quarters **\$76,855.95**

44. \$45,788 at 6% compounded daily (ignoring leap years) for 11 years of 365 days **\$88,585.47**

45. Cathy Wacaser invests a \$25,000 inheritance in a fund paying 5% per year compounded continuously. What will be the amount on deposit after each of the following time periods?
(a) 1 year **(b)** 5 years **(c)** 10 years
(a) \$26,281.78 (b) \$32,100.64 (c) \$41,218.03

46. Linda Youngman, who is self-employed, wants to invest \$60,000 in a pension plan. One investment offers 7% compounded quarterly. Another offers 6.75% compounded continuously. Which investment will earn more interest in 5 years? How much more will the better plan earn? **7% compounded quarterly; \$800.31**

The figure shown here accompanied the article "Is Our World Warming?" which appeared in the October 1990 issue of National Geographic. *It shows projected temperature increases using two graphs: one an exponential-type curve, and the other linear. From the figure, approximate the increase* **(a)** *for the exponential curve, and* **(b)** *for the linear graph for each of the following years.*

47. 2000 **(a) .5°C (b) .35°C**

48. 2010 **(a) 1.0°C (b) .4°C**

49. 2020 **(a) 1.6°C (b) .5°C**

50. 2040 **(a) 3.0°C (b) .7°C**

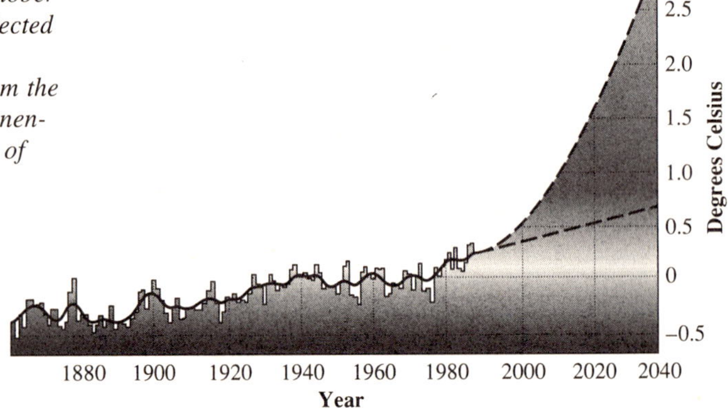

Graph, "Zero Equals Average Global Temperature for the Period 1950–1979" by Dale D. Glasgow from *National Geographic*, October 1990. Reprinted by permission of the National Geographic Society.

Solve the problem. Use a calculator as needed. The equations in Exercises 51 and 52 are models based on the assumption that current growth rates remain the same.

51. The population of Brazil, in millions, is approximated by the function

$$f(x) = 155.3(2)^{.025x},$$

where $x = 0$ corresponds to 1994, $x = 1$ corresponds to 1995, and so on.

(a) What will be the population of Brazil in the year 2000 according to this model?
(b) What will be the population in 2034?
(c) How will the population in 2034 compare to the population in 1994?
(a) 172.3 million (b) 310.6 million
(c) It will be twice as large.

52. The population of Pakistan, in millions, is approximated by the function

$$f(x) = 126.4(2)^{.04x},$$

where $x = 0$ corresponds to 1994, $x = 1$ corresponds to 1995, and so on.

(a) What was the population of Pakistan in the year 1994?

(b) What will the population be in 2019?

(c) How will the population in 2019 compare to the population in 1994?

(a) 126.4 million (b) 252.8 million
(c) It will be twice as large.

53. The amount of radioactive material in an ore sample is given by the function

$$A(t) = 100(3.2)^{-.5t},$$

where $A(t)$ is the amount present, in grams, of the sample t months after the initial measurement.

(a) How much was present at the initial measurement? (*Hint:* $t = 0$.)

(b) How much was present 2 months later?

(c) How much was present 10 months later?

(a) 100 grams (b) 31.25 grams
(c) .30 gram (to the nearest hundredth)

54. A small business estimates that the value $V(t)$ of a copy machine is decreasing according to the function

$$V(t) = 5{,}000(2)^{-.15t},$$

where t is the number of years that have elapsed since the machine was purchased, and $V(t)$ is in dollars.

(a) What was the original value of the machine?

(b) What is the value of the machine 5 years after purchase? Give your answer to the nearest dollar.

(c) What is the value of the machine 10 years after purchase? Give your answer to the nearest dollar.

(a) \$5,000 (b) \$2,973 (c) \$1,768

55. Refer to the function in Exercise 54. When will the value of the machine be \$2,500? (*Hint:* Let $V(t) = 2{,}500$, divide both sides by 5,000, and use the method of Example 7(b).)
6.67 years after it was purchased

56. Refer to the function in Exercise 54. When will the value of the machine be \$1,250?
13.33 years after it was purchased

The bar graph shows the average annual major league baseball player's salary for each year since free agency began. Using a technique from statistics, it was determined that the function

$$S(x) = 74{,}741(1.17)^x$$

approximates the salary, where $x = 0$ corresponds to 1976, and so on, up to $x = 18$ representing 1994. (Salary is in dollars.)

Source: MLBPA

57. Based on this model, what was the average salary in 1986? **about \$360,000**

58. Based on the graph, in what year did the average salary first exceed \$1,000,000? **1992**

59. The accompanying graphing calculator screen shows this function graphed from $x = 0$ to $x = 20$. Interpret the display at the bottom of the screen.

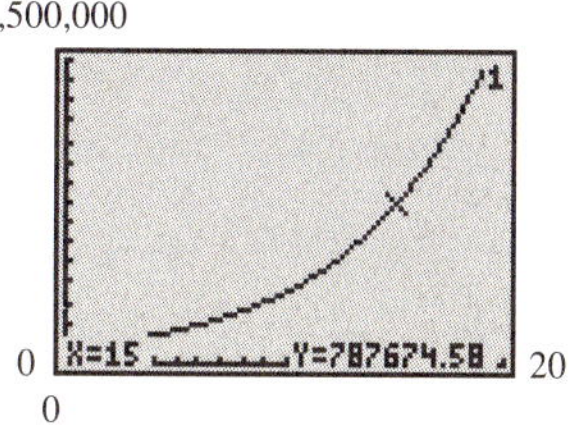

In 1991, the average salary was about \$800,000.

◉ **60.** In the bar graph, we see that the tops of the bars rise from left to right, and in the calculator-generated graph, the curve rises from left to right. What in the model equation $S(x) = 74{,}741(1.17)^x$ indicates that during the time period, baseball salaries were *rising*?
Because the base 1.17 is greater than 1, the exponential function values *increase* as x gets larger.

61. Radioactive strontium decays according to the function

$$y = y_0 e^{-.0239t},$$

where t is time in years.

(a) If an initial sample contains $y_0 = 5$ grams of radioactive strontium, how many grams will be present after 20 years?

(b) How many grams of the initial 5-gram sample will be present after 60 years?

(c) What is the half-life of radioactive strontium?

(a) 3.10 grams (b) 1.19 grams (c) 29 years

62. In the United States, the intensity of an earthquake is rated using the *Richter scale.* The Richter scale rating of an earthquake of intensity x is given by

$$R = \log_{10} \frac{x}{x_0},$$

where x_0 is the intensity of an earthquake of a certain (small) size. The figure shows Richter scale ratings for major Southern California earthquakes since 1920. As the figure indicates, earthquakes "come in bunches" and the 1990s have been an especially busy time.

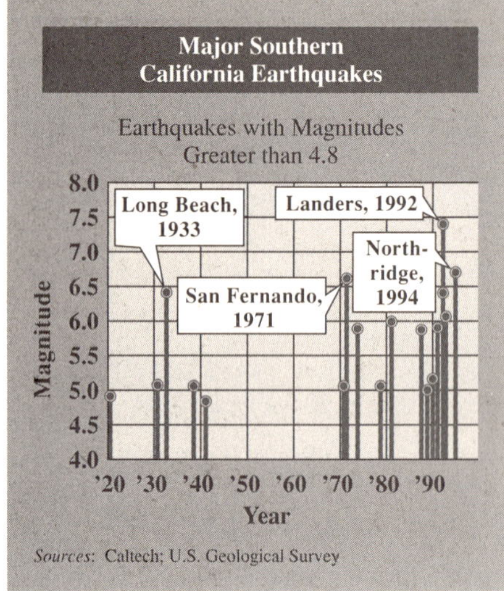

Writing the logarithmic equation just given in exponential form, we get

$$10^R = \frac{x}{x_0} \quad \text{or} \quad x = 10^R x_0.$$

The 1994 Northridge earthquake had a Richter scale rating of 6.7; the Landers earthquake had a rating of 7.3.

(a) How many times as powerful was the Landers earthquake compared to the Northridge earthquake?

(b) Compare the smallest rated earthquake in the figure (at 4.8) with the Landers quake. How many times as powerful was the Landers quake?

(a) almost 4 times as powerful (b) about 300 times as powerful

8.7 Exercises (Text Page 471)

✎ **1.** What is meant by a system of equations?

◉ **2.** A solution of a system of linear equations in two variables is a(n) ______ ______ of real numbers. **ordered pair**

Decide whether the ordered pair is a solution of the given system.

3. $x + y = 6$ (5, 1) **yes**
$x - y = 4$

4. $x - y = 17$ (8, −9) **yes**
$x + y = -1$

5. $2x - y = 8$ (5, 2) **no**
$3x + 2y = 20$

6. $3x - 5y = -12$ (−1, 2) **no**
$x - y = 1$

Solve the system by graphing.

7. $x + y = 4$ $\{(2, 2)\}$
$2x - y = 2$

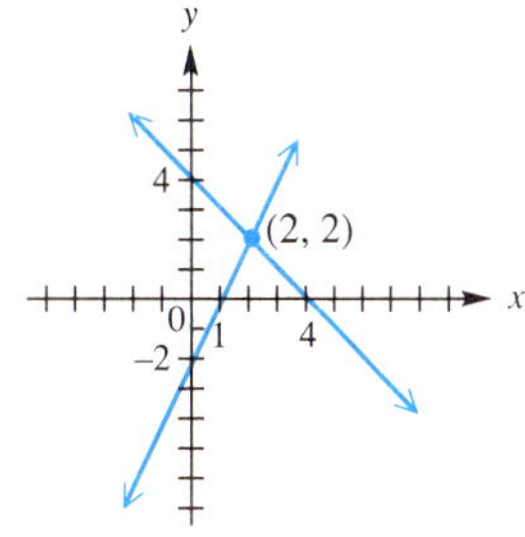

8. $x + y = -5$ $\{(-2, -3)\}$
$-2x + y = 1$

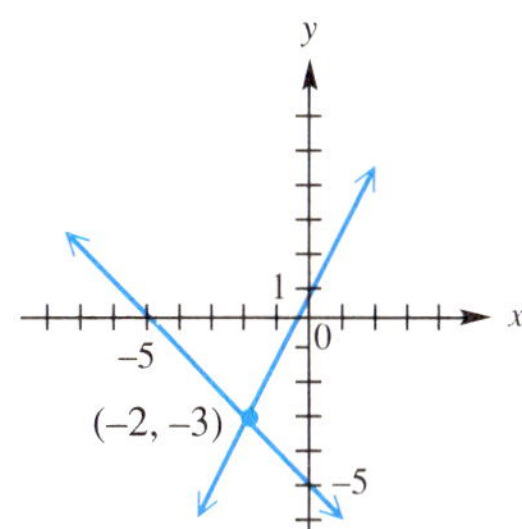

Solve the system by elimination.

9. $2x - 5y = 11$ $\{(3, -1)\}$
$3x + y = 8$

10. $-2x + 3y = 1$ $\{(1, 1)\}$
$-4x + y = -3$

11. $3x + 4y = -6$ $\{(2, -3)\}$
$5x + 3y = 1$

12. $4x + 3y = 1$ $\{(4, -5)\}$
$3x + 2y = 2$

13. $3x + 3y = 0$ $\left\{\left(\frac{3}{2}, -\frac{3}{2}\right)\right\}$
$4x + 2y = 3$

14. $8x + 4y = 0$ $\left\{\left(\frac{1}{4}, -\frac{1}{2}\right)\right\}$
$4x - 2y = 2$

15. $7x + 2y = 6$
$-14x - 4y = -12$
$\left\{\left(\frac{6 - 2y}{7}, y\right)\right\}$

16. $x - 4y = 2$
$4x - 16y = 8$
$\{(2 + 4y, y)\}$

17. $\frac{x}{2} + \frac{y}{3} = -\frac{1}{3}$
$\frac{x}{2} + 2y = -7$
$\{(2, -4)\}$

18. $\frac{x}{5} + y = \frac{6}{5}$
$\frac{x}{10} + \frac{y}{3} = \frac{5}{6}$
$\left\{\left(13, -\frac{7}{5}\right)\right\}$

19. $5x - 5y = 3$
$x - y = 12$
$\emptyset$

20. $2x - 3y = 7$
$-4x + 6y = 14$
$\emptyset$

Solve the system by substitution.

21. $4x + y = 6$
$y = 2x$
$\{(1, 2)\}$

22. $2x - y = 6$
$y = 5x$
$\{(-2, -10)\}$

23. $3x - 4y = -22$
$-3x + y = 0$
$\left\{\left(\frac{22}{9}, \frac{22}{3}\right)\right\}$

24. $-3x + y = -5$
$x + 2y = 0$
$\left\{\left(\frac{10}{7}, -\frac{5}{7}\right)\right\}$

25. $-x - 4y = -14$
$2x = y + 1$
$\{(2, 3)\}$

26. $-3x - 5y = -17$
$4x = y - 8$
$\{(-1, 4)\}$

27. $5x - 4y = 9$
$3 - 2y = -x$
$\{(5, 4)\}$

28. $6x - y = -9$
$4 + 7x = -y$
$\{(-1, 3)\}$

29. $x = 3y + 5$
$x = \frac{3}{2}y$
$\left\{\left(-5, -\frac{10}{3}\right)\right\}$

30. $x = 6y - 2$
$x = \frac{3}{4}y$
$\left\{\left(\frac{2}{7}, \frac{8}{21}\right)\right\}$

31. $\frac{1}{2}x + \frac{1}{3}y = 3$
$y = 3x$
$\{(2, 6)\}$

32. $\frac{1}{4}x - \frac{1}{5}y = 9$
$y = 5x$
$\{(-12, -60)\}$

33. Explain what the following statement means: The solution set of the system
$2x + y + z = 3$
$3x - y + z = -2$ is $\{(-1, 2, 3)\}$.
$4x - y + 2z = 0$

34. Write a system of three linear equations in three variables that has solution set $\{(3, 1, 2)\}$. Then solve the system. (*Hint:* Start with the solution and make up three equations that are satisfied by the solution. There are many ways to do this.)
Answers will vary. One example is
$x + y + z = 6$
$2x + 3y - z = 7$
$3x - y - z = 6.$

Solve the system of equations in three variables.

35. $3x + 2y + z = 8$
$2x - 3y + 2z = -16$
$x + 4y - z = 20$
$\{(1, 4, -3)\}$

36. $-3x + y - z = -10$
$-4x + 2y + 3z = -1$
$2x + 3y - 2z = -5$
$\{(2, -1, 3)\}$

37. $2x + 5y + 2z = 0$
$4x - 7y - 3z = 1$
$3x - 8y - 2z = -6$
$\{(0, 2, -5)\}$

38. $5x - 2y + 3z = -9$
$4x + 3y + 5z = 4$
$2x + 4y - 2z = 14$
$\{(0, 3, -1)\}$

39. $x + y - z = -2$
$2x - y + z = -5$
$-x + 2y - 3z = -4$
$\left\{\left(-\frac{7}{3}, \frac{22}{3}, 7\right)\right\}$

40. $x + 2y + 3z = 1$
$-x - y + 3z = 2$
$-6x + y + z = -2$
$\left\{\left(\frac{20}{59}, -\frac{33}{59}, \frac{35}{59}\right)\right\}$

41. $2x - 3y + 2z = -1$
$x + 2y + z = 17$
$2y - z = 7$
$\{(4, 5, 3)\}$

42. $2x - y + 3z = 6$
$x + 2y - z = 8$
$2y + z = 1$
$\{(5, 1, -1)\}$

43. $4x + 2y - 3z = 6$
$x - 4y + z = -4$
$-x + 2z = 2$
$\{(2, 2, 2)\}$

44. $2x + 3y - 4z = 4$
$x - 6y + z = -16$
$-x + 3z = 8$
$\{(4, 4, 4)\}$

45. $2x + y = 6$
$3y - 2z = -4$
$3x - 5z = -7$
$\left\{\left(\frac{8}{3}, \frac{2}{3}, 3\right)\right\}$

46. $4x - 8y = -7$
$4y + z = 7$
$-8x + z = -4$
$\left\{\left(\frac{3}{4}, \frac{5}{4}, 2\right)\right\}$

47. The accompanying graph shows the trends during the years 1966–1990 relating to bachelor's degrees awarded in the United States. (*Source:* National Science Foundation)

(a) Between what years shown on the horizontal axis did the number of degrees in all fields for men and women reach equal numbers?

(b) When the number of degrees for men and women reached equal numbers, what was that number (approximately)?

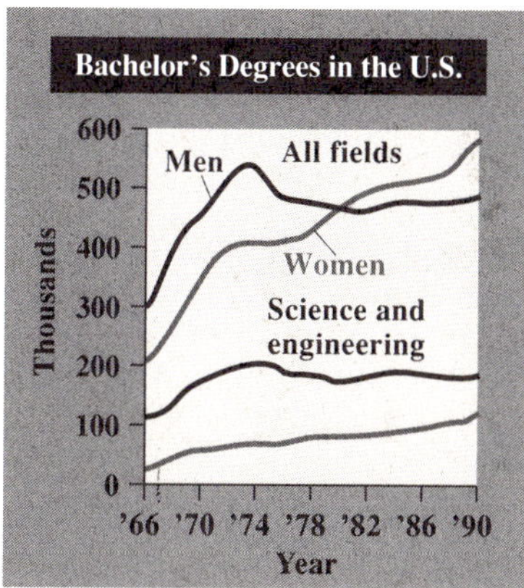

(a) 1978 and 1982 (b) just less than 500,000

48. The accompanying graph shows how the production of vinyl LPs, audiocassettes, and compact discs (CDs) changed over the years from 1983 to 1993. (*Source:* Recording Industry of America)

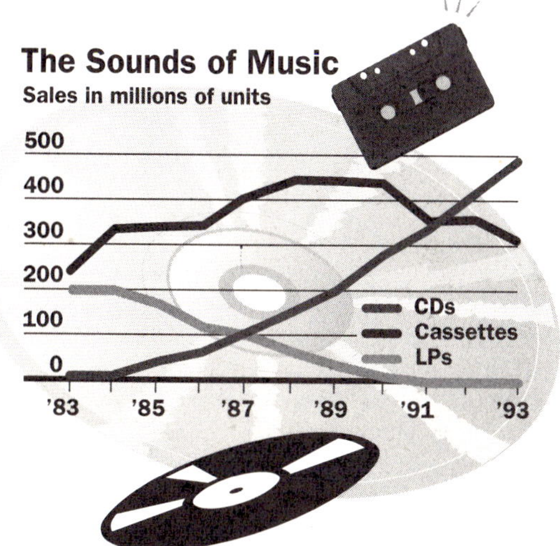

(a) In what year did cassette production and CD production reach equal levels? What was that level?

(b) Express as an ordered pair of the form (year, production level) the point of intersection of the graphs of LP production and CD production.

(a) In 1991 they both reached the level of about 360 million. (b) (1987, 100 million)

Solve the problem by using a system of two equations in two variables.

49. Andre and Monica measured the perimeter of a tennis court and found that it was 42 feet longer than it was wide, and had a perimeter of 228 feet. What were the length and the width of the tennis court?

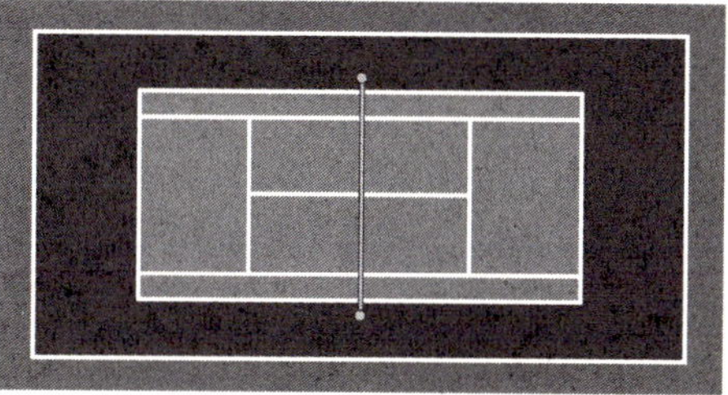

length: 78 feet; width: 36 feet

50. Kareem and Manute found that the width of their basketball court was 44 feet less than the length. If the perimeter was 288 feet, what were the length and the width of their court?

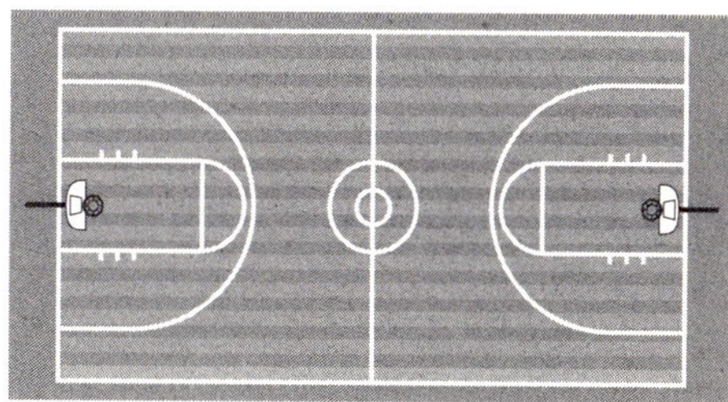

length: 94 feet; width: 50 feet

51. The length of a rectangle is 7 feet more than the width. If the length were decreased by 3 feet and the width were increased by 2 feet, the perimeter would be 32 feet. Find the length and width of the original rectangle.
length: 12 feet; width: 5 feet

52. The side of a square is 4 centimeters longer than the side of an equilateral triangle. The perimeter of the square is 24 centimeters more than the perimeter of the triangle. Find the lengths of a side of the square and a side of the triangle.
square: 12 centimeters; triangle: 8 centimeters

53. How many gallons each of 25% alcohol and 35% alcohol should be mixed to get 20 gallons of 32% alcohol? **6 gallons of 25%; 14 gallons of 35%**

54. How many liters each of 15% acid and 33% acid should be mixed to get 40 liters of 21% acid?
$26\frac{2}{3}$ liters of 15%; $13\frac{1}{3}$ liters of 33%

55. Pure acid is to be added to a 10% acid solution to obtain 27 liters of a 20% acid solution. What amounts of each should be used?
3 liters of pure acid; 24 liters of 10%

56. A truck radiator holds 18 liters of fluid. How much pure antifreeze must be added to a mixture that is 4% antifreeze in order to fill the radiator with a mixture that is 20% antifreeze?
3 liters of pure antifreeze

57. A party mix is made by adding nuts that sell for $2.50 a kilogram to a cereal mixture that sells for $1 a kilogram. How much of each should be added to get 30 kilograms of a mix that will sell for $1.70 a kilogram?
14 kilograms of nuts; 16 kilograms of cereal

58. A popular fruit drink is made by mixing fruit juices. Such a mixture with 50% juice is to be mixed with another mixture that is 30% juice to get 200 liters of a mixture that is 45% juice. How much of each should be used?
150 liters of 50% juice; 50 liters of 30% juice

59. Tickets to a production of *Othello* at Nicholls State University cost $2.50 for general admission or $2.00 with student identification. If 184 people paid to see a performance and $406 was collected, how many of each type of admission were sold?
76 general admission; 108 with student identification

60. A grocer plans to mix candy that sells for $1.20 a pound with candy that sells for $2.40 a pound to get a mixture that he plans to sell for $1.65 a pound. How much of the $1.20 and $2.40 candy should he use if he wants 80 pounds of the mix?
50 pounds of $1.20 candy; 30 pounds of $2.40 candy

61. Angelica Canales has been saving dimes and quarters. She has 94 coins in all. If the total value is $19.30, how many dimes and how many quarters does she have? **28 dimes; 66 quarters**

62. Gayle Sloan is a teller at the Seagull branch of Parish National Bank. She received a checking account deposit in $20 bills and $50 bills. There was a total of 70 bills, and the amount of the deposit was $3,200. How many of each denomination were deposited?
10 twenties; 60 fifties

63. A total of $3,000 is invested, part at 2% simple interest and part at 4%. If the total annual return from the two investments is $100, how much is invested at each rate? **$1,000 at 2%; $2,000 at 4%**

64. A man must invest a total of $15,000 in two accounts, one paying 4% annual simple interest, and the other 3%. If he wants to earn $550 annual interest, how much should he invest at each rate?
$10,000 at 4%; $5,000 at 3%

65. San Jacinto College has decided to supply its mathematics labs with color monitors. A trip to the local electronics outlet leads to the following information: 4 CGA monitors and 6 VGA monitors can be purchased for $4,600, while 6 CGA monitors and 4 VGA monitors will cost $4,400. What are the prices of a single CGA monitor and a single VGA monitor?
CGA monitor: $400; VGA monitor: $500

66. The Cleveland Indians gift shop will sell you 5 New Era baseball caps and 3 Diamond Collection jerseys for $430, or 3 New Era caps and 4 Diamond Collection jerseys for $500. What are the prices of a single cap and a single jersey? **cap: $20; jersey: $110**

67. A factory makes use of two basic machines, A and B, which turn out two different products, yarn and thread. Each unit of yarn requires 1 hour on machine A and 2 hours on machine B, while each unit of thread requires 1 hour on A and 1 hour on B. Machine A runs 8 hours per day, while machine B runs 14 hours per day. How many units each of yarn and thread should the factory make each day to keep its machines running at capacity?
6 units of yarn; 2 units of thread

68. A company that makes personal computers has found that each standard model requires 4 hours to manufacture electronics and 2 hours for the case. The top-of-the-line model requires 5 hours for the electronics and 1.5 hours for the case. On a particular production run, the company has available 200 hours in the electronics department and 76 hours in the cabinet department. How many of each model can be made?
20 standard; 24 top-of-the-line

69. Theodis bought 2 kilograms of dark clay and 3 kilograms of light clay, paying \$22 for the clay. He later needed 1 kilogram of dark clay and 2 kilograms of light clay, costing \$13 altogether. How much did he pay per kilogram for each type of clay?
dark clay: \$5 per kilogram; light clay: \$4 per kilogram

70. A biologist wants to grow two types of algae, green and brown. She has 15 kilograms of nutrient X and 26 kilograms of nutrient Y. A vat of green algae needs 2 kilograms of nutrient X and 3 kilograms of nutrient Y, while a vat of brown algae needs 1 kilogram of nutrient X and 2 kilograms of nutrient Y. How many vats of each type of algae should the biologist grow in order to use all the nutrients?
4 vats of green algae; 7 vats of brown algae

71. A freight train and an express train leave towns 390 kilometers apart, traveling toward one another. The freight train travels 30 kilometers per hour slower than the express train. They pass one another 3 hours later. What are their speeds?
freight train: 50 kilometers per hour; express train: 80 kilometers per hour

72. A train travels 150 kilometers in the same time that a plane covers 400 kilometers. If the speed of the plane is 20 kilometers per hour less than 3 times the speed of the train, find both speeds.
train: 60 kilometers per hour; plane: 160 kilometers per hour

73. In his motorboat, Phan Nguyen travels upstream at top speed to his favorite fishing spot, a distance of 36 miles, in two hours. Returning, he finds that the trip downstream, still at top speed, takes only 1.5 hours. Find the speed of Phan's boat and the speed of the current.
boat: 21 miles per hour; current: 3 miles per hour

74. Traveling for three hours into a steady headwind, a plane makes a trip of 1,650 miles. The pilot determines that flying with the same wind for two hours, he could make a trip of 1,300 miles. What is the speed of the plane and the speed of the wind?
plane: 600 miles per hour; wind: 50 miles per hour

Solve the problem by using a system of three equations in three variables. (In Exercises 75–78 use the fact that the sum of the measures of the angles of a triangle is 180°.)

75. In the figure shown, $z = x + 10$ and $x + y = 100$. Determine a third equation involving x, y, and z, and then find the measures of the three angles.

$x + y + z = 180$; angle measures: 70°, 30°, 80°

76. In the figure shown, x is 10 less than y and 20 less than z. Write a system of equations and find the measures of the three angles.

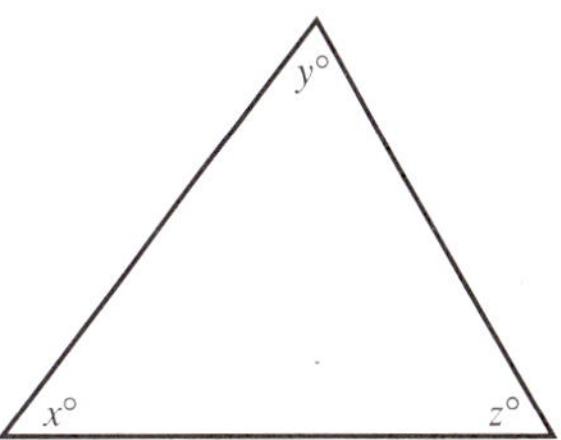

$x = y - 10, x = z - 20, x + y + z = 180$; angle measures: 50°, 60°, 70°

77. In a certain triangle, the measure of the second angle is 10° more than three times the first. The third angle measure is equal to the sum of the measures of the other two. Find the measures of the three angles.
first: 20°; second: 70°; third: 90°

78. The measure of the largest angle of a triangle is 12° less than the sum of the measures of the other two. The smallest angle measures 58° less than the largest. Find the measures of the angles.
largest: 84°; middle: 70°; smallest: 26°

79. The perimeter of a triangle is 70 centimeters. The longest side is 4 centimeters less than the sum of the other two sides. Twice the shortest side is 9 centimeters less than the longest side. Find the length of each side of the triangle.
shortest: 12 centimeters; middle: 25 centimeters; longest: 33 centimeters

80. The perimeter of a triangle is 56 inches. The longest side measures 4 inches less than the sum of the other two sides. Three times the shortest side is 4 inches more than the longest side. Find the lengths of the three sides.
shortest: 10 inches; middle: 20 inches; longest: 26 inches

81. Sharon Dempsey, a Mardi Gras trinket manufacturer, supplies three wholesalers, A, B, and C. The output from a day's production is 320 cases of trinkets. She must send wholesaler A three times as many cases as she sends B, and she must send wholesaler C 160 cases less than she provides A and B together. How many cases should she send to each wholesaler to distribute the entire day's production to them?
A: 180 cases; B: 60 cases; C: 80 cases

82. A hardware supplier manufactures three kinds of clamps, types A, B, and C. Production restrictions require them to make 10 units more type C clamps than the total of the other types and twice as many type B clamps as type A. The shop must produce a total of 490 units of clamps per day. How many units of each type can be made per day?
type A: 80; type B: 160; type C: 250

83. The manager of a candy store wants to feature a special Easter candy mixture of jelly beans, small chocolate eggs, and marshmallow chicks. She plans to make 15 pounds of mix to sell at \$1 a pound. Jelly beans sell for \$.80 a pound, chocolate eggs for \$2 a pound, and marshmallow chicks for \$1 a pound. She will use twice as many pounds of jelly beans as eggs and chicks combined and five times as many pounds of jelly beans as chocolate eggs. How many pounds of each candy should she use?
10 pounds of jelly beans; 2 pounds of chocolate eggs; 3 pounds of marshmallow chicks

84. Three kinds of tickets are available for a Rhonda Rock concert: "up close," "in the middle," and "far out." "Up close" tickets cost \$2 more than "in the middle" tickets, while "in the middle" tickets cost \$1 more than "far out" tickets. Twice the cost of an "up close" ticket is \$1 less than 3 times the cost of a "far out" ticket. Find the price of each kind of ticket.
"up close": \$10; "in the middle": \$8; "far out": \$7

8.8 Exercises (Text Page 479)

Graph the linear inequality.

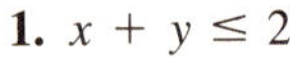
1. $x + y \leq 2$

2. $x - y \geq -3$

3. $4x - y \leq 5$

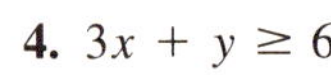
4. $3x + y \geq 6$

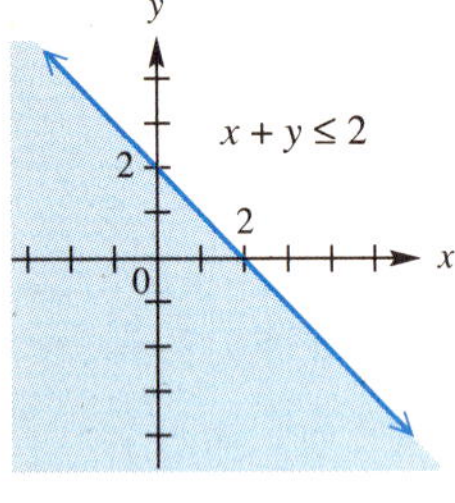

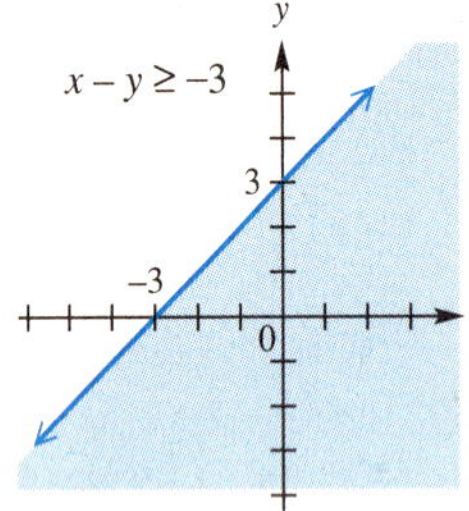

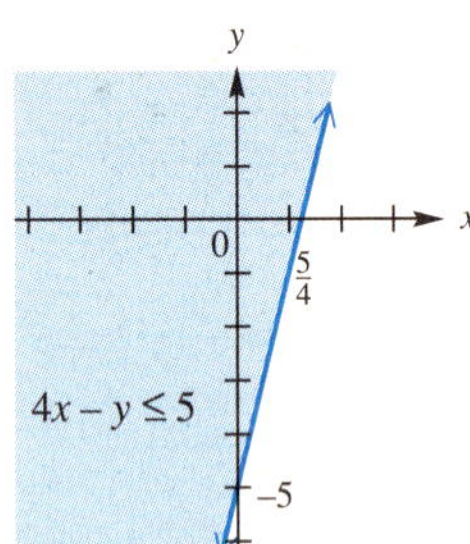

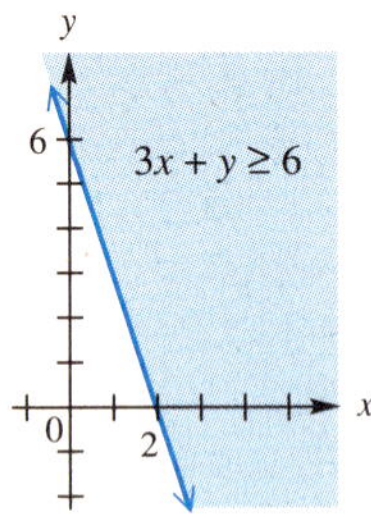

5. $x + 3y \geq -2$

6. $4x + 6y \leq -3$

7. $x + 2y \leq -5$

8. $2x - 4y \leq 3$

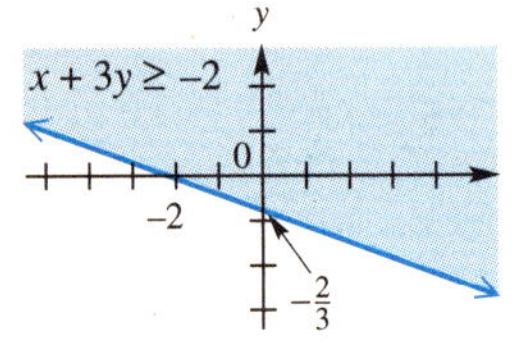

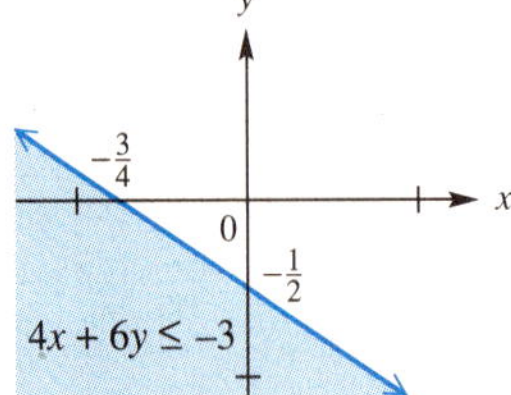

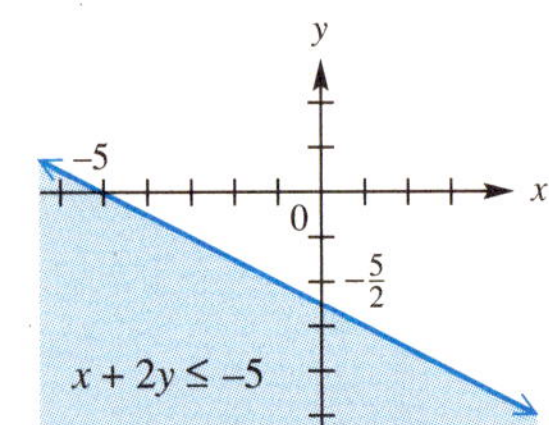

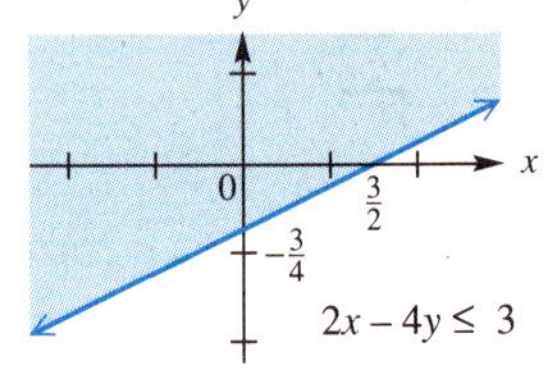

CONCEPTUAL  WRITING  CHALLENGING  SCIENTIFIC CALCULATOR  GRAPHING CALCULATOR

9. $4x - 3y < 12$

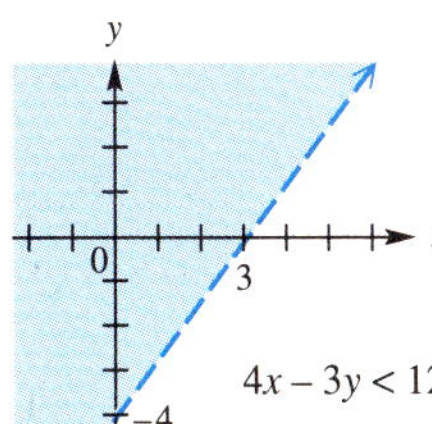

10. $5x + 3y > 15$

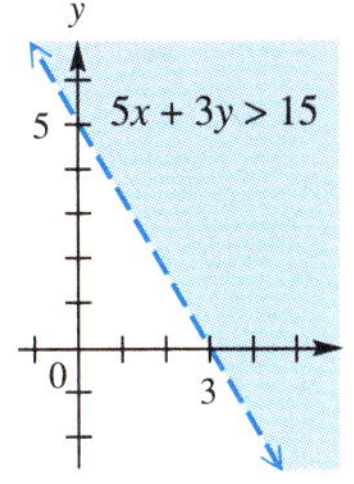

11. $y > -x$

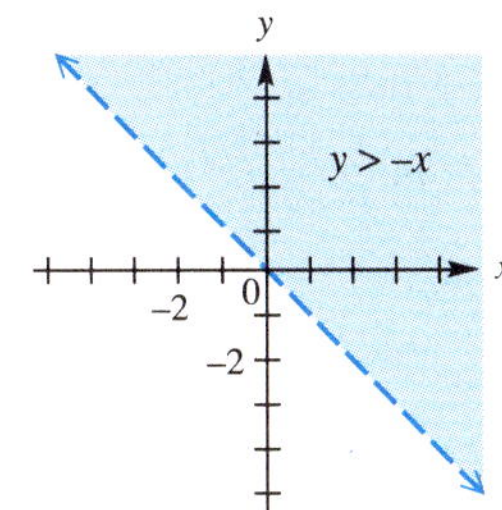

12. $y < x$

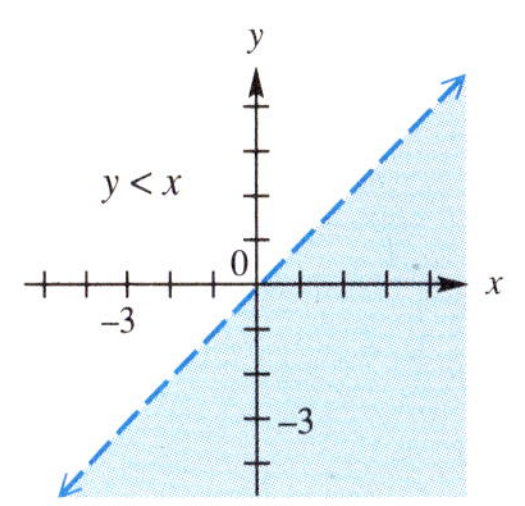

Graph the system of inequalities.

13. $x + y \leq 1$
$x \geq 0$

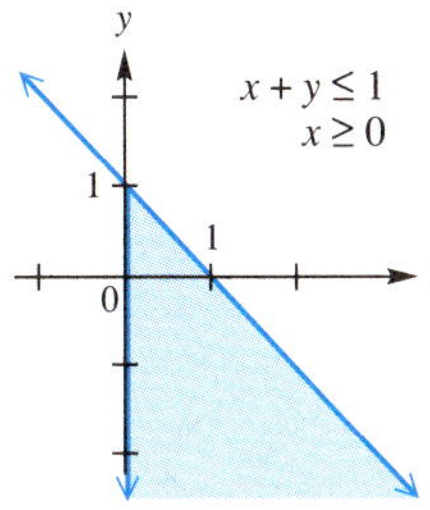

14. $3x - 4y \leq 6$
$y \geq 1$

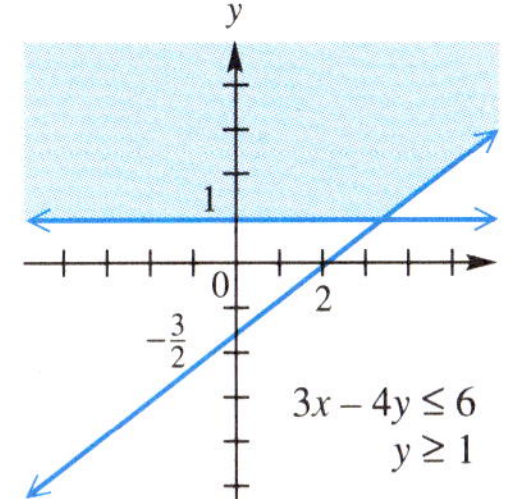

15. $2x - y \geq 1$
$3x + 2y \geq 6$

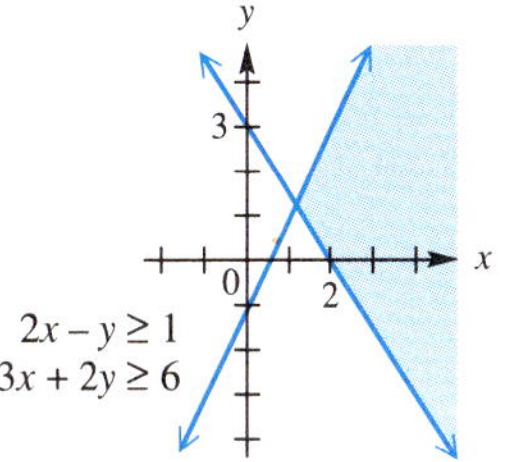

16. $x + 3y \geq 6$
$3x - 4y \leq 12$

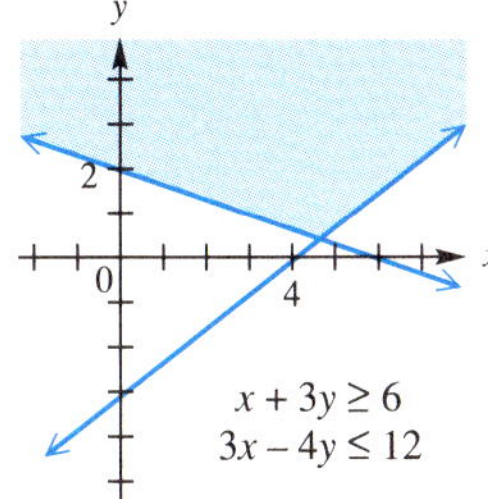

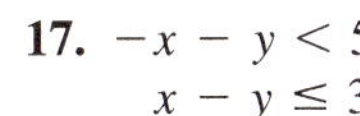
17. $-x - y < 5$
$x - y \leq 3$

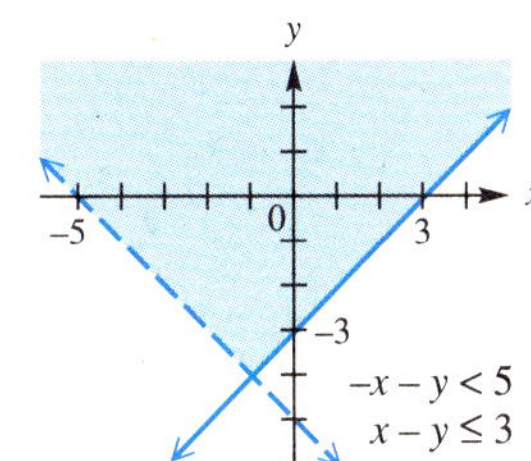

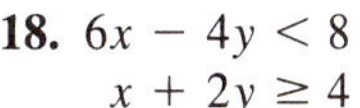
18. $6x - 4y < 8$
$x + 2y \geq 4$

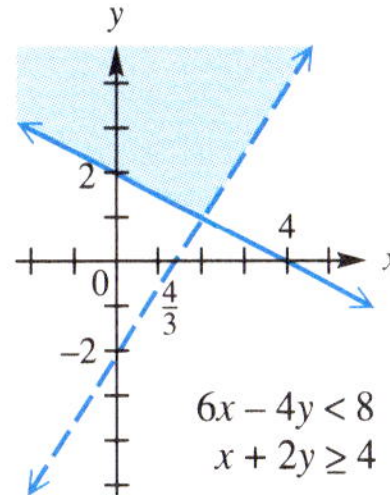

19. Explain how to determine whether a dashed line or a solid line is used for the boundary when graphing a linear inequality in two variables.

20. The graph of $y > -3x + 2$ consists of all points that are __________ (above/below) the line with the equation __________.
above; $y = -3x + 2$

Extension Exercises (Text Page 481)

Exercises 1 and 2 show regions of feasible solutions. Find the maximum and minimum values of the given expressions.

1. $3x + 5y$

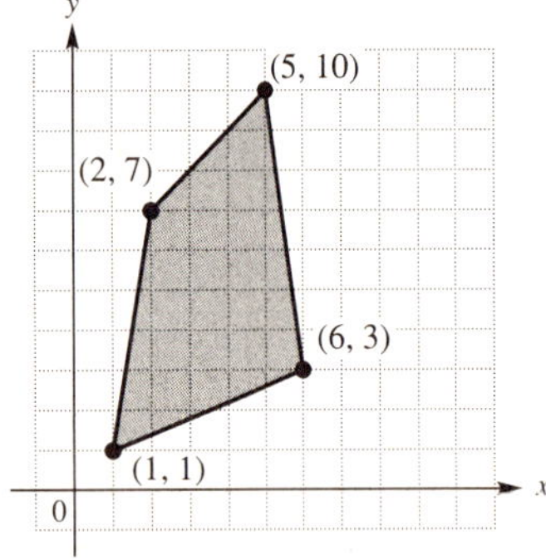

maximum of 65 at (5, 10); minimum of 8 at (1, 1)

2. $40x + 75y$

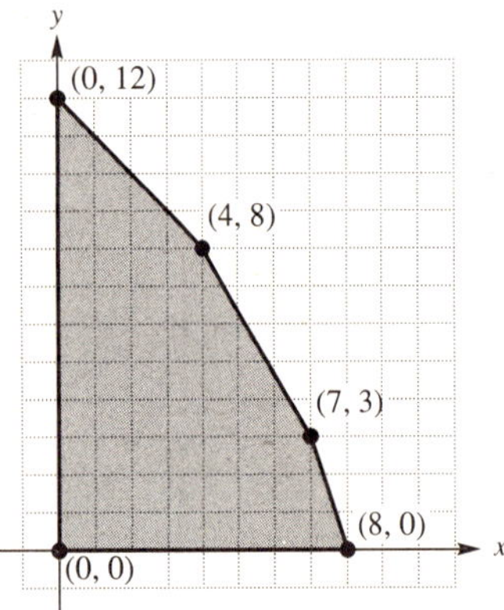

maximum of 900 at (0, 12); minimum of 0 at (0, 0)

In Exercises 3–6, use graphical methods to find values of x and y satisfying the given conditions. (It may be necessary to solve a system of equations in order to find vertices.) Find the value of the maximum or minimum.

3. Find $x \geq 0$ and $y \geq 0$ such that

$$2x + 3y \leq 6$$
$$4x + y \leq 6$$

and $5x + 2y$ is maximized.

$\left(\frac{6}{5}, \frac{6}{5}\right)$; $\frac{42}{5}$

4. Find $x \geq 0$ and $y \geq 0$ such that

$$x + y \leq 10$$
$$5x + 2y \geq 20$$
$$2y \geq x$$

and $x + 3y$ is minimized.

$\left(\frac{10}{3}, \frac{5}{3}\right)$; $\frac{25}{3}$

5. Find $x \geq 2$ and $y \geq 5$ such that

$$3x - y \geq 12$$
$$x + y \leq 15$$

and $2x + y$ is minimized.

$\left(\frac{17}{3}, 5\right)$; $\frac{49}{3}$

6. Find $x \geq 10$ and $y \geq 20$ such that

$$2x + 3y \leq 100$$
$$5x + 4y \leq 200$$

and $x + 3y$ is maximized.

$\left(10, \frac{80}{3}\right)$; 90

Solve each of the following linear programming problems.

7. Gwen, who is dieting, requires two food supplements, I and II. She can get these supplements from two different products, A and B. Product A provides 3 g per serving of supplement I and 2 g per serving of supplement II. Product B provides 2 g per serving of supplement I and 4 g per serving of supplement II. Her dietician, Dr. Shoemake, has recommended that she include at least 15 g of each supplement in her daily diet. If product A costs 25¢ per serving and product B costs 40¢ per serving, how can she satisfy her requirements most economically?

Use $3\frac{3}{4}$ servings of A and $1\frac{7}{8}$ servings of B, for a minimum cost of $1.69.

8. A manufacturer of refrigerators must ship at least 100 refrigerators to its two West coast warehouses. Each warehouse holds a maximum of 100 refrigerators. Warehouse A holds 25 refrigerators already, while warehouse B has 20 on hand. It costs $12 to ship a refrigerator to warehouse A and $10 to ship one to warehouse B. How many refrigerators should

be shipped to each warehouse to minimize cost? What is the minimum cost?
Ship 20 to A and 80 to B, for a minimum cost of $1,040.

9. A machine shop manufactures two types of bolts. Each can be made on any of three groups of machines, but the time required on each group differs, as shown in the table below.

		Machine Groups		
		I	II	III
Bolts	Type 1	.1 min	.1 min	.1 min
	Type 2	.1 min	.4 min	.5 min

Production schedules are made up one day at a time. In a day there are 240, 720, and 160 minutes available, respectively, on these machines. Type 1 bolts sell for 10¢ and type 2 bolts for 12¢. How many of each type of bolt should be manufactured per day to maximize revenue? What is the maximum revenue?
Manufacture 1,600 type 1 and 0 type 2, for a maximum revenue of $160.

10. Karin Wagner takes vitamin pills. Each day, she must have at least 16 units of Vitamin A, at least 5 units of Vitamin B_1, and at least 20 units of Vitamin C. She can choose between red pills costing 10¢ each that contain 8 units of A, 1 of B_1, and 2 of C; and blue pills that cost 20¢ each and contain 2 units of A, 1 of B_1, and 7 of C. How many of each pill should she take in order to minimize her cost and yet fulfill her daily requirements?
Take 3 red pills and 2 blue pills, for a minimum cost of 70¢ per day.

11. A bakery makes both cakes and cookies. Each batch of cakes requires two hours in the oven and three hours in the decorating room. Each batch of cookies needs one and a half hours in the oven and two thirds of an hour in the decorating room. The oven is available no more than 15 hours a day, while the decorating room can be used no more than 13 hours a day. How many batches of cakes and cookies should the bakery make in order to maximize profits if cookies produce a profit of $20 per batch and cakes produce a profit of $30 per batch?
Make 3 batches of cakes and 6 batches of cookies, for a maximum profit of $210.

12. A manufacturing process requires that oil refineries manufacture at least 2 gallons of gasoline for each gallon of fuel oil. To meet the winter demand for fuel oil, at least 3 million gallons a day must be produced. The demand for gasoline is no more than 6.4 million gallons per day. If the price of gasoline is $1.90 per gallon and the price of fuel oil is $1.50 per gallon, how much of each should be produced to maximize revenue?
Produce 6.4 million gallons of gasoline and 3.2 million gallons of fuel oil, for a maximum revenue of $16,960,000.

13. Earthquake victims in China need medical supplies and bottled water. Each medical kit measures 1 cubic foot and weighs 10 pounds. Each container of water is also 1 cubic foot but weighs 20 pounds. The plane can only carry 80,000 pounds with a total volume of 6,000 cubic feet. Each medical kit will aid 4 people, while each container of water will serve 10 people. How many of each should be sent in order to maximize the number of people aided?
Ship no medical kits and 4,000 containers of water.

14. If each medical kit could aid 6 people instead of 4, how would the results in Exercise 13 change?
Ship 4,000 medical kits and 2,000 containers of water.

Chapter 8 Test (Text Page 484)

1. Find the distance between the points $(-3, 5)$ and $(2, 1)$. $\sqrt{41}$

2. Find an equation of the circle whose center has coordinates $(-1, 2)$, with radius 3. Sketch its graph. $(x + 1)^2 + (y - 2)^2 = 9$

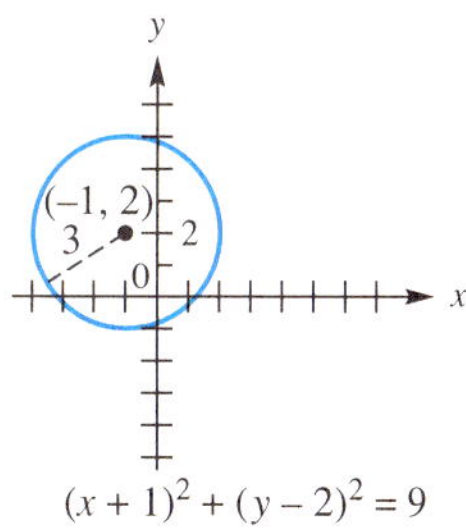

$(x + 1)^2 + (y - 2)^2 = 9$

3. Find the x- and y-intercepts of the graph of $3x - 2y = 8$, and graph the equation.

x-intercept: $\left(\frac{8}{3}, 0\right)$; y-intercept: $(0, -4)$

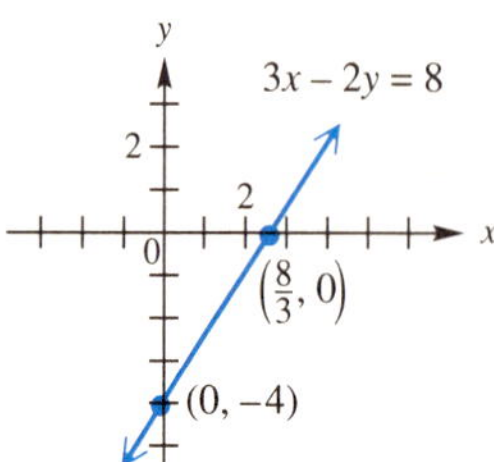

4. Find the slope of the line passing through the points $(6, 4)$ and $(-1, 2)$. $\frac{2}{7}$

5. Find the slope-intercept form of the equation of the line described.

(a) passing through the point $(-1, 3)$, with slope $-2/5$

(b) passing through $(-7, 2)$ and perpendicular to $y = 2x$

(c) the line shown in the figures (Look at the displays at the bottom.)

10

−10 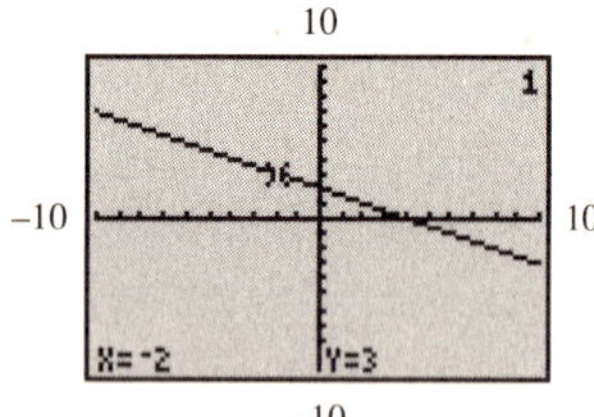 10

−10

10

−10 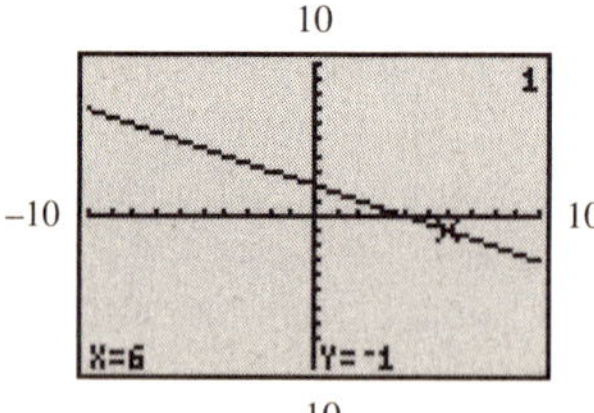 10

−10

(a) $y = -\frac{2}{5}x + \frac{13}{5}$ (b) $y = -\frac{1}{2}x - \frac{3}{2}$ (c) $y = -\frac{1}{2}x + 2$

6. Which one of the following has positive slope and negative y-coordinate for its y-intercept?

(a) 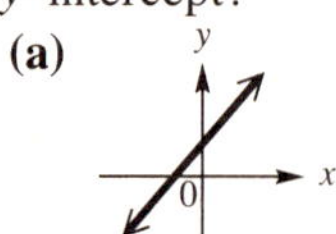

(b) 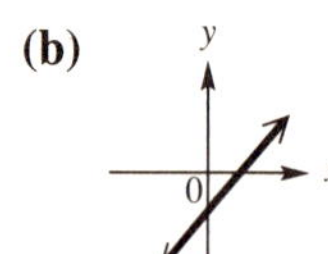

(c) 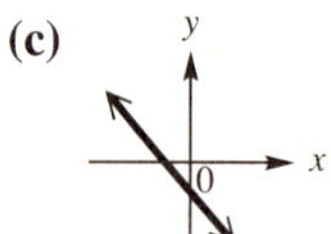

(d) 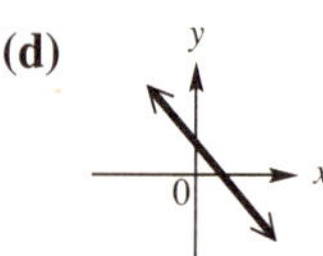

(b)

---

7. The linear equation $y = 825x + 8{,}689$ provides a model for the number of cases served by the Child Support Enforcement program from 1985 to 1990, where $x = 0$ corresponds to 1985, $x = 1$ corresponds to 1986, and so on. Use this model to approximate the number of cases served during 1989. (*Source:* Office of Child Support Enforcement)
11,989

8. What does the number 825 in the equation in Exercise 7 refer to

(a) with respect to the graph?

(b) in the context of the problem?

(a) It is the slope of the line. (b) It is the annual increase in the number of cases served.

CONCEPTUAL  WRITING ▲ CHALLENGING SCIENTIFIC CALCULATOR GRAPHING CALCULATOR

9. The graph indicates the percentage of all U.S. workers without any private or public health insurance. (*Source:* Census Bureau, Employee Benefit Research Institute)

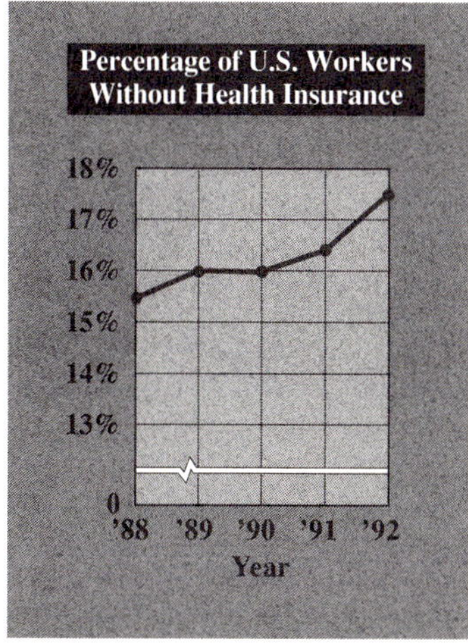

(a) Between which two years was the percentage approximately the same?
(b) Between which two years was the increase the greatest?
(c) In what year was the percent about 16.5%?
(a) 1989 and 1990 (b) 1991 and 1992
(c) 1991

10. For the function $f(x) = x^2 - 3x + 12$,
(a) give its domain; **(b)** find $f(-2)$.
(a) $(-\infty, \infty)$ (b) 22

11. What is the domain of the function

$$g(x) = \sqrt{9 - 4x}\ ?$$

$\left(-\infty, \frac{9}{4}\right]$

12. If the cost to produce x units of calculators is $C(x) = 50x + 5{,}000$ dollars, while the revenue is $R(x) = 60x$ dollars, find the number of units of calculators that must be produced in order to break even. What is the revenue at the break-even point? **500 calculators; $30,000**

13. Graph the quadratic function $f(x) = -(x + 3)^2 + 4$. Give the axis, the vertex, the domain, and the range.

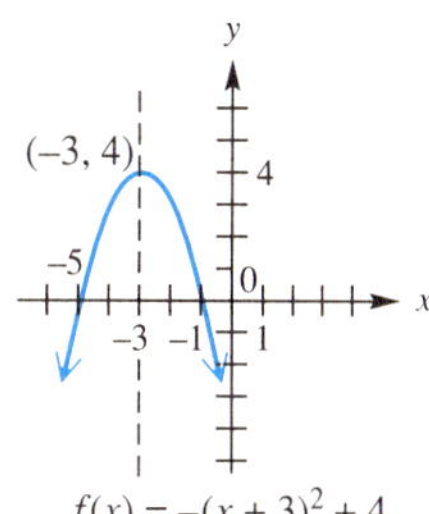

axis: $x = -3$; vertex: $(-3, 4)$; domain: $(-\infty, \infty)$; range: $(-\infty, 4]$

14. Miami-Dade Community College wants to construct a rectangular parking lot on land bordered on one side by a highway. It has 320 ft of fencing with which to fence off the other three sides. What should be the dimensions of the lot if the enclosed area is to be a maximum? **80 feet by 160 feet**

15. Use a scientific calculator to find an approximation of each of the following. Give as many digits as the calculator displays.
(a) $5.1^{4.7}$ **(b)** $e^{-1.85}$ **(c)** ln 23.56
(a) 2116.31264888 (b) .157237166314
(c) 3.15955035878

16. Which one of the following is a false statement?
(a) The domain of the function $f(x) = \log_2 x$ is $(-\infty, \infty)$.
(b) The graph of $F(x) = 3^x$ intersects the y-axis.
(c) The graph of $G(x) = \log_3 x$ intersects the x-axis.
(d) The expression ln x represents the exponent to which e must be raised in order to obtain x.
(a)

17. Suppose that $12,000 is invested in an account that pays 4% annual interest, and is left untouched for 3 years. How much will be in the account if
(a) interest is compounded quarterly (four times per year);
(b) interest is compounded continuously?
(a) $13,521.90 (b) $13,529.96

18. Since 1950, the growth in the world population in millions closely fits the exponential function defined by

$$A(t) = 2{,}600e^{.018t},$$

where t is the number of years since 1950.
(a) The world population was about 3,700 million in 1970. How closely does the function approximate this value?

(b) Use the function to approximate the population in 1990. (The actual 1990 population was about 5,320 million.)

(c) Use the function to estimate the population in the year 2000.

(a) $A(20) \approx 3{,}727$ million, which is off by 27 million. (b) 5,342 million (c) 6,395 million

Solve each system by using elimination, substitution, or a combination of the two methods.

19. $2x + 3y = 2$
$3x - 4y = 20$ $\{(4, -2)\}$

20. $2x + y + z = 3$
$x + 2y - z = 3$
$3x - y + z = 5$ $\{(2, 0, -1)\}$

21. $2x + 3y - 6z = 11$
$x - y + 2z = -2$
$4x + y - 2z = 7$ $\{(1, 2z + 3, z)\}$

Solve the problem by using a system of equations.

22. A manufacturer of portable compact disc players shipped 200 of the players to its two Quebec warehouses. It costs \$3 per unit to ship to Warehouse A, and \$2.50 per unit to ship to Warehouse B. If the total shipping cost was \$537.50, how many were shipped to each warehouse?
75 to A; 125 to B

23. Natalie Jackson sells undeveloped land. On three recent sales, she made a 10% commission, a 6% commission, and a 5% commission. Her total commissions on these sales were \$8,500, and she sold property worth \$140,000. If the 5% sale amounted to the sum of the other two, what were the three sales prices?
\$20,000 at 10%; \$50,000 at 6%; \$70,000 at 5%

24. The accompanying screen was generated by a graphing calculator. Which one of the following linear inequalities in two variables does it illustrate?

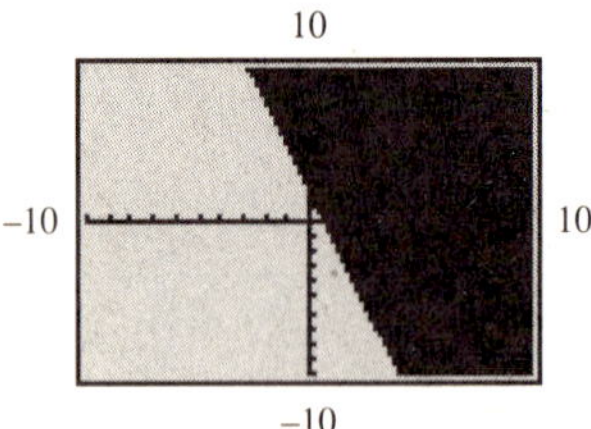

(a) $y > -3x + 2$
(b) $y < -3x + 2$
(c) $y > 3x + 2$
(d) $y < 3x + 2$ (a)

25. Graph the solution of the system of inequalities

$$x + y \le 6$$
$$2x - y \ge 3.$$

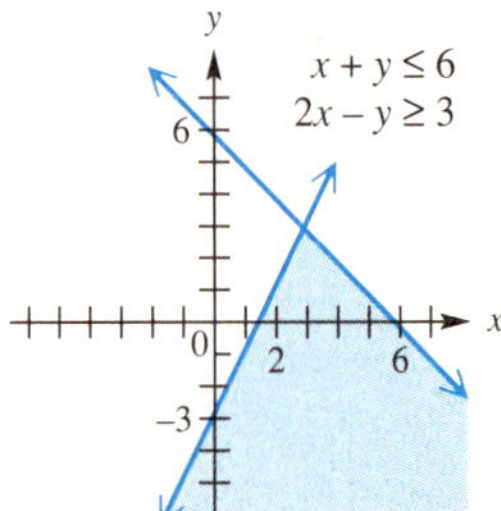

CHAPTER 9 Geometry

9.1 Exercises (Text Page 496)

Decide whether each statement is true *or* false.

1. A line segment has two endpoints. true

2. A ray has one endpoint. true

3. A half-line has one endpoint. false

4. If A and B are distinct points on a line, then ray AB and ray BA represent the same set of points. false

5. If two lines intersect, they lie in the same plane. true

6. If two lines are parallel, they lie in the same plane. true

7. If two lines do not intersect, they must be parallel. false

8. The sum of the measures of two right angles is the measure of a straight angle. true

9. The supplement of an acute angle must be an obtuse angle. true

10. Segment AB and segment BA represent the same set of points. true

11. The sum of the measures of two obtuse angles cannot equal the measure of a straight angle. true

12. There is no angle that is its own complement. false

13. There is no angle that is its own supplement. false

14. The origin of the use of the degree as a unit of measure of an angle goes back to the Egyptians. false

Exercises 15–24 name portions of the line shown. For each exercise, **(a)** *give the symbol that represents the portion of the line named, and* **(b)** *draw a figure showing just the portion named, including all labeled points.*

15. line segment AB
(a) $\overline{AB}$
(b) A B

16. ray BC
(a) $\overrightarrow{BC}$
(b) B C D

17. ray CB
(a) $\overrightarrow{CB}$
(b) A B C

18. line segment AD
(a) $\overline{AD}$
(b) A B C D

19. half-line BC
(a) $\overset{\circ\!\!\longrightarrow}{BC}$
(b) B C D

20. half-line AD
(a) $\overset{\circ\!\!\longrightarrow}{AD}$
(b) A B C D

21. ray BA
(a) $\overrightarrow{BA}$
(b) A B

22. ray DA
(a) $\overrightarrow{DA}$
(b) A B C D

23. line segment CA
(a) $\overline{CA}$
(b) A B C

24. line segment DA
(a) $\overline{DA}$
(b) A B C D

In Exercises 25–32, match the figure in column I with the figure in column II that names the same set of points, based on the given figure.

I		II
25. $\overline{PQ}$	F	**A.** $\overset{\circ\!\!\longrightarrow}{QS}$
26. $\overset{\circ\!\!\longrightarrow}{QR}$	A	**B.** $\overrightarrow{RQ}$
27. $\overrightarrow{QR}$	D	**C.** $\overset{\circ\!\!\longrightarrow}{SR}$
28. $\overleftrightarrow{PQ}$	G	**D.** $\overrightarrow{QS}$
29. $\overrightarrow{RP}$	B	**E.** $\overline{SP}$
30. $\overset{\circ\!\!\longrightarrow}{SQ}$	C	**F.** $\overline{QP}$
31. $\overline{PS}$	E	**G.** $\overleftrightarrow{RS}$
32. $\overset{\circ\!\!\longrightarrow}{PS}$	H	**H.** none of these

Lines, rays, half-lines, and segments may be considered sets of points. The **intersection** *(symbolized ∩) of two sets is composed of all elements common to both sets, while the* **union** *(symbolized ∪) of two sets is composed of all elements found in at least one of the two sets. Based on the figure below, specify each of the sets given in Exercises 33–40 in a simpler way.* **There may be other correct forms of the answers in Exercises 33–40.**

33. $\overline{MN} \cup \overline{NO}$ **$\overline{MO}$** **34.** $\overline{MN} \cap \overline{NO}$ **N** **35.** $\overrightarrow{MO} \cap \overrightarrow{OM}$ **$\overline{MO}$** **36.** $\overrightarrow{MO} \cup \overrightarrow{OM}$ **$\overleftrightarrow{MO}$**

37. $\overset{\circ\!\rightarrow}{OP} \cap O$ **$\emptyset$** **38.** $\overset{\circ\!\rightarrow}{OP} \cup O$ **$\overrightarrow{OP}$** **39.** $\overset{\circ\!\rightarrow}{NP} \cap \overline{OP}$ **$\overset{\circ\!\!-\!\bullet}{OP}$** **40.** $\overset{\circ\!\rightarrow}{NP} \cup \overline{OP}$ **$\overset{\circ\!\rightarrow}{NP}$**

Give the measure of the complement of each angle.

41. 28° **62°** **42.** 32° **58°** **43.** 89° **1°**

44. 45° **45°** **45.** $x°$ **$(90 - x)°$** **46.** $(90 - x)°$ **$x°$**

Give the measure of the supplement of each angle.

47. 132° **48°** **48.** 105° **75°** **49.** 26° **154°**

50. 90° **90°** **51.** $y°$ **$(180 - y)°$** **52.** $(180 - y)°$ **$y°$**

Name all pairs of vertical angles in each figure.

53.

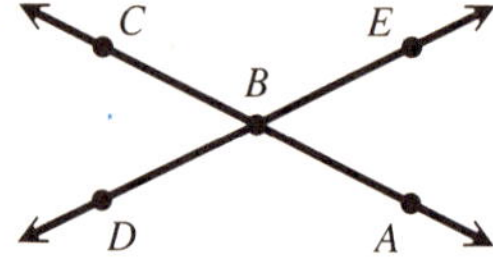

∡CBD and ∡ABE; ∡CBE and ∡DBA

54.

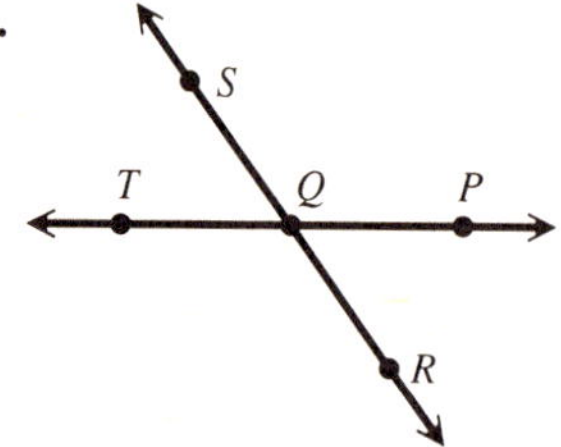

∡TQR and ∡PQS; ∡TQS and ∡PQR

55. In Exercise 53, if ∡*ABE* has a measure of 52°, find the measures of the angles.
(a) ∡*CBD* **52°** **(b)** ∡*CBE* **128°**

56. In Exercise 54, if ∡*SQP* has a measure of 126°, find the measures of the angles.
(a) ∡*TQR* **126°** **(b)** ∡*PQR* **54°**

Find the measure of each marked angle.

57.

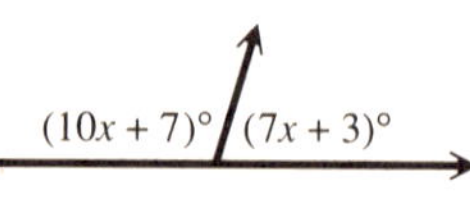

107° and 73°

58.

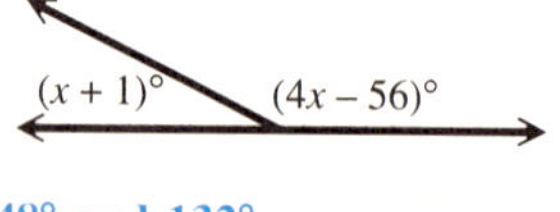

48° and 132°

59.

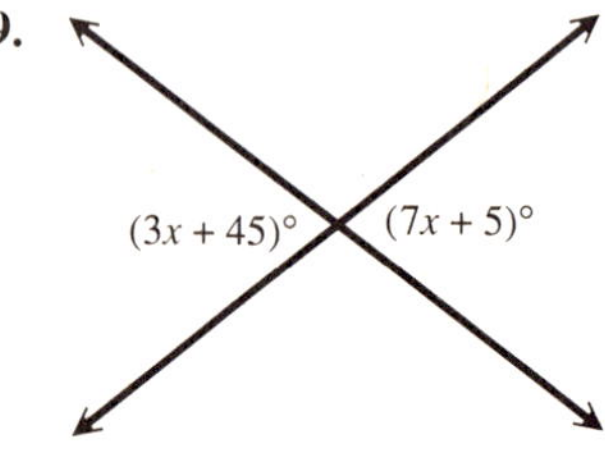

75° and 75°

60.

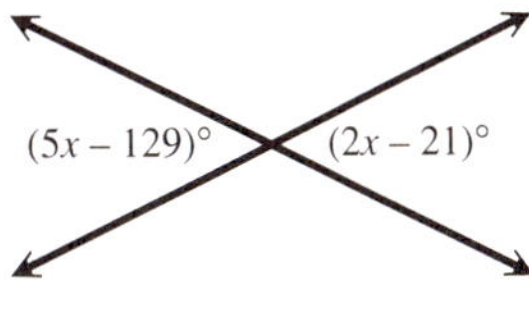

51° and 51°

61.

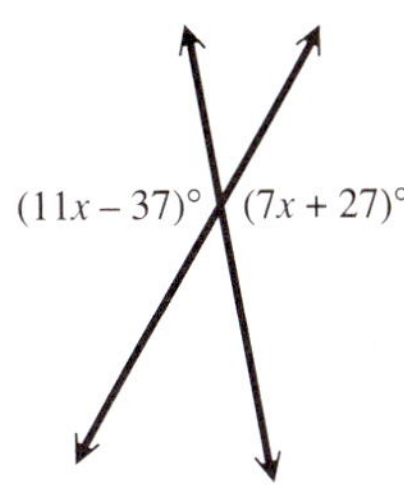

139° and 139°

62.

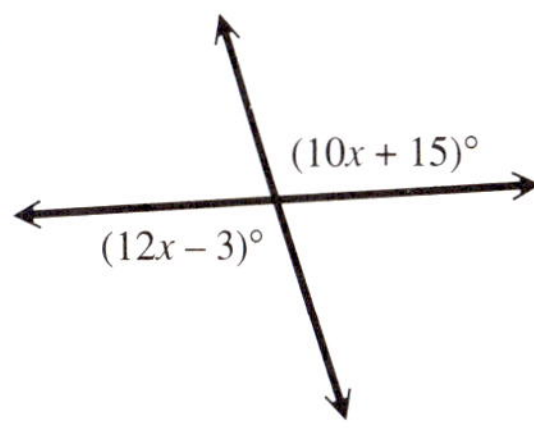

105° and 105°

63.

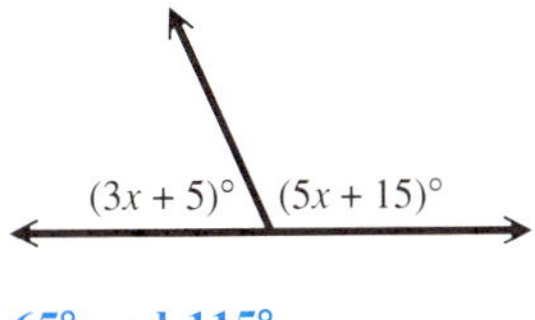

65° and 115°

64.

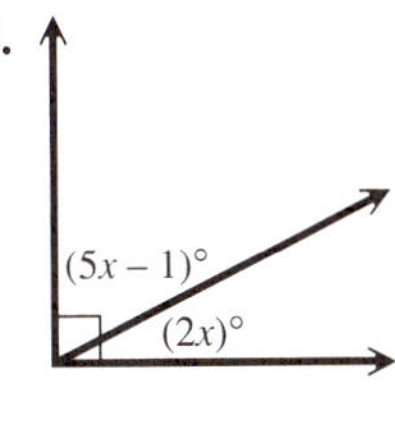

26° and 64°

65.

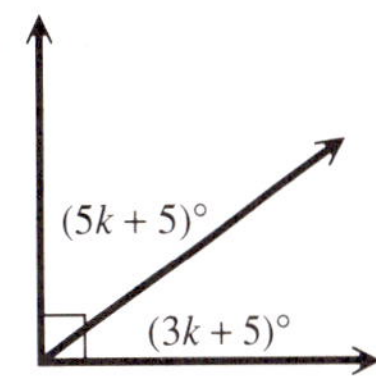

35° and 55°

66. The sketch shows parallel lines m and n cut by a transversal q. (Recall that alternate interior angles have the same measure.) Using the figure, go through the following steps to prove that alternate exterior angles have the same measure.

(a) Measure of ∡2 = measure of ∡ ____, since they are vertical angles. 3

(b) Measure of ∡3 = measure of ∡ ____, since they are alternate interior angles. 6

(c) Measure of ∡6 = measure ∡ ____, since they are vertical angles. 7

(d) By the results of parts (a), (b), and (c), the measure of ∡2 must equal the measure of ∡ ____, showing that alternate ____ angles have equal measures. 7, exterior

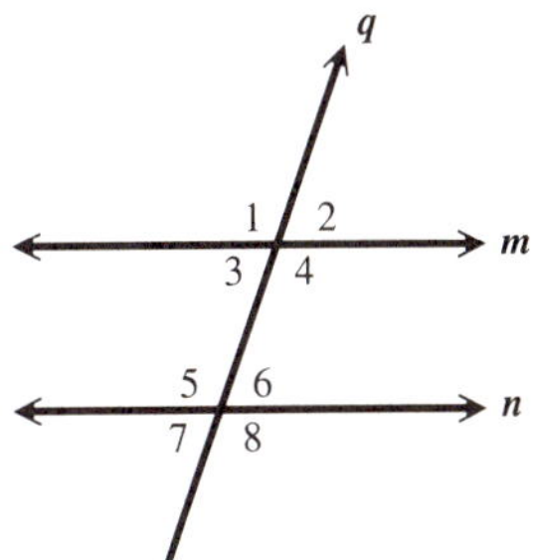

In Exercises 67–70, assume that lines m and n are parallel, and find the measure of each marked angle.

67.

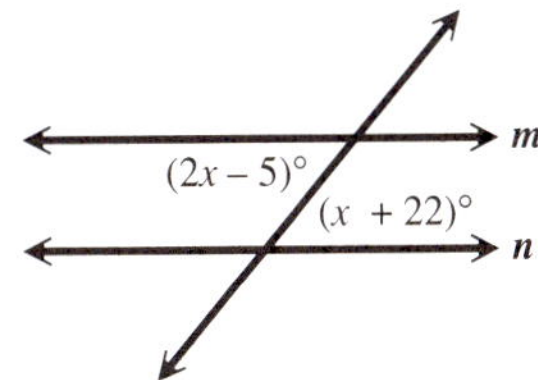

49° and 49°

68.

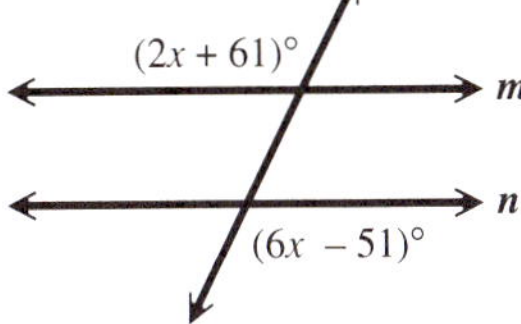

117° and 117°

69.

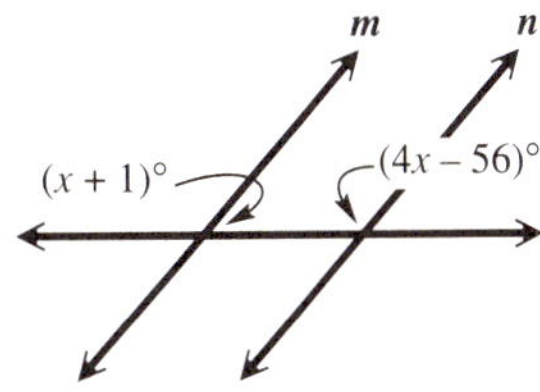

48° and 132°

70.

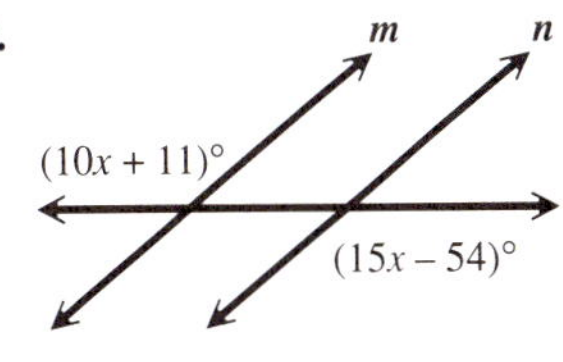

141° and 141°

▲ *Solve each problem in Exercises 71–74.*

71. The supplement of an angle measures 25° more than twice its complement. Find the measure of the angle. 25°

72. The complement of an angle measures 10° less than one-fifth of its supplement. Find the measure of the angle. 80°

73. The supplement of an angle added to the complement of the angle gives 210°. What is the measure of the angle? 30°

74. Half the supplement of an angle is 12° less than twice the complement of the angle. Find the measure of the angle. 52°

75. Write a problem similar to the one in Exercise 71, and solve it.

76. Write a problem similar to the one in Exercise 72, and solve it.

▲ **77.** Use the sketch to find the measure of each numbered angle. Assume that $m \parallel n$.

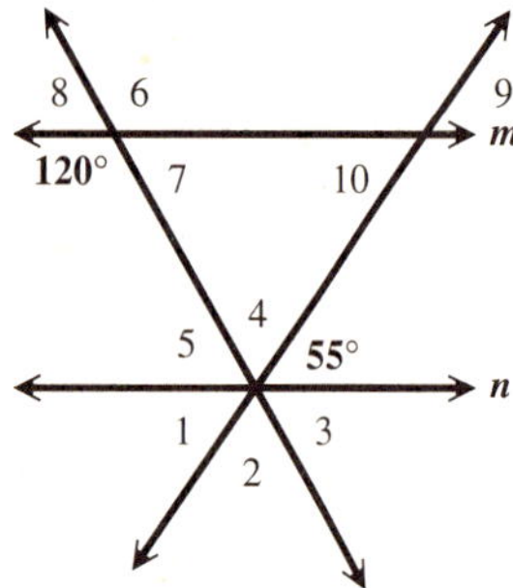

Measures are given in numerical order, starting with angle 1: 55°, 65°, 60°, 65°, 60°, 120°, 60°, 60°, 55°, 55°.

▲ **78.** Complete these steps in the proof that vertical angles have equal measure. In this exercise, m(∡ _____) means "the measure of angle _____." Use the figure.

(a) m(∡1) + m(∡2) = _____ ° 180

(b) m(∡2) + m(∡3) = _____ ° 180

(c) Subtract the equation in part (b) from the equation in part (a) to get [m(∡1) + m(∡2)] − [m(∡2) + m(∡3)] = _____ ° − _____ °. 180, 180

(d) m(∡1) + m(∡2) − m(∡2) − m(∡3) = _____ ° 0

(e) m(∡1) − m(∡3) = _____ ° 0

(f) m(∡1) = m(_____) ∡3

▲ **79.** Use the approach of Exercise 66 to prove that interior angles on the same side of a transversal are supplementary.

▲ **80.** Find the values of x and y in the figure, given that $x + y = 40$. $x = 10, y = 30$

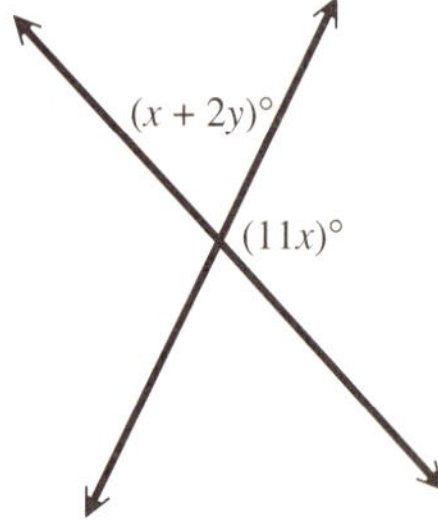

9.2 Exercises (Text Page 505)

◉ *Decide whether each statement in Exercises 1–10 is* true *or* false.

1. A rhombus is an example of a regular polygon. false

2. If a triangle is isosceles, then it is not scalene. true

3. A triangle can have more than one obtuse angle. false

4. A square is both a rectangle and a parallelogram. true

5. A square must be a rhombus. true

6. A rhombus must be a square. false

7. A diameter of a circle is a chord of the circle. true

8. The length of a diameter of a circle is twice the length of a radius of the same circle. true

9. A triangle can have at most one right angle. true

10. A rectangle must be a parallelogram, but a parallelogram might not be a rectangle. true

11. In your own words, explain the distinction between a square and a rhombus.

12. What common traffic sign in the U.S. is in the shape of an octagon? a stop sign

In Exercises 13–20, identify each curve as simple, closed, both, *or* neither.

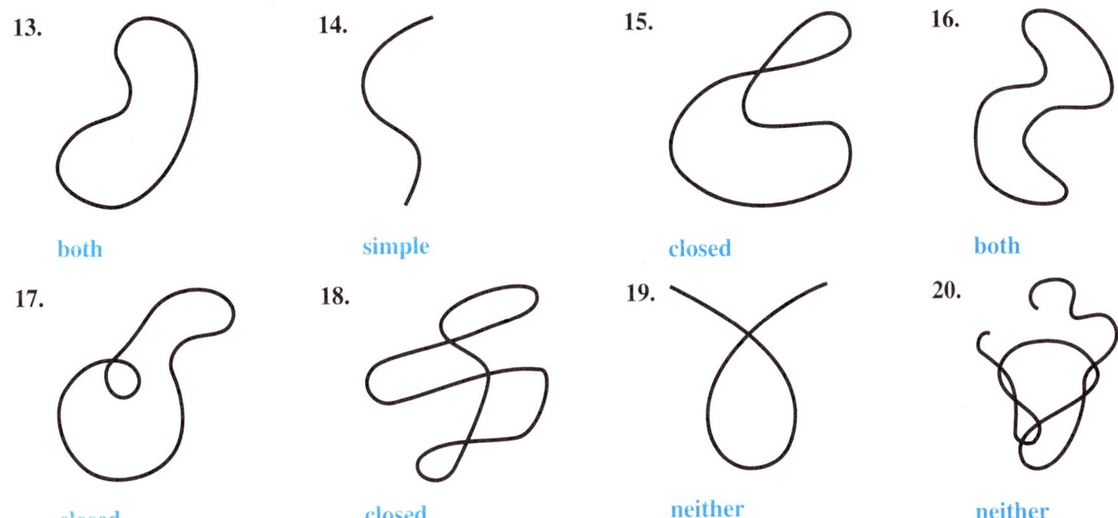

In Exercises 21–26, decide whether each figure is convex *or* not convex.

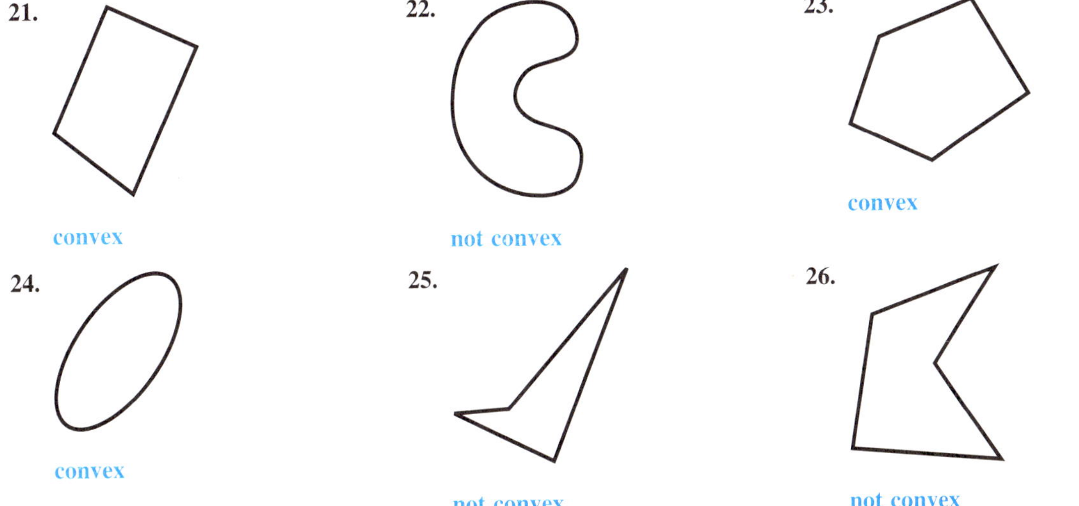

Classify each triangle in Exercises 27–38 as either acute, right, *or* obtuse. *Also classify each as either* equilateral, isosceles, *or* scalene.

27.

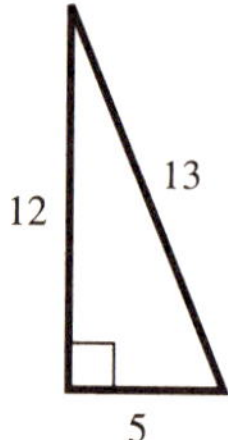

right, scalene

28.

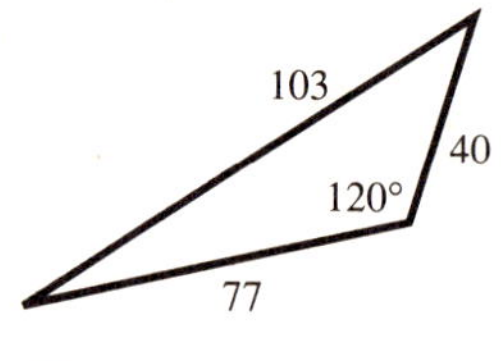

obtuse, scalene

29.

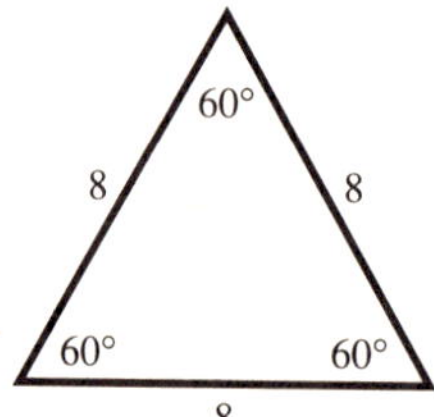

acute, equilateral

30.

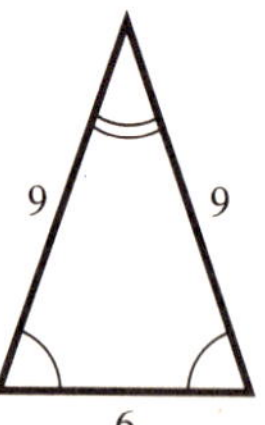

acute, isosceles

31.

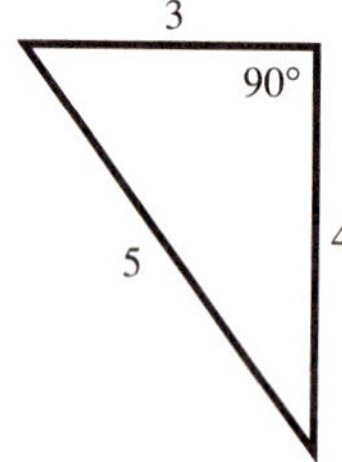

right, scalene

32.

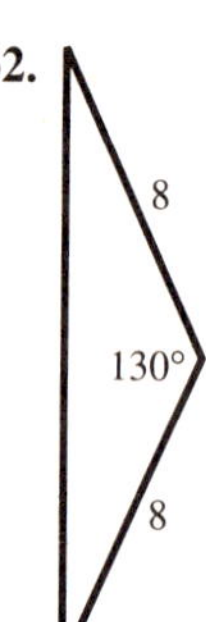

obtuse, isosceles

33.

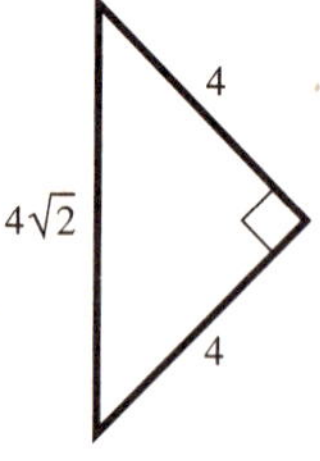

right, isosceles

34.

30°
2
$\sqrt{3}$
60°
1

right, scalene

35.

10
96°
14

obtuse, scalene

36.

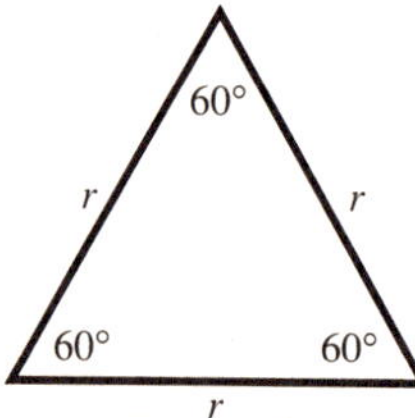

acute, equilateral

37.

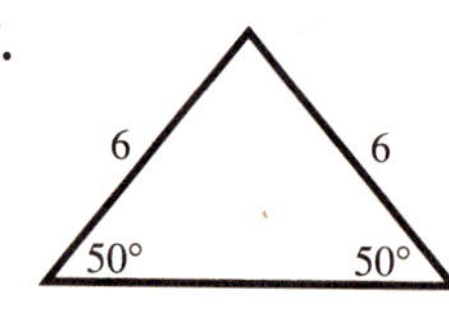

acute, isosceles

38.

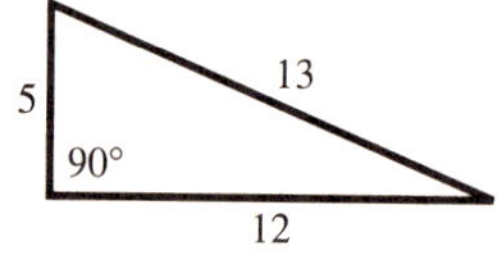

right, scalene

39. Write a definition of *isosceles right triangle.*

40. Explain why the sum of the lengths of any two sides of a triangle must be greater than the length of the third side.

 CONCEPTUAL  WRITING  CHALLENGING  SCIENTIFIC CALCULATOR  GRAPHING CALCULATOR

41. Can a triangle be both right and obtuse? Explain.

42. In the classic 1939 movie *The Wizard of Oz,* the scarecrow, upon getting a brain, says the following: "The sum of the square roots of any two sides of an isosceles triangle is equal to the square root of the remaining side." Give an example to show that his statement is incorrect. **For example, the triangle in Exercise 30 is an isosceles triangle, but $\sqrt{9} + \sqrt{9} \neq \sqrt{6}$, since $3 + 3 = 6 \neq \sqrt{6}$.**

Find the measure of each angle in triangle ABC.

43.

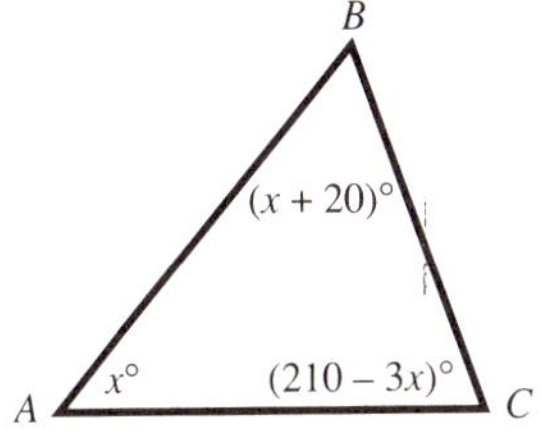

$A = 50°$; $B = 70°$; $C = 60°$

44.

(x + 15)°
B
(x + 5)°
A
C
(10x – 20)°

$A = 130°$; $B = 30°$; $C = 20°$

45.

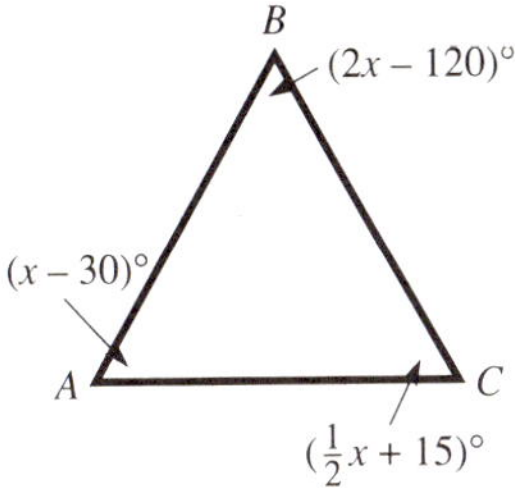

$A = B = C = 60°$

46.

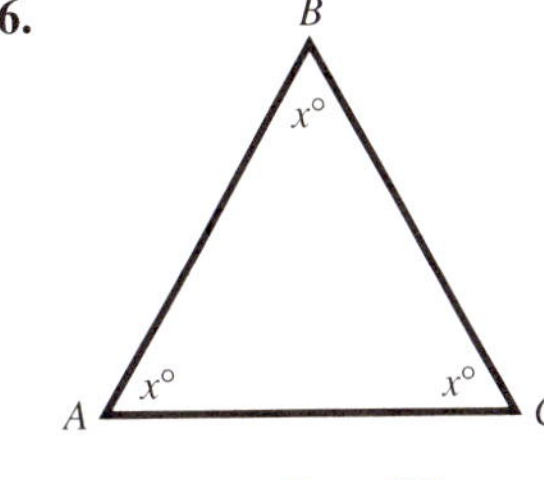

$A = B = C = 60°$

47. In triangle *ABC*, angles *A* and *B* have the same measure, while the measure of angle *C* is 24 degrees larger than the measure of each of *A* and *B*. What are the measures of the three angles? **$A = B = 52°$; $C = 76°$**

48. In triangle *ABC*, the measure of angle *A* is 30 degrees more than the measure of angle *B*. The measure of angle *B* is the same as the measure of angle *C*. Find the measure of each angle. **$B = C = 50°$; $A = 80°$**

In each triangle, find the measure of exterior angle BCD.

49.

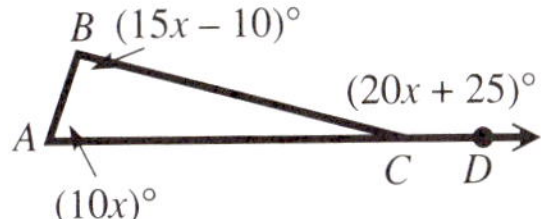

165°

50.

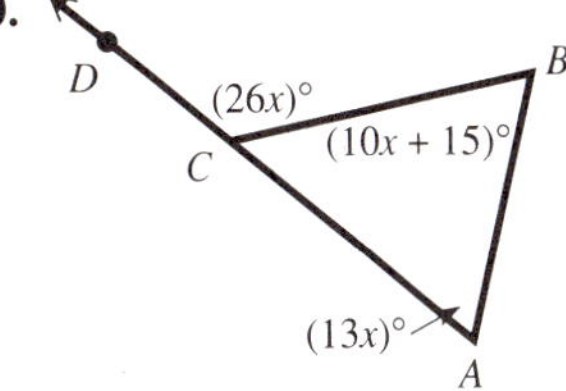

130°

51.

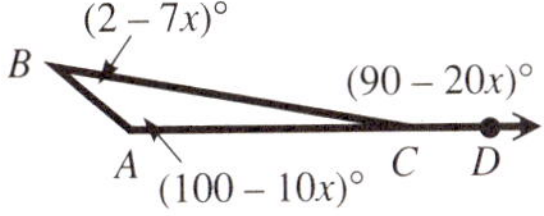

170°

52.

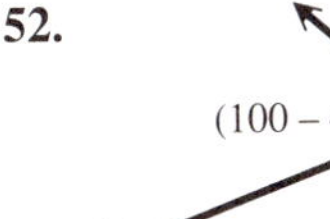
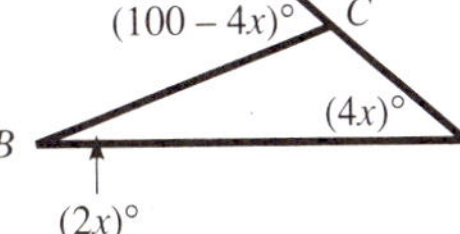

60°

53. Go through the following argument provided by Richard Crouse in a letter to the editor of *Mathematics Teacher* in the February 1988 issue.

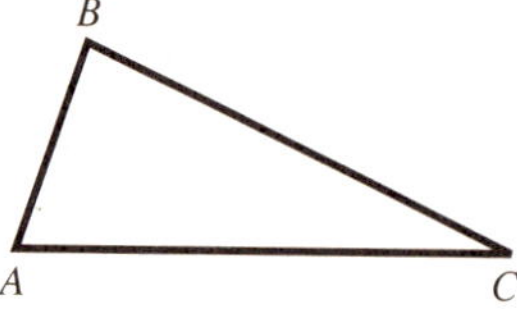

(a) Place the eraser end of a pencil on vertex A of the triangle and let the pencil coincide with side AC of the triangle.
(b) With the eraser fixed at A, rotate the pencil counterclockwise until it coincides with side AB.
(c) With the pencil fixed at point B, rotate eraser end counterclockwise until the pencil coincides with side BC.
(d) With the pencil fixed at point C, rotate the point end of the pencil counterclockwise until the pencil coincides with side AC.
(e) Notice that the pencil is now pointing in the opposite direction. What concept from this section does this exercise reinforce? **The sum of the measures of the angles of a triangle equals 180° (since the pencil has gone through one-half of a complete rotation).**

54. Using the points, segments, and lines in the figure, list all parts of the circle.

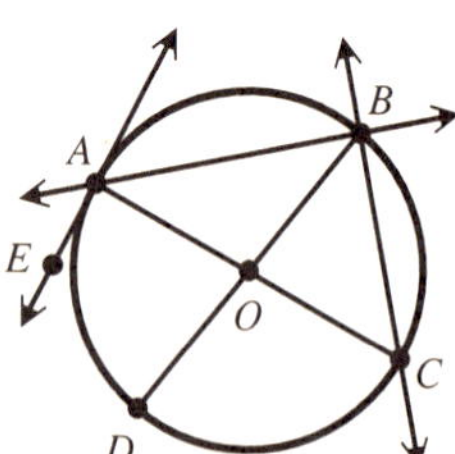

(a) center **O**
(b) radii **$\overline{OA}, \overline{OC}, \overline{OB}, \overline{OD}$**
(c) diameters **$\overline{AC}, \overline{BD}$**
(d) chords **$\overline{AC}, \overline{BD}, \overline{BC}, \overline{AB}$**
(e) secants **$\overleftrightarrow{BC}, \overleftrightarrow{AB}$**
(f) tangents **$\overleftrightarrow{AE}$**

55. Refer to angles 1, 2, and 6 in Figure 19. Prove that the sum of the measures of angles 1 and 2 is equal to the measure of angle 6.

9.3 Exercises (Text Page 516)

Decide whether each statement is true *or* false.

1. The perimeter of an equilateral triangle with side equal to 6 in. is the same as the perimeter of a rectangle with length 7 in. and width 2 in. **true**

2. The area of a circle with radius r inches is greater than the area of a square with side r inches. **true**

3. A square with area 16 cm^2 has a perimeter of 16 cm. **true**

4. If the area of a certain triangle is 24 square units, and the base is 8 units, then the height must be 3 units. **false**

5. The area of a circle with radius r inches is doubled if the radius is doubled. **false**

6. The perimeter of a rectangle is doubled if both the length and the width are doubled. **true**

Use the formulas of this section to find the area of each figure. In Exercises 19–22, use 3.14 as an approximation for π.

7.

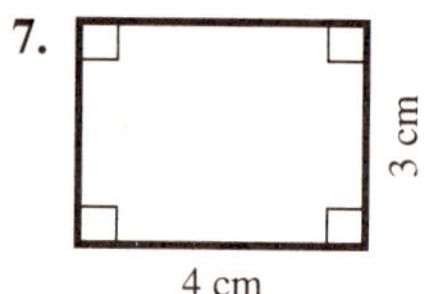

12 cm^2

8.

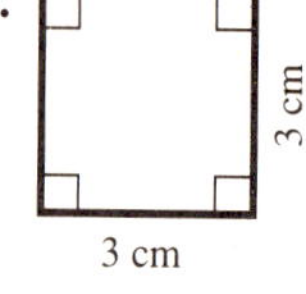

9 cm^2

9.

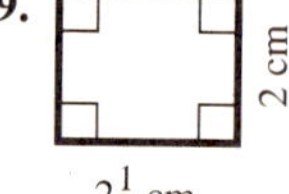

$2\frac{1}{2}$ cm

5 cm^2

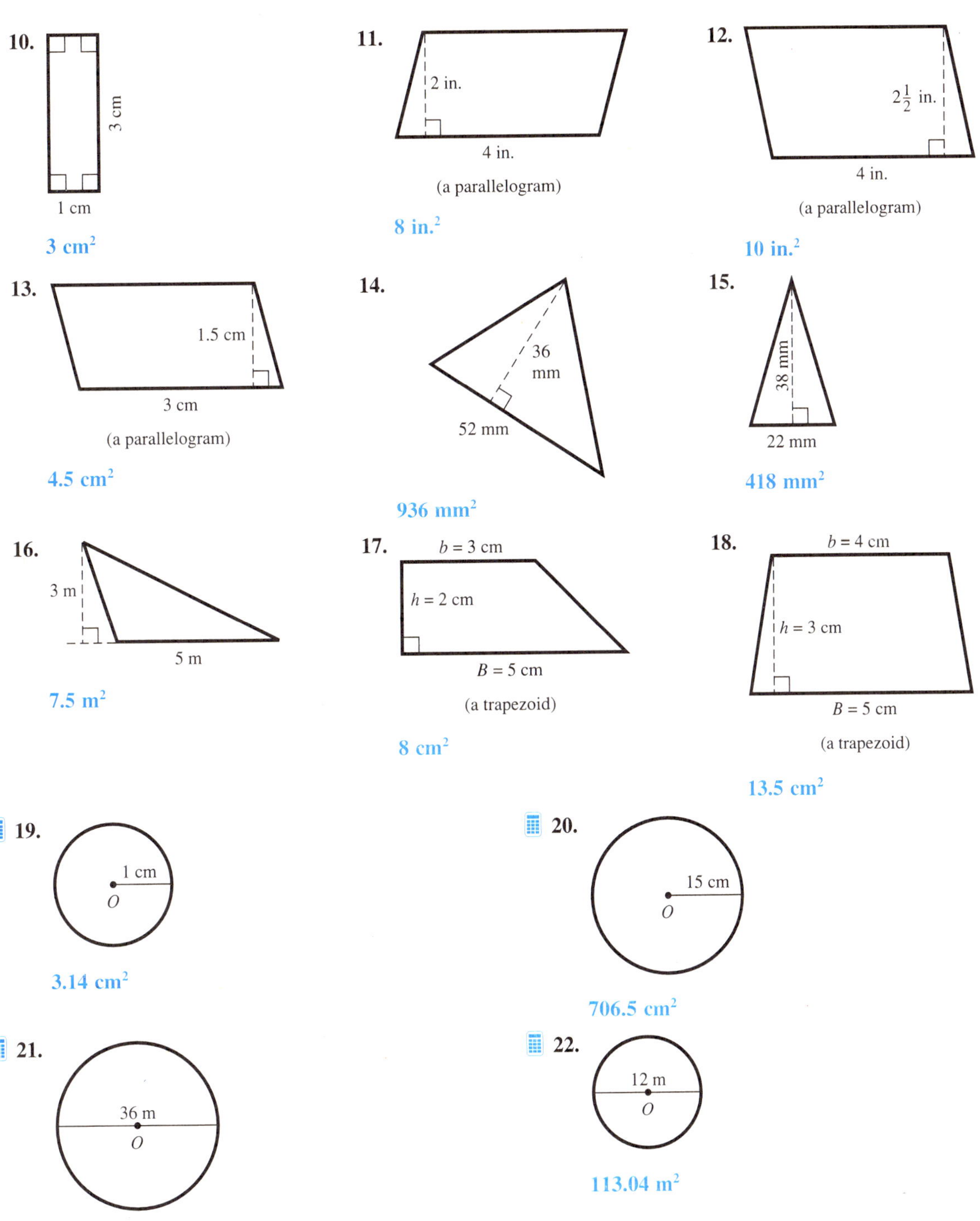

10.
3 cm
1 cm
3 cm²
11.
2 in.
4 in.
(a parallelogram)
8 in.²
12.
$2\frac{1}{2}$ in.
4 in.
(a parallelogram)
10 in.²
13.
1.5 cm
3 cm
(a parallelogram)
4.5 cm²
14.
36 mm
52 mm
936 mm²
15.
38 mm
22 mm
418 mm²
16.
3 m
5 m
7.5 m²
17.
b = 3 cm
h = 2 cm
B = 5 cm
(a trapezoid)
8 cm²
18.
b = 4 cm
h = 3 cm
B = 5 cm
(a trapezoid)
13.5 cm²
19.
1 cm
O
3.14 cm²
20.
15 cm
O
706.5 cm²
21.
36 m
O
1,017.36 m²
22.
12 m
O
113.04 m²

Solve each of the following problems.

23. A stained-glass window in a church is in the shape of a square. The perimeter of the square is 7 times the length of a side in meters, decreased by 12. Find the length of a side of the window.
4 m

24. A video rental establishment displayed a rectangular cardboard standup advertisement for the movie *Forrest Gump*. The length was 20 in. more than the width, and the perimeter was 176 in. What were the dimensions of the rectangle?
34 in. by 54 in.

25. A lot is in the shape of a triangle. One side is 100 ft longer than the shortest side, while the third side is 200 ft longer than the shortest side. The perimeter of the lot is 1,200 ft. Find the lengths of the sides of the lot.
300 ft, 400 ft, 500 ft

26. A wall pennant is in the shape of an isosceles triangle. (Two sides have the same length.) Each of the two equal sides measures 18 in. more than the third side, and the perimeter of the triangle is 54 in. What are the lengths of the sides of the pennant?
6 in., 24 in., 24 in.

27. The Peachtree Plaza Hotel in Atlanta is in the shape of a cylinder, with a circular foundation. The circumference of the foundation is 6 times the radius, increased by 12.88 ft. Find the radius of the circular foundation. (Use 3.14 as an approximation for π.)
46 ft

28. If the radius of a certain circle is tripled, with 8.2 cm then added, the result is the circumference of the circle. Find the radius of the circle. (Use 3.14 as an approximation for π.)
2.5 cm

29. The survey plat in the figure shows two lots that form a trapezoid. The measures of the parallel sides are 115.80 ft and 171.00 ft. The height of the trapezoid is 165.97 ft. Find the combined area of the two lots.

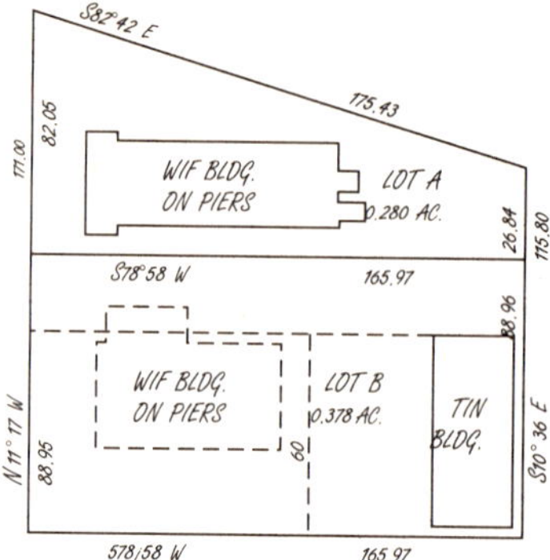

Round your answer to the nearest hundredth of a square foot. **23,800.10 ft²**

30. Lot A in the figure is in the shape of a trapezoid. The parallel sides measure 26.84 ft and 82.05 ft. The height of the trapezoid is 165.97 ft. Find the area of Lot A. Round your answer to the nearest hundredth of a square foot. **9,036.24 ft²**

31. In order to purchase fencing to go around a rectangular yard, would you need to use perimeter or area to decide how much to buy? **perimeter**

32. In order to purchase fertilizer for the lawn of a yard, would you need to use perimeter or area to decide how much to buy? **area**

In the chart below, one of the values r (radius), d (diameter), C (circumference), or A (area) is given for a particular circle. Find the remaining three values. Leave π in your answers.

	r	d	C	A
33.	6 in.	**12 in.**	**12π in.**	**36π in.²**
34.	9 in.	**18 in.**	**18π in.**	**81π in.²**
35.	**5 ft**	10 ft	**10π ft**	**25π ft²**
36.	**20 ft**	40 ft	**40π ft**	**400π ft²**
37.	**6 cm**	**12 cm**	12π cm	**36π cm²**
38.	**9 cm**	**18 cm**	18π cm	**81π cm²**
39.	**10 in.**	**20 in.**	**20π in.**	100π in.²
40.	**16 in.**	**32 in.**	**32π in.**	256π in.²

 CONCEPTUAL  WRITING  CHALLENGING  SCIENTIFIC CALCULATOR  GRAPHING CALCULATOR

Each of the following figures has perimeter as indicated. Find the value of x.

41. $P = 58$

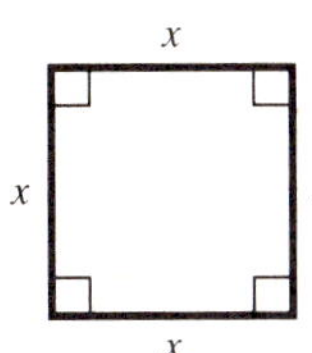

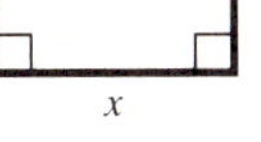

14.5

42. $P = 42$

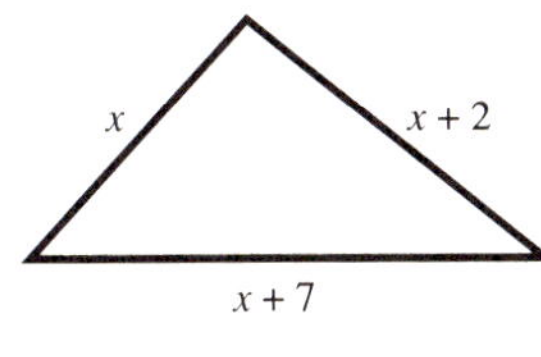

11

43. $P = 38$

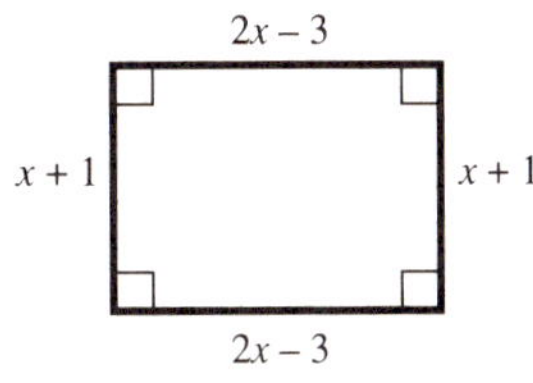

7

44. $P = 278$

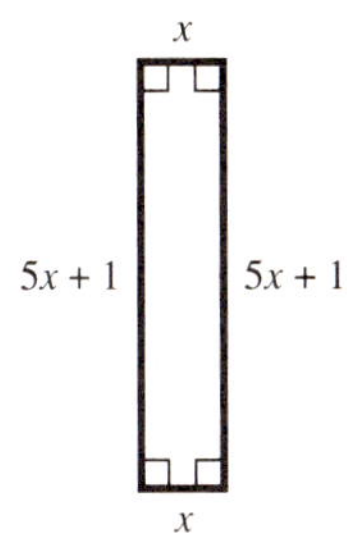

23

Each of the following figures has area as indicated. Find the value of x.

45. $A = 26.01$

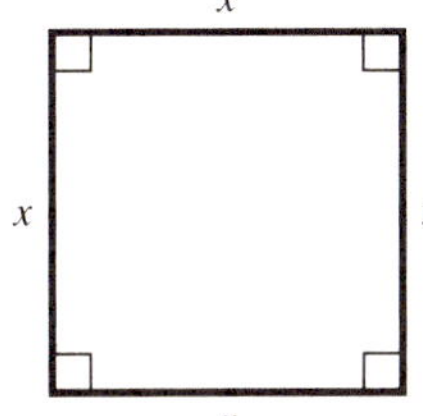

5.1

46. $A = 28$

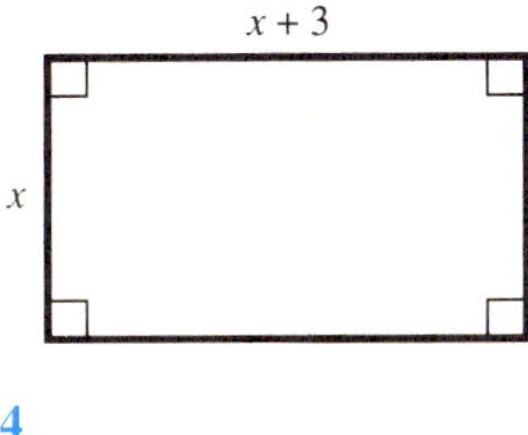

4

47. $A = 15$

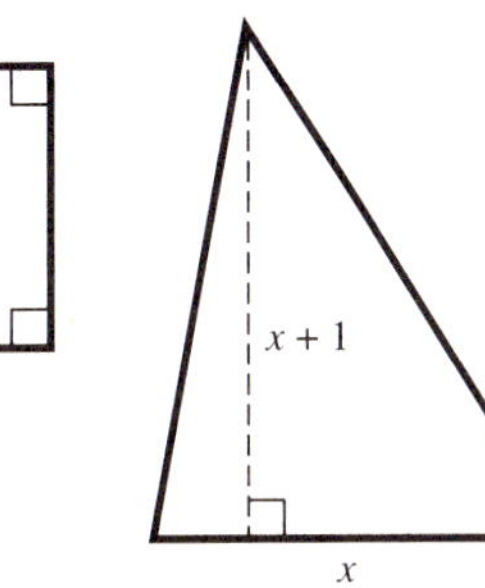

5

48. $A = 30$

(a trapezoid)

8

Each of the following circles has circumference or area as indicated. Find the value of x. Use 3.14 as an approximation for π.

49. $C = 37.68$

5

50. $C = 54.95$

7.5

51. $A = 18.0864$

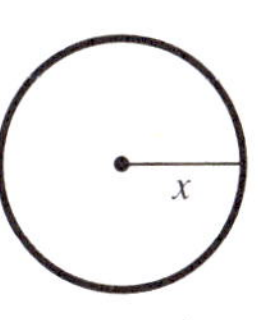

2.4

52. $A = 28.26$

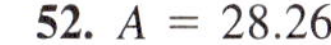

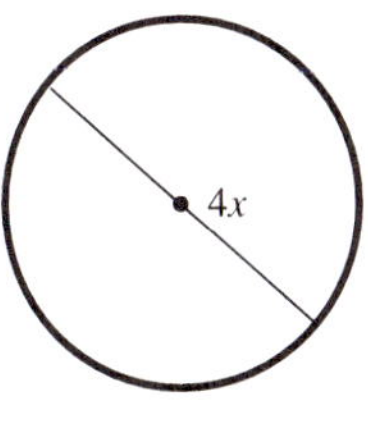

1.5

53. Work through the parts of this exercise in order, and use it to make a generalization concerning areas of rectangles.

(a) Find the area of a rectangle 4 cm by 5 cm. 20 cm^2

(b) Find the area of a rectangle 8 cm by 10 cm. 80 cm^2

(c) Find the area of a rectangle 12 cm by 15 cm. 180 cm^2

(d) Find the area of a rectangle 16 cm by 20 cm. 320 cm^2

(e) The rectangle in part (b) had sides twice as long as the sides of the rectangle in part (a). Divide the larger area by the smaller. By doubling the sides, the area increased _____ times. 4

(f) To get the rectangle in part (c) each side of the rectangle of part (a) was multiplied by _____. This made the larger area _____ times the size of the smaller area. 3; 9

(g) To get the rectangle of part (d) each side of the rectangle of part (a) was multiplied by ______. This made the area increase to ______ times what it was originally.
4; 16

(h) In general, if the length of each side of a rectangle is multiplied by n, the area is multiplied by ______. n^2

Use the results of Exercise 53 to solve each of the following.

54. A ceiling measuring 9 ft by 15 ft can be painted for \$60. How much would it cost to paint a ceiling 18 ft by 30 ft? **\$240**

55. Suppose carpet for a 10 ft by 12 ft room costs \$200. Find the cost to carpet a room 20 ft by 24 ft. **\$800**

56. A carpet cleaner charges \$80 to do an area 31 ft by 31 ft. What would be the charge for an area 93 ft by 93 ft? **\$720**

57. Use the logic of Exercise 53 to answer the following: If the radius of a circle is multiplied by n, then the area of the circle is multiplied by ______. n^2

58. Use the logic of Exercise 53 to answer the following: If the height of a triangle is multiplied by n and the base length remains the same, then the area of the triangle is multiplied by ______. n

By considering total area as the sum of the areas of all of its parts, areas of figures such as those in Exercises 59–62 may be determined. Find the total area of each figure. Use 3.14 as an approximation for π in Exercises 61 and 62.

59.

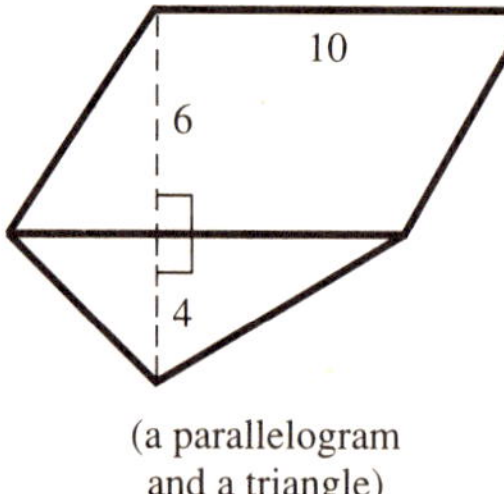

(a parallelogram and a triangle)

80

60.

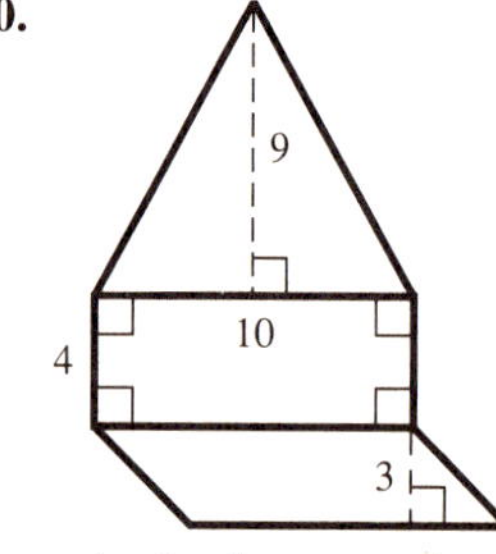

(a triangle, a rectangle, and a parallelogram)

115

61.

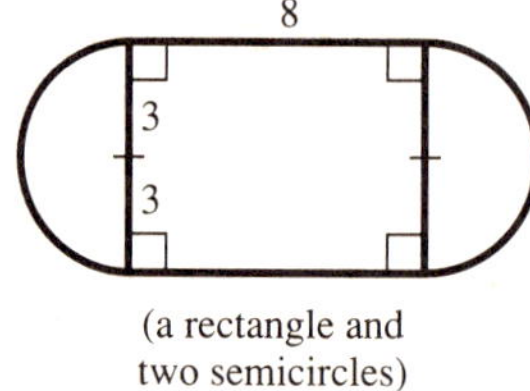

(a rectangle and two semicircles)

76.26

62.

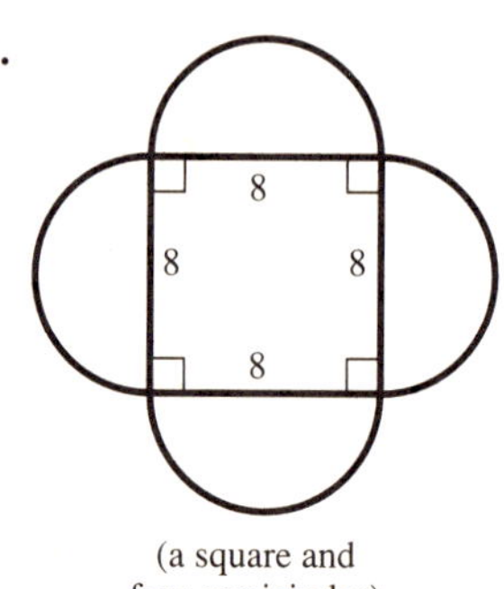

(a square and four semicircles)

164.48

 CONCEPTUAL  WRITING  CHALLENGING  SCIENTIFIC CALCULATOR   GRAPHING CALCULATOR

The shaded areas of the figures in Exercises 63–68 may be found by subtracting the area of the unshaded portion from the total area of the figure. Use this approach to find the area of the shaded portion. Use 3.14 as an approximation for π in Exercises 66–68, and round to the nearest hundredth.

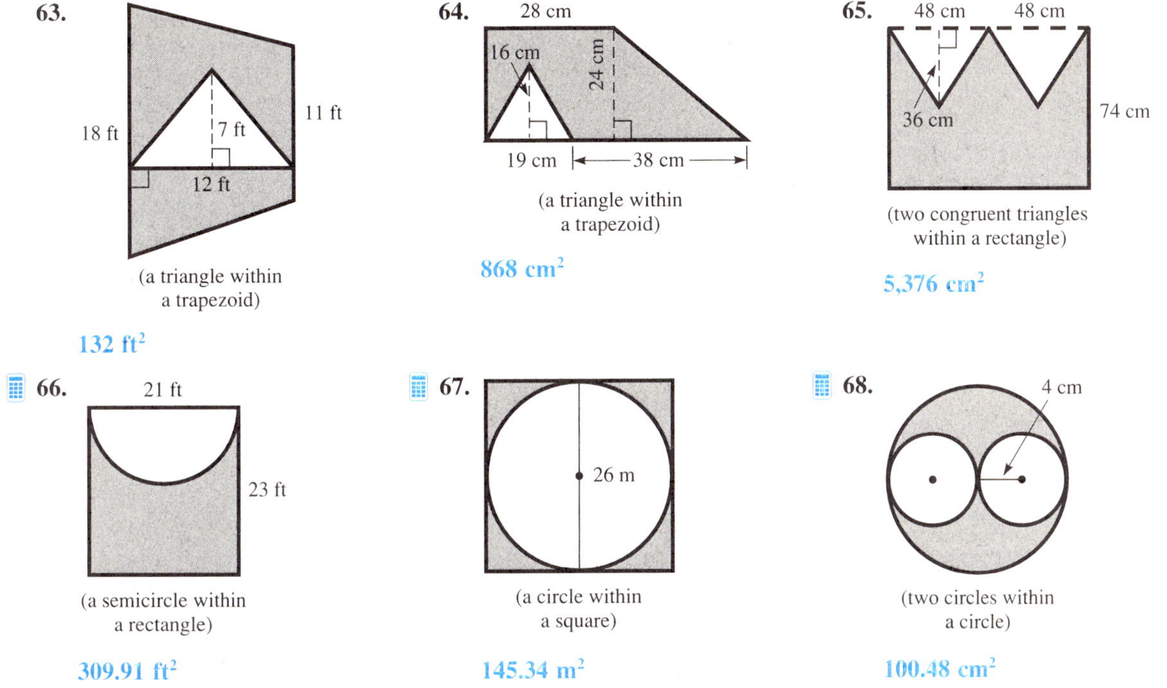

63. (a triangle within a trapezoid)
132 ft^2

64. (a triangle within a trapezoid)
868 cm^2

65. (two congruent triangles within a rectangle)
5,376 cm^2

66. (a semicircle within a rectangle)
309.91 ft^2

67. (a circle within a square)
145.34 m^2

68. (two circles within a circle)
100.48 cm^2

The following exercises show prices actually charged by a local pizzeria. In each case, the dimension is the diameter of the pizza. Find the best buy.

69. Cheese pizza: 10-in. pizza sells for \$5.99, 12-in. pizza sells for \$7.99, 14-in. pizza sells for \$8.99.
14-in. pizza

70. Cheese pizza with two toppings: 10-in. pizza sells for \$7.99, 12-in. pizza sells for \$9.99, 14-in. pizza sells for \$10.99.
14-in. pizza

71. All Feasts pizza: 10-in. pizza sells for \$9.99, 12-in. pizza sells for \$11.99, 14-in. pizza sells for \$12.99.
14-in. pizza

72. Extravaganza pizza: 10-in. pizza sells for \$11.99, 12-in. pizza sells for \$13.99, 14-in. pizza sells for \$14.99.
14-in. pizza

A polygon may be inscribed within a circle or circumscribed about a circle. In the figure, triangle ABC is inscribed within the circle, while square WXYZ is circumscribed about it. These ideas will be used in some of the remaining exercises in this section and later in this chapter.

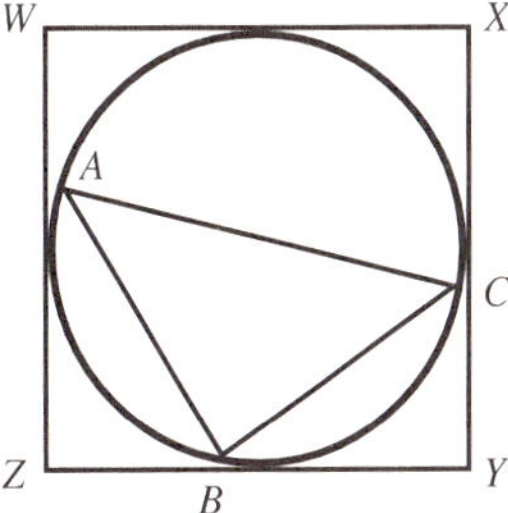

▲ *Exercises 73–80 require some ingenuity, but all may be solved using the concepts presented so far in this chapter.*

73. Given the circle with center O and rectangle $ABCO$, find the diameter of the circle. **26 in.**

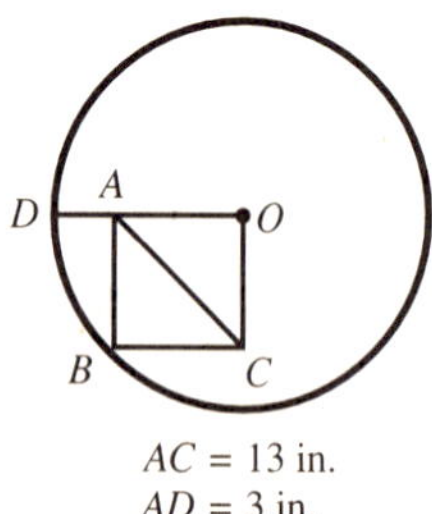

$AC = 13$ in.
$AD = 3$ in.

74. What is the perimeter of $\triangle AEB$, if $AD = 20$ in., $DC = 30$ in., and $AC = 34$ in.? **54 in.**

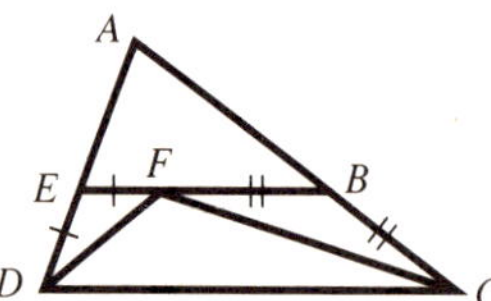

75. The area of square $PQRS$ is 1,250 square feet. T, U, V, and W are the midpoints of $\overleftrightarrow{PQ}$, $\overleftrightarrow{QR}$, $\overleftrightarrow{RS}$, and $\overleftrightarrow{SP}$, respectively. What is the area of square $TUVW$? **625 ft²**

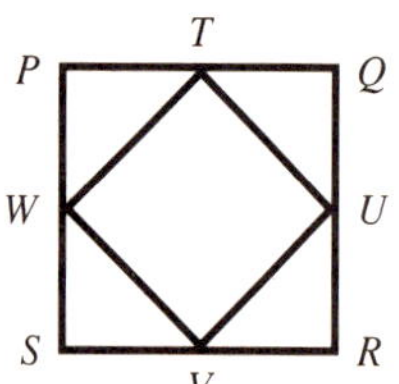

76. The rectangle $ABCD$ has length twice the width. If P, Q, R, and S are the midpoints of the sides, and the perimeter of $ABCD$ is 96 in., what is the area of quadrilateral $PQRS$? **256 in.²**

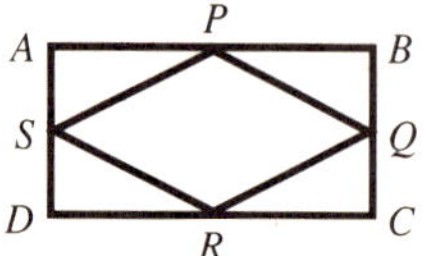

77. If $ABCD$ is a square with each side measuring 36 in., what is the area of the shaded region? **648 in.²**

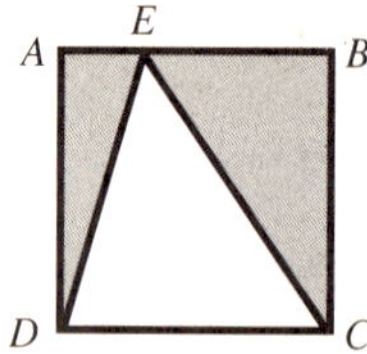

78. Can the perimeter of the polygon shown be determined from the given information? If so, what is the perimeter? **yes; 40 in.**

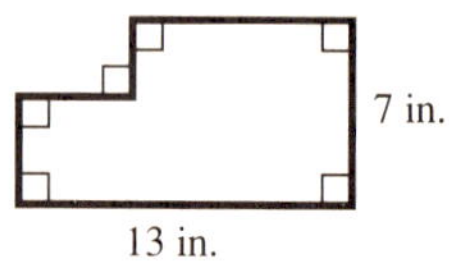

79. Express the area of the shaded region in terms of r, given that the circle is inscribed in the square. $\mathbf{\dfrac{(4 - \pi)r^2}{4}}$

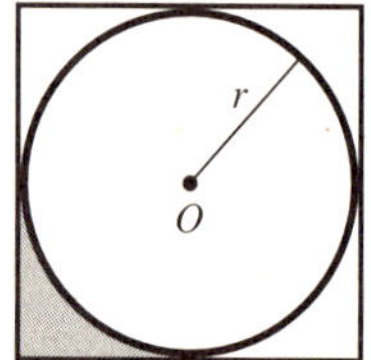

80. Find the area of trapezoid $ABCD$, given that the area of right triangle ABE is 30 in.². **72 in.²**

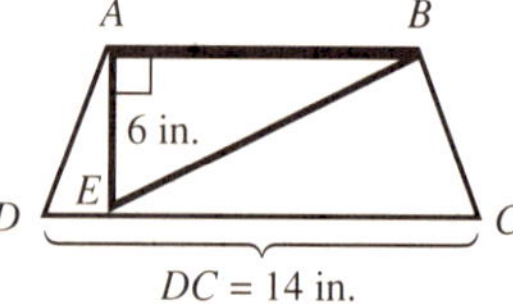

9.4 Exercises (Text Page 530)

In each figure, you are given certain information. Tell which congruence property you would use to verify the congruence required, and give the additional information needed to apply the property.

1. Given: $AC = BD$; $AD = BC$. Show that $\triangle ABD \cong \triangle BAC$.

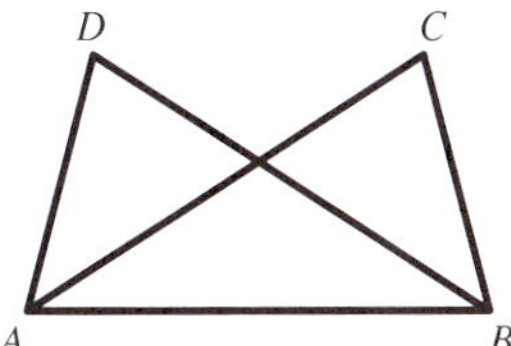

SSS; $AB = AB$, since AB is equal to itself.

2. Given: $AC = BC$; $AD = DB$. Show that $\triangle ADC \cong \triangle BDC$.

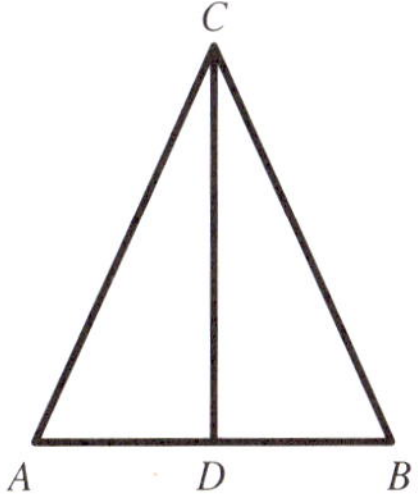

SSS; $CD = CD$, since CD is equal to itself.

3. Given: $\overleftrightarrow{DB}$ is perpendicular to AC; $AB = BC$. Show that $\triangle ABD \cong \triangle CBD$.

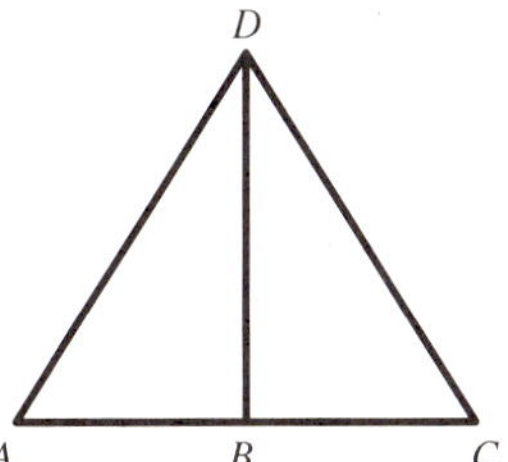

SAS; $DB = DB$, since DB is equal to itself.

4. Given: $BC = BA$; $\measuredangle 1 = \measuredangle 2$. Show that $\triangle DBC \cong \triangle DBA$.

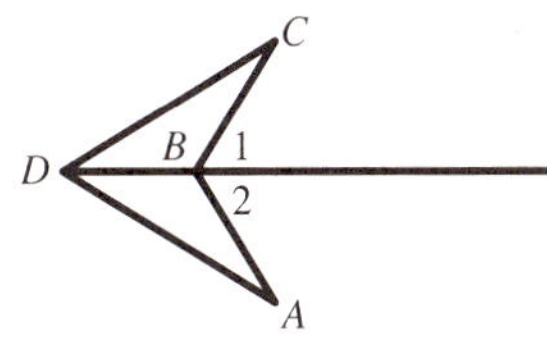

SAS; $DB = DB$ and $\measuredangle DBC = \measuredangle DBA$, since DB is equal to itself, and angles DBC and DBA are supplements of equal angles.

5. Given: $\measuredangle BAC = \measuredangle DAC$; $\measuredangle BCA = \measuredangle DCA$. Show that $\triangle ABC \cong \triangle ADC$.

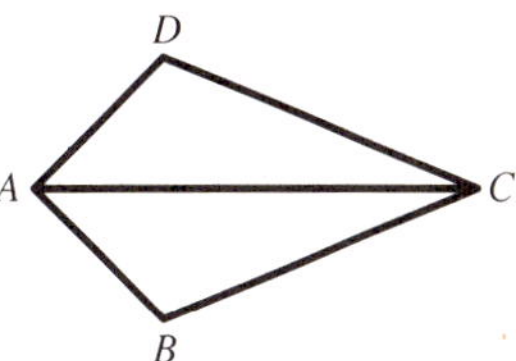

ASA; $AC = AC$, since AC is equal to itself.

6. Given: $BO = OE$; $\overleftrightarrow{OB}$ is perpendicular to $\overleftrightarrow{AC}$; $\overleftrightarrow{OE}$ is perpendicular to $\overleftrightarrow{DF}$. Show that $\triangle AOB \cong \triangle FOE$.

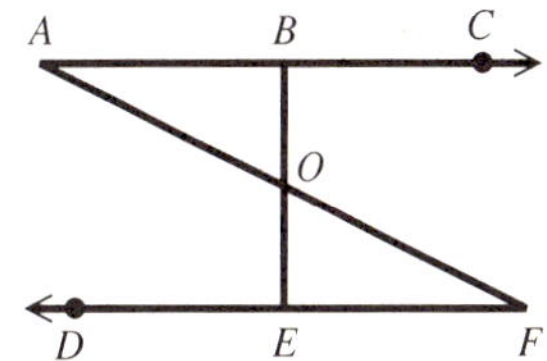

ASA; $\measuredangle AOB = \measuredangle FOE$, since they are vertical angles.

Exercises 7–10 refer to the given figure, an isosceles triangle with $AB = BC$.

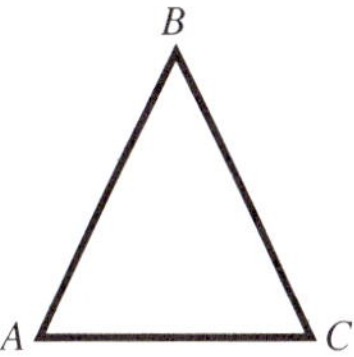

7. If $\measuredangle B$ measures 46°, then $\measuredangle A$ measures ______ and $\measuredangle C$ measures ______. 67°; 67°

8. If $\measuredangle C$ measures 52°, what is the measure of $\measuredangle B$? 76°

9. If $BC = 12$ in., and the perimeter of $\triangle ABC$ is 30 in., what is the length AC? 6 in.

10. If the perimeter of $\triangle ABC = 40$ in., and $AC = 10$ in., what is the length AB? 15 in.

11. Explain why all equilateral triangles must be similar.

12. Explain why two congruent triangles must be similar, but two similar triangles might not be congruent.

Name the corresponding angles and the corresponding sides for each of the following pairs of similar triangles.

13.

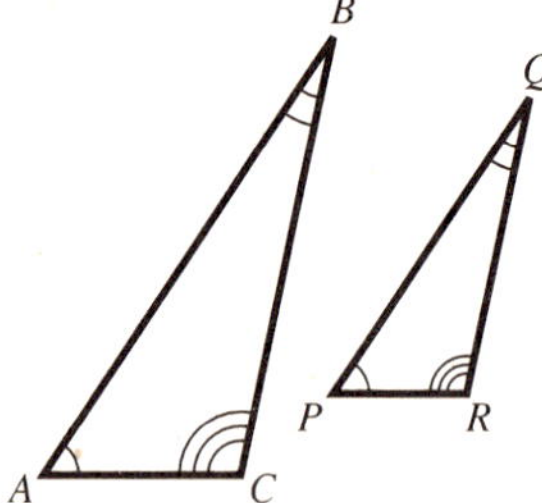

$\measuredangle A$ and $\measuredangle P$; $\measuredangle C$ and $\measuredangle R$; $\measuredangle B$ and $\measuredangle Q$; $\overleftrightarrow{AC}$ and $\overleftrightarrow{PR}$; $\overleftrightarrow{CB}$ and $\overleftrightarrow{RQ}$; $\overleftrightarrow{AB}$ and $\overleftrightarrow{PQ}$

14.

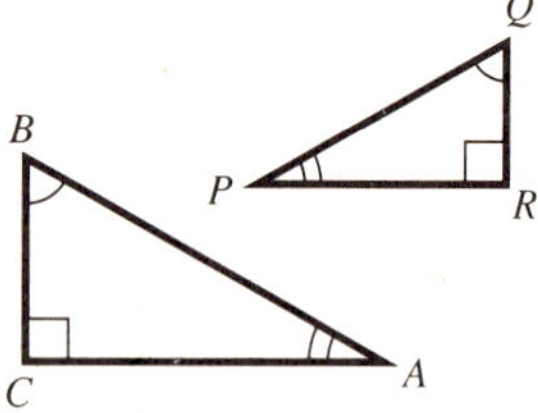

$\measuredangle A$ and $\measuredangle P$; $\measuredangle B$ and $\measuredangle Q$; $\measuredangle C$ and $\measuredangle R$; $\overleftrightarrow{AC}$ and $\overleftrightarrow{PR}$; $\overleftrightarrow{CB}$ and $\overleftrightarrow{RQ}$; $\overleftrightarrow{AB}$ and $\overleftrightarrow{PQ}$

15.

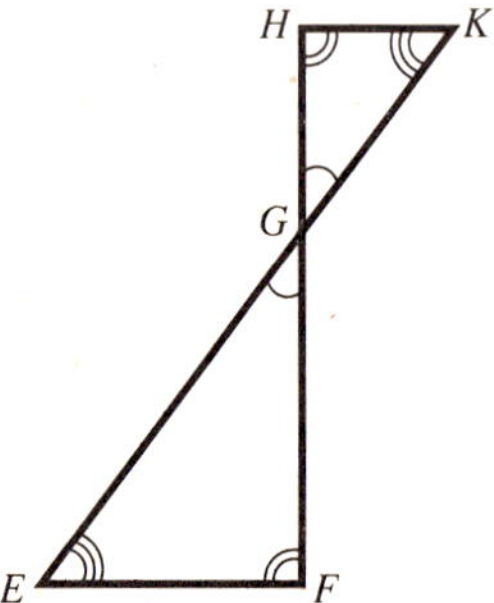

$\measuredangle H$ and $\measuredangle F$; $\measuredangle K$ and $\measuredangle E$; $\measuredangle HGK$ and $\measuredangle FGE$; $\overleftrightarrow{HK}$ and $\overleftrightarrow{FE}$; $\overleftrightarrow{GK}$ and $\overleftrightarrow{GE}$; $\overleftrightarrow{HG}$ and $\overleftrightarrow{FG}$

16.

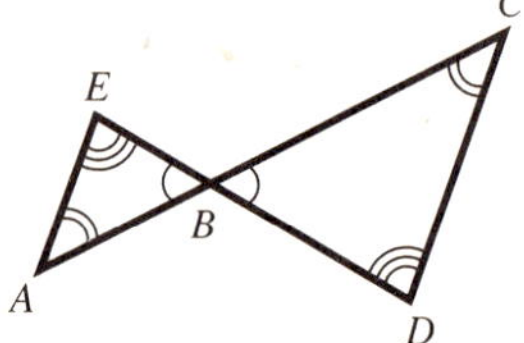

$\measuredangle A$ and $\measuredangle C$; $\measuredangle E$ and $\measuredangle D$; $\measuredangle ABE$ and $\measuredangle CBD$; $\overleftrightarrow{EB}$ and $\overleftrightarrow{DB}$; $\overleftrightarrow{AB}$ and $\overleftrightarrow{CB}$; $\overleftrightarrow{AE}$ and $\overleftrightarrow{CD}$

Find all unknown angle measures in each pair of similar triangles.

17.

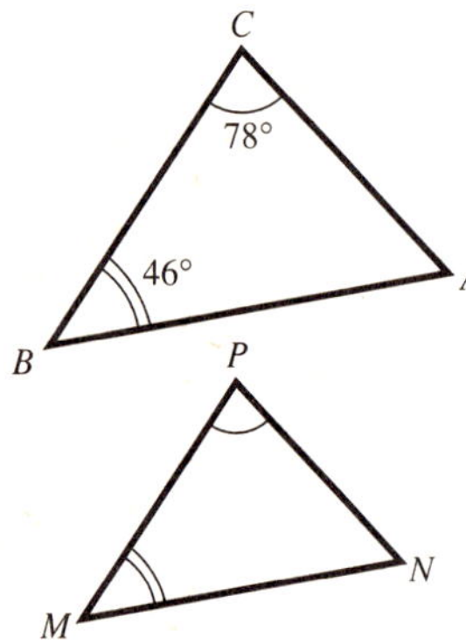

$\measuredangle P = 78°$; $\measuredangle M = 46°$; $\measuredangle A = \measuredangle N = 56°$

18.

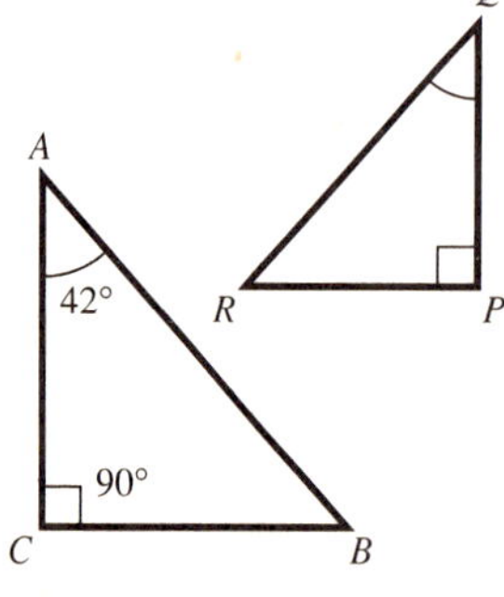

$\measuredangle P = 90°$; $\measuredangle Q = 42°$; $\measuredangle B = \measuredangle R = 48°$

19.

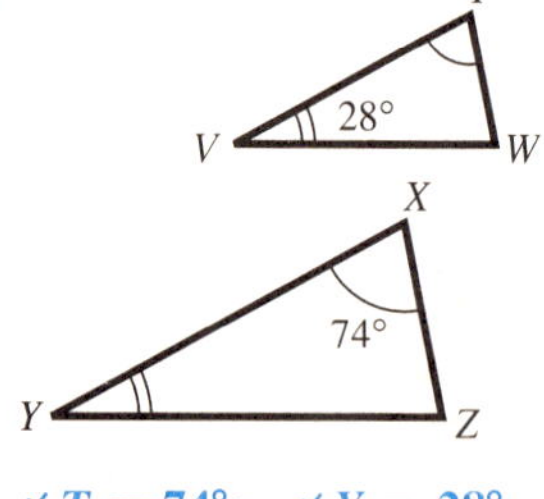

$\measuredangle T = 74°$; $\measuredangle Y = 28°$; $\measuredangle Z = \measuredangle W = 78°$

CONCEPTUAL  WRITING CHALLENGING SCIENTIFIC CALCULATOR GRAPHING CALCULATOR

20. 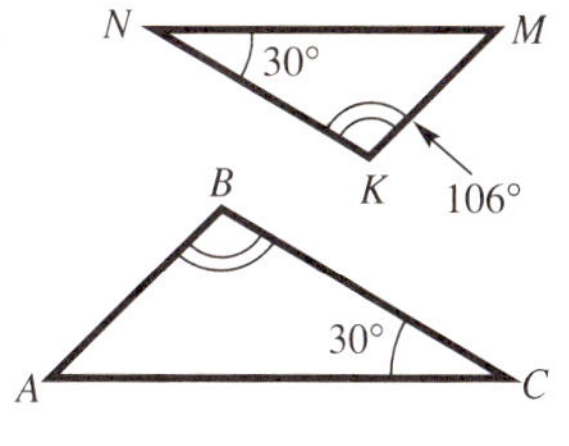

$\measuredangle B = 106°$; $\measuredangle A = \measuredangle M = 44°$

21. 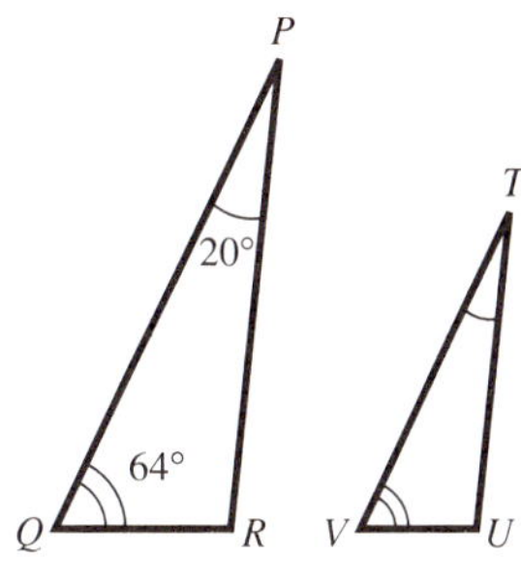

$\measuredangle T = 20°$; $\measuredangle V = 64°$; $\measuredangle R = \measuredangle U = 96°$

22. 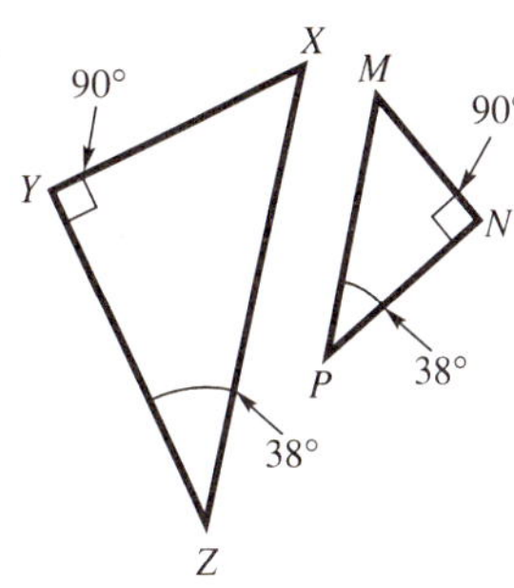

$\measuredangle X = \measuredangle M = 52°$

Find the unknown side lengths in each pair of similar triangles.

23. 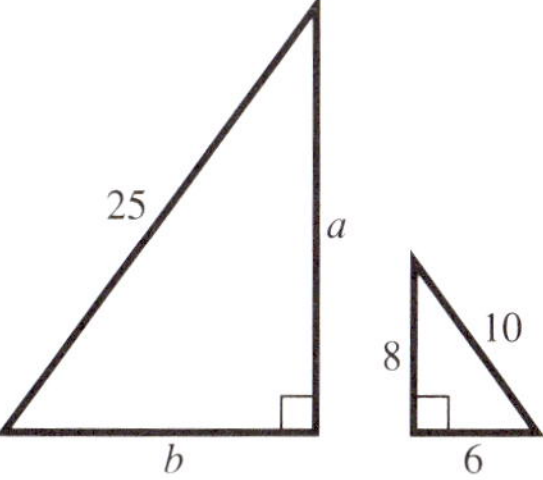

$a = 20$; $b = 15$

24. 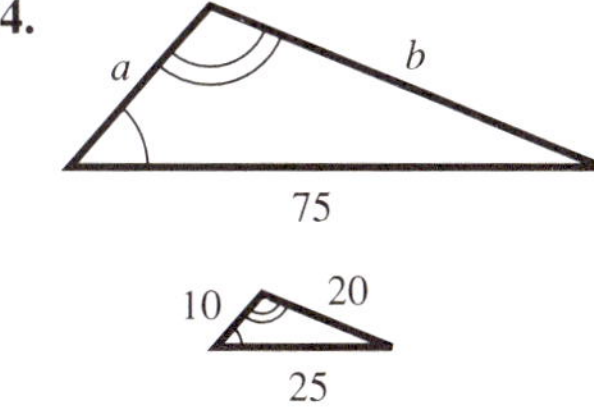

$a = 30$; $b = 60$

25. 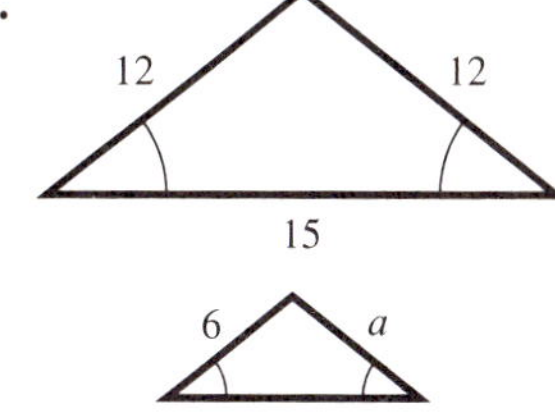

$a = 6$; $b = 15/2$

26. 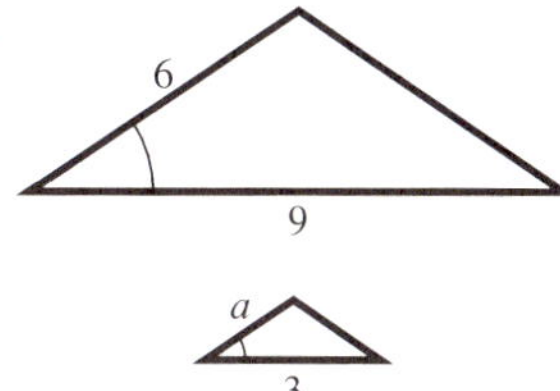

$a = 2$

27. 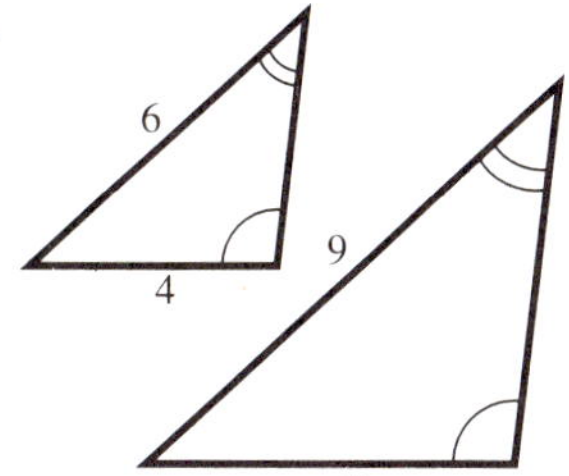

$x = 6$

28. 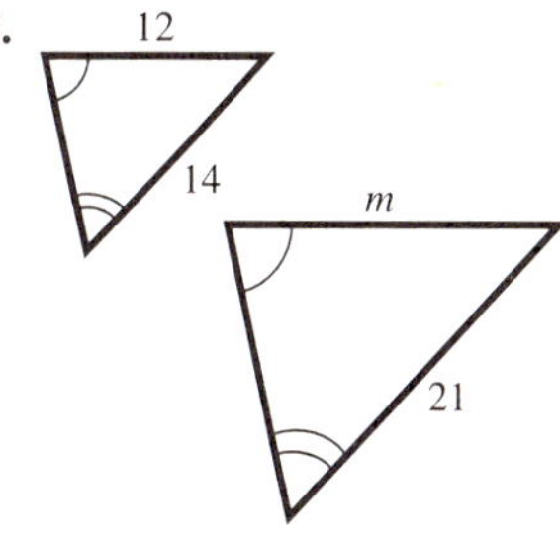

$m = 18$

In each diagram, there are two similar triangles. Find the unknown measurement in each. (Hint: In the figure for Exercise 29, the side of length 100 *in the smaller triangle corresponds to a side of length* 100 + 120 = 220 *in the larger triangle.)*

29. 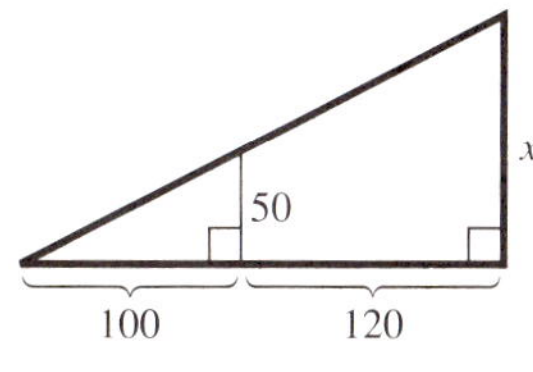

$x = 110$

30. 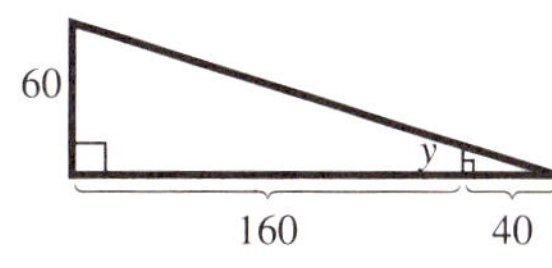

$y = 12$

31.

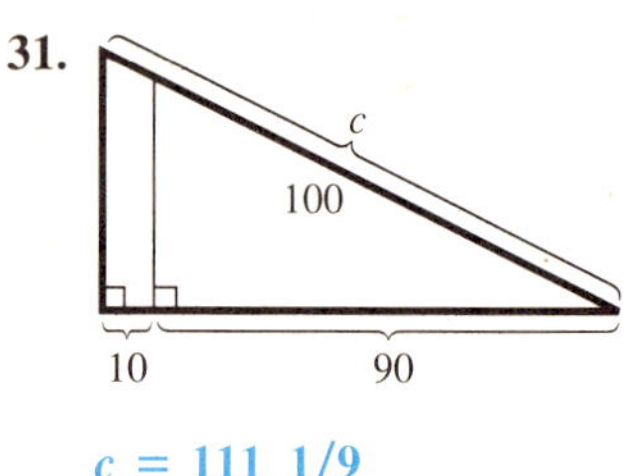

$c = 111\ 1/9$

32.

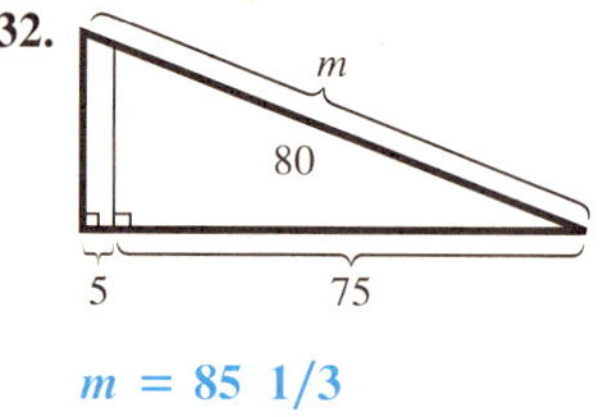

$m = 85\ 1/3$

Solve the following problems.

33. A tree casts a shadow 45 m long. At the same time, the shadow cast by a vertical 2-m stick is 3 m long. Find the height of the tree. **30 m**

34. A forest fire lookout tower casts a shadow 180 ft long at the same time that the shadow of a 9-ft truck is 15 ft long. Find the height of the tower. **108 ft**

35. On a photograph of a triangular piece of land, the lengths of three sides are 4 cm, 5 cm, and 7 cm, respectively. The shortest side of the actual piece of land is 400 m long. Find the lengths of the other two sides. **500 m, 700 m**

36. The Santa Cruz lighthouse is 14 m tall and casts a shadow 28 m long at 7 P.M. At the same time, the shadow of the lighthouse keeper is 3.5 m long. How tall is she? **1.75 m**

37. A house is 15 ft tall. Its shadow is 40 ft long at the same time the shadow of a nearby building is 300 ft long. Find the height of the building. **112.5 ft**

38. By drawing lines on a map, a triangle can be formed by the cities of Phoenix, Tucson, and Yuma. On the map, the distance between Phoenix and Tucson is 8 cm, the distance between Phoenix and Yuma is 12 cm, and the distance between Tucson and Yuma is 17 cm. The actual straight-line distance from Phoenix to Yuma is 230 km. Find the distances between the other pairs of cities. **Phoenix to Tucson: 153.3 km; Yuma to Tucson: 325.8 km**

39. The photograph shows the maintenance of the Mount Rushmore head of Lincoln. Assume that Lincoln was 6 1/3 ft tall and his head 3/4 ft long. Knowing that the carved head of Lincoln is 60 ft tall, find out how tall his entire body would be if it were carved into the mountain. **506 2/3 ft**

40. Robert Wadlow was the tallest human being ever recorded. When a 6-ft stick cast a shadow 24 in., Robert would cast a shadow 35.7 in. How tall was he? **8 ft, 11 in.**

In Exercises 41–46, a and b represent the two legs of a right triangle, while c represents the hypotenuse. Find the lengths of the unknown sides.

41.

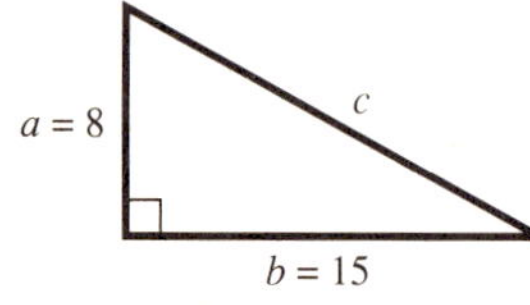

$c = 17$

42.

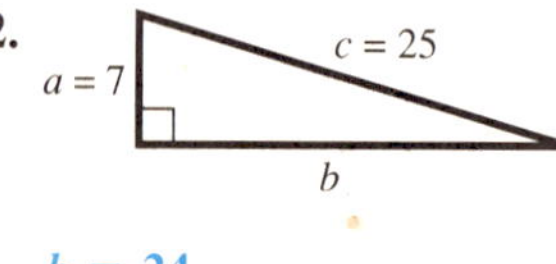

$b = 24$

43.

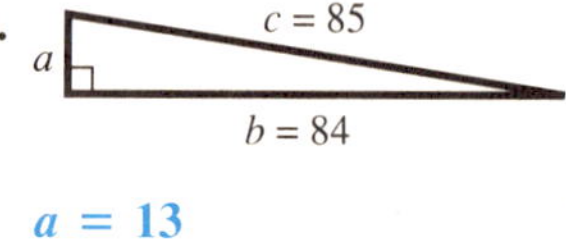

$a = 13$

44. $a = 24$ cm; $c = 25$ cm
$b = 7$ cm

45. $a = 14$ m; $b = 48$ m
$c = 50$ m

46. $a = 28$ km; $c = 100$ km
$b = 96$ km

CONCEPTUAL  WRITING  CHALLENGING  SCIENTIFIC CALCULATOR GRAPHING CALCULATOR

Find the areas of the squares on the sides of the triangles in Exercises 47 and 48 to decide whether the triangle formed is a right triangle.

47.

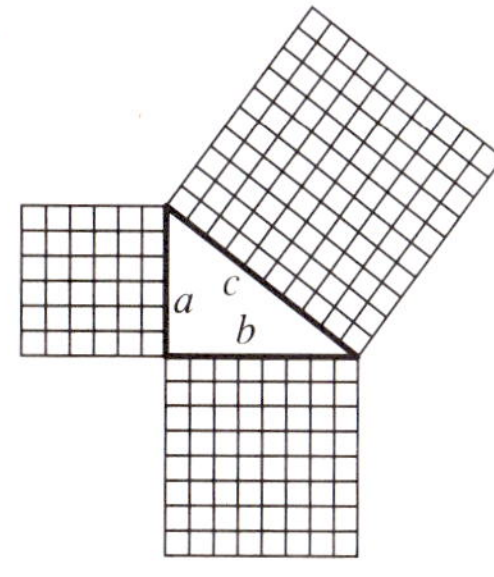

yes

48.

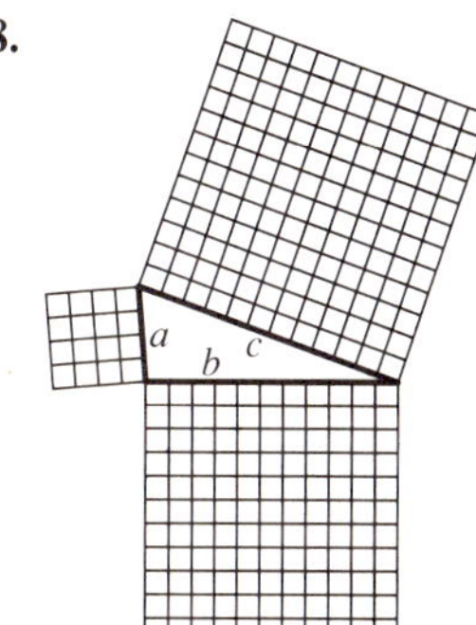

no

49. Refer to Exercise 42 in Section 9.2. Correct the scarecrow's statement.
The sum of the squares of the two shorter sides of a right triangle is equal to the square of the longest side.

50. Show that if $a^2 + b^2 = c^2$, then it is not necessarily true that $a + b = c$.
For example, let $a = 3, b = 4, c = 5$. We have $3^2 + 4^2 = 5^2$ true, but $3 + 4 = 5$ false.

There are various formulas that will generate Pythagorean triples. For example, if we choose positive integers r and s, with r > s, then the set of equations

$$a = r^2 - s^2, \qquad b = 2rs, \qquad c = r^2 + s^2$$

generates a Pythagorean triple (a, b, c). Use the values of r and s given in Exercises 51–56 to generate a Pythagorean triple using this method.

51. $r = 2, s = 1$ **(3, 4, 5)**

52. $r = 3, s = 2$ **(5, 12, 13)**

53. $r = 4, s = 3$ **(7, 24, 25)**

54. $r = 3, s = 1$ **(8, 6, 10)**

55. $r = 4, s = 2$ **(12, 16, 20)**

56. $r = 4, s = 1$ **(15, 8, 17)**

57. Show that the formula given for Exercises 51–56 actually satisfies $a^2 + b^2 = c^2$.

58. It can be shown that if $(x, x + 1, y)$ is a Pythagorean triple, then so is

$$(3x + 2y + 1, 3x + 2y + 2, 4x + 3y + 2).$$

Use this idea to find three more Pythagorean triples, starting with 3, 4, 5. (*Hint:* Here, $x = 3$ and $y = 5$.) **(20, 21, 29); (119, 120, 169); (696, 697, 985)**

If m is an odd positive integer greater than 1, then

$$\left(m, \frac{m^2 - 1}{2}, \frac{m^2 + 1}{2}\right)$$

is a Pythagorean triple. Use this to find the Pythagorean triple generated by each value of m in Exercises 59–62.

59. $m = 3$ **(3, 4, 5)**

60. $m = 5$ **(5, 12, 13)**

61. $m = 7$ **(7, 24, 25)**

62. $m = 9$ **(9, 40, 41)**

63. Show that the expressions in the directions for Exercises 59–62 actually satisfy $a^2 + b^2 = c^2$.

64. Show why (6, 8, 10) is the only Pythagorean triple consisting of consecutive even numbers.

For any integer n greater than 1,

$$(2n, n^2 - 1, n^2 + 1)$$

is a Pythagorean triple. Use this to find the Pythagorean triple generated by each value of n in Exercises 65–68.

65. $n = 2$ (4, 3, 5) **66.** $n = 3$ (6, 8, 10) **67.** $n = 4$ (8, 15, 17) **68.** $n = 5$ (10, 24, 26)

▲ **69.** Show that the expressions in the directions for Exercises 65–68 actually satisfy $a^2 + b^2 = c^2$.

▲ ◉ **70.** Can an isosceles right triangle have sides with integer lengths? Why or why not? No, since if a is the length of one of the equal sides, then $\sqrt{2}a$ must be the length of the hypotenuse.

Solve each problem. (You may wish to review quadratic equations from algebra.)

71. If the hypotenuse of a right triangle is 1 m more than the longer leg, and the shorter leg is 7 m, find the length of the longer leg. 24 m

72. The hypotenuse of a right triangle is 1 cm more than twice the shorter leg, and the longer leg is 9 cm less than three times the shorter leg. Find the lengths of the three sides of the triangle. 8 cm, 15 cm, 17 cm

73. At a point on the ground 30 ft from the base of a tower, the distance to the top of the tower is 2 ft more than twice the height of the tower. Find the height of the tower. 16 ft

74. The length of a rectangle is 2 in. less than twice the width. The diagonal is 5 in. Find the length and width of the rectangle. length: 4 in.; width: 3 in.

75. (Problem of the broken bamboo, from the Chinese work *Arithmetic in Nine Sections* (1261)) There is a bamboo 10 ft high, the upper end of which, being broken, reaches the ground 3 ft from the stem. Find the height of the break. 4.55 ft

76. (Adapted from *Arithmetic in Nine Sections*) There grows in the middle of a circular pond 10 ft in diameter a reed which projects 1 ft out of the water. When it is drawn down it just reaches the edge of the pond. How deep is the water? 12 ft

Imagine that you are a carpenter building the floor of a rectangular room. What must the diagonal of the room measure if your floor is to be squared off properly, given the dimensions in Exercises 77–80? Give your answer to the nearest inch.

77. 12 ft by 15 ft 19 ft, 3 in. **78.** 14 ft by 20 ft 24 ft, 5 in.

79. 16 ft by 24 ft 28 ft, 10 in. **80.** 20 ft by 32 ft 37 ft, 9 in.

▲ **81.** James A. Garfield, the twentieth president of the United States, provided a proof of the Pythagorean theorem using the given figure. Supply the required information in each part below in order to follow his proof.

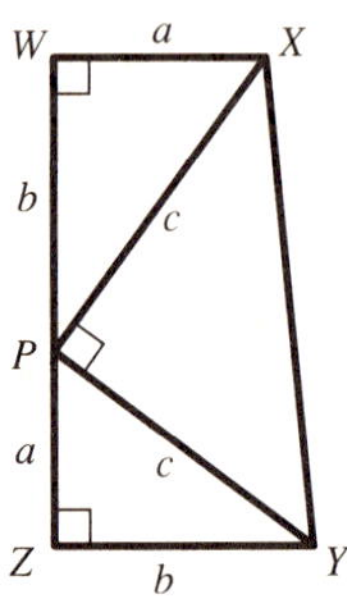

(a) Find the area of the trapezoid $WXYZ$ using the formula for the area of a trapezoid. $\frac{1}{2}(a + b)(a + b)$

(b) Find the area of each of the right triangles PWX, PZY, and PXY. PWX: $\frac{1}{2}ab$; PZY: $\frac{1}{2}ab$; PXY: $\frac{1}{2}c^2$

(c) Since the sum of the areas of the three right triangles must equal the area of the trapezoid, set the expression from part (a) equal to the sum of the three expressions from part (b). Simplify the equation as much as possible. $\frac{1}{2}(a + b)(a + b) = \frac{1}{2}ab + \frac{1}{2}ab + \frac{1}{2}c^2$. When simplified, this gives $a^2 + b^2 = c^2$.

◉ CONCEPTUAL  WRITING ▲ CHALLENGING SCIENTIFIC CALCULATOR GRAPHING CALCULATOR

▲ **82.** In the figure, right triangles ABC, CBD, and ACD are similar. This may be used to prove the Pythagorean theorem. Fill in the blanks with the appropriate responses.

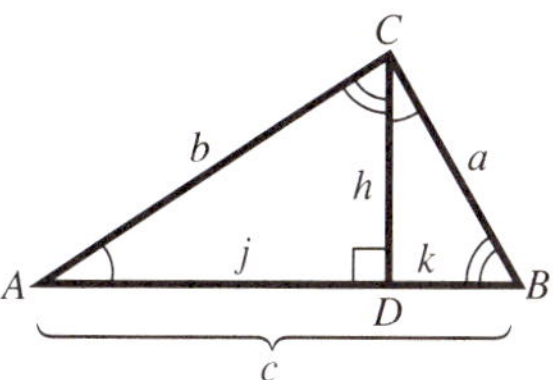

(a) By proportion, we have $c/b =$ ____$/j$. b

(b) By proportion, we also have $c/a = a/$____. k

(c) From part (a), $b^2 =$ ____. cj

(d) From part (b), $a^2 =$ ____. ck

(e) From the results of parts (c) and (d) and factoring, $a^2 + b^2 = c($____$)$. But since ____ $= c$, it follows that ____.
$j + k$; $j + k$; $a^2 + b^2 = c^2$

▲ *Exercises 83–90 require some ingenuity, but all may be solved using the concepts presented so far in this chapter.*

83. Find the area of quadrilateral $ABCD$, if angles A and C are right angles. $24 + 4\sqrt{6}$

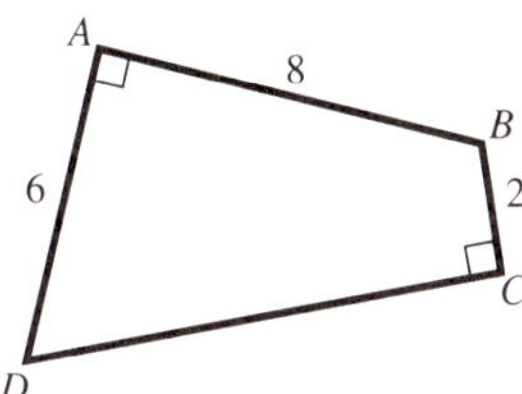

84. The perimeter of the isosceles triangle ABC (with $AB = BC$) is 128 in. The altitude BD is 48 in. What is the area of triangle ABC? 672 in.^2

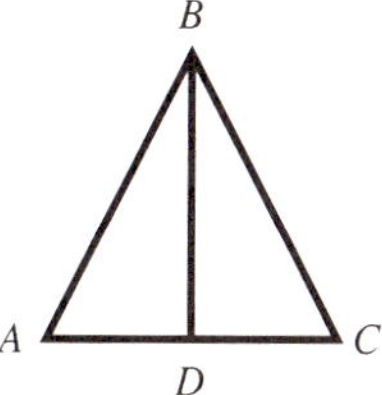

85. An isosceles triangle has a base of 24 and two sides of 13. What other base measure can an isosceles triangle with equal sides of 13 have and still have the same area as the given triangle? 10

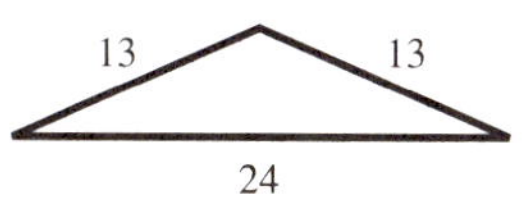

86. In right triangle ABC, if $AD = DB + 8$, what is the value of CD? 9

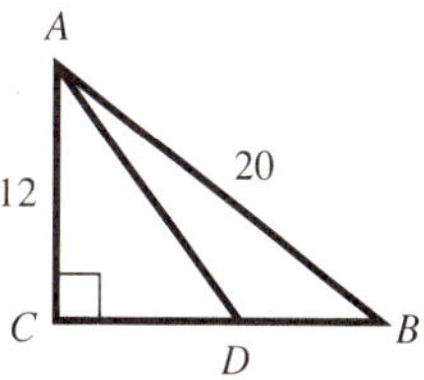

87. In the figure, pentagon $PQRST$ is formed by a square and an equilateral triangle such that $PQ = QR = RS = ST = PT$. The perimeter of the pentagon is 80. Find the area of the pentagon. $256 + 64\sqrt{3}$

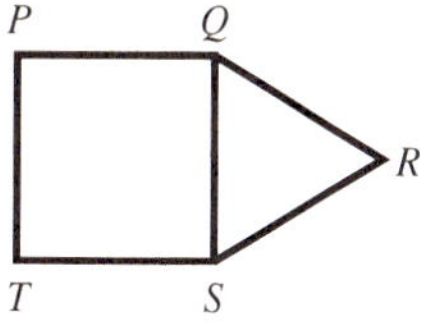

88. (A segment that *bisects* an angle divides the angle into two equal angles.) In the figure, angle A measures 50°. $\overleftrightarrow{OB}$ bisects angle ABC, and $\overleftrightarrow{OC}$ bisects angle ACB. What is the measure of angle BOC?
115°

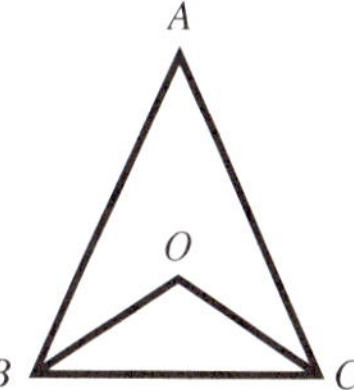

▲ *Exercises 89 and 90 refer to the given figure. The center of the circle is O.*

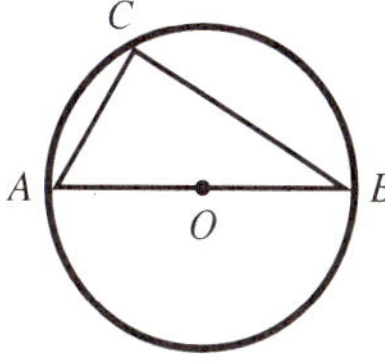

89. If $\overleftrightarrow{AC}$ measures 6 in. and $\overleftrightarrow{BC}$ measures 8 in., what is the radius of the circle?
5 in.

90. If $\overleftrightarrow{AB}$ measures 13 cm, and the length of $\overleftrightarrow{BC}$ is 7 cm more than the length of $\overleftrightarrow{AC}$, what are the lengths of $\overleftrightarrow{BC}$ and $\overleftrightarrow{AC}$?
$\overleftrightarrow{BC}$ measures 12 cm and $\overleftrightarrow{AC}$ measures 5 cm.

Extension Exercises (Text Page 540)

Find sin θ, cos θ, *and* tan θ *for each angle* θ. *(Do not use a calculator.)*

1.

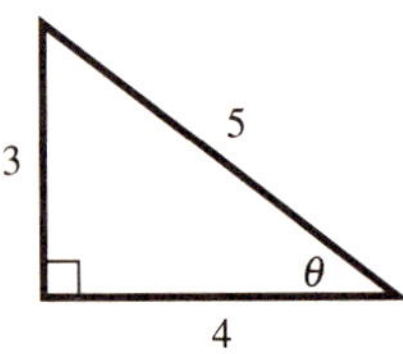

sin $\theta = 3/5$, cos $\theta = 4/5$, tan $\theta = 3/4$

2.

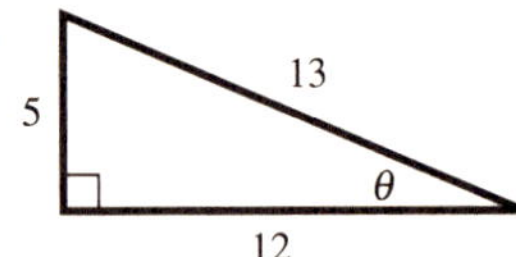

sin $\theta = 5/13$, cos $\theta = 12/13$, tan $\theta = 5/12$

3.

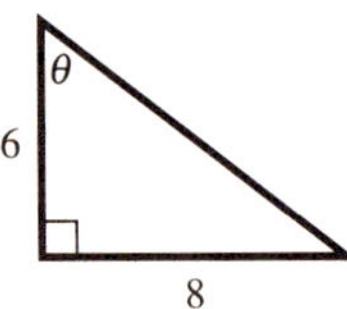

sin $\theta = 4/5$, cos $\theta = 3/5$, tan $\theta = 4/3$

4.

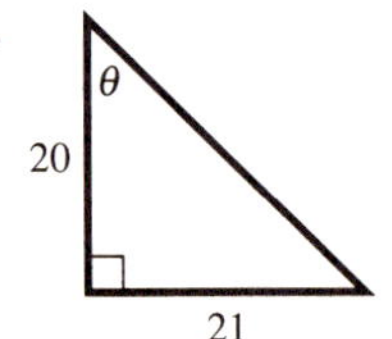

sin $\theta = 21/29$, cos $\theta = 20/29$, tan $\theta = 21/20$

5.

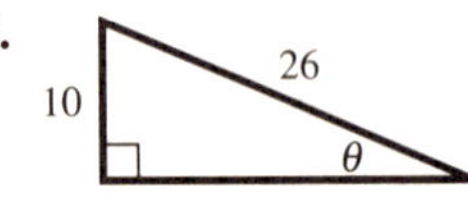

sin $\theta = 5/13$, cos $\theta = 12/13$, tan $\theta = 5/12$

6.

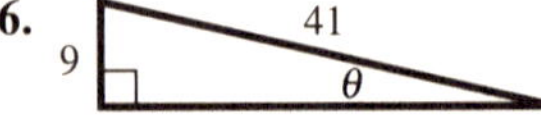

sin $\theta = 9/41$, cos $\theta = 40/41$, tan $\theta = 9/40$

---

7. Refer to the figure to respond to each part of this exercise.

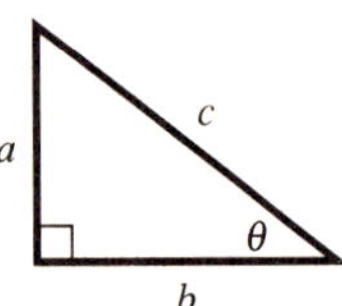

(a) Since the sum of the acute angles of a right triangle is 90°, the two acute angles of a right triangle are _____ of each other.
complements

(b) Express the measure of the unmarked acute angle in terms of θ. $90° - \theta$

(c) Explain why the sine of angle θ is equal to the cosine of the unmarked acute angle of the triangle. Each is equal to a/c.

(d) The sine and cosine ratios are called **cofunctions** (the prefix "co" representing "complementary"). Based on your answers above, what can be concluded about sin θ and cos $(90° - \theta)$?
sin $\theta = \cos(90° - \theta)$

8. Using the accompanying figure, it can be determined that the sine of a 30° angle is $1/2$. Now complete the following statement: In a right triangle with an acute angle 30°, the side opposite the 30° angle measures _____ the hypotenuse of the triangle.

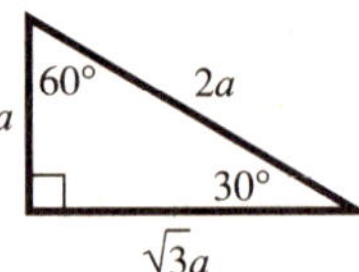

one-half

9. Use the figure in Exercise 8 to find the *exact* values of the sine, cosine, and tangent of a 30° angle and of a 60° angle. (That is, use the definitions, not a calculator.)
$\sin 30° = 1/2$, $\cos 30° = \sqrt{3}/2$, $\tan 30° = 1/\sqrt{3}$ or $\sqrt{3}/3$, $\sin 60° = \sqrt{3}/2$, $\cos 60° = 1/2$, $\tan 60° = \sqrt{3}$

10. Use the figure to find the *exact* values of sin 45°, cos 45°, and tan 45°.

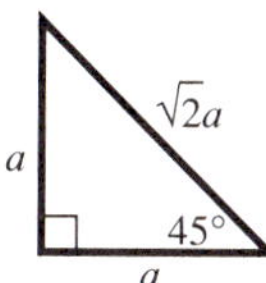

$\sin 45° = 1/\sqrt{2}$ or $\sqrt{2}/2$, $\cos 45° = 1/\sqrt{2}$ or $\sqrt{2}/2$, $\tan 45° = 1$

You will need a scientific calculator for Exercises 11–24.

Use an inverse trigonometric function key to find the measure of each angle θ as described. Express the measure to the nearest tenth of a degree.

11. the angle θ in the figure of Exercise 1 **36.9°**
12. the angle θ in the figure of Exercise 2 **22.6°**
13. the angle θ in the figure of Exercise 4 **46.4°**
14. the angle θ in the figure of Exercise 6 **12.7°**

Solve each problem. Express angle measures to the nearest degree, and lengths to the nearest tenth.

15. The angle of the sun above the horizon is 28°. Find the length of the shadow of a person 4.8 ft tall. (Find x in the figure.) **9.0 ft**

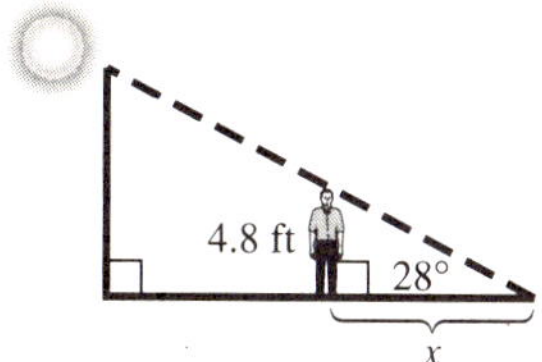

16. Find the angle of the sun above the horizon when a person 5.3 ft tall casts a shadow 8.7 ft long. (Find θ in the figure.) **31°**

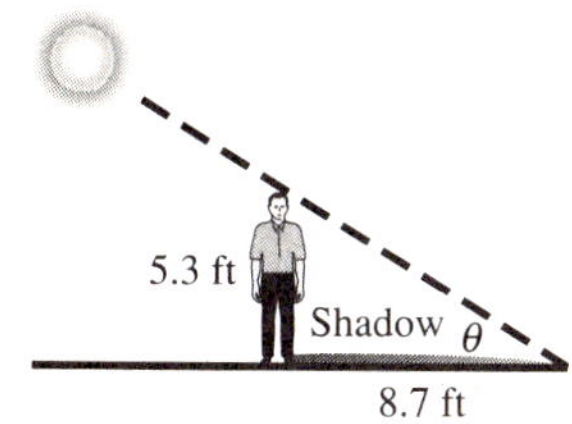

17. In one area, the lowest angle of elevation of the sun in winter is 23°. Find the minimum distance x that a plant needing full sun can be placed from a fence 4.7 ft high. (See the figure.) **11.1 ft**

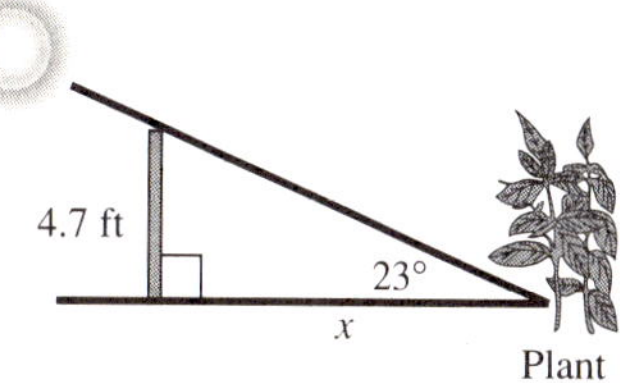

18. The shadow of a vertical tower is 40.6 m long when the angle of elevation of the sun is 35°. Find the height of the tower. **28.4 m**

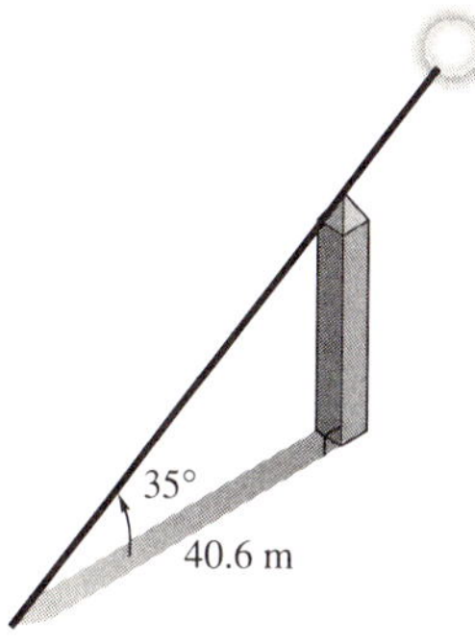

19. To measure the height of a flagpole, Laura Vanek, whose eyes are 6.0 ft above the ground, finds that the angle of elevation measured from eye level to the top of the flagpole is 38°. If Laura is standing at a point 24.8 ft from the base of the flagpole, how tall is the flagpole? **25.4 ft**

20. The angle of depression from the top of a building to a point on the ground is 32°. How far is the point on the ground from the top of the building if the building is 252.0 m high? **475.5 m**

21. A guy wire 77.4 m long is attached to the top of an antenna mast that is 71.3 m high. Find the angle that the wire makes with the ground. **67°**

22. A rectangular piece of land is 528.2 ft by 630.7 ft. Find the acute angles made by the diagonal of the rectangle. **40°, 50°**

23. To find the distance RS across a lake, a surveyor lays off $RT = 53.1$ m, with angle $T = 32°$, and angle $S = 58°$. Find length RS. **33.2 m**

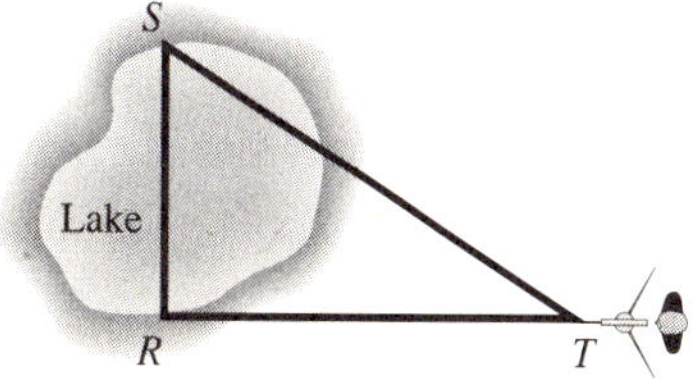

24. A surveyor must find the distance QM across a depressed freeway. She lays off $QN = 769.0$ ft along one side of the freeway, with angle $N = 22°$, and with angle $M = 68°$. Find QM. 310.7 ft

25. The sine and the cosine of an acute angle cannot be greater than or equal to 1. Explain why this is so.

26. Try to calculate $\sin^{-1}(3/2)$ on a calculator. What happens? Why?

9.5 Exercises (Text Page 548)

Decide whether each of the following statements is true *or* false.

1. A cube with volume 64 cubic inches has surface area 96 square inches. true
2. A tetrahedron has the same number of faces as vertices. true
3. A dodecahedron can be used as a model for a calendar for a given year, where each face of the dodecahedron contains a calendar for a single month, and there are no faces left over. true
4. Each face of an octahedron is an octagon. false
5. If you double the length of the edge of a cube, the new cube will have a volume that is twice the volume of the original cube. false
6. The numerical value of the volume of a sphere is $r/3$ times the numerical value of its surface area, where r is the measure of the radius. true

Find **(a)** *the volume and* **(b)** *the surface area of each of the following space figures. When necessary, use 3.14 as an approximation for* π, *and round answers to the nearest hundredth.*

7. $1\frac{1}{4}$ m, 2 m, $1\frac{1}{2}$ m

(a box)

(a) 3 3/4 m^3 (b) 14 3/4 m^2

8. 4 cm, 6 cm, 4 cm

(a box)

(a) 96 cm^3 (b) 128 cm^2

9. 3.2 in., 5 in., 6 in.

(a box)

(a) 96 $in.^3$ (b) 130.4 $in.^2$

10. 7 m

(a sphere)

(a) 1,436.03 m^3 (b) 615.44 m^2

CONCEPTUAL   

SCIENTIFIC CALCULATOR GRAPHING CALCULATOR

11.

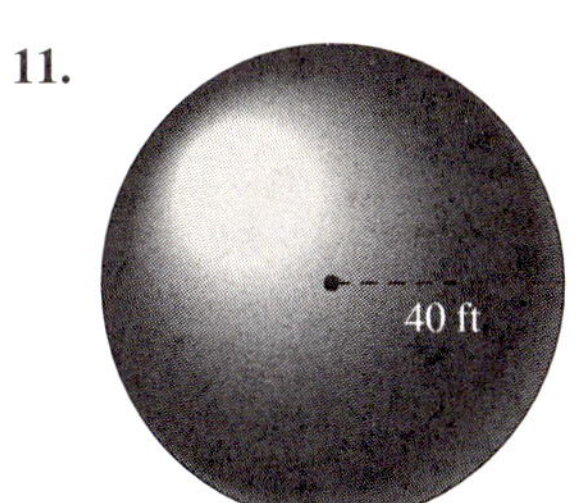

(a sphere)

(a) 267,946.67 ft^3
(b) 20,096 ft^2

12.

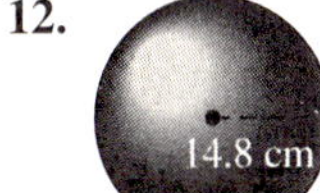

(a sphere)

(a) 1,696.54 cm^3
(b) 687.79 cm^2

13.

5 cm
7 cm

(a right circular cylinder)

(a) 549.5 cm^3
(b) 376.8 cm^2

14.

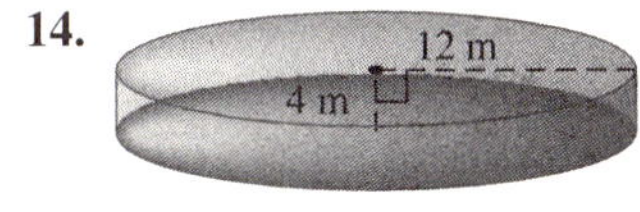

(a right circular cylinder)

(a) 1,808.64 m^3
(b) 1,205.76 m^2

15.

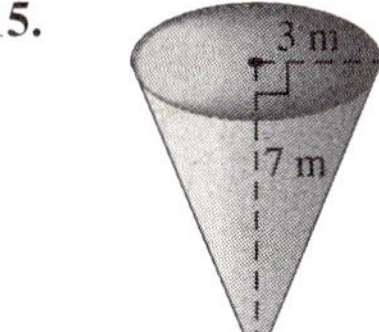

(a right circular cone)

(a) 65.94 m^3
(b) 71.74 m^2

16.

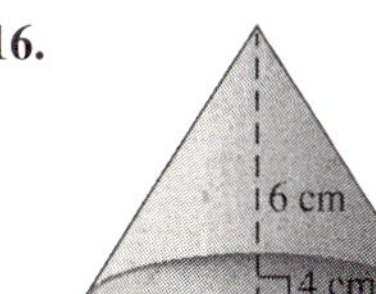

(a right circular cone)

(a) 100.48 cm^3
(b) 90.57 cm^2

Find the volume of each pyramid. In each case, the base is a rectangle.

17.

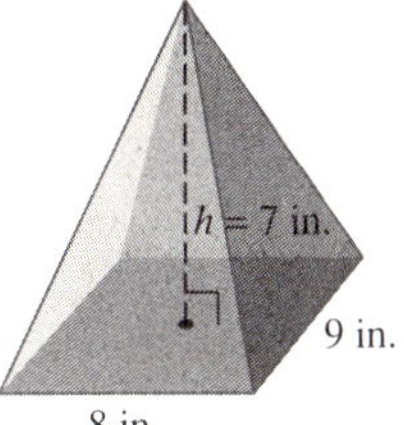

168 in.^{3}

18.

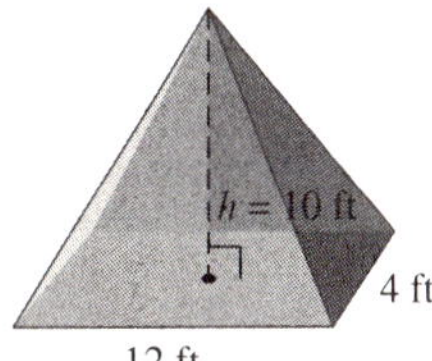

160 ft^3

Find the volume of each of the following objects. Use 3.14 *as an approximation for* π *when necessary.*

19. a coffee can, radius 6.3 cm and height 15.8 cm 1,969.10 cm^3

20. a soup can, radius 3.2 cm and height 9.5 cm 305.46 cm^3

21. a pork-and-beans can, diameter 7.2 cm and height 10.5 cm 427.29 cm^3

22. a cardboard mailing tube, diameter 2 in. and height 40 in. 125.6 in.^{3}

23. a coffee mug, diameter 9 cm and height 8 cm 508.68 cm^3

24. a bottle of typewriter correction fluid, diameter 3 cm and height 4.3 cm 30.38 cm^3

25. the Great Pyramid of Cheops, near Cairo—its base is a square 230 m on a side, while the height is 137 m 2,415,766.67 m^3

26. the Peachtree Plaza Hotel in Atlanta, a cylinder with a base radius of 46 m and a height of 220 m 1,461,732.8 m^3

27. a road construction marker, a cone with height 2 m and base radius 1/2 m .523 m^3

28. the conical portion of a witch's hat for a Halloween costume, with height 12 in. and base radius 4 in. 200.96 in.^{3}

In the chart below, one of the values r (radius), d (diameter), V (volume), or S (surface area) is given for a particular sphere. Find the remaining three values. Leave π in your answers.

	r	d	V	S
29.	6 in.	12 in.	288π in.^{3}	144π in.^{2}
30.	9 in.	18 in.	972π in.^{3}	324π in.^{2}
31.	5 ft	10 ft	$\frac{500}{3}\pi$ ft^3	100π ft^2
32.	20 ft	40 ft	$\frac{32{,}000}{3}\pi$ ft^3	$1{,}600\pi$ ft^2
33.	2 cm	4 cm	$\frac{32}{3}\pi$ cm^3	16π cm^2
34.	4 cm	8 cm	$\frac{256}{3}\pi$ cm^3	64π cm^2
35.	1 m	2 m	$\frac{4}{3}\pi$ m^3	4π m^2
36.	6 m	12 m	288π m^3	144π m^2

37. In order to determine the amount of liquid a spherical tank will hold, would you need to use volume or surface area? **volume**

38. In order to determine the amount of leather it would take to manufacture a basketball, would you need to use volume or surface area? **surface area**

39. One of the three famous construction problems of Greek mathematics required the construction of an edge of a cube with twice the volume of a given cube. If the length of each side of the given cube is x, what would be the length of each side of a cube with twice the original volume? $\sqrt[3]{2}x$

40. Work through the parts of this exercise in order, and use it to make a generalization concerning volumes of spheres. Leave answers in terms of π.

(a) Find the volume of a sphere having a radius of 1 m. $\frac{4}{3}\pi$ m^3

(b) Suppose the radius is doubled to 2 m. What is the volume? $\frac{32}{3}\pi$ m^3

(c) When the radius was doubled, by how many times did the volume increase? (To find out, divide the answer for part (b) by the answer for part (a).) **8 times**

(d) Suppose the radius of the sphere from part (a) is tripled to 3 m. What is the volume? 36π m^3

(e) When the radius was tripled, by how many times did the volume increase? **27 times**

(f) In general, if the radius of a sphere is multiplied by n, the volume is multiplied by ______. n^3

If a spherical tank 2 m in diameter can be filled with a liquid for \$300, find the cost to fill tanks of the following diameters.

41. 6 m **\$8,100** **42.** 8 m **\$19,200**

43. 10 m **\$37,500**

44. Use the logic of Exercise 40 to answer the following: If the radius of a sphere is multiplied by n, then the surface area of the sphere is multiplied by ______. n^2

Each of the following figures has volume as indicated. Find the value of x.

45. $V = 60$

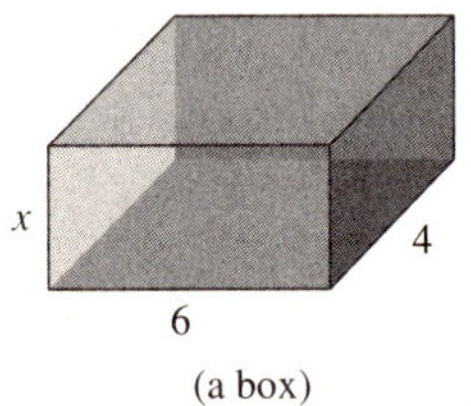

(a box)

2.5

46. $V = 450$

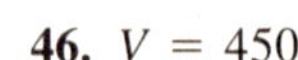

x $x + 1$

$h = 15$ cm
Base is a rectangle.

(a pyramid)

9

47. $V = 36\pi$

(a sphere)

6

48. $V = 245\pi$

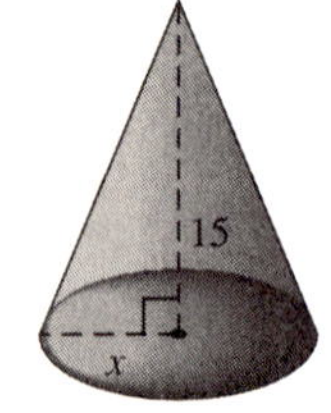

(a right circular cone)

7

▲ *Exercises 49–56 require some ingenuity, but all can be solved using the concepts presented so far in this chapter.*

49. The areas of the sides of a rectangular box are 30 in.^{2}, 35 in.^{2}, and 42 in.^{2}. What is the volume of the box?

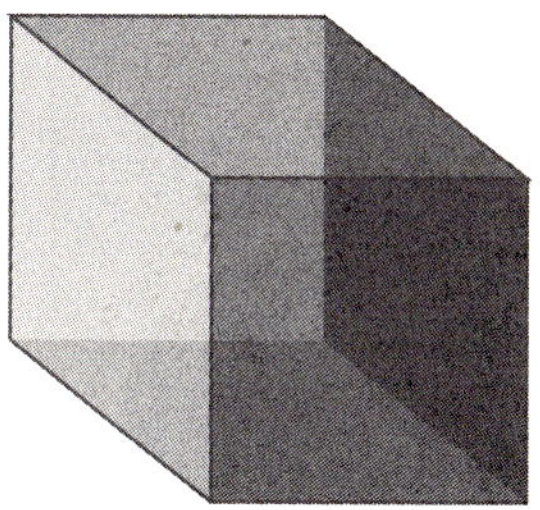

210 in.^{3}

50. In the figure, a right circular cone is inscribed in a hemisphere. What is the ratio of the volume of the cone to the volume of the hemisphere?

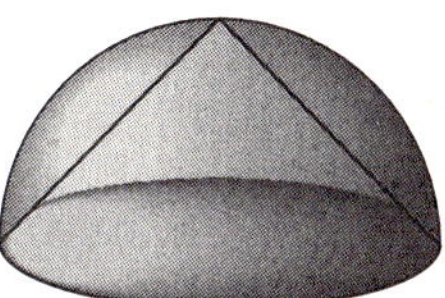

1 to 2

51. A plane intersects a sphere to form a circle as shown in the figure. If the area of the circle formed by the intersection is 576π in.^{2}, what is the volume of the sphere?

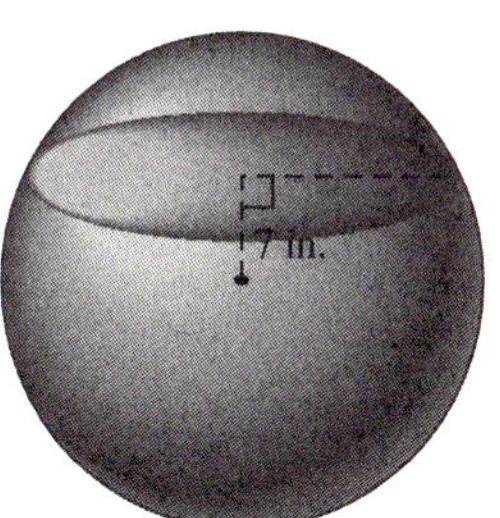

$\frac{62,500}{3}\pi$ in.^{3}

52. If the height of a right circular cylinder is halved and the diameter is tripled, how is the volume changed?

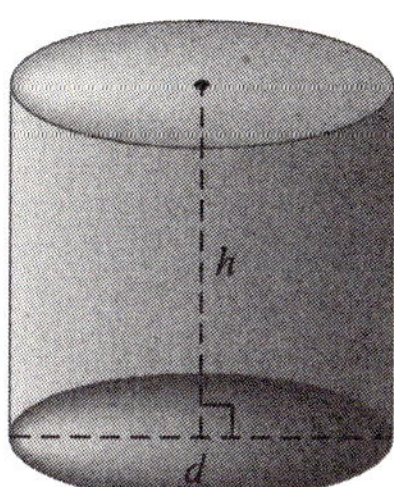

It is multiplied by 4.5.

53. What is the ratio of the area of the circumscribed square to the area of the inscribed square?

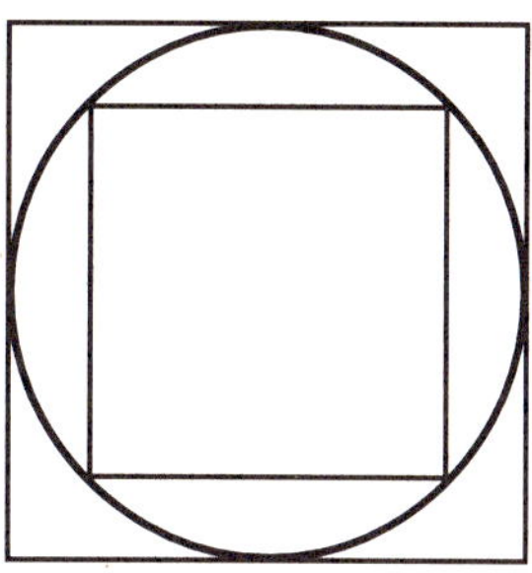

2 to 1

54. The diameter of the circle shown is 8 in. What is the perimeter of the inscribed square *ABCD*?

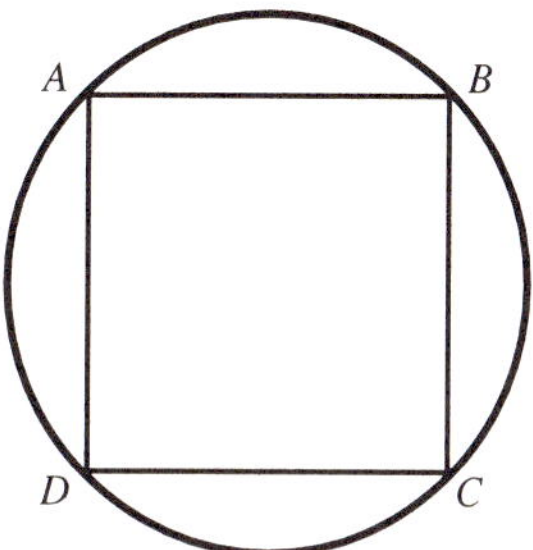

$16\sqrt{2}$ in.

55. In the circle shown with center *O*, the radius is 6. *QTSR* is an inscribed square. Find the value of $PQ^2 + PT^2 + PR^2 + PS^2$.

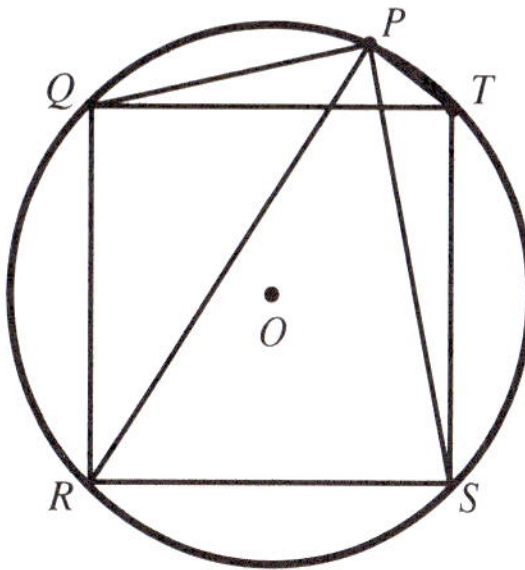

288

56. The square *JOSH* is inscribed in a semicircle. What is the ratio of *x* to *y*?

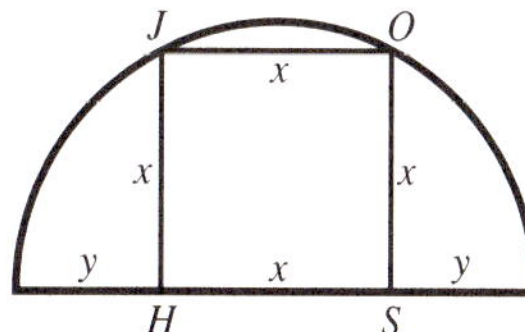

$\frac{1+\sqrt{5}}{2}$ (the golden ratio)

9.6 Exercises (Text Page 560)

The chart that follows characterizes certain properties of Euclidean and non-Euclidean geometries. Study it, and use it to respond to Exercises 1–10.

EUCLIDEAN	**NON-EUCLIDEAN**	
Dates back to about 300 B.C.	**Lobachevskian (about 1830)**	**Riemannian (about 1850)**
Lines have *infinite* length.		Lines have *finite* length.
Geometry on a plane 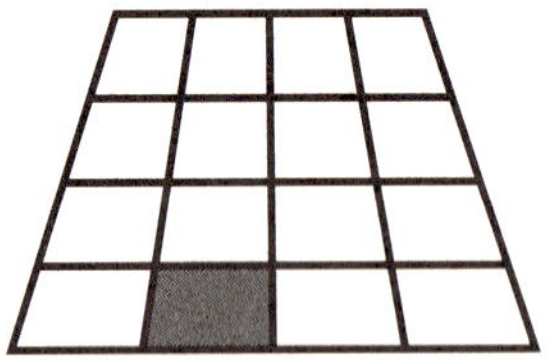	Geometry on a surface like a pseudosphere 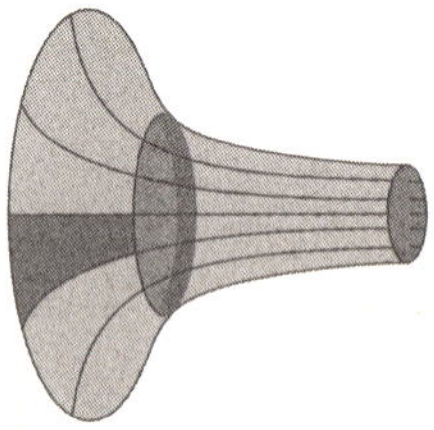	Geometry on a sphere 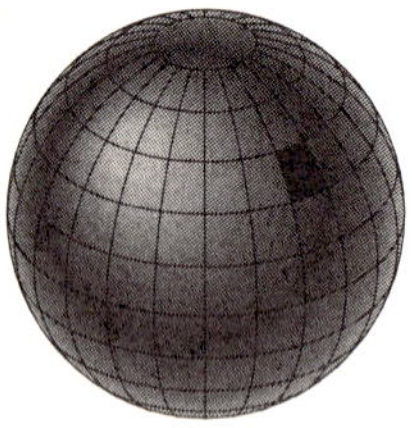
Angles C and D of a Saccheri quadrilateral are *right* angles.	Angles C and D are *acute* angles. 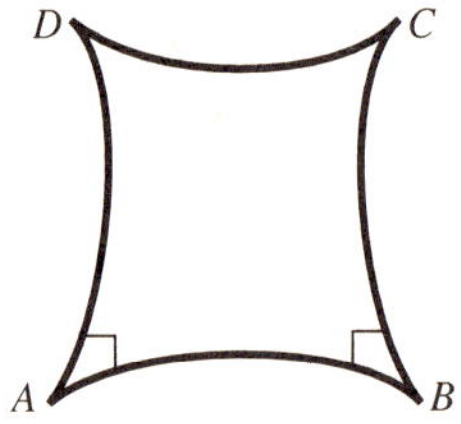	Angles C and D are *obtuse* angles. 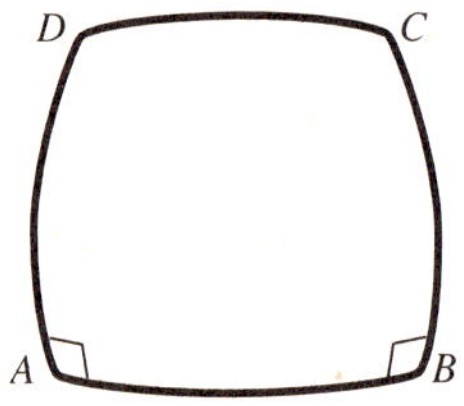
Given point P off line k, exactly *one* line can be drawn through P and parallel to k.	*More than one* line can be drawn through P and parallel to k.	*No* line can be drawn through P and parallel to k.
Typical triangle ABC 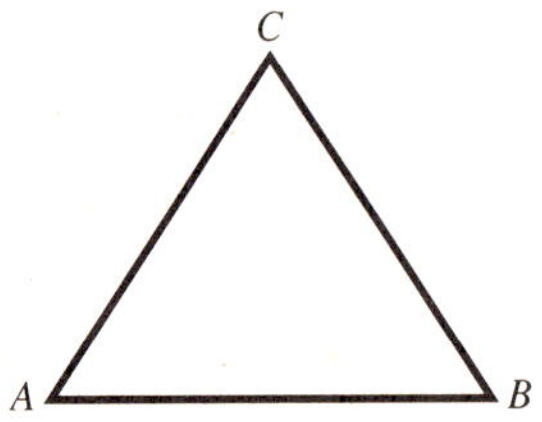	Typical triangle ABC 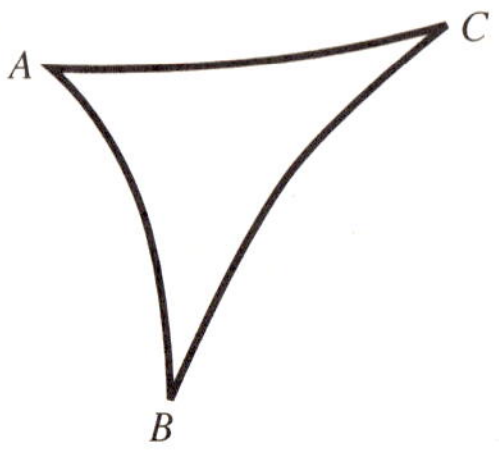	Typical triangle ABC 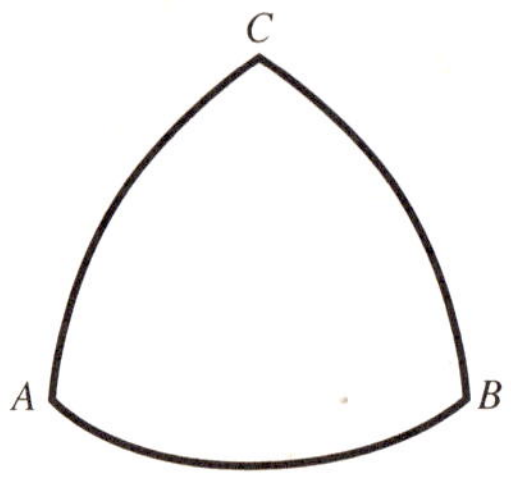
Two triangles can have the same size angles but different size sides (similarity as well as congruence).	Two triangles with the same size angles must have the same size sides (congruence only).	

1. In which geometry is the sum of the measures of the angles of a triangle equal to 180°? **Euclidean**

2. In which geometry is the sum of the measures of the angles of a triangle greater than 180°? **Riemannian**

3. In which geometry is the sum of the measures of the angles of a triangle less than 180°? **Lobachevskian**

4. In a quadrilateral *ABCD* in Lobachevskian geometry, the sum of the measures of the angles must be ________________ 360°. **less than**
(less than/greater than)

5. In a quadrilateral *ABCD* in Riemannian geometry, the sum of the measures of the angles must be ________________ 360°. **greater than**
(less than/greater than)

6. Suppose that *m* and *n* represent lines through *P* that are both parallel to *k*.

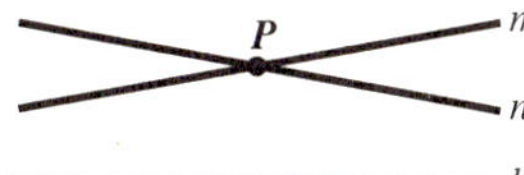

In which geometry is this possible? **Lobachevskian**

7. Suppose that *m* and *n* below *must* meet at a point.

In which geometry is this possible? **Riemannian**

8. A globe representing the earth is a model for a surface in which geometry? **Riemannian**

9. In which geometry is this statement possible? "Triangle *ABC* and triangle *DEF* are such that $\measuredangle A = \measuredangle D$, $\measuredangle B = \measuredangle E$, and $\measuredangle C = \measuredangle F$, and they have different areas." **Euclidean**

10. Draw a figure (on a sheet of paper) as best you can showing the shape formed by the north pole *N* and two points *A* and *B* lying at the equator of a model of the earth.

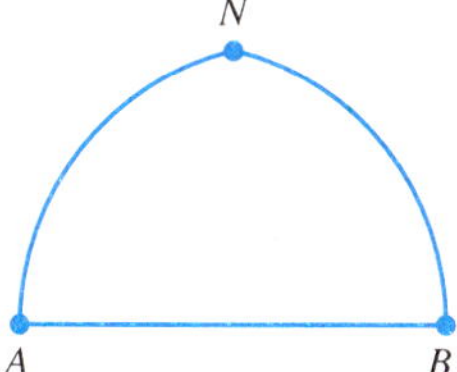

▲ 11. Pappus, a Greek mathematician in Alexandria about 320 A.D., wrote a commentary on the geometry of the times. We will work out a theorem of his about a hexagon inscribed in two intersecting lines.

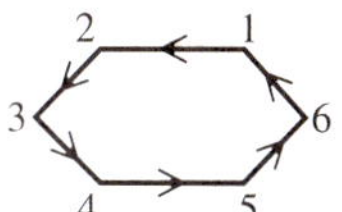

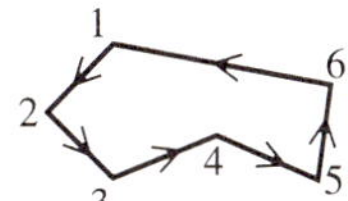

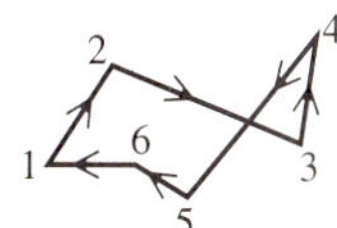

First we need to define an old word in a new way: a **hexagon** consists of any six lines in a plane, no three of which meet in the same point. As the figure shows, the vertices of several hexagons are labeled with numbers. Thus 1–2 represents a line segment joining vertices 1 and 2. Segments 1–2 and 4–5 are opposite sides of a hexagon, as are 2–3 and 5–6, and 3–4 and 1–6.

(a) Draw an angle less than 180°.

(b) Choose three points on one side of the angle. Label them 1, 5, 3 in that order, beginning with the point nearest the vertex.

(c) Choose three points on the other side of the angle. Label them 6, 2, 4 in that order, beginning with the point nearest the vertex.

(d) Draw line segments 1–6 and 3–4. Draw lines through the segments so they extend to meet in a point; call it *N*.

(e) Let lines through 1–2 and 4–5 meet in point *M*.

(f) Let lines through 2–3 and 5–6 meet in *P*.

(g) Draw a straight line through points *M*, *N*, and *P*.

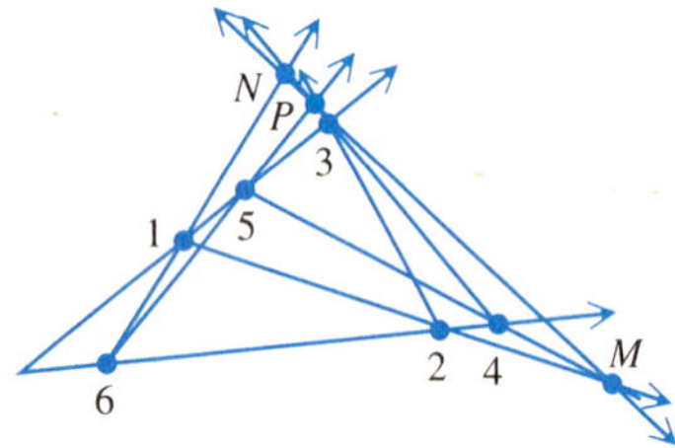

(h) Write in your own words a theorem generalizing your result. **Suppose that a hexagon is inscribed in an angle. Let each pair of opposite sides be extended so as to intersect. Then the three points of intersection thus obtained will lie in a straight line.**

▲ **12.** The following theorem comes from projective geometry:

Theorem of Desargues in a Plane Desargues' theorem states that in a plane, if two triangles are placed so that lines joining corresponding vertices meet in a point, then corresponding sides, when extended, will meet in three collinear points. (*Collinear* points are points lying on the same line.)

Draw a figure that illustrates this theorem.

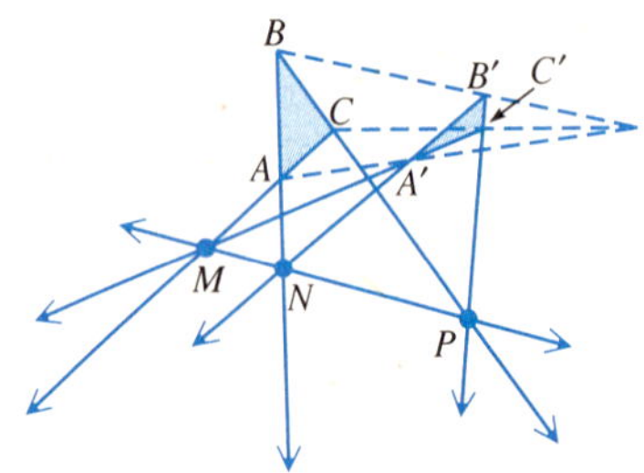

In Exercises 13–20, each figure may be topologically equivalent to none or some of the objects labeled A–E. List all topological equivalences (by letter) for each figure.

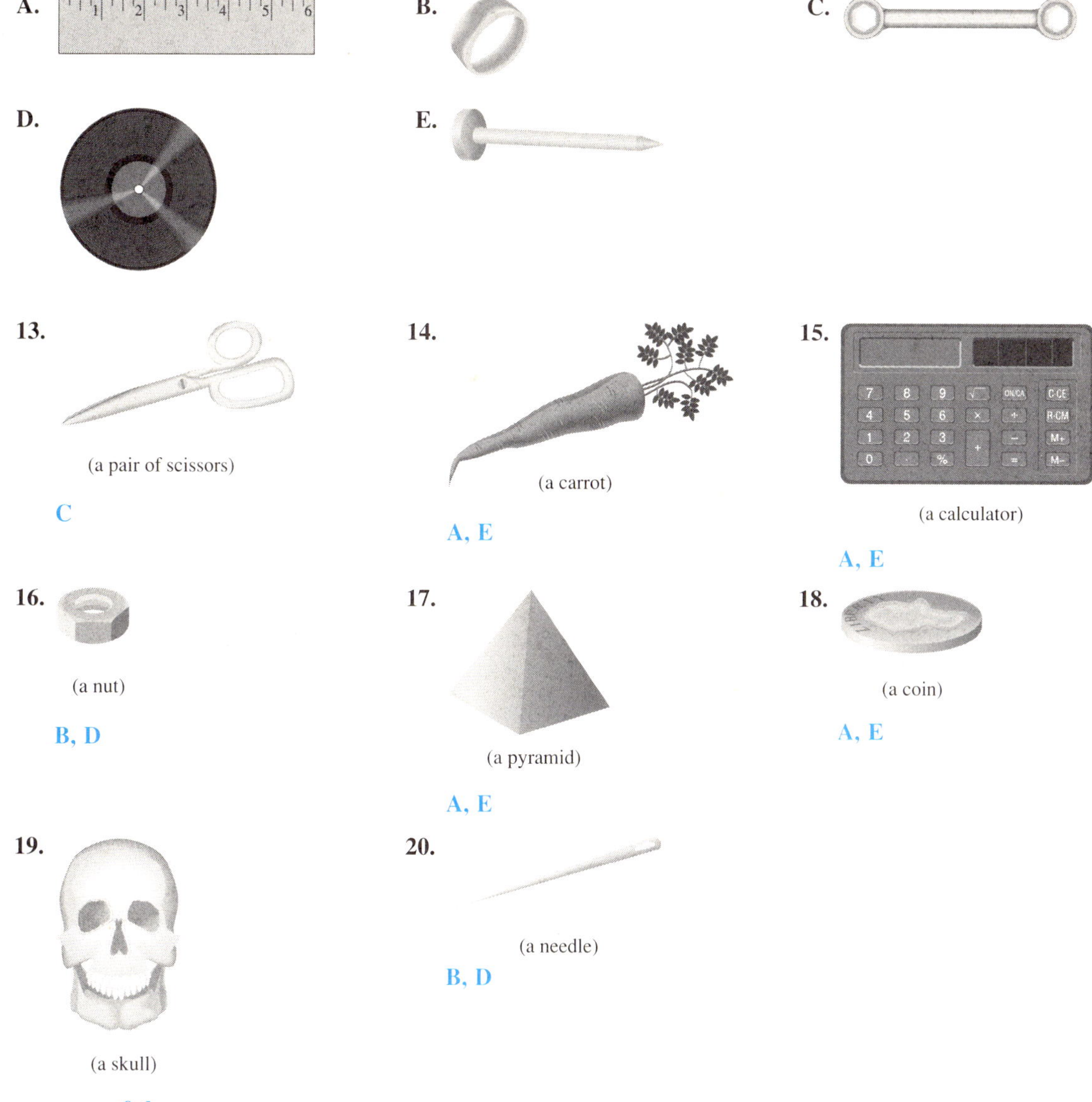

A. **B.** **C.**

D. **E.**

13. (a pair of scissors)

C

14. (a carrot)

A, E

15. (a calculator)

A, E

16. (a nut)

B, D

17. (a pyramid)

A, E

18. (a coin)

A, E

19. (a skull)

none of them

20. (a needle)

B, D

Someone once described a topologist as "a mathematician who doesn't know the difference between a doughnut and a coffee cup." This is due to the fact that both are of genus 1—they are topologically equivalent. Based on this interpretation, would a topologist know the difference between each of the following pairs of objects?

21. a spoon and a fork no

22. a mixing bowl and a colander yes

23. a slice of American cheese and a slice of Swiss cheese yes

24. a penny loafer shoe and a sock no

Give the genus of each of the following objects.

25. a compact disc 1

26. a phonograph record 1

27. a sheet of loose-leaf paper made for a three-ring binder 3

28. a sheet of loose-leaf paper made for a two-ring binder 2

29. a wedding band 1

30. a postage stamp 0

For each of the following networks, decide whether each lettered vertex is even *or* odd.

31.
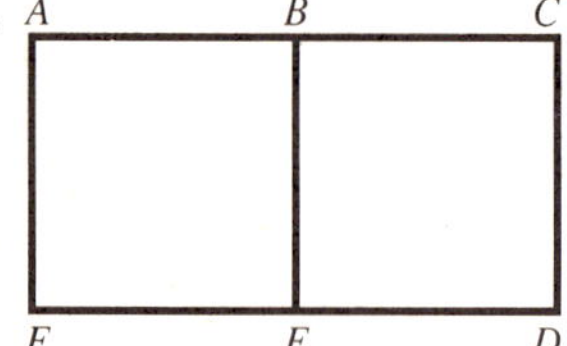

A, C, D, and F are even; B and E are odd.

32.
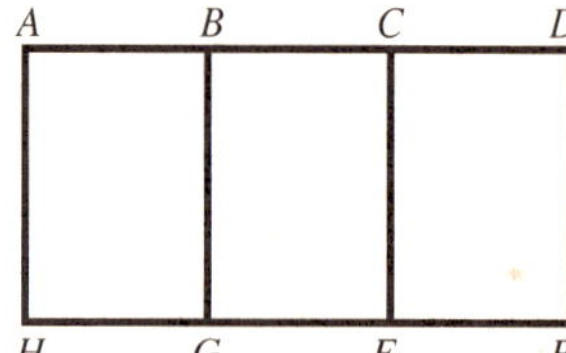

A, D, E, and H are even; B, C, F, and G are odd.

33.
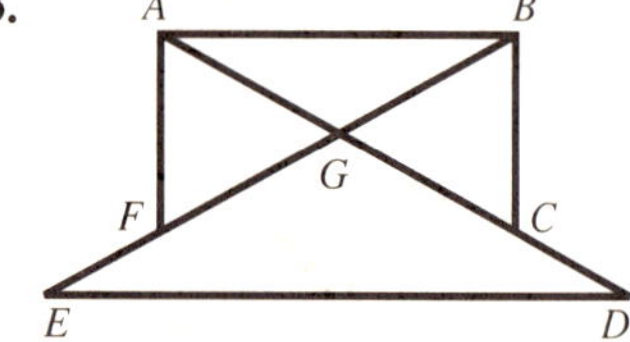

A, B, C, and F are odd; D, E, and G are even.

34.
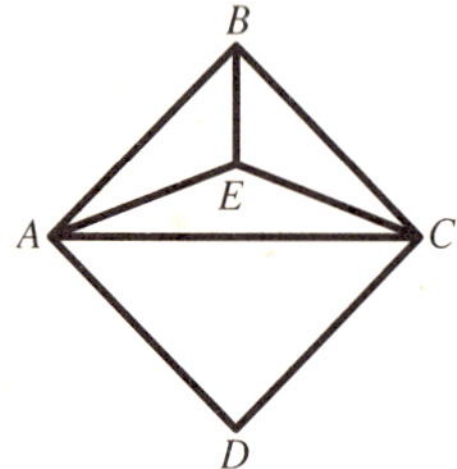

A, C, and D are even; B and E are odd.

35.
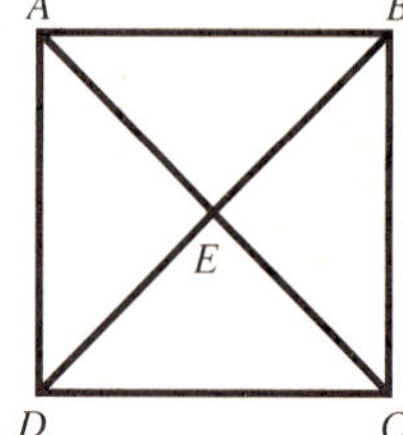

A, B, C, and D are odd; E is even.

36.
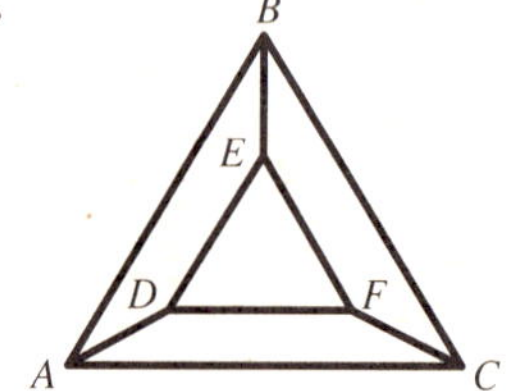

All vertices A–F are odd.

Decide whether or not each network is traversable. If a network is traversable, show how it can be traversed.

37.

Start End

traversable

38.

not traversable

39.

not traversable

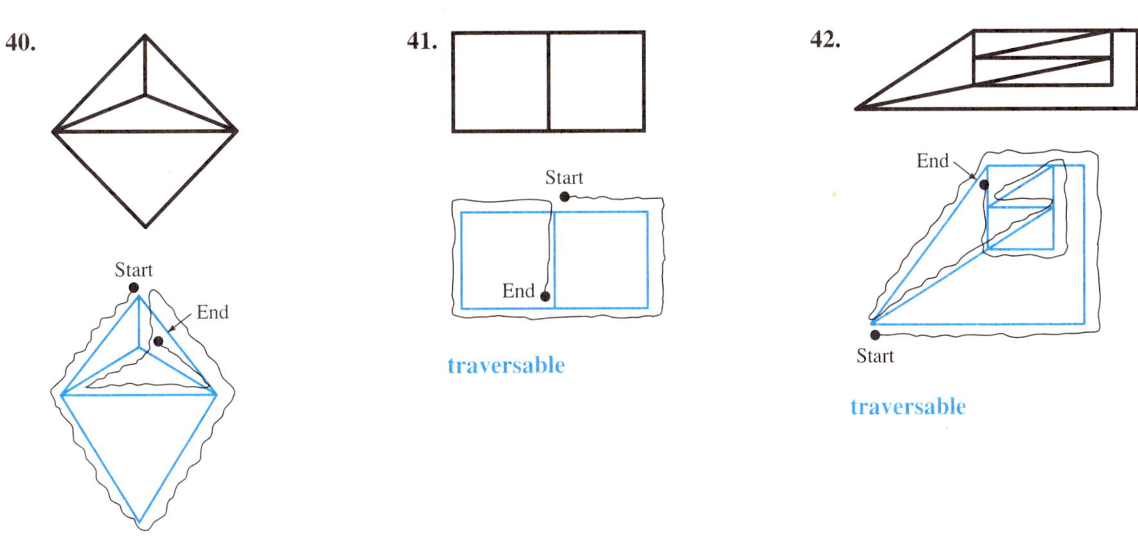

traversable

Is it possible to walk through each door of the following houses exactly once? If the answer is "yes", show how it can be done.

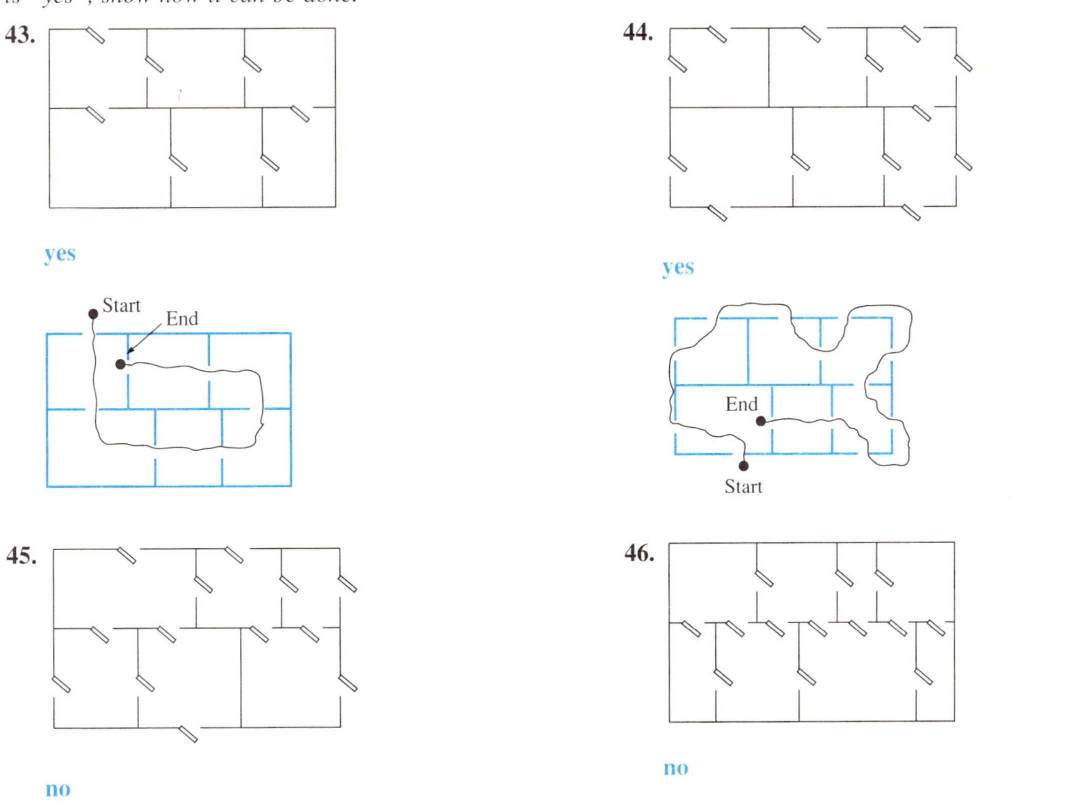

9.7 Exercises (Text Page 568)

Exercises 1–25 are taken from the November 1991 issue of Student Math Notes, *published by the National Council of Teachers of Mathematics. They were written by Tami Martin, School of Education, Boston University, and the authors wish to thank N.C.T.M. and Tami Martin for permission to reproduce this activity. Since the exercises should be done in numerical order, answers to all exercises (both even- and odd-numbered) appear in the answer section of this book.*

Most of the mathematical objects you have studied have dimensions that are whole numbers. For example, such solids as cubes and icosahedrons have dimension three. Squares, triangles, and many other planar figures are two-dimensional. Lines are one-dimensional, and points have dimension zero. Consider a square with side of length one. Gather several of these squares by cutting out or using patterning blocks.

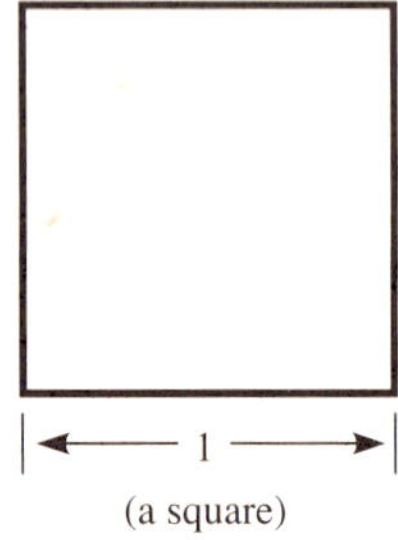

(a square)

1. What is the least number of these squares that can be put together edge to edge to form a larger square? 4

The size *of a figure is calculated by counting the number of* replicas *(small pieces) that make it up. Here, a replica is the original square with edges of length one. The original square is made up of one small square, so its size is one.*

2. What is the size of the new square? 4
3. What is the length of each edge of the new square? 2

Similar figures have the same shape but are not necessarily the same size. The **scale factor** *between two similar figures can be found by calculating the ratio of corresponding edges:*

$$\frac{\text{new length}}{\text{old length}}.$$

4. What is the scale factor between the large square and the small square? 2/1 or 2
5. Find the ratio
$$\frac{\text{new size}}{\text{old size}}$$
for the two squares. 4/1 or 4
6. Form an even larger square that is three units long on each edge. Compare this square to the small square. What is the scale factor between the two squares? What is the ratio of new size to old size? 3/1 or 3; 9/1 or 9
7. Form an even larger square that is four units long on each edge. Compare this square to the small square. What is the scale factor between the two squares? What is the ratio of the new size to the old size? 4/1 or 4; 16/1 or 16
8. Complete the table for squares. 4, 9, 16, 25, 36, 100

Scale Factor	2	3	4	5	6	10
Ratio of new size to old size						

9. How are the two rows in the table related? **Each ratio in the bottom row is the square of the scale factor in the top row.**

Consider an equilateral triangle. The length of an edge of the triangle is one unit. The size of this triangle is one.

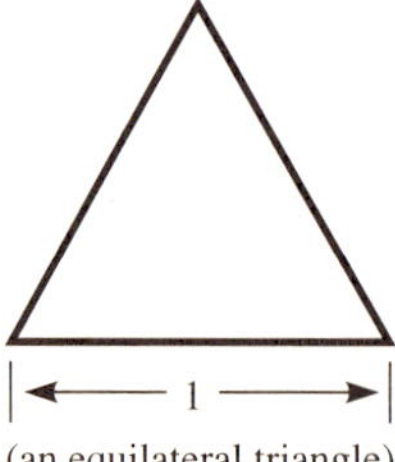

(an equilateral triangle)

10. What is the least number of equilateral triangles that can be put together edge to edge to form a similar larger triangle? 4

CONCEPTUAL WRITING CHALLENGING SCIENTIFIC CALCULATOR GRAPHING CALCULATOR

11. Complete the table for triangles. **4, 9, 16, 25, 36, 100**

Scale Factor	2	3	4	5	6	10
Ratio of new size to old size						

12. How does the relationship between the two rows in this table compare with the one you found in the table for squares? **Each ratio in the bottom row is again the square of the scale factor in the top row.**

One way to define the dimension, d, of a figure relates the scale factor, the new size, and the old size:

$$(\text{scale factor})^d = \frac{\text{new size}}{\text{old size}}.$$

Using a scale factor of two for squares or equilateral triangles, we can see that $2^d = 4/1$; *that is,* $2^d = 4$. *Since* $2^2 = 4$, *the dimension, d, must be two. This definition of dimension confirms what we already know—that squares and equilateral triangles are two-dimensional figures.*

13. Use this definition and your completed tables to confirm that the square and the equilateral triangle are two-dimensional figures for scale factors other than two. **Answers will vary. Some examples are: $3^d = 9$, thus $d = 2$; $5^d = 25$, thus $d = 2$; $4^d = 16$, thus $d = 2$.**

Consider a cube, with edges of length one. Let the size of the cube be one.

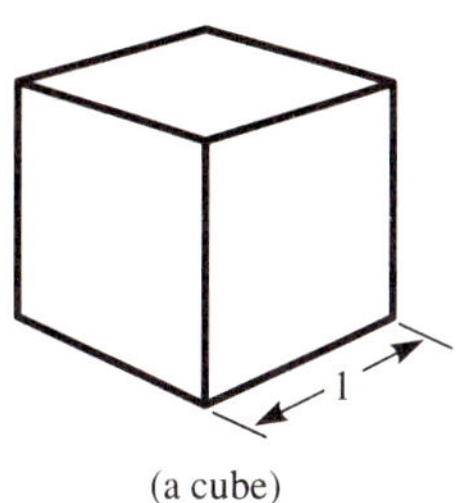

(a cube)

14. What is the least number of these cubes that can be put together face to face to form a larger cube? **8**

15. What is the scale factor between these two cubes? What is the ratio of the new size to the old size for the two cubes? **2/1 or 2; 8/1 or 8**

16. Complete the table for cubes. **8, 27, 64, 125, 216, 1,000**

Scale Factor	2	3	4	5	6	10
Ratio of new size to old size						

17. How are the two rows in the table related? **Each ratio in the bottom row is the cube of the scale factor in the top row.**

18. Use the definition of dimension and a scale factor of two to verify that a cube is a three-dimensional object. **Since $2^3 = 8$, the value of d in $2^d = 8$ must be 3.**

We have explored scale factors and sizes associated with two- and three-dimensional figures. Is it possible for mathematical objects to have fractional dimensions? Consider the following figure formed by replacing the middle third of a line segment of length one by an upside-down V, each of whose two sides are equal in length to the segment removed. The first four stages in the development of this figure are shown.

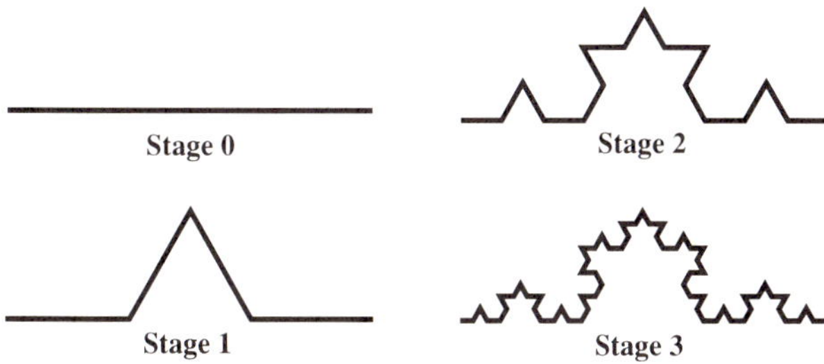

Finding the scale factor for this sequence of figures is difficult because the overall length of the figure remains the same while the number of pieces increases. To simplify the procedure, follow these steps.

Step 1: *Start with any stage (e.g., stage 1).*
Step 2: *Draw the next stage (e.g., stage 2) of the sequence and "blow it up" so that it contains an exact copy of the preceding stage (in this example, stage 1).*

Notice that stage 2 contains four copies, or replicas, of stage 1 and is three times as long as stage 1.

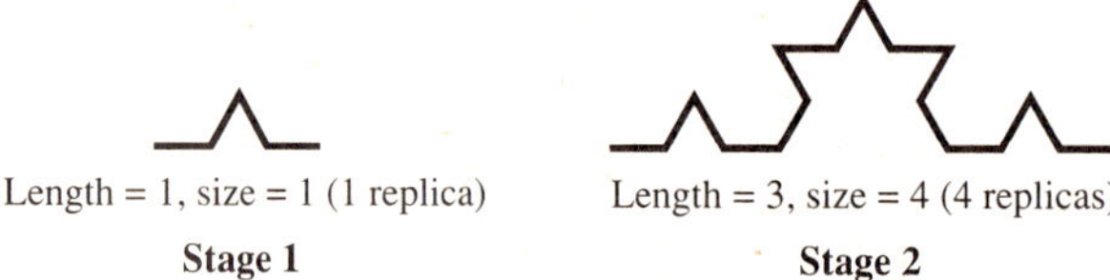

Length = 1, size = 1 (1 replica) — **Stage 1**

Length = 3, size = 4 (4 replicas) — **Stage 2**

19. The scale factor is equal to the ratio

$$\frac{\text{new length}}{\text{old length}}$$

between any two consecutive stages. The scale factor between stage 1 and stage 2 is ______. **3/1 or 3**

20. The size can be determined by counting the number of replicas of stage 1 found in stage 2. Old size = 1, new size = ______. **4**

Use the definition of dimension to compute the dimension, d, of the figure formed by this process: $3^d = 4/1$*; that is,* $3^d = 4$*. Since* $3^1 = 3$ *and* $3^2 = 9$*, for* $3^d = 4$ *the dimension of the figure must be greater than one but less than two:* $1 < d < 2$.

21. Use your calculator to estimate d. Remember that d is the exponent that makes 3^d equal 4. For example, since d must be between 1 and 2, try $d = 1.5$. But $3^{1.5} = 5.196\ldots$, which is greater than 4; thus d must be smaller than 1.5. Continue until you approximate d to three decimal places. (Use logarithms for an exact determination.) **1.262 or ln 4/ln 3**

The original figure was a one-dimensional line segment. By iteratively adding to the line segment, an object of dimension greater than one but less than two was generated. Objects with fractional dimension are known as **fractals.** *Fractals are infinitely self-similar objects formed by repeated additions to, or removals from, a figure. The object attained at the limit of the repeated procedure is the fractal.*

Next consider a two-dimensional object with sections removed iteratively. In each stage of the fractal's development, a triangle is removed from the center of each triangular region.

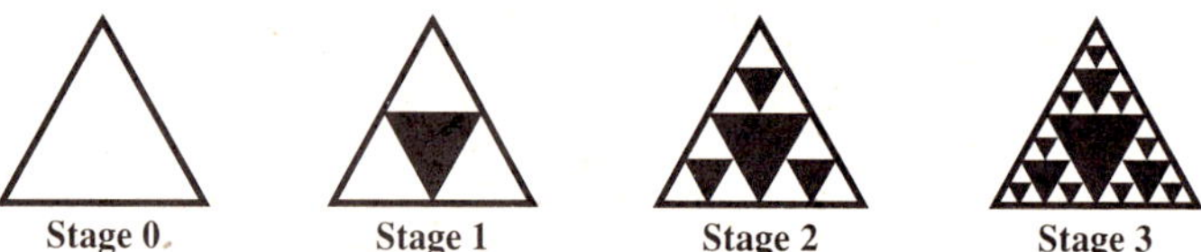

Stage 0 **Stage 1** **Stage 2** **Stage 3**

Use the process from the last example to help answer the following questions.

22. What is the scale factor of the fractal? **2/1 or 2**

23. Old size = 1, new size = ______. **3**

24. The dimension of the fractal is between what two whole number values? **It is between 1 and 2.**

25. Use the definition of dimension and your calculator to approximate the dimension of this fractal to three decimal places. **1.585 or ln 3/ln 2**

Use a calculator to determine the pattern of attractors for the equation $y = kx(1 - x)$ for the given value of k and the given initial value of x.

26. $k = 3.25$, $x = .7$ **.812, .495, .812, .495, ... The two attractors are .812 and .495.**

27. $k = 3.4$, $x = .8$ **.842, .452, .842, .452, ... The two attractors are .842 and .452.**

28. $k = 3.55$, $x = .7$ **.540, .882, .370, .828, .506, .887, .355, .813, ... The eight attractors are these eight numbers.**

Chapter 9 Test (Text Page 572)

1. Consider a 38° angle. Answer each of the following.
- **(a)** What is the measure of its complement? **52°**
- **(b)** What is the measure of its supplement? **142°**
- **(c)** Classify it as acute, obtuse, right, or straight. **acute**

Find the measure of each marked angle.

2.

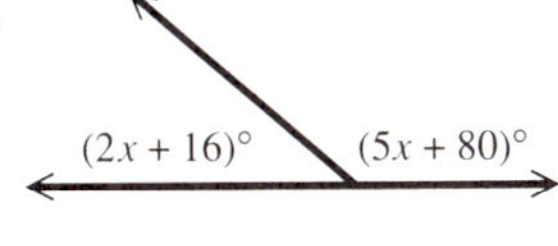

40°, 140°

3.

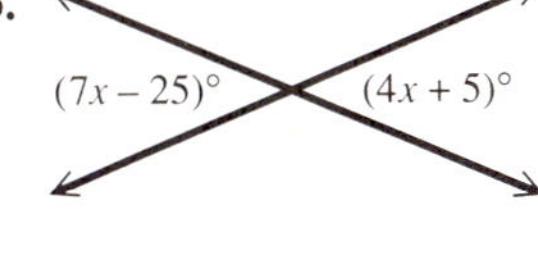

45°, 45°

4.

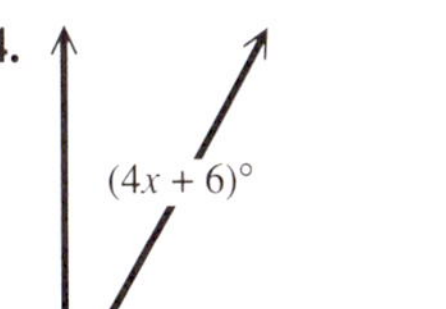

30°, 60°

In Exercises 5 and 6, assume that lines m and n are parallel, and find the measure of each marked angle.

5.

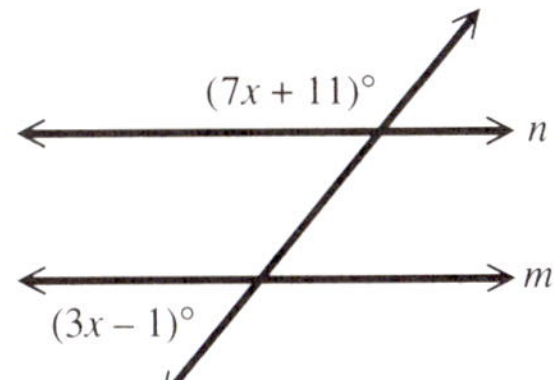

130°, 50°

6.

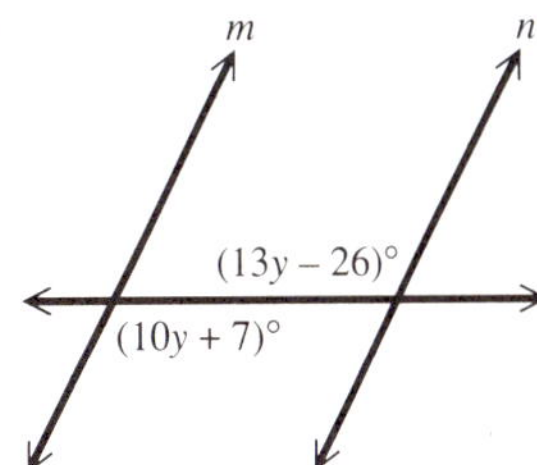

117°, 117°

7. Explain why a rhombus must be a parallelogram, but a parallelogram might not be a rhombus.

8. Which one of the following statements is false?
- **(a)** A square is a rhombus.
- **(b)** The acute angles of a right triangle are complementary.
- **(c)** A triangle may have both a right angle and an obtuse angle.
- **(d)** A trapezoid may have non-parallel sides of the same length. **(c)**

Identify each of the following curves as simple, closed, both, *or* neither.

9.

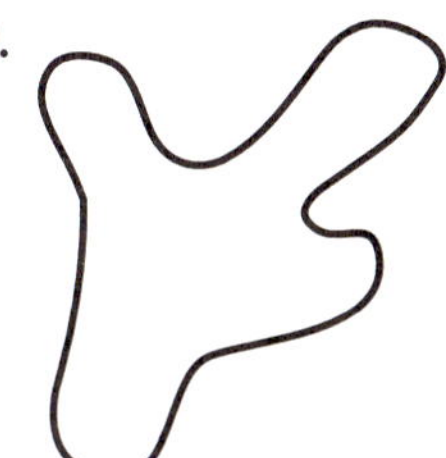

simple, closed

10.

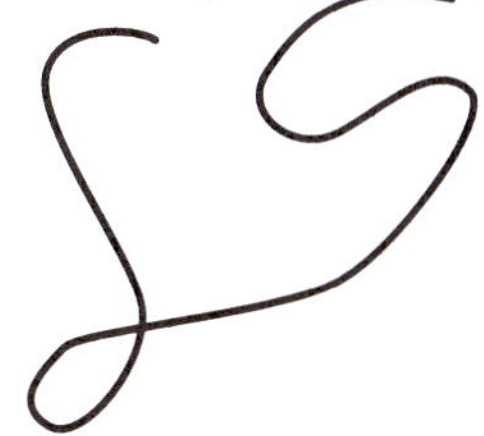

neither

11. Find the measure of each angle in the triangle. **30°, 45°, 105°**

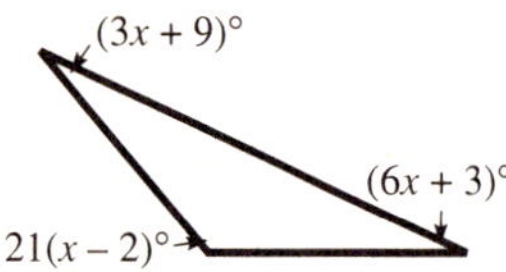

Find the area of each of the following figures.

12.

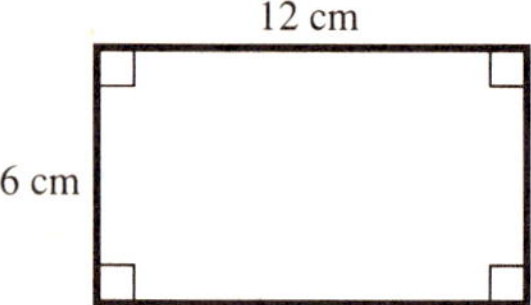

72 cm²

13.

5 in.

12 in.

(a parallelogram)

60 in.²

14.

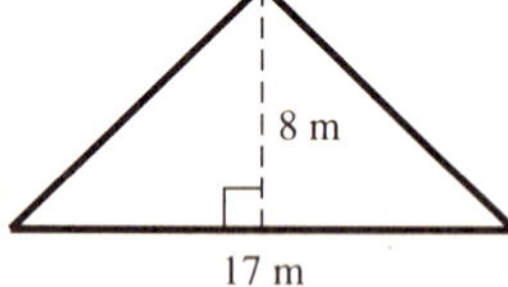

68 m²

15.

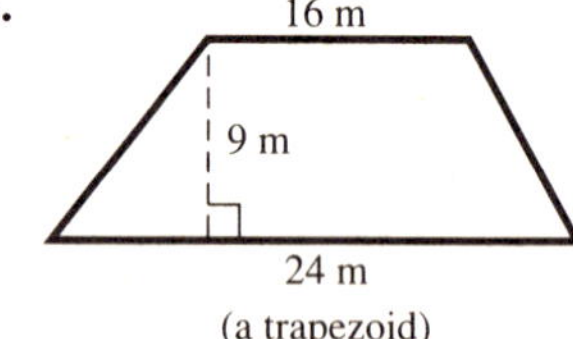

(a trapezoid)

180 m²

---

16. If a circle has area 144π square inches, what is its circumference? **24π in.**

17. The length of a rectangle is 12 in. more than 3 times the width, and the perimeter is 72 in. Find the length and the width of the rectangle.
length: 30 in.; width: 6 in.

18. What is the area of the shaded portion of the figure? Use 3.14 as an approximation for π. **57 cm²**

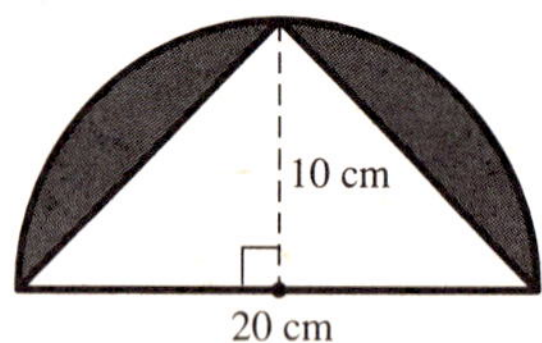

(a triangle within a semicircle)

19. Which congruence property for triangles can be used to show that $\triangle ABD \cong \triangle BAC$, given that $\measuredangle CAB = \measuredangle DBA$, and $DB = CA$? Give the additional information needed to apply the property.

SAS; $AB = AB$, since AB is equal to itself.

20. If a 30-ft pole casts a shadow 45 ft long, how tall is a pole whose shadow is 30 ft long at the same time? **20 ft**

21. What is the measure of a diagonal of a rectangle that has width 20 m and length 21 m? **29 m**

CONCEPTUAL  WRITING CHALLENGING SCIENTIFIC CALCULATOR GRAPHING CALCULATOR

Find **(a)** *the volume and* **(b)** *the surface area of each of the following space figures. When necessary, use 3.14 as an approximation for π.*

22.

(a sphere)

(a) **904.32 in.^{3}**
(b) **452.16 in.^{2}**

23.

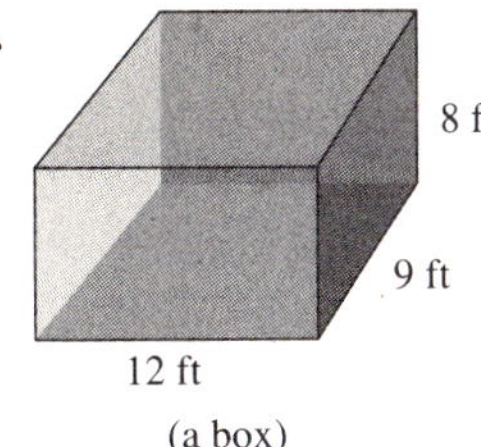

(a box)

(a) **864 ft^3**
(b) **552 ft^2**

24.

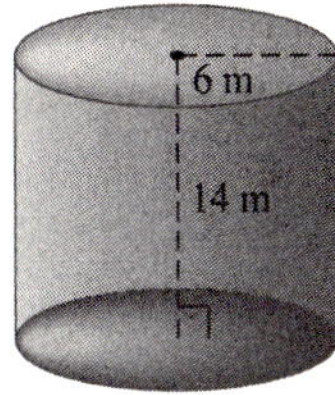

(a right circular cylinder)

(a) **1,582.56 m^3**
(b) **753.60 m^2**

25. List several main distinctions between Euclidean and non-Euclidean geometry.

Are the following pairs of objects topologically equivalent?

26. a page of a book and the cover of the same book **yes**

27. a pair of glasses with the lenses removed, and the Mona Lisa **no**

28. Is it possible to traverse this network? If so, show how it can be done. **yes**

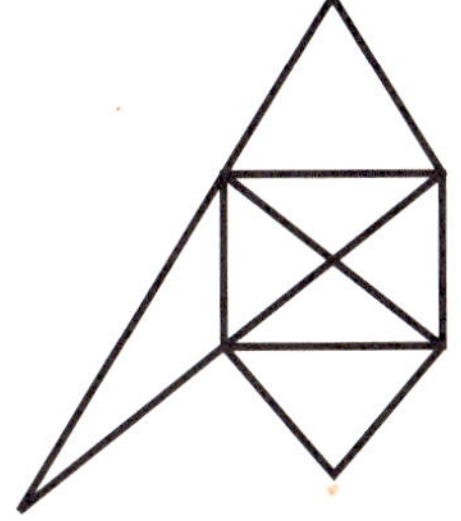

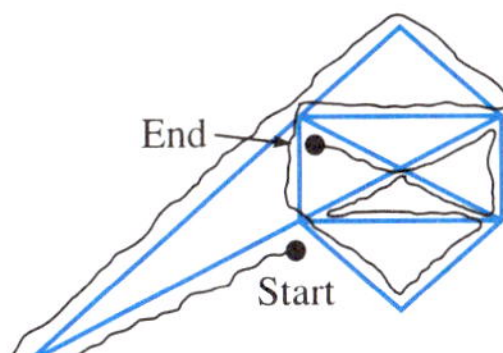

29. Use a calculator to determine the attractors for the sequence generated by the equation $y = 2.1x(1 - x)$, with initial value of $x = .6$. **The only attractor is .524.**

30. Who is Benoit Mandelbrot? **He is the mathematician who developed fractal geometry.**

CHAPTER 10 Counting Methods

10.1 Exercises (Text Page 582)

Refer to Examples 1 and 4, involving the club N = {Andy, Bill, Cathy, David, Evelyn}. *Assuming all members are eligible, but that no one can hold more than one office, list and count the different ways the club could elect the following groups of officers.*

1. a president and a treasurer *AB, AC, AD, AE, BA, BC, BD, BE, CA, CB, CD, CE, DA, DB, DC, DE, EA, EB, EC, ED*; 20 ways

2. a president and a treasurer if the president must be a female *CA, CB, CD, CE, EA, EB, EC, ED*; 8 ways

3. a president and a treasurer if the two officers must not be the same sex *AC, AE, BC, BE, CA, CB, CD, DC, DE, EA, EB, ED*; 12 ways

4. a president, a secretary, and a treasurer, if the president and treasurer must be women *CAE, CBE, CDE, EAC, EBC, EDC*; 6 ways

5. a president, a secretary, and a treasurer, if the president must be a man and the other two must be women *ACE, AEC, BCE, BEC, DCE, DEC*; 6 ways

6. a president, a secretary, and a treasurer, if the secretary must be a woman and the other two must be men *ACB, ACD, AEB, AED, BCA, BCD, BEA, BED, DCA, DCB, DEA, DEB*; 12 ways

List and count the ways club N could appoint a committee of three members under the following conditions.

7. There are no restrictions. *ABC, ABD, ABE, ACD, ACE, ADE, BCD, BCE, BDE, CDE*; 10 ways

8. The committee must include more men than women. *ABC, ABD, ABE, ACD, ADE, BCD, BDE*; 7 ways

Refer to Table 2 (the product table for rolling two dice). Of the 36 possibilities, determine the number for which the sum (for both dice) is the following.

9. 2 1 **10.** 3 2 **11.** 4 3 **12.** 5 4 **13.** 6 5 **14.** 7 6

15. 8 5 **16.** 9 4 **17.** 10 3 **18.** 11 2 **19.** 12 1 **20.** odd 18

21. even 18 **22.** from 6 through 8 inclusive 16

23. between 6 and 10 15 **24.** less than 5 6

25. Construct a product table showing all possible two-digit numbers using digits from the set {1, 2, 3, 4, 5, 6}.

	1	2	3	4	5	6
1	11	12	13	14	15	16
2	21	22	23	24	25	26
3	31	32	33	34	35	36
4	41	42	43	44	45	46
5	51	52	53	54	55	56
6	61	62	63	64	65	66

Of the thirty-six numbers in the product table for Exercise 25, list the ones that belong to each of the following categories.

26. odd numbers **11, 13, 15, 21, 23, 25, 31, 33, 35, 41, 43, 45, 51, 53, 55, 61, 63, 65**

27. numbers with repeating digits **11, 22, 33, 44, 55, 66**

28. multiples of 6 **12, 24, 36, 42, 54, 66**

29. prime numbers **11, 13, 23, 31, 41, 43, 53, 61**

30. triangular numbers **15, 21, 36, 45, 55, 66**

31. square numbers **16, 25, 36, 64**

32. Fibonacci numbers **13, 21, 34, 55**

33. powers of 2 **16, 32, 64**

34. Construct a tree diagram showing all possible results when three fair coins are tossed. Then list the ways of getting the following results.
(a) at least two heads **hhh, hht, hth, thh**
(b) more than two heads **hhh**
(c) no more than two heads **hht, hth, htt, thh, tht, tth, ttt**
(d) fewer than two heads **htt, tht, tth, ttt**

First coin Second coin Third coin Result
h h h hhh
t hht
t h hth
t htt
t h h thh
t tht
t h tth
t ttt

35. Extend the tree diagram of Exercise 34 for four fair coins. Then list the ways of getting the following results.
(a) more than three tails **tttt**
(b) fewer than three tails **hhhh, hhht, hhth, hhtt, hthh, htht, htth, thhh, thht, thth, tthh**
(c) at least three tails **httt, thtt, ttht, ttth, tttt**
(d) no more than three tails **hhhh, hhht, hhth, hhtt, hthh, htht, htth, httt, thhh, thht, thth, thtt, tthh, ttht, ttth**

First coin Second coin Third coin Fourth coin Result

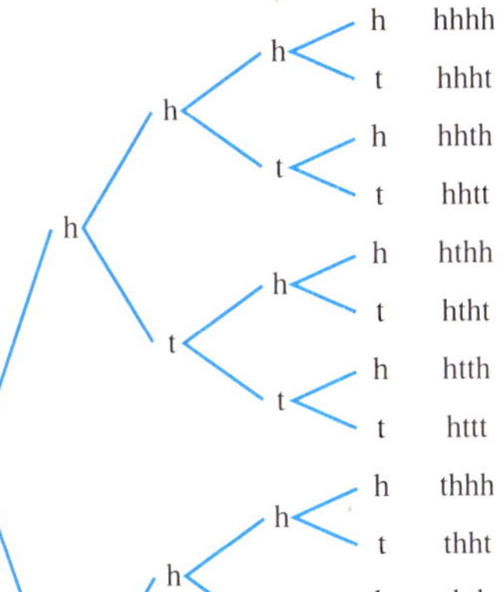

hhhh
hhht
hhth
hhtt
hthh
htht
htth
httt
thhh
thht
thth
thtt
tthh
ttht
ttth
tttt

Determine the number of triangles (of any size) in each of the following figures.

36.

10

37.

16

38.

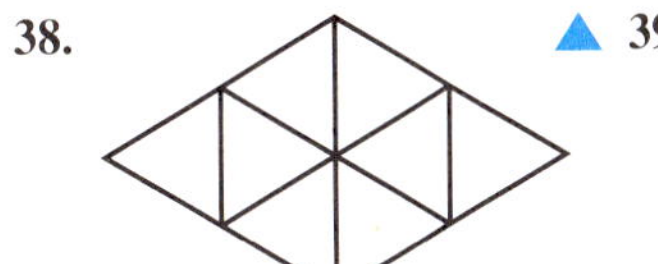

10

▲ **39.**

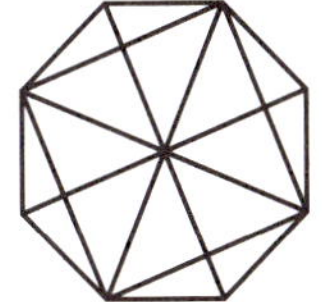

44

Determine the number of squares (of any size) in each of the following figures.

40. 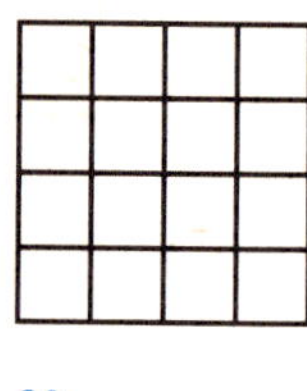

30

41. 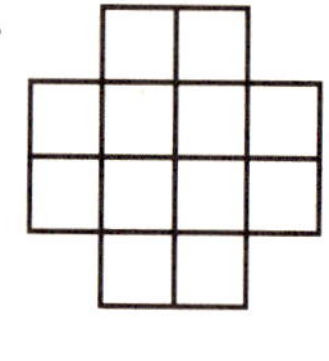

17

▲ **42.** 

19

▲ **43.** 

72

Consider only the smallest individual cubes and assume solid stacks (no gaps). Determine the number of cubes in each stack that are not visible.

44. 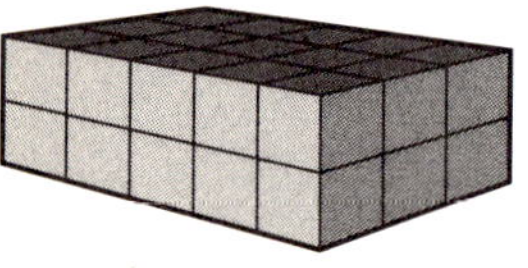

8

45. 

12

46. 

14

47. 

10

---

48. In the plane figure shown here, only movement that tends downward is allowed. Find the total number of paths from A to B. 11

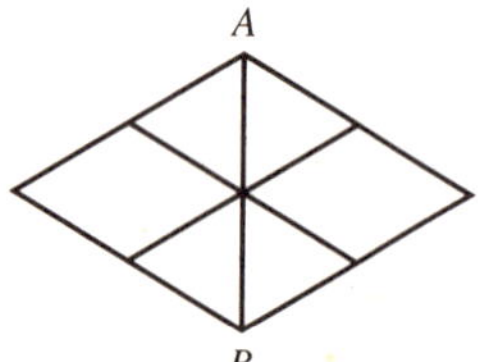

49. Find the number of paths from A to B in the figure shown here if the directions on various segments are restricted as shown. 6

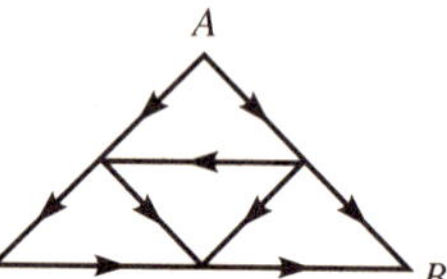

In each of Exercises 50–52, determine the number of different ways the given number can be written as the sum of two primes.

50. 30 **3** **51.** 40 **3** **52.** 95 **none**

53. A group of twelve strangers sat in a circle, and each one got acquainted only with the person to the left and the person to the right. Then all twelve people stood up and each one shook hands (once) with each of the others who was still a stranger. How many handshakes occurred? **54**

54. Fifty people enter a single elimination chess tournament. (If you lose one game, you're out.) Assuming no ties occur, what is the number of games required to determine the tournament champion? **49**

55. How many of the numbers from 10 through 100 have the sum of their digits equal to a perfect square? **18**

▲ **56.** How many three-digit numbers have the sum of their digits equal to 22? **21**

▲ **57.** How many integers between 100 and 400 contain the digit 2? **138**

▲ **58.** Jim Northington and friends are dining at the Bay Steamer Restaurant this evening, where a complete dinner consists of three items: (1) soup (clam chowder or minestrone) or salad (fresh spinach or shrimp), (2) sourdough rolls or bran muffin, and (3) entree (lasagna, lobster, or roast turkey). Jim selects his meal subject to the following restrictions. He cannot stomach more than one kind of seafood at a sitting. Also, whenever he tastes minestrone, he cannot resist having lasagna as well. And he cannot face the teasing he would receive from his companions if he were to order both spinach and bran. Use a tree diagram to determine the number of different choices Jim has. **13**

For Exercises 59–61, refer to Example 7. How many different settings could Gabriella choose in each case?

59. No restrictions apply to adjacent switches. **16**

60. No two adjacent switches can be off *and* no two adjacent switches can be on. **2**

61. There are five switches rather than four, and no two adjacent switches can be on. **13**

62. Determine the number of odd, non-repeating three-digit numbers that can be written using digits from the set {0, 1, 2, 3}. **8**

▲ **63.** A line segment joins the points (8, 12) and (53, 234) in the Cartesian plane. Including its endpoints, how many lattice points does this line segment contain? (A *lattice point* is a point with integer coordinates.) **4**

▲ **64.** In the pattern shown here, dots are one unit apart horizontally and vertically. If a segment can join any two dots, how many segments can be drawn with each of the following lengths?

(a) 1 **40** **(b)** 2 **30** **(c)** 3 **20**
(d) 4 **10** **(e)** 5 **8**

▲ **65.** Uniform length matchsticks are used to build a rectangular grid as shown here. If the grid is 15 matchsticks high and 28 matchsticks wide, how many matchsticks are used? **883**

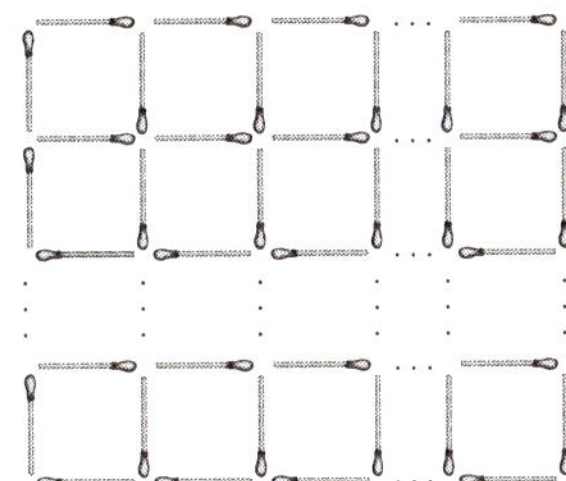

▲ **66.** A square floor is to be tiled with square tiles as shown here with blue tiles on the main diagonals and red tiles everywhere else. (In all cases, both blue and red tiles must be used and the two diagonals must have a common blue tile at the center of the floor.)

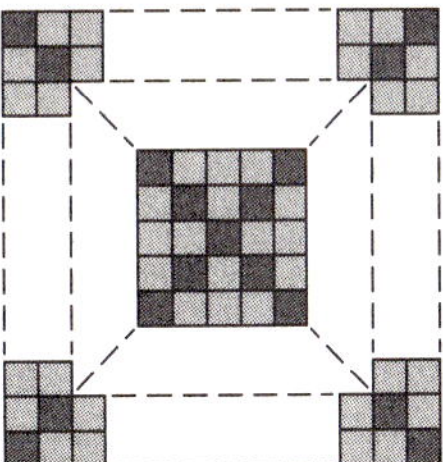

(a) If 81 blue tiles will be used, how many red tiles will be needed? **1,600**

(b) For what numbers in place of 81 would this problem still be solvable? **$4k + 1$ for all positive integers k**

(c) Find a formula expressing the number of red tiles required in general. **$4k^2$**

▲ **67.** The text stated that number chains such as those of Example 10 always terminate with a single-digit number. This follows from the fact that adding the digits of a whole number with more than one digit always produces a sum smaller than the original number.

(a) Show that this last statement is true for any two-digit number. (*Hint:* Express the number as *tu*, where t is the tens digit and u is the units digit. Then the *value* of the number *tu* is $10t + u$.)

(b) Explain how part (a) implies that any two-digit number will eventually lead to a one-digit number.

(c) Show that adding the digits of any three-digit number will produce a sum smaller than the original number.

▲ **68.** Kent Merrill and his son were among four father-and-son pairs who gathered to trade baseball cards. As each person arrived, he shook hands with anyone he had not known previously. Each person ended up making a different number of new acquaintances (0–6), except Kent and his son, who each met the same number of people. How many hands did Kent shake? **3**

10.2 Exercises (Text Page 591)

1. Explain the fundamental counting principle in your own words.

2. Describe how factorials can be used in counting problems.

For Exercises 3 and 4, n and m are counting numbers. Do the following: **(a)** *Say whether or not the given statement is true in general, and* **(b)** *explain your answer, using specific examples.*

3. $(n + m)! = n! + m!$ **(a) no**

4. $(n \cdot m)! = n! \cdot m!$ **(a) no**

Evaluate each expression without using a calculator.

5. $4!$ **24**

6. $7!$ **5,040**

7. $\frac{8!}{5!}$ **336**

8. $\frac{16!}{14!}$ **240**

9. $\frac{5!}{(5-2)!}$ **20**

10. $\frac{6!}{(6-4)!}$ **360**

11. $\frac{9!}{6!\,(6-3)!}$ **84**

12. $\frac{10!}{4!\,(10-4)!}$ **210**

13. $\frac{n!}{(n-r)!}$, where $n = 7$ and $r = 4$ **840**

14. $\frac{n!}{r!\,(n-r)!}$, where $n = 12$ and $r = 4$ **495**

Evaluate each expression using a calculator. (Some answers may not be exact.)

15. $11!$ **39,916,800**

16. $17!$ **$3.556874281 \times 10^{14}$**

17. $\frac{12!}{7!}$ **95,040**

18. $\frac{15!}{9!}$ **3,603,600**

19. $\frac{13!}{(13-3)!}$ **1,716**

20. $\frac{16!}{(16-6)!}$ **5,765,760**

21. $\frac{20!}{10! \cdot 10!}$ **184,756**

22. $\frac{18!}{6! \cdot 12!}$ **18,564**

23. $\frac{n!}{(n-r)!}$, where $n = 23$ and $r = 10$ **$4.151586701 \times 10^{12}$**

24. $\frac{n!}{r!\,(n-r)!}$, where $n = 28$ and $r = 15$ **37,442,160**

CONCEPTUAL  WRITING ▲ CHALLENGING SCIENTIFIC CALCULATOR GRAPHING CALCULATOR

A panel containing three on-off switches in a row is to be set.

25. Assuming no restrictions on individual switches, use the fundamental counting principle to find the total number of possible panel settings. $2 \cdot 2 \cdot 2 = 8$

26. Assuming no restrictions, construct a tree diagram to list all the possible panel settings of Exercise 25.

First switch	Second switch	Third switch	Switch settings
0	0	0	000
		1	001
	1	0	010
		1	011
1	0	0	100
		1	101
	1	0	110
		1	111

27. Now assume that no two adjacent switches can both be off. Explain why the fundamental counting principle does not apply.

28. Construct a tree diagram to list all possible panel settings under the restriction of Exercise 27.

First switch	Second switch	Third switch	Switch settings
0	1	0	010
		1	011
1	0	1	101
	1	0	110
		1	111

29. Table 2 in the previous section shows that there are 36 possible outcomes when two fair dice are rolled. How many would there be if three fair dice were rolled? **216**

30. How many five-digit numbers are there in our system of counting numbers? **90,000**

Recall the club

N = {Andy, Bill, Cathy, David, Evelyn}.

In how many ways could they do each of the following?

31. line up all five members for a photograph
$5! = 120$

32. schedule one member to work in the office on each of five different days, assuming members may work more than one day $5^5 = 3{,}125$

33. select a male and a female to decorate for a party
$3 \cdot 2 = 6$

34. select two members, one to open their next meeting and another to close it, given that Bill will not be present $4 \cdot 3 = 12$

In the following exercises, counting numbers are to be formed using only digits from the set {0, 1, 2, 3, 4}*. Determine the number of different possibilities for each type of number described. (Note that* 201 *is a three-digit number while* 012 *is not.)*

35. three-digit numbers $4 \cdot 5^2 = 100$

36. even four-digit numbers $4 \cdot 5^2 \cdot 3 = 300$

37. five-digit numbers with one pair of adjacent 0s and no other repeated digits $3 \cdot 4 \cdot 3 \cdot 2 = 72$

38. four-digit numbers beginning and ending with 3 and unlimited repetitions allowed $1 \cdot 5^2 \cdot 1 = 25$

39. five-digit multiples of five that contain exactly two 3s and exactly one 0 $6 \cdot 3^2 \cdot 1 = 54$

40. six-digit numbers containing two 0s, two 2s, and two 4s $10 \cdot 6 \cdot 1 = 60$

The Casa Loma Restaurant offers four choices in the soup and salad category (two soups and two salads), two choices in the bread category, and three choices in the entree category. Find the number of dinners available in each of the following cases.

41. One item is to be included from each of the three categories. $4 \cdot 2 \cdot 3 = 24$

42. Only soup and entree are to be included. $2 \cdot 3 = 6$

Determine the number of possible ways to mark your answer sheet (with an answer for each question) for each of the following tests.

43. an eight-question true-or-false test $2^8 = 256$

44. a thirty-question multiple-choice test with four answer choices for each question
$4^{30} = 1.152921505 \times 10^{18}$

Rogers Newman is making up his class schedule for next semester, which must include one class from each of the four categories shown here.

Category	Choices	Number of Choices
English	Medieval Literature Composition Modern Poetry	3
Mathematics	College Algebra Trigonometry	2
Computer Information Science	Introduction to Spreadsheets Advanced Word Processing C Programming BASIC Programming	4
Sociology	Social Problems Sociology of the Middle East Aging in America Minorities in America Women in American Culture	5

For each situation in Exercises 45–50, use the table above to determine the number of different sets of classes Rogers can take.

45. All classes shown are available. $3 \cdot 2 \cdot 4 \cdot 5 = 120$

46. He is not eligible for College Algebra or for BASIC Programming. $3 \cdot 1 \cdot 3 \cdot 5 = 45$

47. All sections of Minorities in America and Women in American Culture are filled already. $3 \cdot 2 \cdot 4 \cdot 3 = 72$

48. He does not have the prerequisites for Medieval Literature, Trigonometry, or C Programming. $2 \cdot 1 \cdot 3 \cdot 5 = 30$

49. Funding has been withdrawn for three of the computer courses and for two of the Sociology courses. $3 \cdot 2 \cdot 1 \cdot 3 = 18$

50. He must complete English Composition and Trigonometry next semester to fulfill his degree requirements. $1 \cdot 1 \cdot 4 \cdot 5 = 20$

---

51. Sean took two pairs of shoes, four pairs of pants, and six shirts on a trip. Assuming all items are compatible, how many different outfits can he wear? $2 \cdot 4 \cdot 6 = 48$

52. A music equipment outlet stocks ten different guitars, four guitar cases, six amplifiers, and three effects processors, with all items mutually compatible and all suitable for beginners. How many different complete setups could Lionel choose to start his musical career? $10 \cdot 4 \cdot 6 \cdot 3 = 720$

53. Tadishi's ZIP code is 95841. How many ZIP codes, altogether, could be formed using all of those same five digits? $5! = 120$

▲ **54.** George Owen keeps four textbooks and three novels on his desk. In how many different ways can he arrange them in a row if

(a) the textbooks must be to the left of the novels? $4! \cdot 3! = 144$

(b) the novels must all be together? $5 \cdot 4! \cdot 3! = 720$

(c) no two novels should be next to each other? $10 \cdot 4! \cdot 3! = 1{,}440$

▲ *Andy, Betty, Clyde, Dawn, Evan, and Felicia have reserved six seats in a row at the theater, starting at an aisle seat. (Refer to Example 7 in this section.)*

55. In how many ways can they arrange themselves? (*Hint:* Divide the task into the series of six parts shown below, performed in order.)

(a) If A is seated first, how many seats are available for him? 6

(b) Now, how many are available for B? 5

(c) Now, how many for C? 4

(d) Now, how many for D? 3

(e) Now, how many for E? 2

(f) Now, how many for F? 1

Now multiply together your six answers above.
$6 \cdot 5 \cdot 4 \cdot 3 \cdot 2 \cdot 1 = 720$

56. In how many ways can they arrange themselves so that Andy and Betty will be next to each other? (*Hint:* First answer the following series of questions, assuming these parts are to be accomplished in order.)

1	2	3	4	5	6
X	X	_	_	_	_
_	X	X	_	_	_
_	_	X	X	_	_
_	_	_	X	X	_
_	_	_	_	X	X

Seats available to A and B

(a) How many pairs of adjacent seats can A and B occupy? 5

(b) Now, given the two seats for A and B, in how many orders can they be seated? 2

(c) Now, how many seats are available for C? 4

(d) Now, how many for D? 3

(e) Now, how many for E? 2

(f) Now, how many for F? 1

$5 \cdot 2 \cdot 4 \cdot 3 \cdot 2 \cdot 1 = 240$

57. In how many ways can they arrange themselves if the men and women are to alternate seats and a man must sit on the aisle? (*Hint:* First answer the following series of questions.)

(a) How many choices are there for the person to occupy the first seat, next to the aisle? (It must be a man.) 3

(b) Now, how many choices of people may occupy the second seat from the aisle? (It must be a woman.) 3

(c) Now, how many for the third seat? (one of the remaining men) 2

(d) Now, how many for the fourth seat? (a woman) 2

(e) Now, how many for the fifth seat? (a man) 1

(f) Now, how many for the sixth seat? (a woman) 1

$3 \cdot 3 \cdot 2 \cdot 2 \cdot 1 \cdot 1 = 36$

58. In how many ways can they arrange themselves if the men and women are to alternate with either a man or a woman on the aisle? (*Hint:* First answer the following series of questions.)

(a) How many choices of people are there for the aisle seat? 6

(b) Now, how many are there for the second seat? (This person may not be of the same sex as the person on the aisle.) 3

(c) Now, how many choices are there for the third seat? 2

(d) Now, how many for the fourth seat? 2

(e) Now, how many for the fifth seat? 1

(f) Now, how many for the sixth seat? 1

$6 \cdot 3 \cdot 2 \cdot 2 \cdot 1 \cdot 1 = 72$

10.3 Exercises (Text Page 602)

Evaluate each of the following expressions.

1. $P(5, 2)$ 20 **2.** $P(8, 3)$ 336

3. $P(10, 0)$ 1 **4.** $P(12, 5)$ 95,040

5. $C(7, 3)$ 35 **6.** $C(10, 6)$ 210

7. $C(8, 4)$ 70 **8.** $C(12, 8)$ 495

Determine the number of permutations (arrangements) of each of the following.

9. 7 things taken 4 at a time 840

10. 8 things taken 5 at a time 6,720

11. 12 things taken 3 at a time 1,320

12. 41 things taken 2 at a time 1,640

Determine the number of combinations (subsets) of each of the following.

13. 5 things taken 4 at a time 5

14. 9 things taken 0 at a time 1

15. 11 things taken 7 at a time 330

16. 14 things taken 3 at a time 364

Use a calculator to evaluate each of these expressions.

17. $P(26, 8)$ 62,990,928,000

18. $C(38, 12)$ 2,707,475,148

In "Super Lotto," a California state lottery game, you select six distinct numbers from the counting numbers 1 *through* 51, *hoping that your selection will match a random list selected by lottery officials.*

19. How many different sets of six numbers can you select? $C(51, 6) = 18{,}009{,}460$

20. Diane Sypniewski always includes her age and her husband's age as two of the numbers in her Super Lotto selections. How many ways can she complete her list of six numbers? $C(49, 4) = 211{,}876$

21. Is it possible to evaluate $P(8, 12)$? Explain.

22. Is it possible to evaluate $C(6, 15)$? Explain.

23. Explain how permutations and combinations differ.

24. Explain how factorials are related to permutations.

25. How many different ways could first, second, and third place winners occur in a race with six runners competing? $P(6, 3) = 120$

26. John Young, a contractor, builds homes of eight different models and presently has five lots to build on. In how many different ways can he place homes on these lots? Assume five different models will be built. $P(8, 5) = 6{,}720$

27. How many different five-member committees could be formed from the 100 U.S. senators? $C(100, 5) = 75{,}287{,}520$

28. If any two points determine a line, how many lines are determined by seven points in a plane, no three of which are collinear? $C(7, 2) = 21$

29. Radio stations in the United States have call letters that begin with K or W (for west or east of the Mississippi River, respectively). Some have three call letters, such as WBZ in Boston, WLS in Chicago, and KGO in San Francisco. Assuming no repetition of letters, how many three-letter sets of call letters are possible? $2 \cdot P(25, 2) = 1{,}200$

30. Most stations that were licensed after 1927 have four call letters starting with K or W, such as WXYZ in Detroit or KRLD in Dallas. Assuming no repetitions, how many four-letter sets are possible? $2 \cdot P(25, 3) = 27{,}600$

31. Subject identification numbers in a certain scientific research project consist of three letters followed by three digits and then three more letters. Assume repetitions are not allowed within any of the three groups, but letters in the first group of three may occur also in the last group of three. How many distinct identification numbers are possible? $P(26, 3) \cdot P(10, 3) \cdot P(26, 3) = 175{,}219{,}200{,}000$

32. How many triangles are determined by twenty points in a plane, no three of which are collinear? $C(20, 3) = 1{,}140$

33. How many ways can a sample of five CD players be selected from a shipment of twenty-four players? $C(24, 5) = 42{,}504$

34. If the shipment of Exercise 33 contains six defective players, how many of the size-five samples would not include any of the defective ones? $C(18, 5) = 8{,}568$

35. In how many ways could twenty-five people be divided into five groups containing, respectively, three, four, five, six, and seven people? $C(25, 3) \cdot C(22, 4) \cdot C(18, 5) \cdot C(13, 6) = 2.47365374256 \times 10^{14}$

36. Larry Sifford and seven of his friends were contemplating the drive back to Denver after a long day of skiing.

(a) If they brought two vehicles, how many choices do they have as to who will do the driving? $C(8, 2) = 28$

(b) Suppose that (due to an insurance limitation) one of the vehicles can only be driven by Larry, his wife, or their son, all of whom are part of the skiing group. Now how many choices of drivers are there? $3 \cdot 7 = 21$

37. Each team in an eight-team basketball league is scheduled to play each other team three times. How many games will be played altogether? $(8 \cdot 21)/2 = 84$

38. The Coyotes, a youth league baseball team, have seven pitchers, who only pitch, and twelve other players, all of whom can play any position other than pitcher. For Saturday's game, the coach has not yet determined which nine players to use nor what the batting order will be, except that the pitcher will bat last. How many different batting orders may occur? $7 \cdot P(12, 8) = 139{,}708{,}800$

CONCEPTUAL    GRAPHING CALCULATOR

39. A music class of eight girls and seven boys is having a recital. If each member is to perform once, how many ways can the program be arranged in each of the following cases?

(a) The girls must all perform first. $8! \cdot 7! = 203{,}212{,}800$

(b) A girl must perform first and a boy must perform last. $8 \cdot 7 \cdot 13! = 348{,}713{,}164{,}800$

(c) Elisa and Doug will perform first and last, respectively. $13! = 6{,}227{,}020{,}800$

(d) The entire program will alternate between the sexes. $8! \cdot 7! = 203{,}212{,}800$

(e) The first, eighth, and fifteenth performers must be girls. $8 \cdot 7 \cdot 6 \cdot 12! = 160{,}944{,}537{,}600$

40. Carole has eight errands to run today, five of them pleasant, but the other three unpleasant. How many ways can she plan her day in each of the following cases?

(a) She decides to put off the unpleasant errands to another day. $5! = 120$

(b) She is determined to complete all eight errands today. $8! = 40{,}320$

(c) She will work up her courage by starting with *at least* two pleasant errands and then completing the rest of the eight. $P(5, 2) \cdot 6! = 14{,}400$

(d) She will begin and end the day with pleasant errands and will accomplish only six altogether. $P(5, 2) \cdot P(6, 4) = 7{,}200$

(e) She will succeed in all eight by facing all three unpleasant errands first. $3! \cdot 5! = 720$

For Exercises 41–46, refer to the standard 52-card deck pictured in the text and notice that the deck contains four aces, twelve face cards, thirteen hearts (all red), thirteen diamonds (all red), thirteen spades (all black), and thirteen clubs (all black). Of the 2,598,960 *different five-card hands possible, decide how many would consist of the following cards.*

41. all diamonds $C(13, 5) = 1{,}287$

42. all black cards $C(26, 5) = 65{,}780$

43. all aces 0 (impossible)

44. four clubs and one non-club $C(13, 4) \cdot 39 = 27{,}885$

45. two face cards and three non-face cards $C(12, 2) \cdot C(40, 3) = 652{,}080$

46. two red cards, two clubs, and a spade $C(26, 2) \cdot C(13, 2) \cdot 13 = 329{,}550$

47. How many different three-number "combinations" are possible on a combination lock having 40 numbers on its dial? (*Hint:* "Combination" is a misleading name for these locks since repetitions are allowed and also order makes a difference.) $40^3 = 64{,}000$

48. In a 7/39 lottery, you select seven distinct numbers from the set 1 through 39, where order makes no difference. How many different ways can you make your selection? $C(39, 7) = 15{,}380{,}937$

John Young (the contractor) is to build six homes on a block in a new subdivision. Overhead expenses have forced him to limit his line to two different models, standard and deluxe. (All standard model homes are the same and all deluxe model homes are the same.)

49. How many different choices does John have in positioning the six houses if he decides to build three standard and three deluxe models? $C(6, 3) = 20$

50. If John builds only two deluxe and four standards, how many different positionings can he use? $C(6, 2) = 15$

Because of his good work, John gets a contract to build homes on three additional blocks in the subdivision, with six homes on each block. He decides to build nine deluxe homes on these three blocks; two on the first block, three on the second, and four on the third. The remaining nine homes will be standard.

51. Altogether on the three-block stretch, how many different choices does John have for positioning the eighteen homes? (*Hint:* Consider the three blocks separately and use the fundamental counting principle.) $C(6, 2) \cdot C(6, 3) \cdot C(6, 4) = 4{,}500$

52. How many choices would he have if he built 2, 3, and 4 deluxe models on the three different blocks as before, but not necessarily on the first, second, and third blocks in that order? $3! \cdot C(6, 2) \cdot C(6, 3) \cdot C(6, 4) = 27{,}000$

53. (a) How many numbers can be formed using all six digits 2, 3, 4, 5, 6, and 7? $6! = 720$

▲ **(b)** Suppose all these numbers were arranged in increasing order: 234,567; 234,576; and so on. Which number would be 363rd in the list? 523,647

54. How many four-digit counting numbers are there whose digits are distinct and have a sum of 10, assuming the digit 0 is not used? $4! = 24$

55. How many paths are possible from A to B if all motion must be to the right or downward?

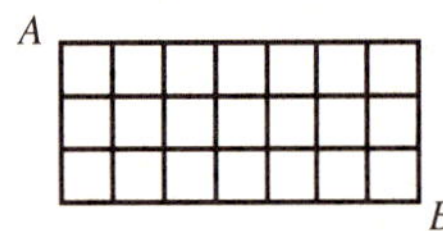

$C(10, 3) = 120$

56. How many cards must be drawn (without replacement) from a standard deck of 52 to guarantee that at least two are from the same suit? 5

57. Judy Zahrndt wants to name her new baby so that her monogram (first, middle, and last initials) will be distinct letters in alphabetical order. How many different monograms could she select?
$C(24, 2) = 300$

▲ **58.** How many pairs of vertical angles are formed by eight distinct lines that all meet at a common point?
$2 \cdot C(8, 2) = 56$

▲ **59.** In how many ways can three men and six women be assigned to two groups so that neither group has more than six people? $C(9, 3) + C(9, 4) = 210$

60. Verify that $C(8, 3) = C(8, 5)$. Each is equal to 56.

61. Verify that $C(12, 9) = C(12, 3)$. Each is equal to 220.

⊙ **62.** Use the factorial formula for combinations to prove that in general, $C(n, r) = C(n, n - r)$.

Each is equal to $\dfrac{n!}{r!\,(n - r)!}$.

10.4 Exercises (Text Page 608)

Read the following combination values directly from Pascal's triangle.

1. $C(4, 3)$ 4 **2.** $C(5, 2)$ 10 **3.** $C(6, 4)$ 15 **4.** $C(7, 5)$ 21

5. $C(8, 2)$ 28 **6.** $C(9, 4)$ 126 **7.** $C(9, 7)$ 36 **8.** $C(10, 6)$ 210

A committee of four Congressmen will be selected from a group of seven Democrats and three Republicans. Find the number of ways of obtaining each of the following.

9. exactly one Democrat $C(7, 1) \cdot C(3, 3) = 7$

10. exactly two Democrats $C(7, 2) \cdot C(3, 2) = 63$

11. exactly three Democrats $C(7, 3) \cdot C(3, 1) = 105$

12. exactly four Democrats $C(7, 4) \cdot C(3, 0) = 35$

Suppose eight fair coins are tossed. Find the number of ways of obtaining each of the following.

13. exactly three heads $C(8, 3) = 56$

14. exactly four heads $C(8, 4) = 70$

15. exactly five heads $C(8, 5) = 56$

16. exactly six heads $C(8, 6) = 28$

Kelly Melcher, searching for an Economics class, knows that it must be in one of nine classrooms. Since the professor does not allow people to enter after the class has begun, and there is very little time left, Kelly decides to try just four of the rooms at random.

17. How many different selections of four rooms are possible? $C(9, 4) = 126$

18. How many of the selections of Exercise 17 will fail to locate the class? $C(8, 4) = 70$

19. How many of the selections of Exercise 17 will succeed in locating the class? 56

20. What fraction of the possible selections will lead to "success"? (Give three decimal places.) $56/126 = .444$

For a set of five elements, find the number of different subsets of each of the following sizes. (Use row 5 of Pascal's triangle to find the answers.)

21. 0 1 **22.** 1 5 **23.** 2 10 **24.** 3 10 **25.** 4 5 **26.** 5 1

27. How many subsets (of any size) are there for a set of five elements? **32**

28. Find and explain the relationship between the row number and row sum in Pascal's triangle.

Over the years, many interesting patterns have been discovered in Pascal's triangle. Exercises 29 and 30 exhibit two such patterns.*

29. Name the next five numbers of the diagonal sequence indicated in the figure at the side. **. . . , 15, 21, 28, 36, 45, . . . ; (These are the triangular numbers.)**

30. Complete the sequence of sums on the diagonals shown in the figure at the side. What pattern do these sums make? What is the name of this important sequence of numbers? The presence of this sequence in the triangle apparently was not recognized by Pascal. **. . . , 8, 13, 21, 34, . . . ; A number in this sequence is the sum of the two preceding terms. This is the Fibonacci sequence.**

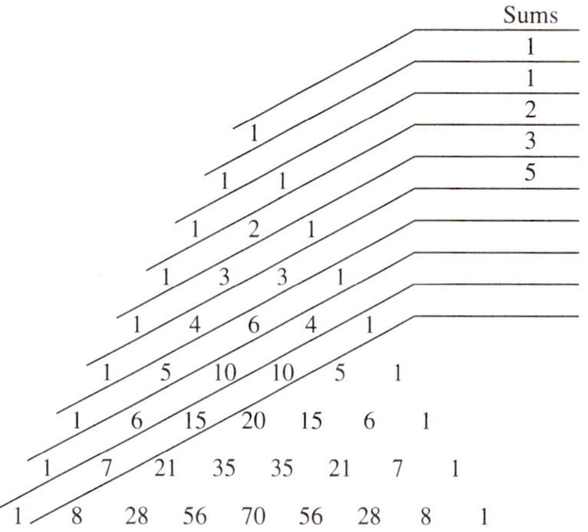

31. More than a century before Pascal's treatise on the "triangle" appeared, another work by the Italian mathematician Niccolo Tartaglia (1506–1559) came out and included the table of numbers shown here.

1	1	1	1	1	1
1	2	3	4	5	6
1	3	6	10	15	21
1	4	10	20	35	56
1	5	15	35	70	126
1	6	21	56	126	252
1	7	28	84	210	462
1	8	36	120	330	792

Explain the connection between Pascal's triangle and Tartaglia's "rectangle." **The rows of Tartaglia's rectangle correspond to the diagonals of Pascal's triangle.**

*For example, see the article "Serendipitous Discovery of Pascal's Triangle" by Francis W. Stanley in *The Mathematics Teacher*, February 1975.

32. Construct another "triangle" by replacing every number in Pascal's triangle (rows **0** through **5**) by its remainder when divided by 2. What special property is shared by rows **2** and **4** of this new triangle?

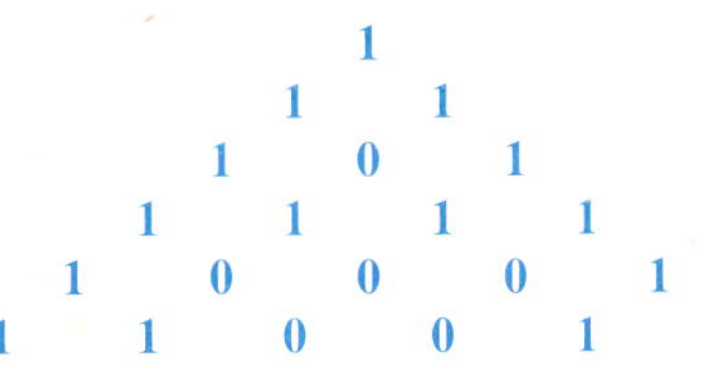

In rows 2 and 4, every entry, except for the beginning and ending 1s, is 0. (This is because the corresponding entries in the original triangle were all even.)

33. What is the next row that would have the same property as rows **2** and **4** in Exercise 32? **row 8**

34. How many even numbers are there in row number **256** of Pascal's triangle? (Do Exercises 32 and 33 first.) **255**

Write out the binomial expansion for each of the following powers.

35. $(x + y)^6$ $x^6 + 6x^5y + 15x^4y^2 + 20x^3y^3 + 15x^2y^4 + 6xy^5 + y^6$

36. $(x + y)^8$ $x^8 + 8x^7y + 28x^6y^2 + 56x^5y^3 + 70x^4y^4 + 56x^3y^5 + 28x^2y^6 + 8xy^7 + y^8$

37. $(z + 2)^3$ $z^3 + 6z^2 + 12z + 8$

38. $(w + 3)^5$ $w^5 + 15w^4 + 90w^3 + 270w^2 + 405w + 243$

39. $(2a + 5b)^4$ $16a^4 + 160a^3b + 600a^2b^2 + 1{,}000ab^3 + 625b^4$

40. $(3d + 5f)^4$ $81d^4 + 540d^3f + 1{,}350d^2f^2 + 1{,}500df^3 + 625f^4$

41. $(b - h)^7$ (*Hint:* First change $b - h$ to $b + (-h)$.)
$b^7 - 7b^6h + 21b^5h^2 - 35b^4h^3 + 35b^3h^4 - 21b^2h^5 + 7bh^6 - h^7$

42. $(2n - 4m)^5$ $32n^5 - 320n^4m + 1{,}280n^3m^2 - 2{,}560n^2m^3 + 2{,}560nm^4 - 1{,}024m^5$

43. How many terms appear in the binomial expansion for $(x + y)^n$? $n + 1$

44. Observe the pattern in this table and fill in the blanks to discover a formula for the rth term (the general term) of the binomial expansion for $(x + y)^n$.

Term Number	Coefficient	Variable Part	Term
1	1	x^n	$1\,x^n$
2	$\frac{n!}{(n-1)!\,1!}$	$x^{n-1}y$	$\frac{n!}{(n-1)!\,1!}x^{n-1}y$
3	$\frac{n!}{(n-2)!\,2!}$	$x^{n-2}y^2$	$\frac{n!}{(n-2)!\,2!}x^{n-2}y^2$
4	$\frac{n!}{(n-3)!\,3!}$	$x^{n-3}y^3$	$\frac{n!}{(n-3)!\,3!}x^{n-3}y^3$
⋮	⋮	⋮	⋮
r	$\frac{n!}{(n-r+1)!(r-1)!}$	______	______

$x^{n-r+1}y^{r-1}$; $\frac{n!}{(n-r+1)!\,(r-1)!}x^{n-r+1}y^{r-1}$

Use the results of Exercise 44 to find the indicated term of each of the following expansions.

45. $(x + y)^{12}$; 5th term $n = 12, r = 5: \dfrac{12!}{8!\,4!}x^8y^4 = 495x^8y^4$

46. $(a + b)^{20}$; 16th term $n = 20, r = 16: \dfrac{20!}{5!\,15!}a^5b^{15} = 15{,}504a^5b^{15}$

10.5 Exercises (Text Page 615)

1. Explain why the complements principle of counting is called an "indirect" method.

2. Explain the difference between the *special* and *general* additive principles of counting.

If you toss seven fair coins, in how many ways can you obtain each of the following results?

3. at least one head ("At least one" is the complement of "none.") $2^7 - 1 = 127$

4. at least two heads ("At least two" is the complement of "zero or one.") $2^7 - (1 + 7) = 120$

5. at least two tails 120

6. at least one of each (a head and a tail) $2^7 - 2 = 126$

If you roll two fair dice (say red and green), in how many ways can you obtain each of the following? (Refer to Table 2 in the first section of this chapter.)

7. a 2 on the red die 6

8. a sum of at least 3 $36 - 1 = 35$

9. a 4 on at least one of the dice $6 + 6 - 1 = 11$

10. a different number on each die $36 - 6 = 30$

Among the 635,013,559,600 *possible bridge hands (13 cards each), how many contain the following cards? (The standard card deck was described in the third section of this chapter.)*

11. at least one card that is not a spade (complement of "all spades")
635,013,559,600 − 1 = 635,013,559,599

12. cards of more than one suit (complement of "all the same suit")
635,013,559,600 − 4 = 635,013,559,596

13. at least one face card (complement of "no face cards")
635,013,559,600 − $C(40, 13)$ = 622,980,336,720

14. at least one diamond, but not all diamonds (complement of "no diamonds or all diamonds")
635,013,559,600 − $[C(39, 13) + 1]$ = 626,891,134,155

How many three-digit counting numbers meet the following requirements?

15. even or a multiple of 5 $9 \cdot 10 \cdot 6 = 540$

16. greater than 600 or a multiple of 10
$(4 \cdot 10^2 - 1) + 9 \cdot 10 \cdot 1 - 39 = 450$

If a given set has twelve elements, how many of its subsets have the given numbers of elements?

17. at most two elements
$C(12, 0) + C(12, 1) + C(12, 2) = 79$

18. at least ten elements
$C(12, 10) + C(12, 11) + C(12, 12) = 79$

19. more than two elements
$2^{12} - 79 = 4{,}017$

20. from three through nine elements
$2^{12} - (79 + 79) = 3{,}938$

Of a group of 50 *students,* 30 *enjoy music,* 15 *enjoy literature, and* 10 *enjoy both music and literature. How many of them enjoy the following?*

21. at least one of these two subjects (general additive principle) $30 + 15 - 10 = 35$

22. neither of these two subjects (complement of "at least one") $50 - 35 = 15$

If a single card is drawn from a standard 52-card deck, in how many ways could it be the following? (Use the general additive principle.)

23. a club or a jack $13 + 4 - 1 = 16$

24. a face card or a black card $12 + 26 - 6 = 32$

25. a diamond or a face card or a denomination greater than 10 (First note that this is the same as "A diamond or a denomination greater than 10" since every face card has denomination greater than 10. Do not consider ace to be greater than 10 in this case.)
$13 + 12 - 3 = 22$

26. a heart or a queen or a red card (First note that this is the same as "A queen or a red card." Why is this true?) $4 + 26 - 2 = 28$

Table 4 in this chapter (For Further Thought) briefly described the various kinds of hands in five-card poker. A "royal flush" is a hand containing 10, *jack, queen, king, and ace all in the same suit. A "straight flush" is any five consecutive denominations in a common suit. The lowest would be ace,* 2, 3, 4, 5, *and the highest is* 9, 10, *jack, queen, king. (An ace high straight flush is not referred to as a straight flush since it has the special name "royal flush.") A "straight" contains five consecutive denominations but not all of the same suit, and a "flush" is any hand with all five cards the same suit, except a royal flush or a straight flush. A "three of a kind" hand contains exactly three of one denomination, and furthermore the remaining two must be of two* additional *denominations. (Why is this?)*

Verify each of the following. (Explain all steps of your argument.)

27. There are four ways to get a royal flush.

28. There are 36 ways to get a straight flush.

▲ **29.** There are 5,108 ways to get a flush.

▲ **30.** There are 10,200 ways to get a straight.

▲ **31.** There are 624 ways to get four of a kind.

▲ **32.** There are 54,912 ways to get three of a kind.

If three-digit numbers are formed using only digits from the set {0, 1, 2, 3, 4, 5, 6}, *how many will belong to the following categories?*

33. even numbers **$6 \cdot 7 \cdot 4 = 168$**

34. multiples of 10 **$6 \cdot 7 \cdot 1 = 42$**

35. multiples of 100 **$6 \cdot 1 \cdot 1 = 6$**

▲ **36.** multiples of 25 **$6 \cdot 1 \cdot 1 = 6$ end in 00; $6 \cdot 1 \cdot 1 = 6$ end in 25; $6 \cdot 1 \cdot 1 = 6$ end in 50; $6 + 6 + 6 = 18$**

37. If license numbers consist of three letters followed by three digits, how many different licenses could be created having at least one letter or digit repeated? (*Hint:* Use the complements principle of counting.) **$26^3 \cdot 10^3 - P(26, 3) \cdot P(10, 3) = 6{,}344{,}000$**

38. If two cards are drawn from a 52-card deck without replacement (that is, the first card is not replaced in the deck before the second card is drawn), in how many different ways is it possible to obtain a king on the first draw and a heart on the second? (*Hint:* Split this event into the two disjoint components "king of hearts and then another heart" and "non-heart king and then heart." Use the fundamental counting principle on each component, then apply the special additive principle.) **$1 \cdot 12 + 3 \cdot 13 = 51$**

39. A committee of four faculty members will be selected from a department of twenty-five which includes professors Fontana and Spradley. In how many ways could the committee include at least one of these two professors? **$C(25, 4) - C(23, 4) = 3{,}795$**

Edward Roberts is planning a long-awaited driving tour, which will take him and his family on the southern route to the West Coast. Ed is interested in seeing the twelve national monuments listed here, but will have to settle for seeing just three of them since some family members are anxious to get to Disneyland.

New Mexico	Arizona	California
Gila Cliff Dwellings	Canyon de Chelly	Devils Postpile
Petroglyph	Organ Pipe Cactus	Joshua Tree
White Sands	Saguaro	Lava Beds
Aztec Ruins		Muir Woods
		Pinnacles

In how many ways could the three monuments chosen include the following?

40. sites in only one state **$C(4, 3) + C(3, 3) + C(5, 3) = 15$**

41. at least one site not in California **$C(12, 3) - C(5, 3) = 210$**

▲ **42.** sites in fewer than all three states **$C(12, 3) - 4 \cdot 3 \cdot 5 = 160$**

▲ **43.** sites in exactly two of the three states **$C(12, 3) - (15 + 60) = 145$**

▲ **44.** In the figure here, a "segment" joins intersections and/or turning points. For example, the path from A straight up to C and then straight across to B consists of six segments. Altogether, how many paths are there from A to B that consist of exactly six segments? **$C(6, 3)$ contain no diagonal segments; 6 contain diagonal segments; $20 + 6 = 26$**

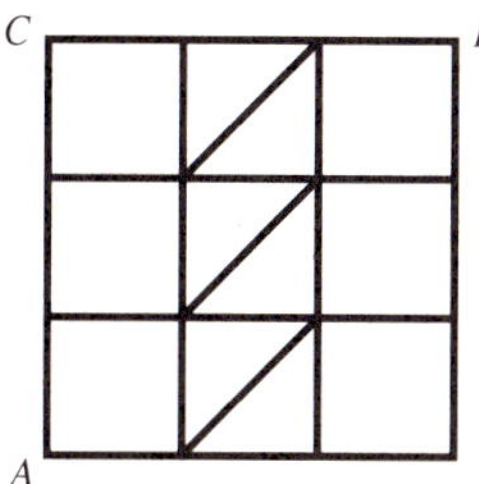

▲ **45.** How many of the counting numbers 1 through 100 can be expressed as a sum of three or fewer powers of three (not necessarily distinct)? For example, 5 is such a number since $5 = 1 + 1 + 3$. (For one solution, see the October 1992 issue of *Mathematics Teacher,* page 497.) **40**

46. Extend the general additive counting principle to three overlapping sets (as in the figure) to show that

$$n(A \cup B \cup C) = n(A) + n(B) + n(C) - n(A \cap B) - n(A \cap C) - n(B \cap C) + n(A \cap B \cap C).$$

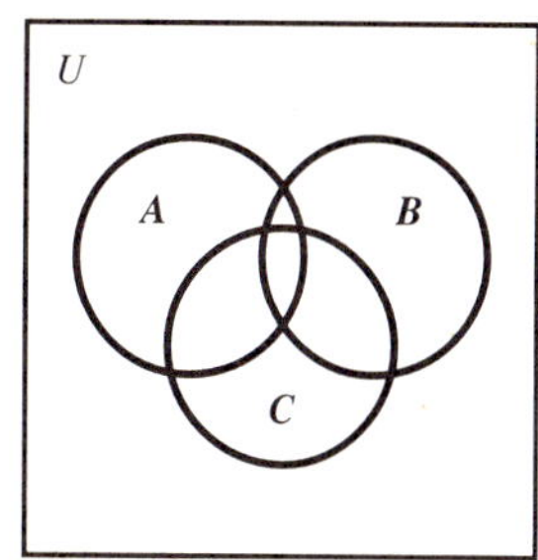

47. How many of the counting numbers 1 through 300 are *not* divisible by 2, 3, or 5? (*Hint:* Use the complements principle and the result of Exercise 46.) **$300 - (150 + 100 + 60 - 50 - 30 - 20 + 10) = 80$**

48. How many three-digit counting numbers do not contain any of the digits 2, 5, 7, or 8? **$5 \cdot 6 \cdot 6 = 180$**

49. Caralee Woods manages the shipping department of a firm that sells gold watches. If the watches can be shipped in packages of four, three, or one, in how many ways can a customer's order for fifteen watches be filled? **15**

A Civil Air Patrol unit of fifteen members includes four officers. In how many ways can four of them be selected for a search and rescue mission in each of the following cases?

50. The search group must include at least one officer. **$C(15, 4) - C(11, 4) = 1{,}035$**

51. The search group must include at least two officers. **$C(15, 4) - [C(11, 4) + C(11, 3) \cdot C(4, 1)] = 375$**

Chapter 10 Test (Text Page 618)

If digits may be used from the set $\{0, 1, 2, 3, 4, 5, 6\}$, *find the number of each of the following.*

1. three-digit numbers **$6 \cdot 7 \cdot 7 = 294$**

2. even three-digit numbers **$6 \cdot 7 \cdot 4 = 168$**

3. three-digit numbers without repeated digits **$6 \cdot 6 \cdot 5 = 180$**

▲ **4.** three-digit multiples of five without repeated digits **$6 \cdot 5 \cdot 1 = 30$ end in 0; $5 \cdot 5 \cdot 1 = 25$ end in 5; $30 + 25 = 55$**

5. Determine the number of triangles (of any size) in the figure shown here. **13**

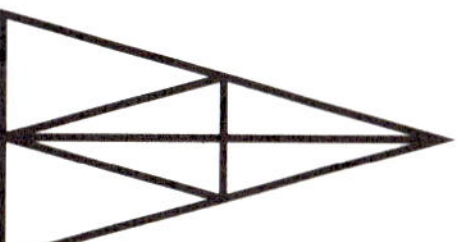

6. Construct a tree diagram showing all possible results when a fair coin is tossed four times, if the third toss must be different than the second.

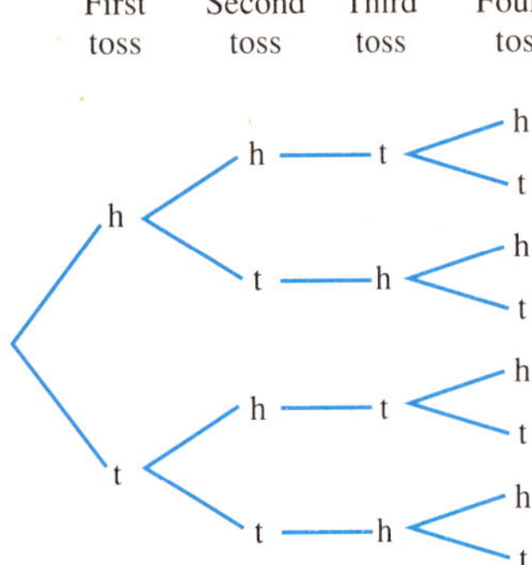

7. How many non-repeating four-digit numbers have the sum of their digits equal to 30? $4! = 24$

8. Using only digits from the set {0, 1, 2}, how many three-digit numbers can be written which have no repeated odd digits? **12**

Evaluate the following expressions.

9. $5!$ **120**

10. $\frac{8!}{5!}$ **336**

11. $P(12, 4)$ **11,880**

12. $C(7, 3)$ **35**

13. How many five-letter "words" without repeated letters are possible using the English alphabet? (Assume that any five letters make a "word.") $P(26, 5) = 7{,}893{,}600$

14. Using the Russian alphabet (which has 32 letters), and allowing repeated letters, how many five-letter "words" are possible? $32^5 = 33{,}554{,}432$

If there are twelve players on a basketball team, find the number of choices the coach has in selecting each of the following.

15. four players to carry the team equipment $C(12, 4) = 495$

16. two players for guard positions and two for forward positions $C(12, 2) \cdot C(10, 2) = 2{,}970$

17. five starters and five subs $C(12, 5) \cdot C(7, 5) = 16{,}632$

18. a set of three or more of the players $2^{12} - [C(12, 0) + C(12, 1) + C(12, 2)] = 4{,}017$

Determine the number of possible settings for a row of four on-off switches under each of the following conditions.

19. There are no restrictions. $2^4 = 16$

20. The first and fourth switches must be on. $2^2 = 4$

21. The first and fourth switches must be set the same. $2 \cdot 2^2 = 8$

22. No two adjacent switches can be off. **8**

23. No two adjacent switches can be set the same. **2**

24. At least two switches must be on. $16 - (1 + 4) = 11$

Four distinct letters are to be chosen from the set {A, B, C, D, E, F, G}. Determine the number of ways to obtain a subset that includes each of the following.

25. the letter D $C(6, 3) = 20$

26. both A and E $C(5, 2) = 10$

27. either A or E, but not both $2 \cdot C(5, 3) = 20$

28. equal numbers of vowels and consonants $C(5, 2) = 10$

29. more consonants than vowels $C(5, 4) + C(2, 1) \cdot C(5, 3) = 25$

30. Write out and simplify the binomial expansion for $(x + 2)^5$. $x^5 + 10x^4 + 40x^3 + 80x^2 + 80x + 32$

31. If $C(n, r) = 495$ and $C(n, r + 1) = 220$, find the value of $C(n + 1, r + 1)$. $495 + 220 = 715$

32. If you write down the second entry of each row of Pascal's triangle (starting with row 1), what sequence of numbers do you obtain? **the counting numbers**

33. Explain why there are $r!$ times as many permutations of n things taken r at a time as there are combinations of n things taken r at a time.

CONCEPTUAL  WRITING  CHALLENGING  SCIENTIFIC CALCULATOR  GRAPHING CALCULATOR

CHAPTER 11 Probability

11.1 Exercises (Text Page 627)

Suppose the spinner shown here is spun once. (Each number corresponds to a 120-degree sector, or one third of the circle.) Find **(a)** *the theoretical probability, and* **(b)** *the odds in favor, of each of the following events. Start by constructing the sample space for the experiment. For each exercise, list the favorable outcomes. Then form the probability fraction.*

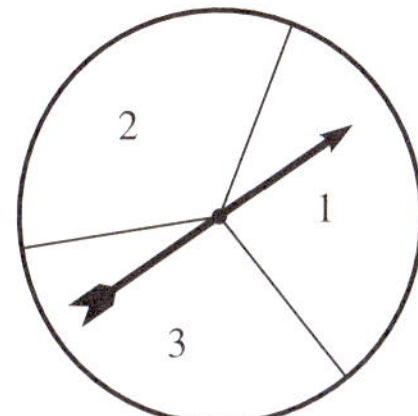

1. It will stop at 1. (a) 1/3 (b) 1 to 2

2. It will stop at an odd number. (a) 2/3 (b) 2 to 1

3. It will stop at an even number.
(a) 1/3 (b) 1 to 2

Suppose the same spinner is spun twice in succession to determine a two-digit number. Construct the sample space and refer to it for Exercises 4–7.

4. How many different two-digit numbers could be determined? $3 \cdot 3 = 9$

Find **(a)** *the probability, and* **(b)** *the odds in favor, of each of the following events.*

5. The resulting number will be even.
(a) 1/3 (b) 1 to 2

6. The resulting number will contain repeated digits.
(a) 1/3 (b) 1 to 2

7. The resulting number will be prime.
(a) 4/9 (b) 4 to 5

For the experiment of tossing two fair coins, say a dime and a nickel, find **(a)** *the probability, and* **(b)** *the odds in favor, of each of the following events. Start by writing out the sample space.*

8. The dime will land heads up.
(a) 1/2 (b) 1 to 1

9. The nickel will land heads up.
(a) 1/2 (b) 1 to 1

10. Two heads will occur.
(a) 1/4 (b) 1 to 3

11. Exactly one head will occur.
(a) 1/2 (b) 1 to 1

If three fair coins are to be tossed, find the probability of each of the following events. Start by writing out the sample space.

12. no heads 1/8

13. exactly one head 3/8

14. exactly two heads 3/8

15. three heads 1/8

The sample space for the rolling of two fair dice appeared in Table 2 of the first section in the previous chapter. Reproduce that table, but replace each of the 36 equally likely ordered pairs with its corresponding sum (for the two dice). Then find the probability of rolling each of the following sums.

16. 2 1/36

17. 3 2/36 = 1/18

18. 4 3/36 = 1/12

19. 5 4/36 = 1/9

20. 6 5/36

21. 7 6/36 = 1/6

22. 8 5/36

23. 9 4/36 = 1/9

24. 10 3/36 = 1/12

25. 11 2/36 = 1/18

26. 12 1/36

In Exercises 27 and 28, compute answers to three decimal places.

27. In a hybrid corn research project, 200 seeds were planted, and 170 of them germinated. Find the empirical probability that any particular seed of this type will germinate. 170/200 = .850

28. In a certain state, 37,052 boys and 35,192 girls were born last year. Find the empirical probability that one of those births, chosen at random, would be
(a) a boy; 37,052/72,244 ≈ .513
(b) a girl. 35,192/72,244 ≈ .487

29. In Example 3, what would be Ina's probability of having exactly two daughters if she were to have four children altogether? (Use a tree diagram to construct the sample space.) 6/16 = 3/8

Mendel found no dominance in snapdragons (in contrast to peas) with respect to red and white flower color. When pure red and pure white parents are crossed (see Table 1), the resulting Rr *combination (one of each gene) produces second generation offspring with* pink *flowers. These second generation pinks, however, still carry one red and one white gene, so when they are crossed the third generation is still governed by Table 2.*

Find the following probabilities for third generation snapdragons.

30. P(red) **1/4** **31.** P(pink) **2/4 = 1/2**

32. P(white) **1/4**

Mendel also investigated various characteristics besides flower color. For example, round peas are dominant over recessive wrinkled peas. First, second, and third generations can again be analyzed using Tables 1 and 2, where R *represents round and* r *represents wrinkled.*

33. Explain why crossing pure round and pure wrinkled first generation parents will always produce round peas in the second generation offspring.

34. When second generation round pea plants (each of which carries both R and r genes) are crossed, find the probability that a third generation offspring will have

(a) round peas; **3/4**

(b) wrinkled peas. **1/4**

Cystic fibrosis *is one of the most common inherited diseases in North America (including the United States), occurring in about 1 of every 2,000 Caucasian births and about 1 of every 250,000 non-Caucasian births. Even with modern treatment, victims usually die from lung damage by their early twenties. If we denote a cystic fibrosis gene with a* c *and a disease-free gene with a* C *(since the disease is recessive), then only a* cc *person will actually have the disease. Such persons would ordinarily die before parenting children, but a child can also inherit the disease from two* Cc *parents (who themselves are healthy, that is, have no symptoms but are "carriers" of the disease). This is like a pea plant inheriting white flowers from two red-flowered parents which both carry genes for white.*

35. Find the empirical probability (to four decimal places) that cystic fibrosis will occur in a randomly selected infant birth among U.S. Caucasians. **1/2,000 = .0005**

36. Find the empirical probability (to six decimal places) that cystic fibrosis will occur in a randomly selected infant birth among U.S. non-Caucasians. **1/250,000 = .000004**

37. Among 150,000 North American Caucasian births, about how many occurrences of cystic fibrosis would you expect? **75**

Suppose that both partners in a marriage are cystic fibrosis carriers (a rare occurrence). Construct a chart similar to Table 2 and determine the probability of each of the following events.

38. Their first child will have the disease. **1/4**

39. Their first child will be a carrier. **2/4 = 1/2**

40. Their first child will neither have nor carry the disease. **1/4**

Suppose a child is born to one cystic fibrosis carrier parent and one non-carrier parent. Find the probability of each of the following events.

41. The child will have cystic fibrosis. **0**

42. The child will be a healthy cystic fibrosis carrier. **2/4 = 1/2**

43. The child will neither have nor carry the disease. **2/4 = 1/2**

Sickle-cell anemia *occurs in about 1 of every 500 black baby births and about 1 of every 160,000 non-black baby births. It is ordinarily fatal in early childhood. There is a test to identify carriers. Unlike cystic fibrosis, which is recessive, sickle-cell anemia is* **codominant.** *This means that inheriting two sickle-cell genes causes the disease, while inheriting just one sickle-cell gene causes a mild (non-fatal) version (which is called* **sickle-cell trait***). This is like a snapdragon plant manifesting pink flowers by inheriting one red gene and one white gene.*

Find the empirical probability of each of the following events (Exercises 44 and 45).

44. A randomly selected black baby will have sickle-cell anemia. (Give your answer to three decimal places.) **1/500 = .002**

45. A randomly selected non-black baby will have sickle-cell anemia. (Give your answer to six decimal places.) **1/160,000 ≈ .000006**

CONCEPTUAL WRITING CHALLENGING SCIENTIFIC CALCULATOR GRAPHING CALCULATOR

46. Among 80,000 births of black babies, about how many occurrences of sickle-cell anemia would you expect? about 160

Find the theoretical probability of each of the following conditions in a child, both of whose parents have sickle-cell trait. (Exercises 47–49)

47. The child will have sickle-cell anemia. 1/4

48. The child will have sickle-cell trait. 2/4 = 1/2

49. The child will be healthy. 1/4

50. In the history of track and field, no woman has broken the 10-second barrier in the 100-meter run.

(a) From the statement above, find the empirical probability that a woman runner will break the 10-second barrier next year. 0

(b) Can you find the theoretical probability for the event of part (a)? no

(c) Is it possible that the event of part (a) will occur? yes

51. Is there any way a coin could fail to be "fair"? Explain.

52. On page 27 of their book *Descartes' Dream,* Philip Davis and Reuben Hersh ask the question, "Is probability real or is it just a cover-up for ignorance?" What do you think? Are some things truly random, or is everything potentially deterministic?

Three married couples arrange themselves randomly in six consecutive seats in a row. Find the probability of each event in Exercises 53–56. (Hint: In each case the denominator of the probability fraction will be 720.)

53. Each man will sit immediately to the left of his wife. $3 \cdot 1 \cdot 2 \cdot 1 \cdot 1 \cdot 1 = 6$; $6/720 = 1/120 \approx .0083$

54. Each man will sit immediately to the left of a woman. $3 \cdot 3 \cdot 2 \cdot 2 \cdot 1 \cdot 1 = 36$; $36/720 = 1/20 = .05$

55. The women will be in three adjacent seats. $4 \cdot 3! \cdot 3! = 144$; $144/720 = 1/5 = .2$

56. The women will be in three adjacent seats, as will the men.
$2 \cdot 3! \cdot 3! = 72$; $72/720 = 1/10 = .1$

57. If two distinct numbers are chosen randomly from the set $\{-2, -4/3, -1/2, 0, 1/2, 3/4, 3\}$, find the probability that they will be the slopes of two perpendicular lines. $2/C(7, 2) = 2/21 \approx .095$

58. At most horse-racing tracks, the "trifecta" is a particular race where you win if you correctly pick the "win," "place," and "show" horses (the first, second, and third place winners), in their proper order. If five horses of equal ability are entered in today's trifecta race, and you select an entry, what is the probability that you will be a winner?
$1/P(5, 3) = 1/60 \approx .017$

59. If a dart hits the square target shown here at random, what is the probability that it will hit in a colored region? (*Hint:* Compare the area of the colored regions to the total area of the target.)

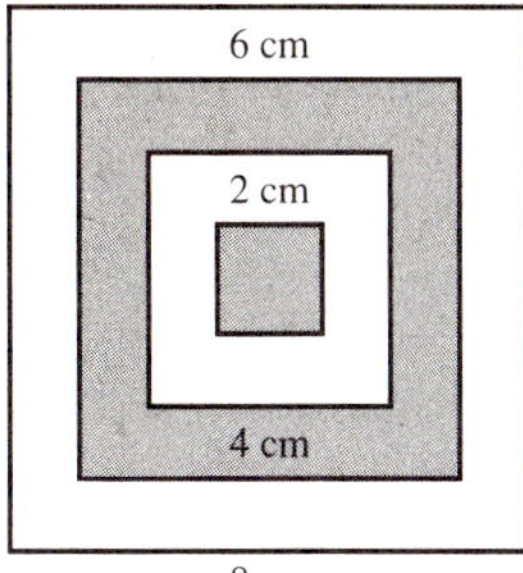

$(6^2 - 4^2 + 2^2)/8^2 = 3/8 = .375$

60. Suppose you plan to take four courses next term. If you select them randomly from a listing of ten courses, six of which are science courses, what is the probability that all four courses you select will be science courses?
$C(6, 4)/C(10, 4) = 1/14 \approx .071$

Assuming that Marcy, Todd, and Jennifer are three of the 32 members of the class, and that three of the class members will be chosen randomly to deliver their reports during the next class meeting, find the probability (to six decimal places) of each of the following events (Exercises 61 and 62).

61. Marcy, Todd, and Jennifer are selected, in that order. $1/P(32, 3) \approx .000034$

62. Marcy, Todd, and Jennifer are selected, in any order. $1/C(32, 3) \approx .000202$

63. Theresa Piselli randomly selects two symphony performances to attend this season, choosing from a schedule of eight performances, two of which will feature works by Mozart. Find the probability that Theresa will select both of the Mozart programs.
$1/C(8, 2) = 1/28 \approx .036$

64. When drawing cards without replacement from a standard 52-card deck, find the highest number of cards you could possibly draw and still get

(a) fewer than three black cards, 28

(b) fewer than six spades, 44

(c) fewer than four face cards, 43

(d) fewer than two kings. 49

In 5-card poker, find the probability of being dealt each of the following (Exercises 65–70). Give each answer to eight decimal places. (Refer to Table 3.)

65. a straight flush $36/2{,}598{,}960 \approx .00001385$

66. four of a kind $624/2{,}598{,}960 \approx .00024010$

67. four aces $48/2{,}598{,}960 \approx .00001847$

68. one pair $1{,}098{,}240/2{,}598{,}960 \approx .42256903$

69. a pair of 7s $84{,}480/2{,}598{,}960 \approx .03250531$

70. two pairs $123{,}552/2{,}598{,}960 \approx .04753902$

71. If two distinct prime numbers are randomly selected from among the first eight prime numbers, what is the probability that their sum will be 24? $3/28 \approx .107$

72. The digits 1, 2, 3, 4, and 5 are randomly arranged to form a five-digit number. Find the probability of each of the following events.

(a) The number is even. $2 \cdot 4!/5! = 2/5$

(b) The first and last digits of the number are both even. $2 \cdot 3!/5! = 1/10$

73. Two integers are randomly selected from the set $\{1, 2, 3, 4, 5, 6, 7, 8, 9\}$ and are added together. Find the probability that their sum is 11 if they are selected

(a) with replacement; $8/9^2 = 8/81 \approx .099$

(b) without replacement. $4/C(9, 2) = 1/9 \approx .111$

Numbers that are ***palindromes*** *read the same forward and backward. For example, 30203 is a five-digit palindrome. If a single number is chosen randomly from each of the following sets, find the probability that it will be palindromic.*

74. the set of all two-digit numbers $9/(9 \cdot 10) = 1/10$

75. the set of all three-digit numbers $(9 \cdot 10)/(9 \cdot 10^2) = 1/10$

11.2 Exercises (Text Page 636)

1. Patricia Quinlin has three office assistants. If A is the event that at least two of them are men and B is the event that at least two of them are women, are A and B mutually exclusive? yes

2. Vicki Valerin earned her college degree several years ago. Consider the following four events.

Her alma mater is in the East.
Her alma mater is a private college.
Her alma mater is in the Northwest.
Her alma mater is in the South.

Are these events all mutually exclusive of one another? no

3. Explain the difference between the "general" and "special" addition rules of probability, illustrating each one with an appropriate example.

For the experiment of rolling a single fair die, find the probability of each of the following events.

4. not less than 2 $5/6$

5. not prime $1/2$

6. odd or less than 5 $5/6$

7. even or prime $5/6$

8. odd or even 1

9. less than 3 or greater than 4 $2/3$

For the experiment of drawing a single card from a standard 52-card deck, find **(a)** *the probability, and* **(b)** *the odds in favor, of each of the following events.*

10. not an ace (a) $12/13$ (b) 12 to 1

11. king or queen (a) $2/13$ (b) 2 to 11

12. club or heart (a) $1/2$ (b) 1 to 1

13. spade or face card (a) $11/26$ (b) 11 to 15

14. not a heart, or a 7 (a) $10/13$ (b) 10 to 3

15. neither a heart nor a 7 (a) $9/13$ (b) 9 to 4

CONCEPTUAL WRITING CHALLENGING SCIENTIFIC CALCULATOR GRAPHING CALCULATOR

For the experiment of rolling an ordinary pair of dice, find the probability that the sum will be each of the following. (You may want to construct a table showing the sum for each of the 36 equally likely outcomes.)

16. 11 or 12 **1/12**

17. even or a multiple of 3 **2/3**

18. odd or greater than 9 **11/18**

19. less than 3 or greater than 9 **7/36**

20. Find the probability of getting a prime number in each of the following cases.
(a) A number is chosen randomly from the set $\{1, 2, 3, 4, \ldots, 12\}$. **5/12**
(b) Two dice are rolled and the sum is observed. **5/12**

21. Suppose, for a given experiment, A, B, C, and D are events, all mutually exclusive of one another, such that $A \cup B \cup C \cup D = S$ (the sample space). By extending the special addition rule of probability to this case, and utilizing probability property 3, what statement can you make? $P(A) + P(B) + P(C) + P(D) = 1$

If you are dealt a 5-card hand (this implies without replacement) from a standard 52-card deck, find the probability of getting each of the following. Refer to Table 3 of the previous section, and give answers to six decimal places.

22. a flush or three of a kind **.023094**

23. a full house or a straight **.005365**

24. a black flush or two pairs **.048522**

25. nothing any better than two pairs **.971285**

The table at the side gives golfer Jennifer Twomey's probabilities of scoring in various ranges on a par-70 course. In a given round, find the probability that her score will be each of the following (Exercises 26–30).

x	$P(x)$
Below 60	.04
60–64	.06
65–69	.14
70–74	.30
75–79	.23
80–84	.09
85–89	.06
90–94	.04
95–99	.03
100 or above	.01

26. 95 or higher **.04**

27. par or above **.76**

28. in the 80s **.15**

29. not in the 90s **.93**

30. not in the 70s, 80s, or 90s **.25**

31. What are the odds of Jennifer's shooting below par? **6 to 19**

32. Anne Kelly randomly chooses a single ball from the urn shown here, and x represents the color of the ball chosen. Construct a complete probability distribution for the random variable x.

x	$P(x)$
red	4/11
yellow	5/11
blue	2/11

33. Let x denote the sum of two distinct numbers selected randomly from the set $\{1, 2, 3, 4, 5\}$. Construct the probability distribution for the random variable x.

x	$P(x)$
3	.1
4	.1
5	.2
6	.2
7	.2
8	.1
9	.1

34. Toss a pair of dice 50 times, keeping track of the number of times the sum is "less than 3 or greater than 9" (that is, 2, 10, 11, or 12).

(a) From your results, calculate an empirical probability for the event "less than 3 or greater than 9." **Answers will vary.**

(b) By how much does your answer differ from the *theoretical* probability of Exercise 19? **Answers will vary.**

◉ *For Exercises 35–38, let A be an event within the sample space S, and let* $n(A) = a$ *and* $n(S) = s$.

35. Use the complements principle of counting to evaluate $n(A')$. $n(A') = s - a$

36. Use the theoretical probability formula to express $P(A)$ and $P(A')$.

$P(A) = \frac{a}{s}$; $P(A') = \frac{s - a}{s}$

37. Evaluate, and simplify, $P(A) + P(A')$. $P(A) + P(A') = 1$

38. What rule have you proved? **complements rule of probability**

Suppose we want to form three-digit numbers using the set of digits {0, 1, 2, 3, 4, 5}. *For example,* 501 *and* 224 *are such numbers but* 035 *is not.*

39. How many such numbers are possible? **180**

40. How many of these numbers are multiples of 5? **60**

41. If one three-digit number is chosen at random from all those that can be made from the above set of digits, find the probability that the one chosen is not a multiple of 5. **2/3**

▲ **42.** An experiment consists of spinning both spinners shown here and multiplying the resulting numbers together. Find the probability that the resulting product will be even.

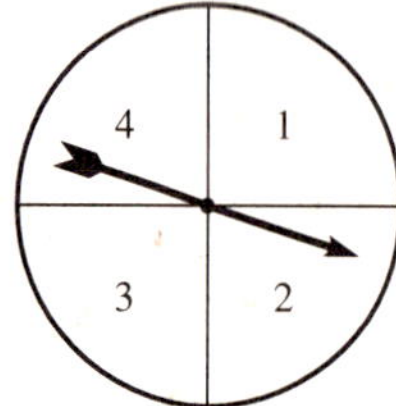

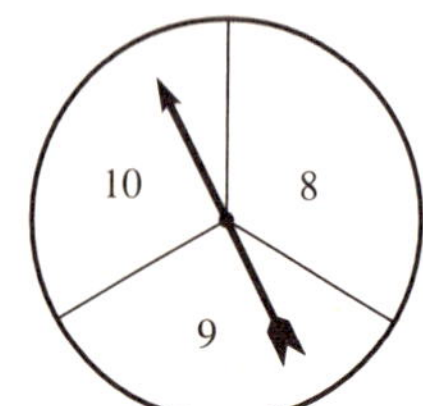

5/6

▲ **43.** A bag contains fifty blue and fifty green marbles. Two marbles at a time are randomly selected. If they are both green, they are placed in box A; if both blue, in box B; if one is green and the other blue, in box C. After all marbles are drawn, what is the probability that the numbers of marbles in box A and box B are the same? (This problem is borrowed with permission from the December 1992 issue of *Mathematics Teacher*, page 736.) **1**

11.3 Exercises (Text Page 645)

For each of the following experiments, determine whether the two given events are independent.

1. A fair coin is tossed twice. The events are "head on the first" and "head on the second." independent
2. A pair of dice are rolled. The events are "even on the first" and "odd on the second." independent
3. Two cards are drawn, with replacement, from a standard 52-card deck. The events are "first card red" and "second card a club." independent
4. Two cards are drawn, without replacement, from a standard deck. The events are "first card an ace" and "second card a face card." not independent
5. The answers are all guessed on a twenty-question multiple-choice test. The events are "first answer correct" and "last answer correct." independent
6. A committee of five is randomly selected from the 100 U.S. Senators. The events are "first member selected is a Republican" and "second member selected is a Republican." (Assume that there are both Republicans and non-Republicans in the Senate.) not independent

One hundred college seniors attending a career fair at a major northeastern university were categorized according to sex and according to primary career motivation, as summarized here.

		Primary Career Motivation			
		Money	**Allowed to Be Creative**	**Sense of Giving to Society**	**Total**
Sex	**Male**	15	22	18	55
	Female	10	15	20	45
	Total	25	37	38	100

If one of these students is to be selected at random, find the probability that the student selected will satisfy each of the following conditions.

7. female $45/100 = 9/20$
8. motivated primarily by creativity $37/100$
9. not motivated primarily by money $75/100 = 3/4$
10. male and motivated primarily by money $15/100 = 3/20$
11. male, given that primary motivation is a sense of giving to society $18/38 = 9/19$
12. motivated primarily by money or creativity, given that the student is a female $25/45 = 5/9$

Suppose two cards are drawn with replacement *from a standard 52-card deck. Find the probability of each of the following events.*

13. a red card on the first draw, and a heart on the second $(26/52) \cdot (13/52) = 1/8$
14. both red and neither one a heart $(13/52) \cdot (13/52) = 1/16$
15. a 5 on the first draw, and an ace on the second $(4/52) \cdot (4/52) = 1/169$
16. one 5 and one ace $2 \cdot (1/169) = 2/169$

Let two cards be dealt successively, without replacement, *from a standard 52-card deck. Find the probability of each of the following events (Exercises 17–27).*

17. spade second, given spade first 12/51 = 4/17

18. club second, given diamond first 13/51

19. ace second, given face card first 4/51

20. two face cards (12/52) · (11/51) = 11/221

21. spade first, then red card
(13/52) · (26/51) = 13/102

22. a jack, then a queen (4/52) · (4/51) = 4/663

23. one jack and one queen 2 · (4/663) = 8/663

24. no face cards (40/52) · (39/51) = 10/17

25. The first card dealt is a jack and the second is a face card. (4/52) · (11/51) = 11/663

26. The first card dealt is a face card and the second is a jack.
(4/52) · (3/51) + (8/52) · (4/51) = 11/663

27. The first card dealt is a club and the second is a king.
(1/52) · (3/51) + (12/52) · (4/51) = 1/52

28. Given events A and B within the sample space S, the following sequence of steps establishes formulas that can be used to compute conditional probabilities. (The formula of step (b) involves probabilities, while that of step (d) involves cardinal numbers.) Justify each statement.

(a) $P(A \text{ and } B) = P(A) \cdot P(B \mid A)$
general multiplication rule of probability

(b) Therefore, $P(B \mid A) = \dfrac{P(A \text{ and } B)}{P(A)}$.
Divide both sides of the equation in (a) by $P(A)$.

(c) Therefore, $P(B \mid A) = \dfrac{n(A \text{ and } B)/n(S)}{n(A)/n(S)}$.
theoretical probability formula

(d) Therefore, $P(B \mid A) = \dfrac{n(A \text{ and } B)}{n(A)}$.
Multiply the numerator and denominator of the expression in (c) by $n(S)$.

Use the results of Exercise 28 to find each of the following probabilities when a single card is drawn from a standard 52-card deck.

29. $P(\text{queen} \mid \text{face card})$ 1/3

30. $P(\text{face card} \mid \text{queen})$ 1

31. $P(\text{red} \mid \text{diamond})$ 1

32. $P(\text{diamond} \mid \text{red})$ 1/2

33. If one number is chosen randomly from the integers 1 through 10, the probability of getting a number that is *odd and prime,* by the general multiplication rule, is

$$P(\text{odd}) \cdot P(\text{prime} \mid \text{odd}) = \frac{5}{10} \cdot \frac{3}{5} = \frac{3}{10}.$$

Compute, and compare, the product $P(\text{prime}) \cdot P(\text{odd} \mid \text{prime})$. 3/10 (the same)

34. What does Exercise 33 imply, in general, about the probability of an event of the form *A and B*?

Sarah Devaney manages a discount supermarket which encourages volume shopping on the part of its customers. Sarah has discovered that, on any given weekday, 60 percent of the market's sales amount to more than \$50. That is, any given sale on such a day has a probability of .60 of being for more than \$50. Find the probability of each of the following events. (Give answers to two decimal places.)

35. The first two sales on Wednesday are both for more than \$50. .36

36. The first three sales on Wednesday are all for more than \$50. .22

37. None of the first three sales on Wednesday is for more than \$50. .06

38. Exactly one of the first three sales on Wednesday is for more than \$50. .29

One problem encountered by developers of the space shuttle program is air pollution in the area surrounding the launch site. A certain direction from the launch site is considered critical in terms of hydrogen chloride pollution from the exhaust cloud. It has been determined that weather conditions would cause emission cloud movement in the critical direction only 5% of the time.

Find probabilities for the following events (Exercises 39–42), assuming that probabilities for a particular launch in no way depend on the probabilities for other launches. (Give answers to two decimal places.)

39. A given launch will not result in cloud movement in the critical direction. .95

40. No cloud movement in the critical direction will occur during any of 5 launches. .77

41. Any 5 launches will result in at least one cloud movement in the critical direction. .23

42. Any 10 launches will result in at least one cloud movement in the critical direction. .40

43. One of the authors of this text has three sons and no daughters. Find the probability that a couple having three children will have all boys. 1/8

44. In Example 7, where Anne draws three balls without replacement, what would be her probability of getting one of each color, where the order does not matter? 8/33

Five men and three women are waiting to be interviewed for jobs. If they are all selected in random order, find the probability of each of the following events.

45. All the women will be interviewed first. 1/56

46. All the men will be interviewed first. 1/56

47. The first person interviewed will be a woman. 3/8

48. The second person interviewed will be a woman. 3/8

49. The last person interviewed will be a woman. 3/8

Let A and B be events, neither one certain and neither one impossible. For each statement in Exercises 50 and 51, **(a)** *determine whether it is true, and* **(b)** *explain why or why not.*

50. If A and B are mutually exclusive, then A and B are also independent. (a) not true

51. If A and B are independent, then A and B are also mutually exclusive. (a) not true

52. Roll a pair of dice until a sum of seven appears, keeping track of how many rolls it took. Repeat the process a total of 50 times, each time recording the number of rolls it took to get a sum of seven.

(a) Use your experimental data to compute an empirical probability (to two decimal places) that it would take at least three rolls to get a sum of seven. Answers will vary.

(b) Find the theoretical probability (to two decimal places) that it would take at least three rolls to obtain a sum of seven. $25/36 \approx .69$

53. Assuming boy and girl babies are equally likely, find the probability that it would take

(a) at least three births to obtain two girls, 3/4

(b) at least four births to obtain two girls, 1/2

(c) at least five births to obtain two girls. 5/16

54. Cards are drawn, without replacement, from an ordinary 52-card deck.

(a) How many must be drawn before the probability of obtaining at least one face card is greater than 1/2? 3

(b) How many must be drawn before the probability of obtaining at least one king is greater than 1/2? 9

Many everyday decisions, like who will drive to lunch, or who will pay for the coffee, are made by the toss of a (presumably fair) coin and using the criterion "heads, you will; tails, I will." This criterion is not quite fair, however, if the coin is biased (perhaps due to slightly irregular construction or wear). John von Neumann suggested a way to make perfectly fair decisions even with a possibly biased coin. If a coin, biased so that $P(\text{h}) = .5200$ *and* $P(\text{t}) = .4800$, *is tossed twice, find the following probabilities. (Give answers to four decimal places.)*

55. $P(\text{hh})$ .2704 **56.** $P(\text{ht})$ .2496

57. $P(\text{th})$ .2496 **58.** $P(\text{tt})$ .2304

59. Having completed Exercises 55–58, what do you think was von Neumann's scheme?

A certain brand of automatic garage door opener utilizes a transmitter control with six independent switches, each one set on or off. The receiver (wired to the door) must be set with the same pattern as the transmitter. Exercises 60–63 are based on ideas similar to those of the "birthday problem" in the "For Further Thought" feature in this section.*

60. How many different ways can the owner of one of these garage door openers set the switches? $2^6 = 64$

61. If two residents in the same neighborhood each have one of this brand of opener, and both set the switches randomly, what is the probability to four decimal places that they will be opening each other's garage doors? $1/64 \approx .0156$

62. If five neighbors with the same type of opener set their switches independently, what is the probability of at least one pair of neighbors using the same settings? (Give your answer to four decimal places.) .1479

* For more information, see "Matching Garage-Door Openers," by Bonnie H. Litwiller and David R. Duncan in the March 1992 issue of *Mathematics Teacher*, page 217.

63. What is the minimum number of neighbors who must use this brand of opener before the probability of at least one duplication of settings is greater than 1/2? 10

64. There are three cards, one that is green on both sides, one that is red on both sides, and one that is green on one side and red on the other. One of the three cards is selected randomly and laid on the table, with a red side up. What is the probability that it is also red on the other side? 2/3

In November, the rain in a certain valley tends to fall in storms of several days' duration. The unconditional probability of rain on any given day of the month is 0.500. But the probability of rain on a day that follows a rainy day is 0.700, and the probability of rain on a day following a nonrainy day is 0.400. Find the probability of each of the following events. Give answers to three decimal places.

65. rain on two consecutive days in November .350

66. rain on three consecutive days in November .245

67. rain on November 1 and 2, but not on November 3 .105

68. rain on the first four days of November, given that October 31 was clear all day .137

69. rain throughout the first week of November, given that there was no rain on October 31 .047

70. no rain in the first week of November, given that there was rain on October 31 .014

In a certain four-engine vintage aircraft, now quite unreliable, each engine has a 10 percent chance of failure on any flight, as long as it is carrying its one-fourth share of the load. But if one engine fails, then the chance of failure increases to 20 percent for each of the other three engines. And if a second engine fails, each of the remaining two has a 30 percent chance of failure. Assuming that no two engines ever fail simultaneously, and that the aircraft can continue flying with as few as two operating engines, find each of the following for a given flight of this aircraft. (Give answers to four decimal places.)

71. the probability of no engine failures
$(.90)^4 = .6561$

72. the probability of exactly one engine failure
$C(4, 1) \cdot (.10) \cdot (.80)^3 = .2048$

73. the probability of exactly two engine failures
$C(4, 2) \cdot (.10) \cdot (.20) \cdot (.70)^2 = .0588$

74. the probability of a failed flight
$1 - (.6561 + .2048 + .0588) = .0803$

In basketball "one-and-one" foul shooting is done as follows: if the player makes the first shot (1 point), he is given a second shot. If he misses the first shot, he is not given a second shot (see the tree diagram).

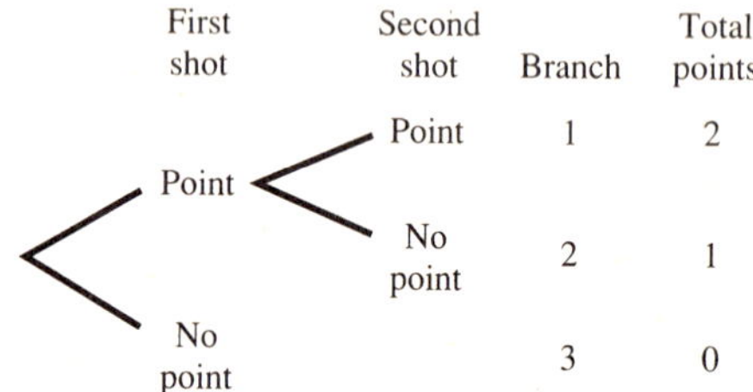

Karin Wagner, a basketball player, has a 60% foul shot record. (She makes 60% of her foul shots.) Find the probability that, on a given one-and-one foul shooting opportunity, Karin will score the following numbers of points.

75. no points .40

76. one point .24

77. two points .36

11.4 Exercises (Text Page 652)

For Exercises 1–24, give all numerical answers as common fractions reduced to lowest terms. For Exercises 25–59, give all numerical answers to three decimal places.

If three fair coins are tossed, find the probability of each of the following numbers of heads.

1. 0 1/8
2. 1 3/8
3. 2 3/8
4. 3 1/8
5. 1 or 2 3/4
6. at least 1 7/8
7. no more than 1 1/2
8. fewer than 3 7/8

9. Assuming boy and girl babies are equally likely, find the probability that a family with three children will have exactly two boys. 3/8

10. Pascal's triangle was shown in Table 5 of Section 4 in the previous chapter. Explain how the probabilities in Exercises 1–4 above relate to row 3 of the "triangle." (Recall that we referred to the topmost row of the triangle as "row number 0," and to the leftmost entry of each row as "entry number 0.")

11. Generalize the pattern of Exercise 10 to complete the following statement. If n fair coins are tossed, the probability of exactly x heads is the fraction whose numerator is entry number ____ of row number ____ in Pascal's triangle, and whose denominator is the sum of row number ____. x; n; n

Use the pattern noted in Exercises 10 and 11 to find the probabilities of the following numbers of heads when seven fair coins are tossed.

12. 0 1/128 **13.** 1 7/128 **14.** 2 21/128 **15.** 3 35/128
16. 4 35/128 **17.** 5 21/128 **18.** 6 7/128 **19.** 7 1/128

A fair die is rolled three times. A 4 is considered "success," while all other outcomes are "failures." Find the probability of each of the following numbers of successes.

20. 0 125/216 **21.** 1 25/72 **22.** 2 5/72 **23.** 3 1/216

24. Exercises 10 and 11 established a way of using Pascal's triangle rather than the binomial probability formula to find probabilities of different numbers of successes in coin tossing experiments. Explain why the same process would not work for Exercises 20–23.

In the remaining exercises of this section, give all numerical answers to three decimal places.

For n repeated independent trials, with constant probability of success p for all trials, find the probability of exactly x successes in each of the following cases.

25. $n = 5$, $p = 1/3$, $x = 4$ .041

26. $n = 10$, $p = .7$, $x = 5$ .103

27. $n = 20$, $p = 1/8$, $x = 2$ .268

28. $n = 30$, $p = .6$, $x = 22$ .050

For Exercises 29–31, refer to Example 4 in the text.

29. Does Rudy's probability of a hit really remain constant at exactly .300 through all ten times at bat? Explain your reasoning.

30. If Rudy's batting average is exactly .300 going into the series, and that value is based on exactly 1,200 career hits out of 4,000 previous times at bat, what is the greatest his average could possibly be (to three decimal places) when he goes up to bat the tenth time of the series? What is the least his average could possibly be when he goes up to bat the tenth time of the series? .302; .299

31. Do you think the use of the binomial probability formula was justified in Example 4, even though p is not strictly constant? Explain your reasoning.

Robin Strang is taking a ten-question multiple-choice test for which each question has three answer choices, only one of which is correct. Robin decides on answers by rolling a fair die and marking the first answer choice if the die shows 1 or 2, the second if it shows 3 or 4, and the third if it shows 5 or 6. Find the probability of each of the following events.

32. exactly four correct answers .228

33. exactly seven correct answers .016

34. fewer than three correct answers .299

35. at least seven correct answers .020

It is known that a certain prescription drug produces undesirable side effects in 30 percent of all patients who use it. Among a random sample of eight patients using the drug, find the probability of each of the following events.

36. None have undesirable side effects. .058

37. Exactly one has undesirable side effects. .198

38. Exactly two have undesirable side effects. .296

39. More than two have undesirable side effects. .448

In a certain state, it has been shown that only 50 percent of the high school graduates who are capable of college work actually enroll in colleges. Find the probability that, among nine capable high school graduates in this state, each of the following numbers will enroll in college (Exercises 40–43).

40. exactly 4 .246

41. from 4 through 6 .656

42. none .002

43. all 9 .002

44. At a large midwestern university, 40 percent of all students have their own personal computers. If five students at that university are selected at random, find the probability that exactly three of them have their own computers. .230

45. If it is known that 65 percent of all orange trees will survive a hard frost, then what is the probability that at least half of a group of six trees will survive such a frost? .883

46. An extensive survey revealed that, during a certain Presidential election campaign, 44 percent of the political columns in a certain group of major newspapers were favorable to the incumbent President. If a sample of fifteen of these columns is selected at random, what is the probability that exactly six of them will be favorable? .197

In the case of n independent repeated Bernoulli trials, the formula developed in this section gives the probability of exactly x successes. Sometimes, however, we are interested not in the event that exactly x successes will occur in n trials, but rather the event that the first success will occur on the xth trial. For example, consider the probability that, in a series of coin tosses, the first success (head) will occur on the fourth toss. This implies a failure first, then a second failure, then a third failure, and finally a success. Symbolically, the event is F_1 and F_2 and F_3 and S_4. The probability of this sequence of outcomes is $q \cdot q \cdot q \cdot p$, or $q^3 \cdot p$. In general, if the probability of success stays constant at p (which implies a probability of failure of $q = 1 - p$), then the probability that the first success will occur on the xth trial can be computed as follows.

$$P(F_1 \text{ and } F_2 \text{ and } \ldots \text{ and } F_{x-1} \text{ and } S_x) = q^{x-1} \cdot p.$$

47. Explain why, in the formula above, there is no combination factor, such as the $C(n, x)$ in the binomial probability formula.

48. Assuming male and female babies are equally likely, find the probability that a family's fourth child will be their first daughter. .063

49. If the probability of getting caught when you exceed the speed limit on a certain stretch of highway is .35, find the probability that the first time you will get caught is the fifth time that you exceed the speed limit. .062

50. If 20 percent of all workers in a certain industry are union members, and workers in this industry are selected successively at random, find the probability that the first union member to occur will be on the sixth selection. .066

51. If a certain type of rocket always has a three percent chance of an aborted launching, find the probability that the first launch to be aborted is the 25th launch. .014

Harvey is standing on the corner tossing a coin. He decides he will toss it 10 times, each time walking 1 block north if it lands heads up, and 1 block south if it lands tails up. In each of the following exercises, find the probability that he will end up in the indicated location. (In each case, ask how many successes, say heads, would be required and use the binomial formula. Some ending positions may not be possible with 10 tosses.) The random process involved here illustrates what we call a **random walk.** *It is a simplified model of Brownian motion, mentioned in the chapter introduction. Further applications of the idea of a random walk are found in the section on simulation at the end of this chapter.*

52. 10 blocks north of his corner $1/1{,}024 \approx .001$

53. 6 blocks north of his corner $45/1{,}024 \approx .044$

54. 6 blocks south of his corner $45/1{,}024 \approx .044$

55. 5 blocks south of his corner 0

56. 2 blocks north of his corner $210/1{,}024 = 105/512 \approx .205$

57. at least 2 blocks north of his corner $(210 + 120 + 45 + 10 + 1)/1{,}024 = 193/512 \approx .377$

58. at least 2 blocks from his corner $(1{,}024 - 252)/1{,}024 = 193/256 \approx .754$

59. on his corner $252/1{,}024 = 63/256 \approx .246$

11.5 Exercises (Text Page 660)

1. A couple who are planning their future say, "We expect to have 2.5 daughters." Explain what this statement means.

A certain game consists of rolling a single fair die and pays off as follows: \$3 *for a* 6, \$2 *for a* 5, \$1 *for a* 4, *and no payoff otherwise.*

2. Find the expected winnings for this game. \$1

3. What is a fair price to pay to play this game? \$1

CONCEPTUAL WRITING CHALLENGING SCIENTIFIC CALCULATOR GRAPHING CALCULATOR

For Exercises 4 and 5, consider a game consisting of rolling a single fair die, with payoffs as follows. If an even number of spots turns up, you receive that many dollars. But if an odd number of spots turns up, you must pay that many dollars.

4. Find the expected net winnings of this game. 50¢

5. Is this game fair, or unfair against the player, or unfair in favor of the player? **unfair in favor of the player**

6. A certain game involves tossing 3 fair coins, and it pays 10¢ for 3 heads, 5¢ for 2 heads, and 3¢ for 1 head. Is 5¢ a fair price to pay to play this game? (That is, does the 5¢ cost to play make the game fair?) **no (Expected net winnings: −3/4¢)**

7. In a form of roulette slightly different from that in Example 4, a more generous management supplies a wheel having only thirty-seven compartments, with eighteen red, eighteen black, and one zero. Find the expected net winnings if you bet on red in this game. $-\$1/37 \approx -2.7$¢

If two cards are drawn from a standard 52-card deck, find the expected number of spades in each of the following cases (Exercises 8 and 9).

8. The drawing is done with replacement. 1/2

9. The drawing is done without replacement. 1/2

10. In a certain mathematics class, the probabilities have been empirically determined for various numbers of absences on any given day. These values are shown in the table below. Find the expected number of absences on a given day. (Give the answer to two decimal places.) 1.93

Number absent	0	1	2	3	4
Probability	.12	.27	.30	.18	.13

11. An insurance company will insure a $100,000 home for its total value for an annual premium of $300. If the probability of total loss for such a home in a given year is .002, and you assume that either total loss or no loss will occur, what is the company's expected annual gain (or profit) on each such policy? $100

A college foundation raises funds by selling raffle tickets for a new car worth $25,000.

12. If 500 tickets are sold for $100 each, determine
(a) the expected *net* winnings of a person buying one of the tickets, −$50
(b) the total profit for the foundation, assuming they had to purchase the car, $25,000
(c) the total profit for the foundation, assuming the car was donated. $50,000

13. If 750 tickets are sold for $100 each, determine
(a) the expected *net* winnings of a person buying one of the tickets, −$66.67
(b) the total profit for the foundation, assuming they had to purchase the car, $50,000
(c) the total profit for the foundation, assuming the car was donated. $75,000

Five thousand raffle tickets are sold. One first prize of $1,000, two second prizes of $500 each, and five third prizes of $100 each are to be awarded, with all winners selected randomly.

14. If you purchased one ticket, what are your expected winnings? 50¢

15. If you purchased two tickets, what are your expected winnings? $1

16. If the tickets were sold for $1 each, how much profit goes to the raffle sponsor? $2,500

17. An amusement park, considering adding some new attractions, conducted a study over several typical days and found that, of 10,000 families entering the park, 830 brought just one child (defined as younger than age twelve), 2,370 brought two children, 4,980 brought three children, 1,210 brought four children, 260 brought five children, 180 brought six children, and 170 brought no children at all. Find the expected number of children per family attending this park. (Round your answer to the nearest tenth.) 2.8

18. Five cards are numbered 1 through 5. Two of these cards are chosen randomly (without replacement), and the numbers on them are added. Find the expected value of this sum. 6

19. Most members of the Mathematical Association of America in a recent year paid for membership dues and regular association publications according to the table below, where M indicates the *American Mathematical Monthly*, G indicates *Mathematics Magazine*, and J indicates the *College Mathematics Journal*. The third row of the table shows how many members in a survey of 1,000 fell into each category. For those members surveyed, find the expected amount paid (to the nearest cent). **$118.72**

Membership category	M	G	J	M + G	M + J	G + J	M + G + J
Amount paid	$108	$90	$94	$126	$130	$112	$148
Number in category	97	144	106	51	195	186	221

20. In a certain California city, projections for the next year are that there is a 30% chance that electronics jobs will increase by 1,200, a 50% chance that they will increase by 500, and a 20% chance that they will *decrease* by 800. What is the expected change in the number of electronics jobs in that city in the next year? **an increase of 450**

21. In one version of the game *keno*, the house has a pot containing 80 balls, numbered 1 through 80. A player buys a ticket for $1 and marks one number on it (from 1 to 80). The house then selects 20 of the 80 numbers at random. If the number selected by the player is among the 20 selected by the management, the player is paid $3.20. Find the expected net winnings for this game. **−20¢**

A game show contestant is offered the option of receiving a set of new kitchen appliances worth $1,800, or accepting a chance to win either a luxury vacation worth $4,000 or a boat worth $5,000. The contestant's probabilities of winning the vacation or the boat are .25 and .15, respectively.

22. If the contestant were to turn down the appliances and go for one of the other prizes, what would be the expected winnings? **$1,750**

▲ 23. Purely in terms of monetary value, what is the contestant's wiser choice? **Accept the appliances (since $1,800 > $1,750).**

*The table below illustrates how a salesman for Levi Strauss & Co. rates his accounts by considering the existing volume of each account plus potential additional volume.**

1	2	3	4	5	6	7
Account Number	**Existing Volume**	**Potential Additional Volume**	**Probability of Getting Additional Volume**	**Expected Value of Additional Volume**	**Existing Volume plus Expected Value of Additional Volume**	**Classification**
1	$15,000	$ 10,000	.25	$2,500	$17,500	
2	40,000	0	—	—	40,000	
3	20,000	10,000	.20	2,000		
4	50,000	10,000	.10	1,000		
5	5,000	50,000	.50			
6	0	100,000	.60			
7	30,000	20,000	.80			

* This information was provided by James McDonald of Levi Strauss & Co., San Francisco.

▲ *Use the table to work Exercises 24–27.*

24. Compute the missing expected values in column 5. **column 5: 25,000, 60,000, 16,000**

25. Compute the missing amounts in column 6. **column 6: 22,000, 51,000, 30,000, 60,000, 46,000**

26. In column 7, classify each account according to this scheme: Class A if the column 6 value is $55,000 or more; Class B if the column 6 value is at least $45,000 but less than $55,000; Class C if the column 6 value is less than $45,000. **column 7: C, C, C, B, C, A, B**

27. Considering all seven of this salesman's accounts, compute the total additional volume he can "expect" to get. **$106,500**

▲ **28.** Recall that in the game keno of Exercise 21, the house randomly selects 20 numbers from the counting numbers 1–80. In the variation called 6-spot keno, the player pays 60¢ for his ticket and marks 6 numbers of his choice. If the 20 numbers selected by the house contain at least 3 of those chosen by the player, he gets a payoff according to this scheme.

3 of the player's numbers among the 20	$ .35
4 of the player's numbers among the 20	2.00
5 of the player's numbers among the 20	60.00
6 of the player's numbers among the 20	1,250.00

Find the player's expected net winnings in this game. **about −15¢**

11.6 Exercises (Text Page 665)

1. Explain why, in Example 1, fifty tosses of the coins produced only 48 sets of three successive offspring.

2. Use the sequence of flower colors of Example 1 to approximate the probability that *four* successive offspring will all have red flowers. **15/47 ≈ .319**

3. Should the probability of two successive girl births be any different from that of two successive boy births? **no**

4. Simulate 40 births by tossing coins yourself, and obtain an empirical probability for two successive girls. **Answers will vary.**

5. Use Table 10 to simulate fifty families with three children. Let 0–4 correspond to boys and 5–9 to girls, and use the middle grouping of three digits (159, 787, 650, and so on). Estimate the probability of exactly two boys in a family of three children. Compare the estimation with the theoretical probability of exactly two boys in a family of three children, which is 3/8 = .375. **18/50 = .36 (This is quite close to .375, the theoretical value.)**

Exercises 75–77 of Section 3 involved one-and-one foul shooting in basketball. Karin, who had a 60% foul-shooting record, had probabilities of scoring 0, 1, *or* 2 *points of* .40, .24, *and* .36, *respectively. Use Table 10 (with digits 0–5 representing hit and 6–9 representing miss) to simulate 50 one-and-one shooting opportunities for Karin. Begin at the top of the first column (5, 7, 3, etc., to the bottom), then move to the second column (1, 7, 6, etc.), going until 50 one-and-one opportunities are obtained. (Notice that some "opportunities" involve one shot while others involve two shots.) Keep a tally of the numbers of times* 0, 1, *and* 2 *points are scored. From the tally, find the empirical probability that, on a given opportunity, Karin will score as follows. (Round your answers to two decimal places.)*

Number of Points	Tally
0	
1	
2	

6. no points **18/50 = .36**

7. 1 point **13/50 = .26**

8. 2 points **19/50 = .38**

Exercises 52–59 of Section 4 illustrated a simple version of the idea of a "random walk." Atomic particles released in nuclear fission also move in a random fashion. During World War II, John von Neumann and Stanislaw Ulam used simulation with random numbers to study particle motion in nuclear reactions. Von Neumann coined the name "Monte Carlo" for the methods used, and since then the terms "Monte Carlo methods" and "simulation methods" have often been used with very little distinction.

The figure below suggests a model for random motion in two dimensions. Assume that a particle moves in a series of 1-unit "jumps," each one in a random direction, any one of 12 *equally likely possibilities. One way to choose directions is to roll a fair die and toss a fair coin. The die determines one of the directions 1–6, coupled with heads on the coin. Tails on the coin reverses the direction of the die, so that the die coupled with tails gives directions 7–12. Thus* 3h *(meaning* 3 *with the die and heads with the coin) gives direction* 3; 3t *gives direction* 9 *(opposite to* 3*); and so on.*

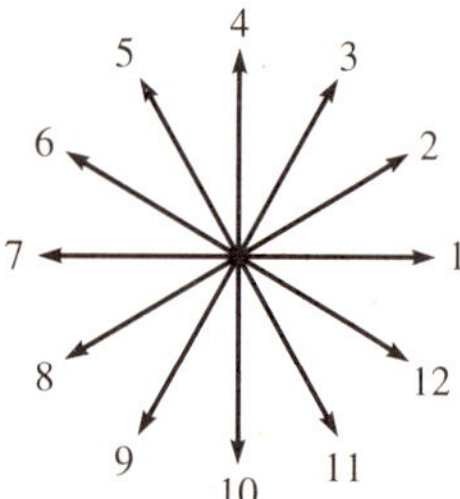

9. Simulate the motion described above with 10 rolls of a die (and tosses of a coin). Draw the 10-jump path you get. Make your drawing accurate enough so you can estimate (by measuring) how far from its starting point the particle ends up. **Answers will vary.**

10. Repeat the experiment of Exercise 9 four more times. Measure distance from start to finish for each of the 5 "random trips." Add these 5 distances and divide the sum by 5, to arrive at an "expected net distance" for such a trip. **Answers will vary.**

Consider another two-dimensional random walk governed by the following conditions.

1. *Start out from a given street corner, and travel one block north. At each intersection:*
2. *Turn left with probability 1/6.*
3. *Go straight with probability 2/6 (= 1/3).*
4. *Turn right with probability 3/6 (= 1/2).*

(Never turn around.)

11. Explain how a fair die could be used to simulate this random walk.

12. Use Table 10 to simulate this random walk. For every 1 encountered in the table, turn left and proceed for another block. For every 2 or 3, go straight and proceed for another block. For every 4, 5, or 6, turn right and proceed for another block. Disregard all other digits. (Do you see how this scheme satisfies the probabilities given above?) This time begin at the upper right corner of the table, running down the column 9, 7, 7, and so on, to the bottom. Then start at the top of the next column to the left, 1, 0, 0, and so on, to the bottom. When these two columns of digits are used up, stop the "walk." Describe, in terms of distance and direction, where you have ended up relative to your starting point. **The walk ends $3\sqrt{2}$ blocks northwest of the starting point (that is, 3 blocks west and 3 blocks north).**

Chapter 11 Test (Text Page 667)

1. Explain the difference between *empirical* and *theoretical* probability.

2. State the *law of large numbers*, and use coin tossing to illustrate it.

A single card is chosen at random from a standard 52-card deck. Find the odds against its being each of the following.

3. a heart **3 to 1** **4.** a red queen **25 to 1**

5. a king or a black face card **11 to 2**

CONCEPTUAL  WRITING  CHALLENGING  SCIENTIFIC CALCULATOR  GRAPHING CALCULATOR

The chart below represents genetic transmission of cystic fibrosis. C *denotes a normal gene while* c *denotes a cystic fibrosis gene. (Normal is dominant.) Both parents in this case are* Cc, *which means that they inherited one of each gene, and are therefore carriers but do not have the disease.*

		Second Parent	
		C	c
First Parent	C		Cc
	c		

6. Complete the chart, showing all four equally likely gene combinations.

	C	c
C	CC	Cc
c	cC	cc

7. Find the probability that a child of these parents will also be a carrier without the disease. 1/2

8. What are the odds against a child of these parents actually having cystic fibrosis? 3 to 1

The manager of a pizza parlor (which operates seven days a week) allows each of three employees to select one day off next week. Assuming the selection is done randomly, find the probability of each of the following events.

9. All three select different days.
$(7/7) \cdot (6/7) \cdot (5/7) = 30/49$

10. All three select the same day.
$(7/7) \cdot (1/7) \cdot (1/7) = 1/49$

▲ **11.** Exactly two of them select the same day.
$1 - (30/49 + 1/49) = 18/49$

Two distinct numbers are randomly selected from the set {1, 2, 3, 4, 5}. *Find the probability of each of the following events.*

12. Both numbers are even. $C(2, 2)/C(5, 2) = 1/10$

13. Both numbers are prime. $C(3, 2)/C(5, 2) = 3/10$

14. The sum of the two numbers is odd. $6/10 = 3/5$

15. The product of the two numbers is odd. 3/10

A three-member committee is selected randomly from a group consisting of three men and two women.

▲ **16.** Let x denote the number of men on the committee, and complete the probability distribution table.

x	$P(x)$
0	0
1	
2	
3	

x	$P(x)$
0	0
1	3/10
2	6/10
3	1/10

17. Find the probability that the committee members are not all men. 9/10

18. Find the expected number of men on the committee.
$18/10 = 9/5$

A pair of dice are rolled. Find the following.

19. the odds against "doubles" (the same number on both dice) 5 to 1

20. the probability of a sum greater than 2
$1 - 1/36 = 35/36$

21. the odds against a sum of "7 or 11" 7 to 2

22. the probability of a sum that is even and less than 5
$4/36 = 1/9$

Jeff Cole has an .82 chance of making par on each hole of golf that he plays. Today he plans to play just three holes. Find the probability of each of the following events. Round answers to three decimal places.

23. He makes par on all three holes. $(.82)^3 \approx .551$

24. He makes par on exactly two of the three holes.
$C(3, 2)(.82)^2(.18) \approx .363$

25. He makes par on at least one of the three holes.
$1 - (.18)^3 \approx .994$

26. He makes par on the first and third holes but not on the second. $(.82)(.18)(.82) \approx .121$

Two cards are drawn, without replacement, from a standard 52-card deck. Find the probability of each of the following events.

27. Both cards are red. 25/102

28. Both cards are the same color. 25/51

29. The second card is a queen, given that the first card is an ace. 4/51

30. The first card is a face card and the second is black. 3/26

Refer to the sequence of 40 coin tosses in Example 2 of Section 6.

31. How many triples of 3 successive births are listed? 38

32. How many of those triples consist of all girls? 1

33. What is the empirical probability that 3 successive births will be all girls? .026

CHAPTER 12 Statistics

12.1 Exercises (Text Page 678)

In Exercises 1 and 2, use the given data to do the following:

(a) *Construct frequency and relative frequency distributions, in a table similar to Table 2.*
(b) *Construct a histogram.*
(c) *Construct a frequency polygon.*

1. The following data are the numbers of questions omitted on a math exam by the 30 members of the class.

1 0 3 0 0 2 1 2 2 0 0 5 1 1 3
4 2 0 2 0 1 0 1 2 3 3 4 0 1 0

(a)

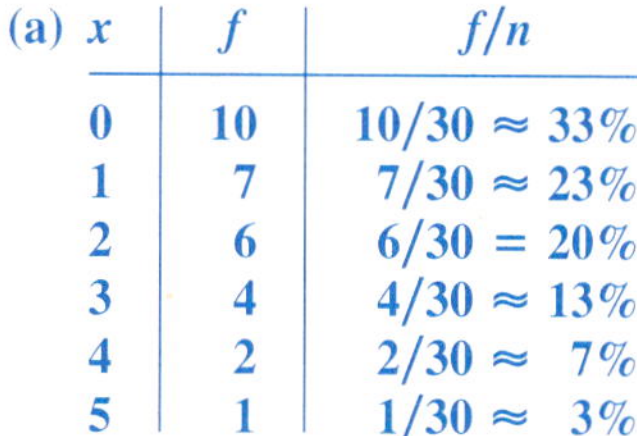

x	f	f/n
0	10	$10/30 \approx 33\%$
1	7	$7/30 \approx 23\%$
2	6	$6/30 = 20\%$
3	4	$4/30 \approx 13\%$
4	2	$2/30 \approx 7\%$
5	1	$1/30 \approx 3\%$

(b)

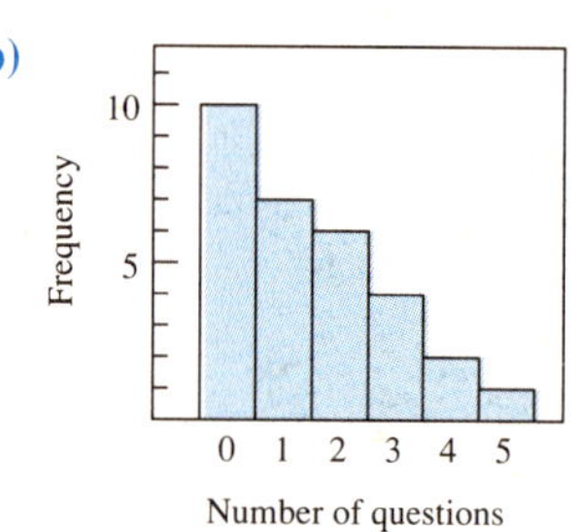

(c)

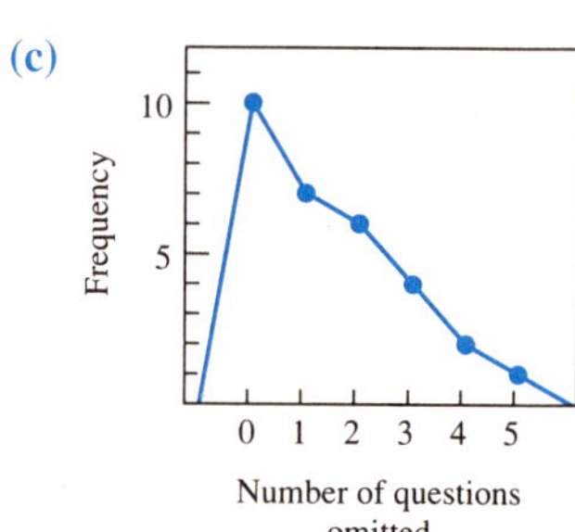

2. The following data are quiz scores for the members of an economics class.

8 5 6 10 4 7 2 7 6 3 1 7 4 9
5 9 2 6 5 4 6 6 8 4 10 8 9 7

(a)

x	f	f/n
1	1	$1/28 \approx 4\%$
2	2	$2/28 \approx 7\%$
3	1	$1/28 \approx 4\%$
4	4	$4/28 \approx 14\%$
5	3	$3/28 \approx 11\%$
6	5	$5/28 \approx 18\%$
7	4	$4/28 \approx 14\%$
8	3	$3/28 \approx 11\%$
9	3	$3/28 \approx 11\%$
10	2	$2/28 \approx 7\%$

(b)

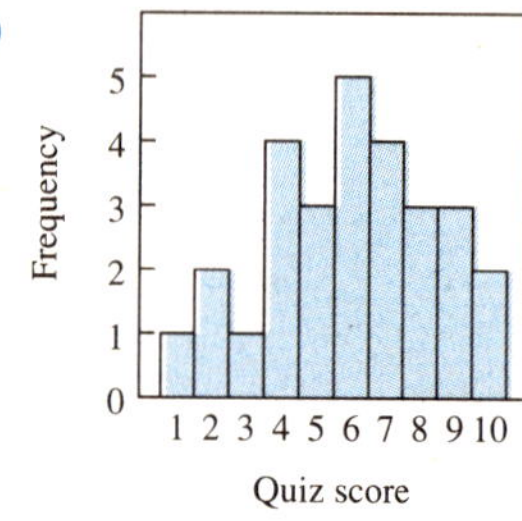

(c)

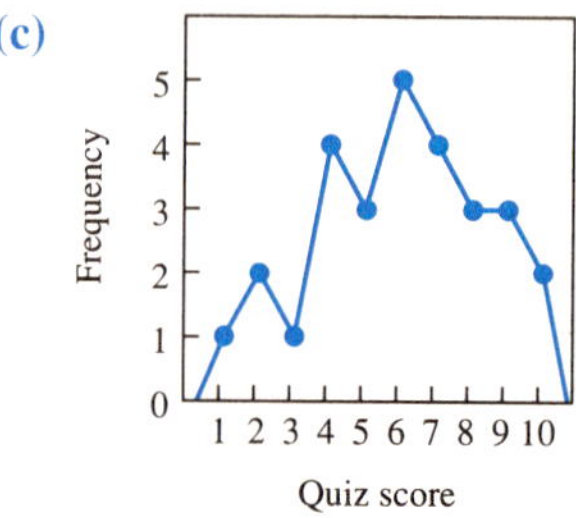

In each of Exercises 3–6, use the given data to do the following:

(a) *Construct grouped frequency and relative frequency distributions, in a table similar to Table 3. (In each case, follow the suggested guidelines for class limits and class width.)*
(b) *Construct a histogram.*
(c) *Construct a frequency polygon.*

3. The heights (in inches) of the 54 starting players in a baseball tournament were as follows.

53	51	65	62	61	55	59	52	62
64	48	54	64	57	51	67	60	49
49	59	54	52	53	60	58	60	64
52	56	56	58	66	59	62	50	58
60	63	64	52	60	58	63	53	56
58	61	55	50	65	56	61	55	54

Use five classes with a uniform class width of 5 inches, and use a lower limit of 45 inches for the first class.

(a)

Class Limits	Tally	Frequency f	Relative Frequency f/n
45–49	|||	3	$3/54 \approx 5.6\%$
50–54	𝍸 𝍸 ||||	14	$14/54 \approx 25.9\%$
55–59	𝍸 𝍸 𝍸 |	16	$16/54 \approx 29.6\%$
60–64	𝍸 𝍸 𝍸 ||	17	$17/54 \approx 31.5\%$
65–69	||||	4	$4/54 \approx 7.4\%$
	Total:	$n = 54$	

(b)

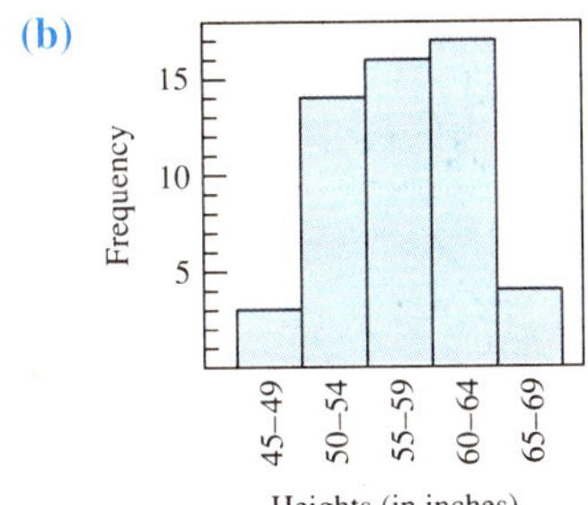

(c)

15
10
5
Frequency
47 52 57 62 67
Heights (in inches)

4. The following raw data represent the monthly account balances (to the nearest dollar) for a sample of 50 brand new charge card users.

138	78	175	46	79	118	90	163	88	107
126	154	85	60	42	54	62	128	114	73
129	130	81	67	119	116	145	105	96	71
100	145	117	60	125	130	94	88	136	112
118	84	74	62	81	110	108	71	85	165

Use seven classes with a uniform width of 20 dollars, where the lower limit of the first class is 40 dollars.

(a)

Class Limits	Tally	Frequency f	Relative Frequency f/n
40–59	|||	3	$3/50 = 6\%$
60–79	~~||||~~ ~~||||~~		11	$11/50 = 22\%$
80–99	~~||||~~ ~~||||~~	10	$10/50 = 20\%$
100–119	~~||||~~ ~~||||~~ ||	12	$12/50 = 24\%$
120–139	~~||||~~ |||	8	$8/50 = 16\%$
140–159	|||	3	$3/50 = 6\%$
160–179	|||	3	$3/50 = 6\%$

Total: $n = 50$

(b)

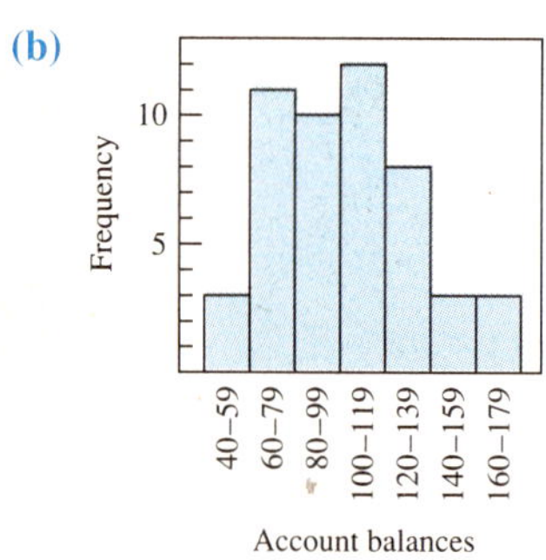

(c)

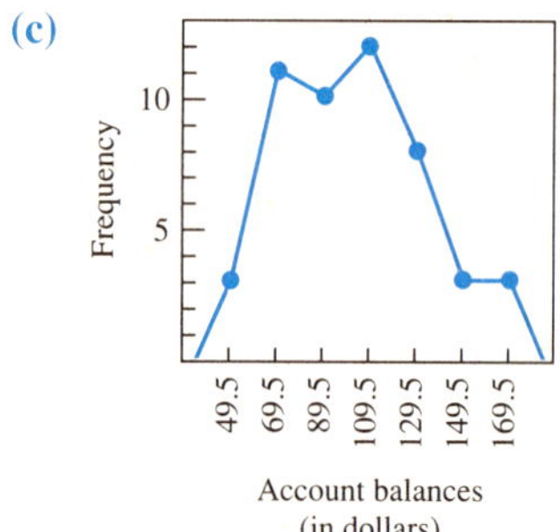

5. The following data represent the daily high temperatures (in degrees Fahrenheit) for the month of June in a southwestern U.S. city.

79	84	88	96	102	104	99	97	92	94
85	92	100	99	101	104	110	108	106	106
90	82	74	72	83	107	111	102	97	94

Use nine classes with a uniform width of 5 degrees, where the lower limit of the first class is 70 degrees.

(a)

Class Limits	Tally	Frequency f	Relative Frequency f/n
70–74	||	2	$2/30 \approx 6.7\%$
75–79	|	1	$1/30 \approx 3.3\%$
80–84	|||	3	$3/30 = 10.0\%$
85–89	||	2	$2/30 \approx 6.7\%$
90–94	~~||||~~	5	$5/30 \approx 16.7\%$
95–99	~~||||~~	5	$5/30 \approx 16.7\%$
100–104	~~||||~~ |	6	$6/30 = 20.0\%$
105–109	||||	4	$4/30 \approx 13.3\%$
110–114	||	2	$2/30 \approx 6.7\%$

Total: $n = 30$

(b)

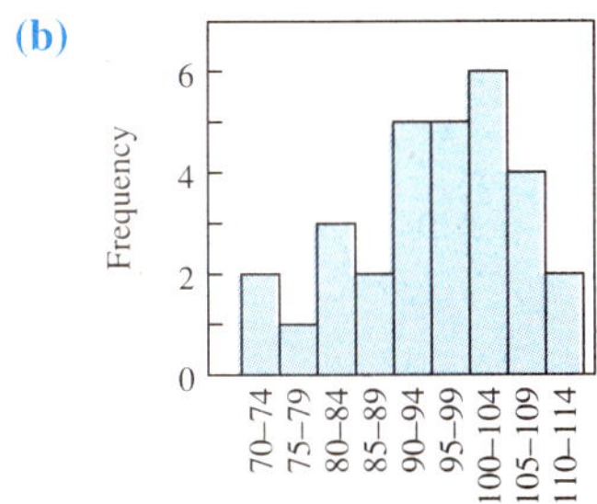

(c)

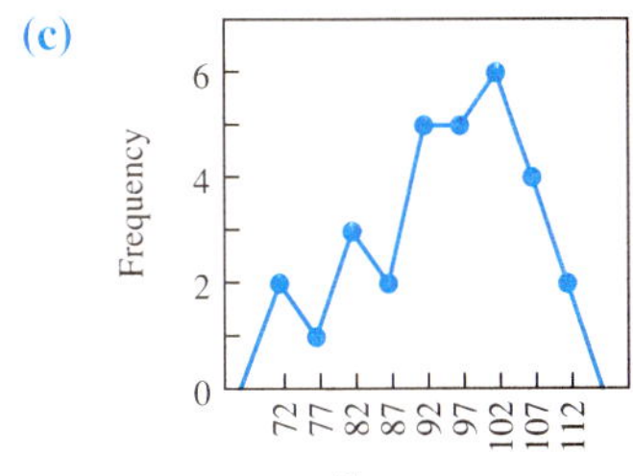

6. The following data represent IQ scores of a group of 50 tenth graders.

113	109	118	92	130	112	114	117	122	115
127	107	108	113	124	112	111	106	116	118
121	107	118	118	110	124	115	103	100	114
104	124	116	123	104	135	121	126	116	111
96	134	98	129	102	103	107	113	117	112

Use nine classes with a uniform width of 5, where the lower limit of the first class is 91.

(a)

Class Limits	Tally	Frequency f	Relative Frequency f/n
91– 95	|	1	1/50 = 2%
96–100	|||	3	3/50 = 6%
101–105	~~||||~~	5	5/50 = 10%
106–110	~~||||~~ ||	7	7/50 = 14%
111–115	~~||||~~ ~~||||~~ ||	12	12/50 = 24%
116–120	~~||||~~ ||||	9	9/50 = 18%
121–125	~~||||~~ ||	7	7/50 = 14%
126–130	||||	4	4/50 = 8%
131–135	||	2	2/50 = 4%
	Total:	$n = 50$	

(b)

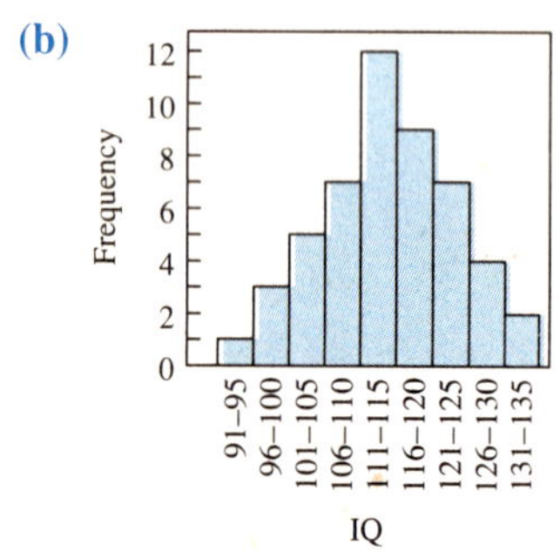

(c)

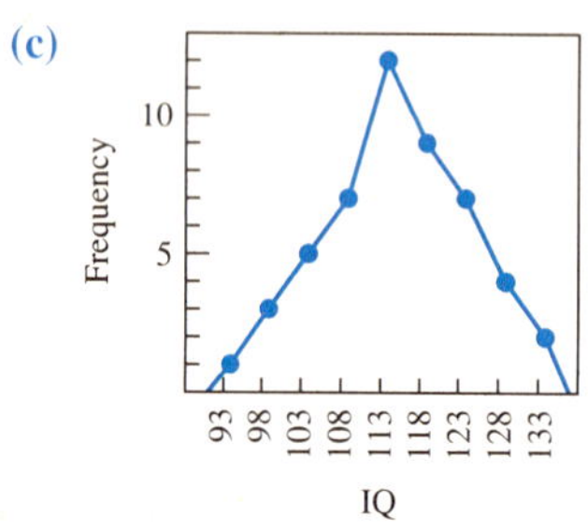

In each of Exercises 7–10, construct a stem-and-leaf display for the given data. In each case, treat the ones digits as the leaves. For any single-digit data, use a stem of 0.

7. On a certain date approaching midseason, the teams in the National Basketball Association had won the following numbers of games.

20	29	11	26	11	12	7	26	18
19	14	13	22	9	25	11	10	15
10	22	23	31	8	24	15	24	15

Stem	Leaves
0	7 9 8
1	1 1 2 8 9 4 3 1 0 5 0 5 5
2	0 9 6 6 2 5 2 3 4 4
3	1

8. The students in a calculus class were asked how many college units they had accumulated to date. Their responses were as follows.

12	4	13	12	21	22	15	17	33	24
32	42	26	11	53	62	42	25	13	8
54	18	21	14	19	17	38	17	20	10

Stem	Leaves
0	4 8
1	2 3 2 5 7 1 3 8 4 9 7 7 0
2	1 2 4 6 5 1 0
3	3 2 8
4	2 2
5	3 4
6	2

9. Following are the daily round-trip distances to school (in miles) for 30 randomly chosen students at a community college in California.

12	30	10	11	18	26	34	18	8	12
26	14	5	22	4	25	9	10	6	21
18	18	9	16	44	23	4	13	36	8

Stem	Leaves
0	8 5 4 9 6 9 4 8
1	2 0 1 8 8 2 4 0 8 8 6 3
2	6 6 2 5 1 3
3	0 4 6
4	4

10. The following data represent net yards gained per game by National Football League running backs who played during a given week of the season.

28	19	36	73	37	88	67	33	54	123	79
12	39	45	22	58	7	73	30	43	24	36
51	43	33	55	40	29	112	60	94	86	62
42	29	18	25	41	3	49	102	16	32	46

Stem	Leaves
0	7 3
1	9 2 8 6
2	8 2 4 9 9 5
3	6 7 3 9 0 6 3 2
4	5 3 3 0 2 1 9 6
5	4 8 1 5
6	7 0 2
7	3 9 3
8	8 6
9	4
10	2
11	2
12	3

Bar graphs (as well as other kinds of graphs) are often drawn in special ways in order to catch attention or to emphasize certain information. The example shown here pertains to the changing nature of U.S. cities throughout this century. (The picket fence is a common symbol of the suburban "American dream.")

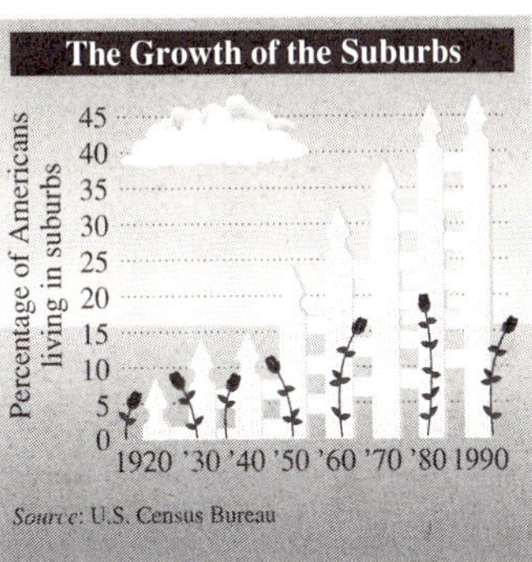

Refer to the graph for Exercises 11–13.

11. About what percentage of Americans lived in suburbs in the following years?
  (a) 1920 about 8%
  (b) 1950 about 24%
  (c) 1990 about 48%

12. In which decade was there an increase of about 10%? the 1940s

13. In which decade was the percentage closest to 50%? the 1980s

Example 5 used a line graph to compare two quantities. Bar graphs are often used for comparison purposes as well.

The bar graphs here show "balance of trade" data for the three years leading up to the North American Free Trade Agreement (NAFTA) of 1992 (ratified in 1993). Refer to these bar graphs for Exercises 14–21. Note that a country's "trade deficit" is the amount by which imports exceed exports, while a "trade surplus" is the amount by which exports exceed imports.

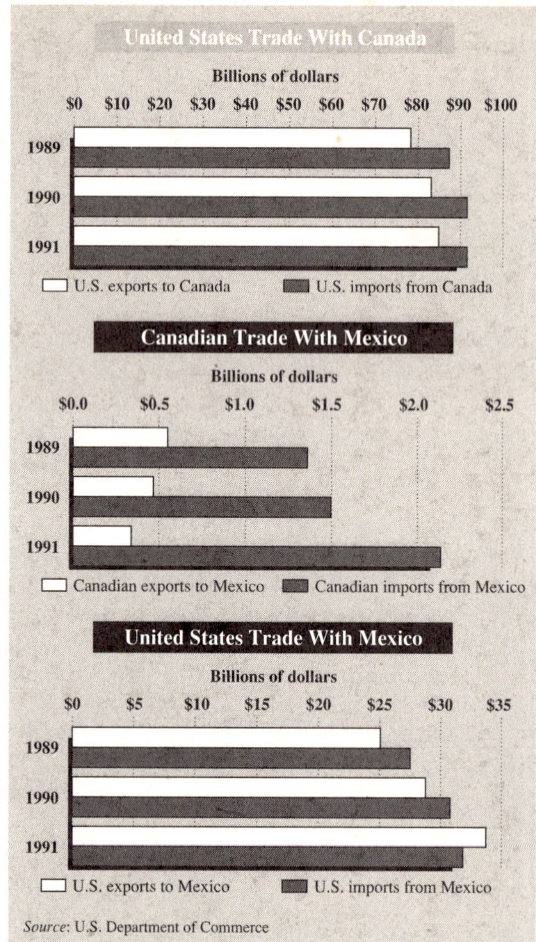

14. What was the dollar amount of Canadian imports from Mexico in 1990? **$1.5 billion**

15. What was the approximate value of U.S. exports to Canada in 1991? **about $85 billion**

16. What trade category declined over the 1989–1991 period? **Canadian exports to Mexico**

17. Over the three-year period, which country was able to steadily increase its trade surplus over which other country? **Mexico over Canada**

18. Over the three-year period, which country was able to steadily decrease (but not eliminate) its trade deficit with which other country?
U.S. with Canada

19. Over the three-year period, which country changed a deficit to a surplus (with which other country)?
U.S. with Mexico

20. In 1990, what was the approximate U.S. net combined trade deficit with both Canada and Mexico?
about $8 billion + $3 billion = $11 billion

21. **(a)** If the period 1991–1998 saw a 50% increase in U.S. exports to Mexico, what would be the approximate value of those exports in 1998?
about $50 billion

(b) Find out the most recent data on U.S. exports to Mexico, and discuss how close the scenario in part (a) is to reality.

 CONCEPTUAL  WRITING ▲ CHALLENGING  SCIENTIFIC CALCULATOR  GRAPHING CALCULATOR

The two circle graphs here show revenue sources for U.S. cities in 1980 and 1990. Refer to them for Exercises 22 and 23.

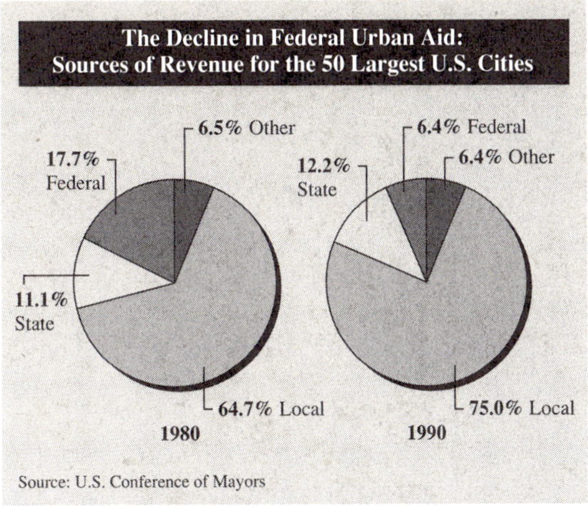

22. How much less was the federal support percentage in 1990 than in 1980? 11.3%

23. How much more was the local support percentage in 1990 than in 1980? 10.3%

24. A survey by the Bureau of Labor Statistics asked American workers how they were trained for their jobs. The percentages who responded in various categories are shown in the table below. Use the information in the table to draw a circle graph.

Principal Source of Training	**Approximate Percentage of Workers**
Trained in school	33%
Informal on-the-job training	25
Formal training from employers	12
Trained in military, or correspondence or other courses	10
No particular training, or could not identify any	20

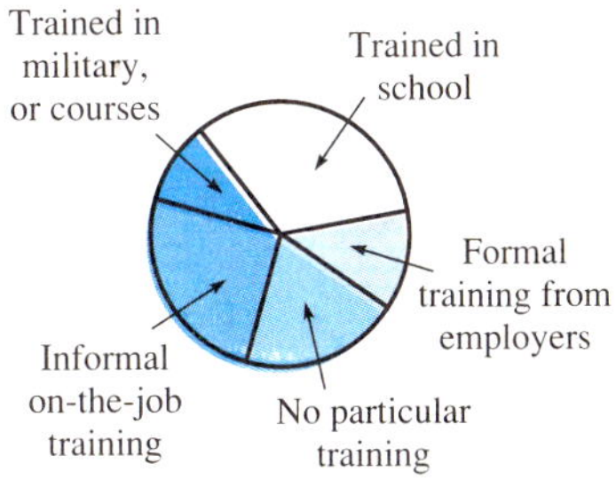

25. Bureau of Labor Statistics data for a recent year showed that the average annual earnings of American workers corresponded to educational level as shown in the table below. Draw a bar graph that shows this information.

Educational Level	Average Annual Earnings
Less than 4 years of high school	$19,168
High school graduate	24,308
Four years of college	38,620

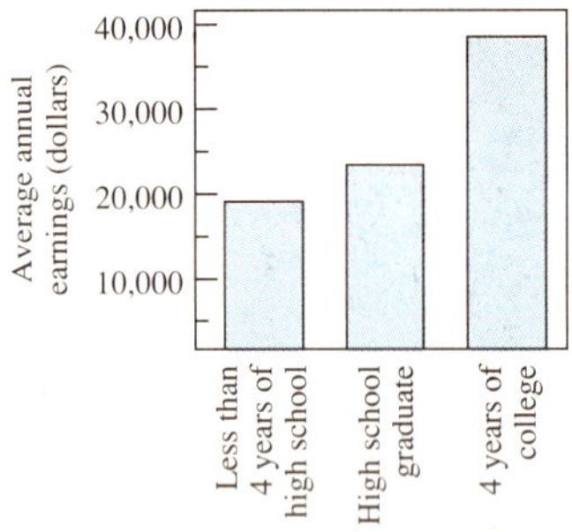

Dick Stratton, wishing to retire at age 60, is studying the comparison line graph here, which shows (under certain assumptions) how the net worth of his retirement savings (initially $200,000 at age 60) will change as he gets older, and as he withdraws living expenses from savings. Refer to the graph for Exercises 26–29.

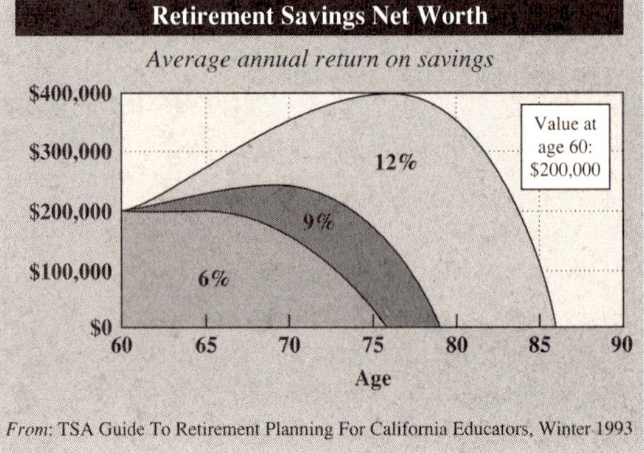

From: TSA Guide To Retirement Planning For California Educators, Winter 1993

26. Assuming Dick can maintain an average annual return of 9%, how old will he be when his money runs out? **about 79 years**

27. If he could earn an average of 12% annually, what maximum net worth would Dick achieve? At about what age would the maximum occur? **about $400,000; about 76 years**

28. Suppose Dick reaches age 70, in good health, and the average annual return has proved to be 6%.
   (a) About how much longer can he expect his money to last? **about 6 years**
   (b) What options might he consider, in order to extend that time?

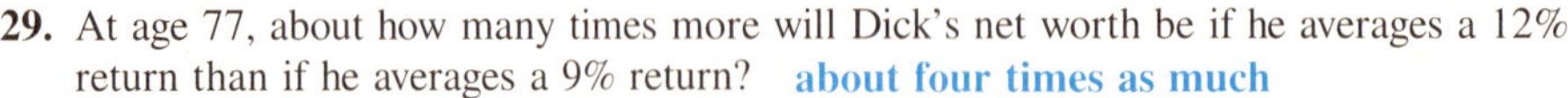
29. At age 77, about how many times more will Dick's net worth be if he averages a 12% return than if he averages a 9% return? **about four times as much**

 CONCEPTUAL  WRITING ▲ CHALLENGING SCIENTIFIC CALCULATOR GRAPHING CALCULATOR

Stem-and-leaf displays can be modified in various ways in order to obtain a reasonable number of stems. For example, the following data, representing the measured masses (in grams) of thirty mineral samples in a geology lab, are shown in a ***double-stem*** *display in Table 8.*

60.7	41.4	50.6	39.5	46.4	58.1	49.7	38.8	61.6	55.2
47.3	52.7	62.4	59.0	44.9	35.6	36.2	40.6	56.9	42.6
34.7	48.3	55.8	54.2	33.8	51.3	50.1	57.0	42.8	43.7

TABLE 8 Stem-And-Leaf Display for Mineral Sample Masses

(30–34)	3	4.7	3.8				
(35–39)	3	9.5	8.8	5.6	6.2		
(40–44)	4	1.4	4.9	0.6	2.6	2.8	3.7
(45–49)	4	6.4	9.7	7.3	8.3		
(50–54)	5	0.6	2.7	4.2	1.3	0.1	
(55–59)	5	8.1	5.2	9.0	6.9	5.8	7.0
(60–64)	6	0.7	1.6	2.4			

30. Describe how the stem-and-leaf display of Table 8 was constructed.

31. Explain why Table 8 is called a "double stem" display.

32. In general, how many stems (total) are appropriate for a stem-and-leaf display? Explain your reasoning.

33. According to the *1995 Information Please Almanac,* the highest temperatures (in degrees Fahrenheit) ever recorded in the 50 states were as follows.

112	100	127	120	134	118	105	110	109	113
100	118	117	116	118	121	114	114	105	109
107	112	114	115	118	117	118	122	106	110
116	108	110	121	113	120	119	111	104	111
120	113	120	117	105	110	118	112	114	114

Present these data in a double stem display.

(100–104)	10	0	0	4												
(105–109)	10	5	9	5	9	7	6	8	5							
(110–114)	11	2	0	3	4	4	2	4	0	0	3	1	1	3	0	2	4	4
(115–119)	11	8	8	7	6	8	5	8	7	8	6	9	7	8		
(120–124)	12	0	1	2	1	0	0	0								
(125–129)	12	7														
(130–134)	13	4														

34. The table here shows commonly accepted percentages of occurrence for the various letters in English language usage. (Code breakers have carefully analyzed these percentages as an aid in deciphering secret codes.) For example, notice that E is the most commonly occurring letter, followed by T, A, O, N, and so on. The letters Q and Z occur least often. Referring to Figure 5 in the text, would you say that the relative frequencies of occurrence of the vowels in the associated paragraph were typical or unusual? Explain your reasoning.

Letter	Percent	Letter	Percent
E	13	L	3 1/2
T	9	C, M, U	3
A, O	8	F, P, Y	2
N	7	W, G, B	1 1/2
I, R	6 1/2	V	1
S, H	6	K, X, J	1/2
D	4	Q, Z	1/5

The percentages shown in Exercise 34 are based on a very large sampling of English language text. Since they are based upon experiment, they are "empirical" rather than "theoretical." By converting each percent in that table to a decimal fraction, you can produce an ***empirical probability distribution.*** *For example, if a single letter is randomly selected from a randomly selected passage of text, the probability that it will be an* E *is* .13. *The probability that a randomly selected letter would be a vowel* (A, E, I, O, *or* U) *is* .385 (.08 + .13 + .065 + .08 + .03).

35. Rewrite the distribution shown in Exercise 34 as an empirical probability distribution. Give values to three decimal places. Note that the 26 probabilities in this distribution—one for each letter of the alphabet—should add up to 1 (except for, perhaps, a slight round-off error).

Letter	Probability
E	.13
T	.09
A, O	.08
N	.07
I, R	.065
S, H	.06
D	.04
L	.035
C, M, U	.03
F, P, Y	.02
W, G, B	.015
V	.01
K, X, J	.005
Q, Z	.002

36. From your distribution of Exercise 35, construct an empirical probability distribution just for the vowels A, E, I, O, and U. (*Hint:* Divide each vowel's probability, from Exercise 35, by .385 to obtain a distribution whose five values add up to 1.) Give values to three decimal places.

Letter	Probability
A	.208
E	.338
I	.169
O	.208
U	.078

37. Construct an appropriately labeled bar chart from your distribution of Exercise 36.

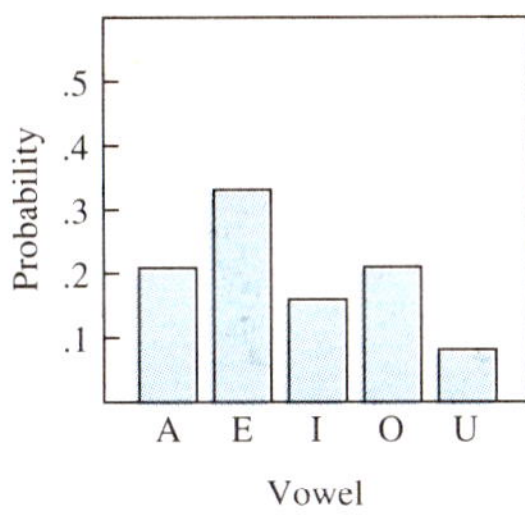

38. Based on the occurrences of vowels in the paragraph represented by Figure 5, construct a probability distribution for the vowels. The frequencies are: A–31, E–34, I–20, O–23, U–10. Give probabilities to three decimal places.

Letter	Probability
A	.263
E	.288
I	.169
O	.195
U	.085

39. Is the probability distribution of Exercise 38 theoretical or empirical? Is it different from the distribution of Exercise 36? Which one is more accurate? Explain your reasoning.

40. Convert the grouped frequency distribution of Table 3 to an empirical probability distribution, using the same classes and giving probability values to three decimal places.

Class Limits	Probability
10–19	.150
20–29	.275
30–39	.225
40–49	.175
50–59	.100
60–69	.050
70–79	.025

41. Recall that the distribution of Exercise 40 was based on weekly study times for a sample of 40 students. Suppose one of those students was chosen randomly. Using your distribution, find the probability that the study time in the past week for the student selected would have been in each of the following ranges.
(a) 30–39 hours .225 **(b)** 40–59 hours .275 **(c)** fewer than 30 hours .425
(d) at least 50 hours .175

The 40 members of a recreation class were asked to name their favorite sport. The table below shows the numbers who responded in various ways.

Sport	Number of Class Members
Sailing	9
Hang gliding	5
Bungee jumping	7
Sky diving	3
Canoeing	12
Rafting	4

Use this information in Exercises 42–45.

42. If a member of this class is selected at random, what is the probability that the favorite sport of the person selected is bungee jumping? 7/40

43. Based on the data above, construct a probability distribution, giving probabilities to three decimal places.

Sport	Probability
Sailing	.225
Hang gliding	.125
Bungee jumping	.175
Sky diving	.075
Canoeing	.300
Rafting	.100

44. **(a)** Is the distribution of Exercise 43 theoretical or is it empirical? **empirical**
(b) Explain your answer to part (a).

45. Explain why a frequency polygon trails down to the axis at both ends while a line graph ordinarily does not.

12.2 Exercises (Text Page 694)

For each of the following lists of data, calculate **(a)** *the mean,* **(b)** *the median, and* **(c)** *the mode or modes (if any). Round mean values to the nearest tenth.*

1. 3, 7, 12, 16, 23
(a) 12.2 (b) 12 (c) none

2. 21, 25, 32, 48, 53, 62
(a) 40.2 (b) 40 (c) none

3. 128, 230, 196, 224, 196, 233
(a) 201.2 (b) 210 (c) 196

4. 26, 31, 46, 31, 26, 29, 31
(a) 31.4 (b) 31 (c) 31

5. 3.1, 4.5, 6.2, 7.1, 4.5, 3.8, 6.2, 6.3
(a) 5.2 (b) 5.35 (c) 4.5 and 6.2

6. 14,320, 16,950, 17,330, 15,470
(a) 16,017.5 (b) 16,210 (c) none

7. .78, .93, .66, .94, .87, .62, .74, .81
(a) .8 (b) .795 (c) none

8. .53, .03, .28, .18, .39, .28, .14, .22, .04
(a) .2 (b) .22 (c) .28

9. 12.3, 45.6, 78.9, 1.2, 34.5, 67.8, 90.1
(a) 47.2 (b) 45.6 (c) none

10. .3, .8, .4, .3, .7, .9, .2, .1, .5, .9, .6
(a) .5 (b) .5 (c) .3 and .9

11. 128, 131, 136, 125, 132, 128, 125, 127
(a) 129 (b) 128 (c) 125 and 128

12. 8.97, 5.64, 2.31, 1.02, 4.35, 7.68
(a) 5.0 (b) 4.995 (c) none

CONCEPTUAL WRITING CHALLENGING SCIENTIFIC CALCULATOR GRAPHING CALCULATOR

The tables here show population figures for the world's twelve most populous countries (1994) and the twelve most populous cities (midyear 1995 estimates). "City" here means a population cluster, which may consist of a city or a city with adjoining cities or clusters if separated by less than one mile. (Source of data: 1995 Information Please Almanac, *pages 133 and 130)*

Most Populous Countries

Rank	**Country**	**Population (in millions)**
1	China	1,192.0
2	India	911.6
3	United States	260.8
4	Indonesia	199.7
5	Brazil	155.3
6	Russia	147.8
7	Pakistan	126.4
8	Japan	125.0
9	Bangladesh	116.6
10	Nigeria	98.1
11	Mexico	91.8
12	Germany	81.2

Most Populous Cities

Rank	**City**	**Population (in thousands)**
1	Tokyo-Yokohama, Japan	28,447
2	Mexico City, Mexico	23,913
3	São Paulo, Brazil	21,539
4	Seoul, South Korea	19,065
5	New York, United States	14,638
6	Osaka-Kobe-Kyoto, Japan	14,060
7	Bombay, India	13,532
8	Calcutta, India	12,885
9	Rio de Janeiro, Brazil	12,788
10	Buenos Aires, Argentina	12,232
11	Manila, Philippines	11,342
12	Moscow, Russia	10,769

Find **(a)** *the mean population, and* **(b)** *the median population for each of the following.*

13. the 5 most populous countries **(a) 543.9 million (b) 260.8 million**

14. the 12 most populous countries **(a) 292.2 million (b) 137.1 million**

15. the 5 most populous cities **(a) 21,520,000 (b) 21,539,000**

16. the 12 most populous cities **(a) 16,268,000 (b) 13,796,000**

Amid concerns about global warming and rain forest destruction, the World Resources Institute reported the five countries with the greatest average yearly forest loss for a recent five-year period, as well as the five greatest emitters of chlorofluorocarbons in a one-year period. The estimates are shown here.

Country	**Average Yearly Forest Loss (in square miles)**
Brazil	9,700
Colombia	3,400
Indonesia	2,400
Mexico	2,375
Ivory Coast	2,000

Country	**Chlorofluorocarbon Emissions (in metric tons)**
United States	130,000,000
Japan	95,000,000
C.I.S.	67,000,000
Germany	34,000,000
United Kingdom	25,000,000

17. Find the mean value of yearly forest loss per country for the five countries listed. Give your answer to the nearest 100 square miles. **4,000 square miles**

18. Find the mean value of chlorofluorocarbon emissions per country for the five countries listed. Give your answer to the nearest one million metric tons.
70,000,000 metric tons

While doing an experiment, a physics student recorded the following sequence of elapsed times (in seconds) in a lab notebook: 2.16, 22.2, 2.96, 2.20, 2.73, 2.28, 2.39.

19. Find the mean. 5.27 sec

20. Find the median. 2.39 sec

The student, when reviewing the calculations later, decided that the entry 22.2 *should have been recorded as* 2.22, *and made that change in the listing.*

21. Find the mean for the new list. 2.42 sec

22. Find the median for the new list. 2.28 sec

23. Which measure, the mean or the median, was affected more by correcting the error? the mean

24. In general, which measure, mean or median, is affected less by the presence of an extreme value in the data? the median

Luis Rebonne earned the following scores on his six math tests last semester.

88, 95, 91, 32, 90, 95

25. Find the mean, the median, and the mode for Luis's scores.
mean = 81.8; median = 90.5; mode = 95

26. Which of the three averages probably is the best indicator of Luis's ability?
the median

27. If Luis's instructor gives him a chance to replace his score of 32 by taking a "make-up" exam, what must he score on the make-up to get an overall average (mean) of 93? 99

Exercises 28–30 give frequency distributions for sets of data values. For each set find the **(a)** *mean (to the nearest hundredth),* **(b)** *median, and* **(c)** *mode or modes (if any).*

28.

Value	Frequency
2	5
4	1
6	8
8	4

(a) 5.22 (b) 6 (c) 6

29.

Value	Frequency
25	4
28	7
30	10
31	3
32	15

(a) 29.97 (b) 30 (c) 32

30.

Value	Frequency
603	13
597	8
589	9
598	12
601	6
592	4

(a) 597.42 (b) 598 (c) 603

31. A company has five employees with a salary of $17,500, eight with a salary of $18,000, four with a salary of $20,300, two with a salary of $24,500, seven with a salary of $26,900, and one with a salary of $115,500. Find the mean salary for the employees (to the nearest hundred dollars). $24,600

Find the grade point average for each of the following students. Assume A = 4, B = 3, C = 2, D = 1, F = 0. *Round to the nearest hundredth.*

32.

Units	Grade
4	C
7	B
3	A
3	F

2.41

33.

Units	Grade
2	A
6	B
5	C

2.77

The table below gives the land area and population of the eleven founding members of the Commonwealth of Independent States. (All were former republics of the Soviet Union.) Use the information from the table for Exercises 34–37.

Name	Area (square miles)	Population
Armenia	11,506	3,373,000
Azerbaijan	33,436	7,222,000
Belarus	80,155	10,480,000
Kazakhstan	1,049,156	16,992,000
Kyrgystan	76,641	4,409,000
Moldova	13,012	4,460,000
Russia	6,592,850	151,436,000
Tajikistan	55,251	5,252,000
Turkmenistan	188,456	3,631,000
Ukraine	233,090	53,125,000
Uzbekistan	172,742	20,453,000

34. Find the mean area (to the nearest square mile) for these 11 states.
773,300 square miles

35. Discuss the meaningfulness of the average calculated in Exercise 34.

36. Find the mean population (to the nearest 1,000) for the 11 states.
25,530,000

37. Discuss the meaningfulness of the average calculated in Exercise 36.

The table below lists the ten most common languages (other than English) spoken at home by Americans. The numbers in the right column indicate the change (increase or decrease) from 1980 to 1990 in the number of Americans speaking the different languages. (Source of data: 1995 Information Please Almanac, *page 835) Use the table for Exercises 38–41. (The rankings are given for information only. They will not enter into your calculations.)*

Most Common Non-English Languages Among Americans

Rank	Language	Change in Number of People
1	Spanish	5,789,839
2	French	129,901
3	German	−59,644
4	Italian	−324,631
5	Chinese	617,476
6	Tagalog	391,289
7	Polish	−102,667
8	Korean	350,766
9	Vietnamese	303,801
10	Portuguese	68,759

38. Find the mean change per language for all ten languages. **716,489**

39. Find the mean change per language for the seven categories with increases.
1,093,119

40. Find the mean change per language for the three categories with decreases.
−162,314

41. Use your answers for Exercises 39 and 40 to find the weighted mean for the change per language for all ten categories. Compare with the value calculated in Exercise 38.
[(1,093,119 × 7) + (−162,314 × 3)]/(7 + 3) = 716,489. This is the same as Exercise 38.

42. The table below shows the "value to the winner" for each of the three so-called "triple crown" horse races in 1994. Find the average value to winner for the three races.

Race	Value to Winner
Belmont Stakes	$392,280
Kentucky Derby	$628,800
Preakness Stakes	$447,720

$489,600

The table at the right shows the medal standings for the 1994 Winter Olympics in Lillehammer, Norway. Use the given information for Exercises 43–46.

Medal Standings for the 1994 Winter Olympics

Nation	Gold	Silver	Bronze	Total
Norway	10	11	5	26
Germany	9	7	8	24
Russia	11	8	4	23
Italy	7	5	8	20
United States	6	5	2	13
Canada	3	6	4	13
Switzerland	3	4	2	9
Austria	2	3	4	9
South Korea	4	1	1	6
Finland	0	1	5	6

Calculate each of the following for all nations shown. For all calculations of mean, round answers to one decimal place.

43. the mean number of gold medals **5.5**

44. the median number of silver medals **5**

45. the mode, or modes, for the number of bronze medals **4**

46. each of the following for the total number of medals
(a) mean **14.9** **(b)** median **13**
(c) mode or modes **none**

In each of Exercises 47 and 48, use the given stem-and-leaf display to identify **(a)** *the mean,* **(b)** *the median, and* **(c)** *the mode (if any) for the data represented.*

47. The display here represents prices (to the nearest dollar) charged by 23 different auto repair shops for a new alternator (installed). Give answers to the nearest cent.

9	7
10	2 4
10	5 7 9
11	1 3 4 4
11	5 5 8 8 9
12	0 4 4
12	5 7 7 9
13	8

(a) $116.30 **(b) $115.00** **(c) none**

48. The display here represents scores achieved on a 100-point biology midterm exam by the 34 members of the class.

4	7
5	1 3 6
6	2 5 5 6 7 8 8
7	0 4 5 6 7 7 8 8 8 8 9
8	0 1 1 3 4 5 5
9	0 0 0 1 6

(a) 74.8 (b) 77.5 (c) 78

49. Yolanda Tubalinal's Business professor lost his grade book, which contained Yolanda's five test scores for the course. A summary of the scores (each of which was an integer from 0 to 100) indicates that the mean was 88, the median was 87, and the mode was 92. (The data set was not bimodal.) What is the lowest possible number among the missing scores? 83

50. Explain what an "outlier" is and how it affects measures of central tendency.

51. A food processing company that packages individual cups of instant soup wishes to find out the best number of cups to include in a package. In a survey of 22 consumers, they found that five prefer a package of 1, five prefer a package of 2, three prefer a package of 3, six prefer a package of 4, and three prefer a package of 6.

(a) Calculate the mean, median, and mode values for preferred package size.
mean = 3; median = 3; mode = 4

(b) Which measure in part (a) should the food processing company use?
mode

(c) Explain your answer to part (b).

52. The following are scores earned by 15 college students on a 20-point math quiz.

0, 1, 3, 14, 14, 15, 16, 16, 17, 17, 18, 18, 18, 19, 20

(a) Calculate the mean, median, and mode values for these scores.
mean = 13.7; median = 16; mode = 18

(b) Which measure in part (a) is most representative of the data?
median

(c) Explain your answer to part (b).

In each of Exercises 53–56, begin a list of the given numbers, in order, starting with the smallest one. Continue the list only until the median of the listed numbers is a multiple of 4. Stop at that point and find **(a)** *the number of numbers listed, and* **(b)** *the mean of the listed numbers (to two decimal places).*

53. counting numbers (a) 7 (b) 4

54. prime numbers (a) 4 (b) 4.25

55. Fibonacci numbers (a) 8 (b) 6.75

56. triangular numbers (a) 6 (b) 9.33

57. Seven consecutive whole numbers add up to 147. What is the result when their mean is subtracted from their median? **0**

58. If the mean, median, and mode are all equal for the set $\{70, 110, 80, 60, x\}$, find the value of x. **80**

59. Vince Straub wants to include a fifth number, n, along with the numbers 2, 5, 8, and 9 so that the mean and median of the five numbers will be equal. How many choices does Vince have for the number n, and what are those choices?
three choices: 1, 6, 16

60. Refer to the salary data of Example 2. Explain what is wrong with simply calculating the mean salary as follows.

$$\begin{aligned}\bar{x} &= \frac{\Sigma x}{n} \\ &= \frac{\$12{,}000 + \$16{,}000 + \$18{,}500 + \$21{,}000 + \$34{,}000 + \$50{,}000}{6} \\ &= \$25{,}250\end{aligned}$$

For Exercises 61–63, refer to the grouped frequency distribution shown here.

Class Limits	Frequency f
21–25	5
26–30	3
31–35	8
36–40	12
41–45	21
46–50	38
51–55	35
56–60	20

61. Can you identify any specific data items that occurred in this sample? **no**

62. Can you compute the actual mean for this sample? **no**

63. Describe how you might approximate the mean for this sample. Justify your procedure.

12.3 Exercises (Text Page 705)

1. If your calculator finds both kinds of standard deviation, the sample standard deviation and the population standard deviation, which of the two will be a larger number for a given set of data? (*Hint:* Recall the difference between how the two standard deviations are calculated.) **the sample standard deviation**

2. If your calculator finds only one kind of standard deviation, explain how you would determine whether it is sample or population standard deviation (assuming your calculator manual is not available).

Find **(a)** *the range, and* **(b)** *the standard deviation for each sample (Exercises 3–12). Round fractional answers to the nearest hundredth.*

3. 2, 5, 6, 8, 9, 11, 15, 19
(a) 17 (b) 5.53

4. 8, 5, 12, 8, 9, 15, 21, 16, 3
(a) 18 (b) 5.74

5. 25, 34, 22, 41, 30, 27, 31
(a) 19 (b) 6.27

6. 67, 83, 55, 68, 77, 63, 84, 72, 65
(a) 29 (b) 9.54

7. 318, 326, 331, 308, 316, 322, 310, 319, 324, 330 (a) 23 (b) 7.75

8. 5.7, 8.3, 7.4, 6.6, 7.4, 6.8, 7.1, 8.0, 8.5, 7.9, 7.1, 7.4, 6.9, 8.2 (a) 2.8 (b) .76

9. 84.53, 84.60, 84.58, 84.48, 84.72, 85.62, 85.03, 85.10, 84.96 (a) 1.14 (b) .37

10. 206.3, 210.4, 209.3, 211.1, 210.8, 213.5, 212.6, 210.5, 211.0, 214.2 (a) 7.9 (b) 2.23

11.

Value	Frequency
9	3
7	4
5	7
3	5
1	2

(a) 8 (b) 2.41

12.

Value	Frequency
14	8
16	12
18	15
20	14
22	10
24	6
26	3

(a) 12 (b) 3.29

Use Chebyshev's theorem for Exercises 13–28.

Find the least possible fraction of the numbers in a data set lying within the given number of standard deviations of the mean. Give answers as standard fractions reduced to lowest terms.

13. 2 3/4

14. 4 15/16

15. 7/2 45/49

16. 11/4 105/121

Find the least possible percentage (to the nearest tenth of a percent) of the items in a distribution lying within the given number of standard deviations of the mean.

17. 3 88.9%

18. 5 96%

19. 5/3 64%

20. 5/2 84%

▲ *In a certain distribution of numbers, the mean is* 70 *and the standard deviation is* 8. *At least what fraction of the numbers are between the following pairs of numbers? Give answers as standard fractions reduced to lowest terms.*

21. 54 and 86 3/4

22. 46 and 94 8/9

23. 38 and 102 15/16

24. 30 and 110 24/25

▲ *In the same distribution (mean* 70 *and standard deviation* 8*), find the largest fraction of the numbers that could meet the following requirements. Give answers as standard fractions reduced to lowest terms.*

25. less than 54 or more than 86 1/4

26. less than 50 or more than 90 4/25

27. less than 42 or more than 98 4/49

28. less than 52 or more than 88 16/81

Katherine Steinbacher owns a minor league baseball team. Each time the team wins a game, Katherine pays the nine starting players, the manager and two coaches bonuses, which are certain percentages of their regular salaries. The amounts paid are listed here.

$80, $105, $120, $175, $185, $190, $205, $210, $215, $300, $320, $325

Use the distribution of bonuses above for Exercises 29–34.

29. Find the mean of the distribution. **$202.50**

30. Find the standard deviation of the distribution. **$80.38**

31. How many of the bonus amounts are within one standard deviation of the mean? **six**

32. How many of the bonus amounts are within two standard deviations of the mean? **all twelve**

33. What does Chebyshev's theorem say about the number of the amounts that are within two standard deviations of the mean? **at least nine**

34. Explain any discrepancy between your answers for Exercises 32 and 33.

William Tobey manages the service department of a trucking company. Each truck in the fleet utilizes an electronic engine control module, which must be replaced when it fails. Long-lasting modules are desirable, of course, but a preventive replacement program can also avoid costly breakdowns on the highway. For this purpose it is desirable that the modules be fairly consistent in their lifetimes; that is, they should all last about the same number of miles before failure, so that the timing of preventive replacements can be done accurately. William tested a sample of 20 Brand A modules, and they lasted 43,560 highway miles on the average (mean), with a standard deviation of 2,116 miles. The listing below shows how long each of another sample of 20 Brand B modules lasted.

50,660, 41,300, 45,680, 48,840, 47,300,
51,220, 49,100, 48,660, 47,790, 47,210,
50,050, 49,920, 47,420, 45,880, 50,110,
49,910, 47,930, 48,800, 46,690, 49,040

Use the data above for Exercises 35–37.

35. According to the sampling that was done, which brand of module has the longer average life (in highway miles)? **Brand *B* ($\bar{x}_B$ = 48,176 > 43,560)**

36. Which brand of module apparently has a more consistent (or uniform) length of life (in highway miles)? **Brand *A* (s_B = 2,235 > 2,116)**

37. If Brands *A* and *B* are the only modules available, which one should William purchase for his maintenance program? Explain your reasoning.

Utilize the following sample for Exercises 38–43.

13, 14, 16, 18, 20, 22, 25

38. Compute the mean and standard deviation for the sample (each to the nearest hundredth). **18.29; 4.35**

39. Now add 5 to each item of the sample above and compute the mean and standard deviation for the new sample. **23.29; 4.35**

 CONCEPTUAL  WRITING  CHALLENGING  SCIENTIFIC CALCULATOR GRAPHING CALCULATOR

40. Go back to the original sample. This time subtract 10 from each item, and compute the mean and standard deviation of the new sample. 8.29; 4.35

41. Based on your answers for Exercises 38–40, make conjectures about what happens to the mean and standard deviation when all items of the sample have the same constant k added or subtracted.

42. Go back to the original sample again. This time multiply each item by 3, and compute the mean and standard deviation of the new sample. 54.86; 13.04

43. Based on your answers for Exercises 38 and 42, make conjectures about what happens to the mean and standard deviation when all items of the sample are multiplied by the same constant k.

44. In Section 2 we showed that the mean, as a measure of central tendency, is highly sensitive to extreme values. Which measure of dispersion, covered in this section, would be more sensitive to extreme values? Illustrate your answer with one or more examples.

45. The Quaker Oats Company conducted a survey to determine whether or not a proposed premium to be included in boxes of their cereal was appealing enough to generate new sales. Four cities were used as test markets, where the cereal was distributed with the premium, and four cities as control markets, where the cereal was distributed without the premium. The eight cities were chosen on the basis of their similarity in terms of population, per capita income, and total cereal purchase volume. The results follow.

		Percent Change in Average Market Share per Month
Test Cities	1	+18
	2	+15
	3	+7
	4	+10
Control Cities	1	+1
	2	−8
	3	−5
	4	0

(a) Find the mean of the change in market share for the four test cities. 12.5

(b) Find the mean of the change in market share for the four control cities. −3.0

(c) Find the standard deviation of the change in market share for the test cities. 4.9

(d) Find the standard deviation of the change in market share for the control cities. 4.2

(e) Find the difference between the means of (a) and (b). This difference represents the estimate of the percent change in sales due to the premium. 15.5

(f) The two standard deviations from (c) and (d) were used to calculate an "error" of ±7.95 for the estimate in (e). With this amount of error, what are the smallest and largest estimates of the increase in sales? 7.55 and 23.45

On the basis of the interval estimate of part (f) the company decided to mass produce the premium and distribute it nationally.

◉ *In a skewed distribution, the mean will be farther out toward the tail than the median, as shown in the sketch.*

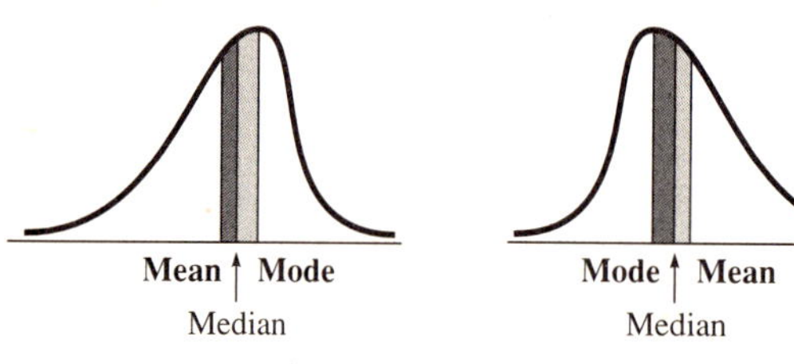

Skewed to the left　　Skewed to the right

A common way of measuring the degree of skewness, which involves both central tendency and dispersion, is with the **skewness coefficient,** *calculated as follows.*

$$\frac{3 \times (\text{mean} - \text{median})}{\text{standard deviation}}$$

46. Under what conditions would the skewness coefficient be each of the following?
(a) positive　**(b)** negative
(a) when the distribution is skewed to the right　(b) when the distribution is skewed to the left

47. Explain why the mean of a skewed distribution is always farther out toward the tail than the median.

48. Suppose that the mean length of patient stay (in days) in U.S. hospitals is 2.7 days, with a standard deviation of 7.1 days. Make a sketch of how this distribution may look.

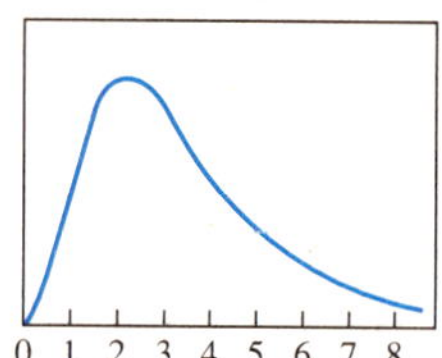

◉ *For Exercises 49–51, refer to the grouped frequency distribution shown here. (Also refer to Exercises 61–63 in the previous section.)*

Class Limits	Frequency f
21–25	5
26–30	3
31–35	8
36–40	12
41–45	21
46–50	38
51–55	35
56–60	20

49. Can you identify any specific data items that occurred in this sample? **no**

50. Can you compute the actual standard deviation for this sample? **no**

51. Describe how you might approximate the standard deviation for this sample. Justify your procedure.

52. Suppose the frequency distribution of Example 3 involved 50 or 100 (or even more) distinct data values, rather than just four. Explain why the procedure of that example would then be very inefficient.

53. A "J-shaped" distribution can be skewed to the right as well as to the left.
(a) In a J-shaped distribution skewed to the right, which data item would be the mode, the largest or the smallest item? smallest
(b) In a J-shaped distribution skewed to the left, which data item would be the mode, the largest or the smallest item? largest
(c) Explain why the mode is a weak measure of central tendency for a J-shaped distribution.

12.4 Exercises (Text Page 713)

For each of Exercises 1–4, make use of z-scores.

1. In a calculus class, Emily Francke scored 5 on a quiz for which the class mean and standard deviation were 4.6 and 2.1, respectively. Charles Gomez scored 6 on another quiz for which the class mean and standard deviation were 4.9 and 2.3, respectively. Relatively speaking, which student did better? Charles (since $z = .48 > .19$)

2. In Saturday's track meet, Gregory Syftestad, a high jumper, jumped 6 feet 3 inches. Conference high jump marks for the past season had a mean of 6 feet even and a standard deviation of 3.5 inches. Alexander Szabo, Gregory's teammate, achieved 18 feet 4 inches in the long jump. In that event the conference's season average (mean) and standard deviation were 16 feet 6 inches and 1 foot 10 inches, respectively. Relative to this past season in this conference, which athlete had a better performance on Saturday? Alexander (since $z = 1.00 > .86$)

3. The lifetimes of Brand A tires are distributed with mean 45,000 miles and standard deviation 4,500 miles, while Brand B tires last for only 38,000 miles on the average (mean) with standard deviation 2,080 miles. Jutta's Brand A tires lasted 37,000 miles and Arvind's Brand B tires lasted 35,000 miles. Relatively speaking, within their own brands, which driver got the better wear? Arvind (since $z = -1.44 > -1.78$)

4. In a certain lake, the trout average 12 inches in length with a standard deviation of 2.75 inches. The bass average 4 pounds in weight with a standard deviation of .8 pound. If Imelda caught an 18-inch trout and Timothy caught a 6-pound bass, then relatively speaking, which catch was the better trophy?
Timothy's bass (since $z = 2.50 > 2.18$)

Refer to the dinner customers data of Example 2. Approximate each of the following. (Use the methods illustrated in this section.)

5. the fifteenth percentile 58

6. the seventy-fifth percentile 70

7. the third decile 62

8. the eighth decile 71

For Exercises 9–20, refer to the top 20 scorers data for the 1993–1994 season of the National Basketball Association.

Leading Scorers	**Games Played**	**Total Points**	**Average Points per Game**
David Robinson, San Antonio	80	2,383	29.8
Shaquille O'Neal, Orlando	81	2,377	29.4
Hakeem Olajuwon, Houston	80	2,184	27.3
Dominique Wilkins, L.A. Clippers	74	1,923	26.0
Karl Malone, Utah	82	2,039	24.9
Patrick Ewing, N.Y. Knicks	79	1,939	24.5
Mitch Richmond, Sacramento	78	1,823	23.4
Scottie Pippen, Chicago	72	1,587	22.0
Charles Barkley, Phoenix	65	1,402	21.6
Alonzo Mourning, Charlotte	60	1,287	21.5
Glen Rice, Miami	81	1,708	21.1
Latrell Sprewell, Golden State	82	1,720	21.0
Danny Manning, Atlanta	68	1,403	20.6
Joe Dumars, Detroit	69	1,410	20.4
Derrick Coleman, New Jersey	77	1,559	20.3
Ron Harper, L.A. Clippers	75	1,508	20.1
Clifford Robinson, Portland	82	1,647	20.1
Kevin Johnson, Phoenix	67	1,340	20.0
Reggie Miller, Indiana	79	1,574	19.9
Jim Jackson, Dallas	82	1,576	19.2

Compute z-scores for Exercises 9–12.

9. Scottie Pippen's games played **−.55**

10. Mitch Richmond's total points **.31**

11. Shaquille O'Neal's average points per game **2.07**

12. David Robinson's total points **2.02**

Determine who occupied each of the following positions.

13. the eighty-fifth percentile in average points per game **Hakeem Olajuwon**

14. the forty-fifth percentile in games played **Mitch Richmond**

15. the eighth decile in total points scored **Karl Malone**

16. the ninth decile in average points per game **Shaquille O'Neal**

Determine who was relatively highest in the groupings of Exercises 17 and 18.

17. Charles Barkley in total points, Danny Manning in average points per game, or Joe Dumars in games played
Danny Manning (since $z = -.63$ is greater than either $-.97$ or -1.00)

18. Reggie Miller in total points, Kevin Johnson in games played, or Ron Harper in average points per game
Reggie Miller (since $z = -.44$ is greater than either -1.30 or $-.79$)

19. Construct a stem-and-leaf plot with double stems for the games played data.

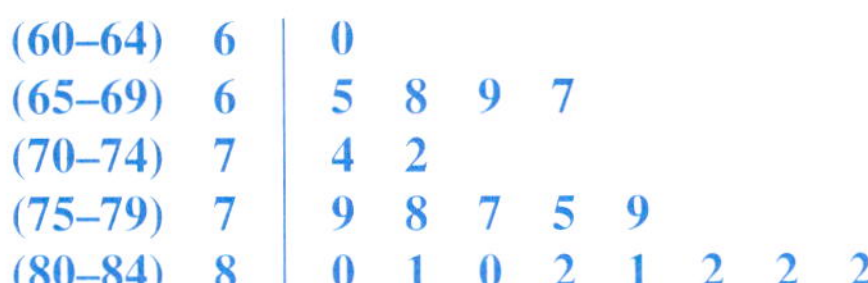

	Stem	Leaves
(60–64)	6	0
(65–69)	6	5 8 9 7
(70–74)	7	4 2
(75–79)	7	9 8 7 5 9
(80–84)	8	0 1 0 2 1 2 2 2

20. Construct a box plot for the average points per game data.

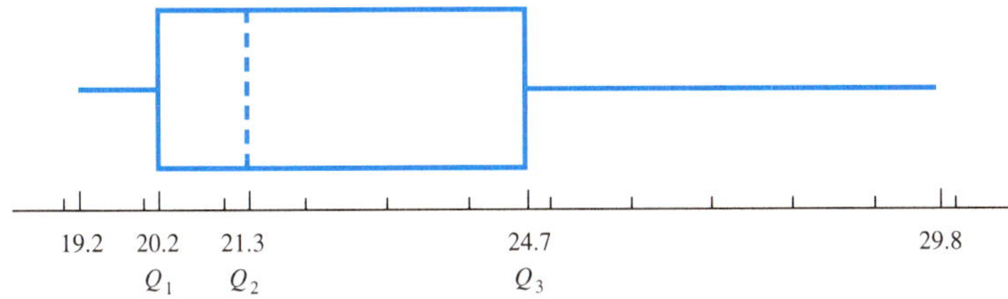

21. What does your box plot of Exercise 20 indicate about each of the following?
(a) the central tendency of these data **The median is 21.3.**
(b) the dispersion of these data **The range is 10.6.**
(c) the skewness of these data **The data are skewed to the right.**

22. The text stated that, for *any* distribution of data, at least 89% of the items will be within three standard deviations of the mean. Why couldn't we just move some items farther out from the mean to obtain a new distribution that would violate this condition?

23. Describe the basic difference between a measure of central tendency and a measure of position.

This chapter has introduced three major characteristics: central tendency, dispersion, and position, and has developed various ways of measuring them in numerical data. In each of Exercises 24–28, a new measure is described. Explain in each case which of the three characteristics you think it would measure and why.

24. Midrange $= \dfrac{\text{minimum item} + \text{maximum item}}{2}$

25. Midquartile $= \dfrac{Q_1 + Q_3}{2}$

26. Coefficient of Variation $= \dfrac{s}{\bar{x}} \cdot 100$

27. Interquartile range $= Q_3 - Q_1$

28. Semi-interquartile range $= \dfrac{Q_3 - Q_1}{2}$

29. The "skewness coefficient" was defined in the exercises of the previous section to be

$$\frac{3 \times (\bar{x} - Q_2)}{s}.$$

Is this a measure of individual data items or of the overall distribution?
the overall distribution

30. In a national standardized test, Jennifer scored at the ninety-second percentile. If 67,500 individuals took the test, about how many scored higher than Jennifer did?
5,400

31. Let the three quartiles (from smallest to largest) for a large population of scores be denoted Q_1, Q_2, and Q_3.

(a) Is it necessarily true that $Q_2 - Q_1 = Q_3 - Q_2$? **no**

(b) Explain your answer to part (a).

In Exercises 32–35, answer yes *or* no *and explain your answer. (Consult the exercises above for definitions of some of these measures.)*

32. Is the midquartile necessarily the same as the median?

33. Is the midquartile necessarily the same as the midrange?

34. Is the interquartile range necessarily half the range?

35. Is the semi-interquartile range necessarily half the interquartile range?

36. Omer and Alessandro participated in the standardization process for a new state-wide chemistry test. Within the large group participating, their raw scores and corresponding z-scores were as shown here.

	Raw score	z-score
Omer	60	.69
Alessandro	72	1.67

Find the overall mean and standard deviation of the distribution of scores. (Give answers to two decimal places.)
mean = 51.55; standard deviation = 12.24

*Since the National Football League began keeping official statistics in 1932, the passing effectiveness of quarterbacks has been rated by several different methods. The current system, adopted in 1973, is based on four performance components considered most important: completions, touchdowns, yards gained, and interceptions, as percentages of the number of passes attempted. The actual computation of the ratings seems to be poorly understood by fans and media sportscasters alike, but can be accomplished with the following formula.**

* Joe Farabee of American River College (a long-time friend and colleague of author Heeren) derived the formula after sifting through a variety of descriptions of the rating system in sports columns and NFL publications.

$$\text{Rating} = \frac{\left(250 \times \frac{C}{A}\right) + \left(1{,}000 \times \frac{T}{A}\right) + \left(12.5 \times \frac{Y}{A}\right) + 6.25 - \left(1{,}250 \times \frac{I}{A}\right)}{3},$$

where A = attempted passes
C = completed passes
T = touchdown passes
Y = yards gained
I = interceptions

In addition to the weighting factors (coefficients) appearing in the formula, the four category ratios are limited to the following maximums.

$$.775 \text{ for } \frac{C}{A}, \qquad .11875 \text{ for } \frac{T}{A}, \qquad 12.5 \text{ for } \frac{Y}{A}, \qquad .095 \text{ for } \frac{I}{A}$$

These limitations are intended to prevent any one component of performance from having an undue effect on the overall rating. They are not often invoked, but in special cases can have a significant effect. (See Exercises 48–51.)

The formula above rates all passers against the same performance standard and is applied, for example, after a single game, an entire season, or a career.

Based on the 1994 regular-season statistics shown here for Steve Young of the San Francisco 49ers and Troy Aikman of the Dallas Cowboys, compute each of the following (to one decimal place).

	Young	Aikman
Attempts	461	361
Completions	324	233
Yards	3,969	2,676
TDs	35	13
Interceptions	10	12

37. Young's season rating 112.8

38. Aikman's season rating 84.9

The 1994 regular season ratings are shown here (ranked by conference) for the fifteen leading passers in the National Conference (N.F.C.) and the fourteen in the American Conference (A.F.C.). Find the requested measures in Exercises 39–42.

N.F.C. Passer	Rating Points	A.F.C. Passer	Rating Points
Young, San Francisco	112.8	Marino, Miami	89.2
Favre, Green Bay	90.7	Elway, Denver	85.7
Everett, New Orleans	84.9	Kelly, Buffalo	84.6
Aikman, Dallas	84.9	Montana, Kansas City	83.6
George, Atlanta	83.3	Humphries, San Diego	81.6
Erickson, Tampa Bay	82.5	Hostetler, Raiders	81.0
Moon, Minnesota	79.9	O'Donnell, Pittsburgh	78.9
Walsh, Chicago	77.9	Esiason, New York Jets	77.3
Cunningham, Philadelphia	74.4	Blake, Cincinnati	76.9
Miller, Rams	73.6	Bledsoe, New England	73.6
Brown, New York Giants	72.5	Testaverde, Cleveland	70.7
Schroeder, Arizona	68.4	Mirer, Seattle	70.2
Mitchell, Detroit	62.0	Klingler, Cincinnati	65.7
Beuerlein, Arizona	61.6	Tolliver, Houston	62.6
Shuler, Washington	59.6		

39. the three quartiles for the N.F.C. ratings $Q_1 = 68.4$; $Q_2 = 77.9$; $Q_3 = 84.9$

40. the three quartiles for the A.F.C. ratings $Q_1 = 70.7$; $Q_2 = 78.1$; $Q_3 = 83.6$

41. the three quartiles for all ratings (the N.F.C. and A.F.C. figures combined)
$Q_1 = 70.45$; $Q_2 = 77.9$; $Q_3 = 84.1$

42. the ninetieth percentile for the combined figures 89.2

43. Construct box plots for both conferences, one above the other in the same drawing.

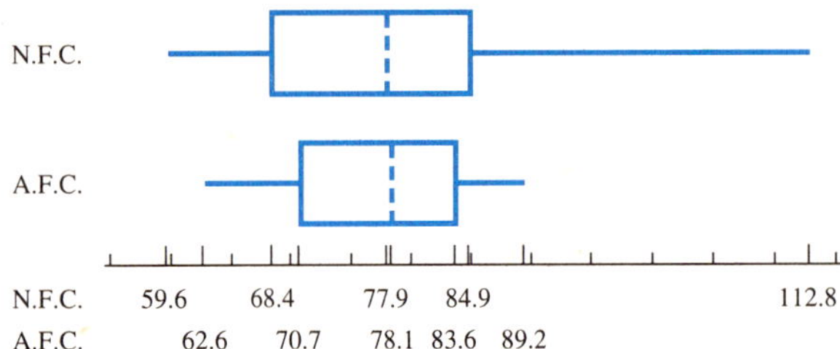

Refer to the box plots of Exercise 43 for Exercises 44–47. In each case, identify the conference for which the given measure is greater, and estimate about how much greater it is.

44. the median **A.F.C.; slightly greater**

45. the midquartile **A.F.C.; about 1/2 unit**

46. the range **N.F.C.; about 27 units**

47. the interquartile range **N.F.C.; about 4 units**

Sid Luckman of the Chicago Bears, the passing champion of the 1943 season, had the following pertinent statistics in that year.

202 attempts, 110 completions, 28 touchdowns, 2,194 yards, 12 interceptions

Use these numbers in the rating formula for the following exercises.

48. Compute Luckman's 1943 rating, being careful to replace each ratio in the formula with its allowed maximum if necessary. **107.5**

49. Compute Luckman's rating assuming no restrictions on the ratios. **114.2**

50. Which component of Luckman's passing game that year was apparently very unusual, even among high caliber passers?
high percentage of touchdown passes (about 13.9%)

51. Compare the two ratings of Exercises 48 and 49 with Steve Young's all-time record season rating of 112.8 (achieved in 1994).

12.5 Exercises (Text Page 725)

Identify each of the following variable quantities as discrete *or* continuous.

1. the number of heads in 30 tossed coins **discrete**
2. the number of babies born in one day at a certain hospital **discrete**
3. the average weight of babies born in a week **continuous**
4. the heights of seedling fir trees at six weeks of age **continuous**
5. the time as shown on a digital watch **discrete**
6. the time as shown on a watch with a sweep hand **continuous**

Suppose 100 geology students measure the mass of an ore sample. Due to human error and limitations in the reliability of the balance, not all the readings are equal. The results are found to closely approximate a normal curve, with mean 37 g and standard deviation 1 g.

Use the symmetry of the normal curve and the empirical rule to estimate the number of students reporting readings in the following ranges.

7. more than 37 g **50**
8. more than 36 g **84**
9. between 36 and 38 g **68**
10. between 36 and 39 g **81 or 82**

On standard IQ tests, the mean is 100, with a standard deviation of 15. The results come very close to fitting a normal curve. Suppose an IQ test is given to a very large group of people. Find the percent of people whose IQ scores fall into the following categories.

11. less than 100 **50%**

12. greater than 115 **16%**

13. between 70 and 130 **95%**

14. more than 145 **.15%**

Find the percent of area under a normal curve between the mean and the given number of standard deviations from the mean. (Note that positive indicates above the mean, while negative indicates below the mean.)

15. 2.50 **49.4%** **16.** .81 **29.1%**

17. −1.71 **45.6%** **18.** −2.04 **47.9%**

Find the percent of the total area under the normal curve between the given values of z.

19. $z = 1.41$ and $z = 2.83$ **7.7%**

20. $z = -1.74$ and $z = -1.02$ **11.3%**

21. $z = -3.11$ and $z = 1.44$ **92.4%**

22. $z = -1.98$ and $z = 1.98$ **95.2%**

▲ *Find a value of z such that the following conditions are met.*

23. 5% of the total area is to the right of z. **1.64**

24. 1% of the total area is to the left of z. **any of these: −2.34, −2.33, −2.32, −2.31 (A more accurate table of normal curve areas would enable us to be more specific.)**

25. 15% of the total area is to the left of z. **−1.03 or −1.04**

26. 25% of the total area is to the right of z. **.67**

The Alva light bulb has an average life of 500 hr, with a standard deviation of 100 hr. The length of life of the bulb can be closely approximated by a normal curve. An amusement park buys and installs 10,000 such bulbs. Find the total number that can be expected to last the following amounts of time.

27. at least 500 hr **5,000**

28. between 500 and 650 hr **4,330**

29. between 650 and 780 hr **640**

30. between 290 and 540 hr **6,370**

31. less than 740 hr **9,920**

32. less than 410 hr **1,840**

The chickens at Ben and Ann Rice's farm have a mean weight of 1,850 g with a standard deviation of 150 g. The weights of the chickens are closely approximated by a normal curve. Find the percent of all chickens having the following weights.

33. more than 1,700 g **84.1%**

34. less than 1,800 g **37.1%**

35. between 1,750 and 1,900 g **37.8%**

36. between 1,600 and 2,000 g **79.4%**

A box of oatmeal must contain 16 oz. The machine that fills the oatmeal boxes is set so that, on the average, a box contains 16.5 oz. The boxes filled by the machine have weights that can be closely approximated by a normal curve. What fraction of the boxes filled by the machine are underweight if the standard deviation is as follows?

37. .5 oz **.159 or 15.9%** **38.** .3 oz **.047 or 4.7%**

39. .2 oz **.006 or .6%** **40.** .1 oz **0%**

41. In nutrition, the recommended daily allowance of vitamins is a number set by the government to guide an individual's daily vitamin intake. Actually, vitamin needs vary drastically from person to person, but the needs are closely approximated by a normal curve. To calculate the recommended daily allowance, the government first finds the average need for vitamins among people in the population and the standard deviation. The **recommended daily allowance** is then defined as the mean plus 2.5 times the standard deviation. What fraction of the population will receive adequate amounts of vitamins under this plan? **.994 or 99.4%**

Find the recommended daily allowance for each vitamin if the mean need and standard deviation are as follows. (See Exercise 41.)

42. mean need = 1,800 units;
standard deviation = 140 units
2,150 units

43. mean need = 159 units;
standard deviation = 12 units **189 units**

Assume the following distributions are all normal, and use the areas under the normal curve given in Table 10 to find the appropriate areas.

44. A machine that fills quart milk cartons is set up to average 32.2 oz per carton, with a standard deviation of 1.2 oz. What is the probability that a filled carton will contain less than 32 oz of milk? **.432**

45. The mean clotting time of blood is 7.45 sec, with a standard deviation of 3.6 sec. What is the probability that an individual's blood-clotting time will be less than 7 sec or greater than 8 sec? .888

46. The average size of the fish caught in Lake Amotan is 12.3 in., with a standard deviation of 4.1 in. Find the probability that a fish caught there will be longer than 18 in. .082

47. To be graded extra large, an egg must weigh at least 2.2 oz. If the average weight for an egg is 1.5 oz, with a standard deviation of .4 oz, how many of five dozen eggs would you expect to grade extra large? about 2 eggs

Kimberly Workman teaches a course in marketing. She uses the following system for assigning grades to her students.

Grade	Score in Class
A	Greater than $\bar{x} + (3/2)s$
B	$\bar{x} + (1/2)s$ to $\bar{x} + (3/2)s$
C	$\bar{x} - (1/2)s$ to $\bar{x} + (1/2)s$
D	$\bar{x} - (3/2)s$ to $\bar{x} - (1/2)s$
F	Below $\bar{x} - (3/2)s$

What percent of the students receive the following grades?

48. A 6.7% 49. B 24.1% 50. C 38.4%

 51. Do you think this system would be more likely to be fair in a large freshman class in psychology or in a graduate seminar of five students? Why?

Extension Exercises (Text Page 731)

The Norwegian stamp at the side features two graphs.

1. What does the solid line represent? We have no way of telling.

2. Has it increased much? We can't tell; there is no scale.

3. What can you tell about the dashed line? We can tell only that it rises and then falls.

4. Does the graph represent a long period of time or a brief period of time? Again, there is no scale.

▲ *A teacher gives a test to a large group of students. The results are closely approximated by a normal curve. The mean is 74 with a standard deviation of 6. The teacher wishes to give As to the top 8% of the students and Fs to the bottom 8%. A grade of B is given to the next 15%, with Ds given similarly. All other students get Cs. Find the bottom cutoff (rounded to the nearest whole number) for the following grades. (Hint: read Table 10 backwards.)*

52. A 82 53. B 78

54. C 70 55. D 66

A normal distribution has mean 94.2 and standard deviation 7.68. Follow the method of Example 6 and find data values corresponding to the following values of z. Round to the nearest tenth.

56. $z = .72$ 99.7 57. $z = 1.44$ 105.3

58. $z = -2.39$ 75.8 59. $z = -3.87$ 64.5

60. What percentage of the items lie within 1.25 standard deviations of the mean
(a) in any distribution (by Chebyshev's theorem)?
(b) in a normal distribution (by Table 10)?
(a) at least 36% (b) 78.8%

61. Explain the difference between the answers to parts (a) and (b) in Exercise 60.

The illustration at the side shows the decline in the value of the British pound for a certain period.

$2.40

5. Calculate the percent of decrease in the value by using the formula

$$\text{Percent of decrease} = \frac{\text{old value} - \text{new value}}{\text{old value}}.$$

28%

$1.72

6. We estimate that the smaller banknote shown has about 50% less area than the larger one. Do you think this is close enough? **The actual decrease ($2.40 to $1.72) is about 28%, but the artist for the magazine reduced *each* dimension of the original figure by 28%. This causes the area to decrease by about 50%, thus giving a false impression.**

Several advertising claims are given below. Decide what further information you might need before deciding to accept the claim.

7. 98% of all Toyotas ever sold in the United States are still on the road. **How long have Toyotas been sold in the United States? How do other makes compare?**

8. Sir Walter Raleigh pipe tobacco is 44% fresher. **The tobacco is 44% fresher than what?**

9. Eight of 10 dentists responding to a survey preferred Trident Sugarless Gum. **The dentists preferred Trident Sugarless Gum to what? Which and how many dentists were surveyed? What percentage responded?**

10. A Volvo has 2/3 the turning radius of a Continental. **So what—a Volvo is a smaller car than a Continental and should have a smaller turning radius.**

11. A Ford LTD is as quiet as a glider. **Just how quiet *is* a glider, really?**

12. *Wall Street Journal* circulation has increased, as shown by the graph below.

There is no scale. We can't tell if the increase is substantial or not.

Exercises 13–16 come from Huff's book. Decide how these exercises describe possibly misleading uses of numbers.

13. Each of the following maps shows what portion of our national income is now being taken and spent by the federal government. The map on the left does this by shading the areas of most of the states *west* of the Mississippi to indicate that federal spending has become equal to the total incomes of the people of those states. The map on the right does this for states *east* of the Mississippi.

The Darkening Shadow
(Western Style)

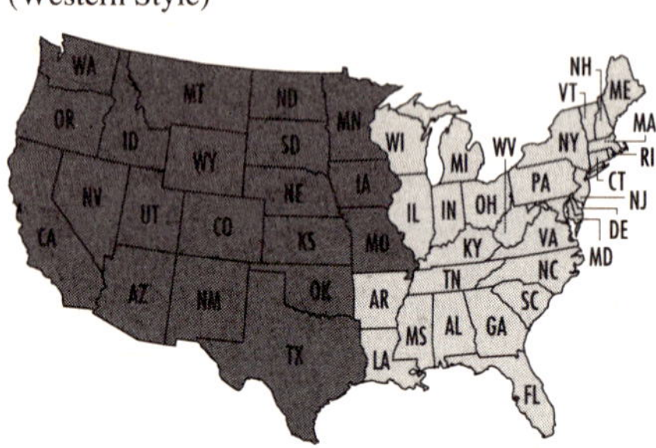

(Eastern Style)

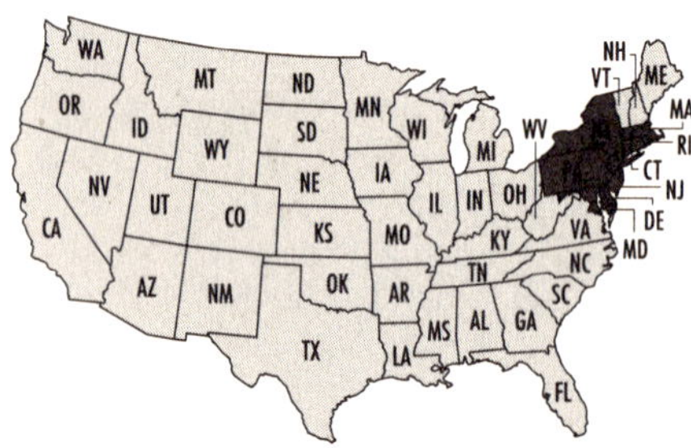

To show we aren't cheating, we added MD, DE, and RI for good measure

The maps convey their impressions in terms of *area* distribution, whereas personal income distribution may be quite different. The map on the left probably implies too high a level of government spending, while that on the right implies too low a level.

14. Long ago, when Johns Hopkins University had just begun to admit women students, someone not particularly enamored of coeducation reported a real shocker: Thirty three and one-third percent of the women at Hopkins had married faculty members! **It turns out that there were *three* women students, and just *one* had married a faculty member.**

15. The death rate in the Navy during the Spanish-American War was nine per thousand. For civilians in New York City during the same period it was sixteen per thousand. Navy recruiters later used these figures to show that it was safer to be in the Navy than out of it. **By the time the figures were used, circumstances may have changed greatly. (The Navy was much larger.) Also, New York City was most likely not typical of the nation as a whole.**

16. If you should look up the latest available figures on influenza and pneumonia, you might come to the strange conclusion that these ailments are practically confined to three southern states, which account for about eighty percent of the reported cases. **When Huff's book was written, those three states had the best system for reporting these diseases.**

When a sample is selected from a population, it is important to decide whether the sample is reasonably representative of the entire population. For example, a sample of the general population that is only 20% women should make us suspicious. The same is true of a questionnaire asking, "Do you like to answer questionnaires?"

In each exercise, pick the choice that gives the most representative sample. (Sampling techniques are a major branch of inferential statistics.*)*

17. A factory has 10% management employees, 30% clerical employees, and 60% assembly-line workers. A sample of 50 is chosen to discuss parking.
(a) 4 management, 21 clerical, 25 assembly-line
(b) 6 management, 15 clerical, 29 assembly-line
(c) 8 management, 9 clerical, 33 assembly-line (b)

18. A college has 35% freshmen, 28% sophomores, 21% juniors, and 16% seniors. A sample of 80 is chosen to discuss methods of electing student officers.
(a) 22 freshmen, 22 sophomores, 24 juniors, 12 seniors
(b) 24 freshmen, 20 sophomores, 22 juniors, 14 seniors
(c) 28 freshmen, 23 sophomores, 16 juniors, 13 seniors (c)

19. A computer company has plants in Boca Raton, Jacksonville, and Tampa. The plant in Boca Raton produces 42% of all the company's output, with 27% and 31% coming from Jacksonville and Tampa, respectively. A sample of 120 parts is chosen for quality testing.
(a) 38 from Boca Raton, 39 from Jacksonville, 43 from Tampa
(b) 43 from Boca Raton, 37 from Jacksonville, 40 from Tampa
(c) 50 from Boca Raton, 31 from Jacksonville, 39 from Tampa (c)

20. At one resort, 56% of all guests come from the Northeast, 29% from the Midwest, and 15% from Texas. A sample of 75 guests is chosen to discuss the dinner menu.
(a) 41 from the Northeast, 21 from the Midwest, 13 from Texas
(b) 45 from the Northeast, 18 from the Midwest, 12 from Texas
(c) 47 from the Northeast, 20 from the Midwest, 8 from Texas (a)

▲ *Use the information supplied in the following problems to solve for the given variables.*

21. An insurance agency has 7 managers, 25 agents, and 18 clerical employees. A sample of 10 is chosen.
(a) Let m be the number of managers in the sample, a the number of agents, and c the number of clerical employees. Find m, a, and c, if

$$c = m + 2$$
$$a = 2c.$$

$m = 1, a = 6, c = 3$

(b) To check that your answer is reasonable, calculate the numbers of the office staff that *should* be in the sample, if all groups are represented proportionately. **There should be 1.4 managers, 5 agents, and 3.6 clerical employees.**

22. A small college has 12 deans, 24 full professors, 39 associate professors, and 45 assistant professors. A sample of 20 employees is chosen to discuss the graduation speaker.
(a) Let d be the number of deans in the sample, f the number of full professors, a the number of associate professors, and s the number of assistant professors. Find d, f, a, and s if

$$f = 2d$$
$$a = f + d + 1$$
$$a = s.$$

(*Hint:* there are 2 deans.)
$d = 2, f = 4, a = 7, s = 7$

(b) To check that your answer is reasonable, calculate the number of each type of employee that *should* be in the sample, if all groups are represented proportionately. **There should be 2 deans, 4 full professors, 6.5 associate professors, and 7.5 assistant professors.**

12.6 Exercises (Text Page 740)

1. In a study to determine the linear relationship between the length (in decimeters) of an ear of corn (y) and the amount (in tons per acre) of fertilizer used (x), the following data were collected.

$$n = 10 \qquad \Sigma xy = 75$$
$$\Sigma x = 30 \qquad \Sigma x^2 = 100$$
$$\Sigma y = 24 \qquad \Sigma y^2 = 80$$

(a) Find an equation for the least squares line. $y' = .3x + 1.5$

(b) Find the coefficient of correlation. $r = .20$

(c) If 3 tons per acre of fertilizer are used, what length (in decimeters) would the equation in (a) predict for an ear of corn? **2.4 decimeters**

2. In an experiment to determine the linear relationship between temperatures on the Celsius scale (y) and on the Fahrenheit scale (x), a student got the following results.

$$n = 5 \qquad \Sigma xy = 28{,}050$$
$$\Sigma x = 376 \qquad \Sigma x^2 = 62{,}522$$
$$\Sigma y = 120 \qquad \Sigma y^2 = 13{,}450$$

(a) Find an equation for the least squares line. $y' = .556x - 17.8$

(b) Find the reading on the Celsius scale that corresponds to a reading of 120° Fahrenheit, using the equation of part (a). **48.9°**

(c) Find the coefficient of correlation. $r = 1$ **(The experimental points must lie perfectly along a line.)**

3. A sample of 10 adult men gave the following data on their heights and weights.

Height (inches) (x)	62	62	63	65	66	67	68	68	70	72
Weight (pounds) (y)	120	140	130	150	142	130	135	175	149	168

(a) Find the equation of the least squares line. $y' = 3.35x - 78.4$

(b) Using the results of (a), predict the weight of a man whose height is 60 inches. **123 lb**

(c) What would be the predicted weight of a man whose height is 70 inches? **156 lb**

(d) Compute the coefficient of correlation. $r = .66$

4. The table below gives reading ability scores and IQs for a group of 10 individuals.

Reading (x)	83	76	75	85	74	90	75	78	95	80
IQ (y)	120	104	98	115	87	127	90	110	134	119

(a) Plot a scatter diagram with reading on the horizontal axis.

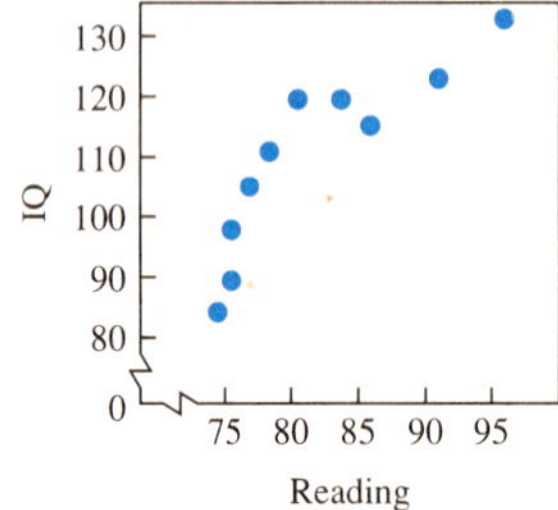

(b) Find the equation of a regression line. **$y' = 2x - 51$**
(c) Use the regression line equation to estimate the IQ of a person with a reading score of 65. **79**

5. Sales, in thousands of dollars, of a certain company are shown here.

Year (x)	0	1	2	3	4	5
Sales (y)	48	59	66	75	80	90

Find the equation of the least squares line. Find the coefficient of correlation.
$y' = 8.06x + 49.52$; $r = .996$

The data for the remaining exercises were adapted from the 1995 Information Please Almanac.

In each case, obtain the least squares regression line and the linear correlation coefficient, and make a statement describing the linear correlation (for example, as strong, moderate, *or* weak*).*

6. Largest U.S. Cities

Rank	City	Population (x)	Mayor's Salary (y)
1	New York	7,311,966	$130,000
2	Los Angeles	3,489,779	123,778
3	Chicago	2,768,483	115,000
4	Houston	1,690,180	133,000
5	Philadelphia	1,552,572	110,000
6	San Diego	1,148,851	65,300
7	Dallas	1,022,497	2,600
8	Phoenix	1,012,230	37,500
9	Detroit	1,012,110	117,000
10	San Antonio	966,437	3,000

$y' = .0139x + 53{,}270$; $r = .528$; The linear correlation is moderate.

7. World's Most Populous Cities

Rank	City	Population, x (in thousands)	Area, y (in square miles)
1	Tokyo-Yokohama, Japan	28,447	1,089
2	Mexico City, Mexico	23,913	522
3	Saõ Paulo, Brazil	21,539	451
4	Seoul, South Korea	19,065	342
5	New York, United States	14,638	1,274
6	Osaka-Kobe-Kyoto, Japan	14,060	495
7	Bombay, India	13,532	95
8	Calcutta, India	12,885	209
9	Rio de Janeiro, Brazil	12,788	260
10	Buenos Aires, Argentina	12,232	535

$y' = .0276x + 50.1$; $r = .413$; The linear correlation is moderate.

8.

Animal	Average Gestation or Incubation Period, x (days)	Record Life Span, y (years)
Cat	63	26
Dog	63	24
Duck	28	15
Elephant	624	71
Goat	151	17
Guinea pig	68	6
Hippopotamus	240	49
Horse	336	50
Lion	108	29
Parakeet	18	12
Pig	115	22
Rabbit	31	15
Sheep	151	16

$y' = .101x + 11.6$; $r = .909$; The linear correlation is fairly strong.

9. In this case, relate **(a)** period of revolution (y) to mean distance from sun (x), and **(b)** period of rotation (y) to mean distance from sun (x).

Planet	Period of Revolution (days)	Period of Rotation (hours)	Mean Distance From Sun (millions of miles)
Mercury	88	1,416	36
Venus	225	5,832	67
Earth	365	24	93
Mars	687	25	142
Jupiter	4,329	10	484
Saturn	10,753	11	887
Uranus	30,660	17	1,784
Neptune	60,225	16	2,796
Pluto	90,520	153	3,666

(a) $y' = 24.02x - 4{,}582$; $r = .989$; The linear correlation is strong.
(b) $y' = -.4972x + 1{,}384$; $r = -.347$; The linear correlation is moderate at best.

Chapter 12 Test (Text Page 743)

The bar graphs here show trends in several economic indicators over the period 1989–1994. Refer to these graphs for Exercises 1–4.

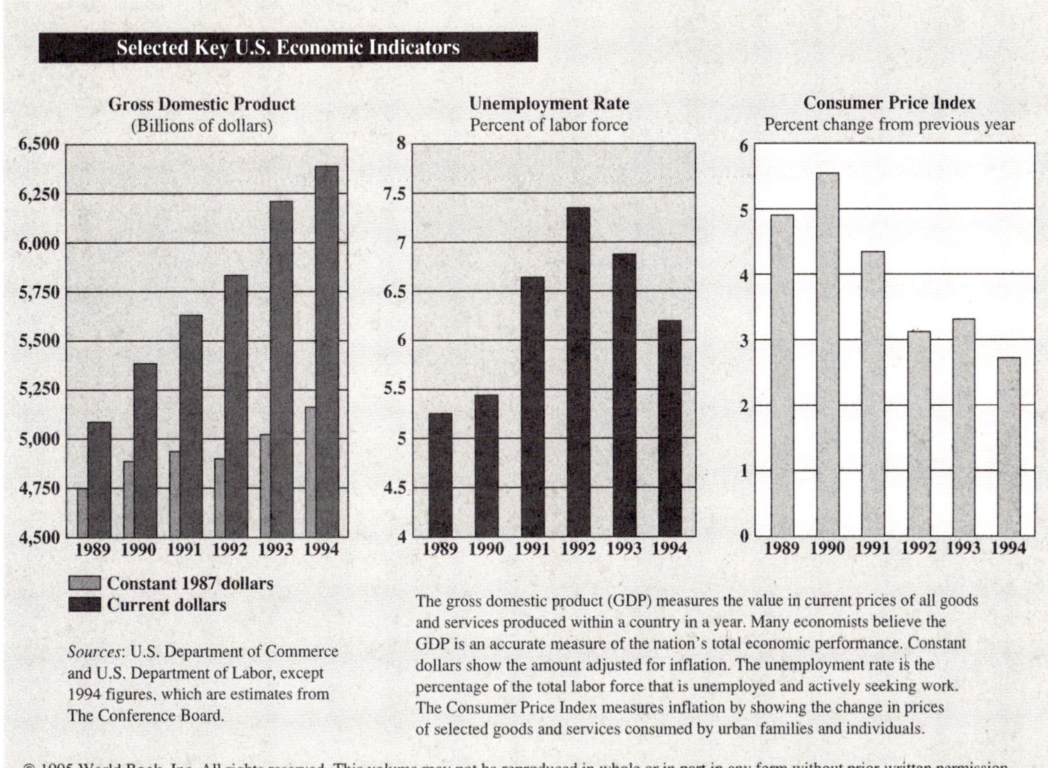

1. About what was the gross domestic product (current dollars) in 1992?
$5,840 billion

2. Over the six-year period, about what was the lowest percent change in consumer price index, and what year did it occur? 2.7%; 1994

3. About what was the highest unemployment rate, and what year did it occur?
7.3%; 1992

4. Describe the apparent trends in both constant dollar and current dollar gross domestic product. Describe how these two quantities seem to relate to one another over the years, and explain why this is so.

Nannette Williams, a sales representative for a publishing company, recorded the following numbers of client contacts for the twenty-two days that she was on the road in the month of March. Use the given data for Exercises 5–7.

12	8	15	11	20	18	14	22	13	26	17
19	16	25	19	10	7	18	24	15	30	24

5. Construct grouped frequency and relative frequency distributions. Use five uniform classes of width 5 where the first class has a lower limit of 6. (Round relative frequencies to two decimal places.)

Class Limits	Frequency f	Relative Frequency f/n
6–10	3	$3/22 \approx .14$
11–15	6	$6/22 \approx .27$
16–20	7	$7/22 \approx .32$
21–25	4	$4/22 \approx .18$
26–30	2	$2/22 \approx .09$

6. From your frequency distribution of Exercise 5, construct **(a)** a histogram and **(b)** a frequency polygon. Use appropriate scales and labels.

(a)

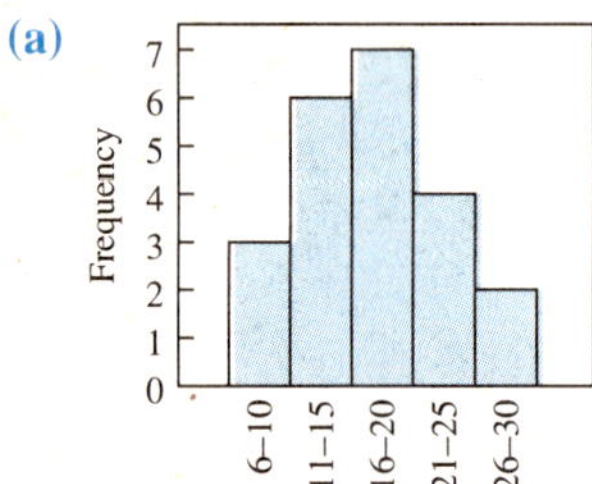

(b)

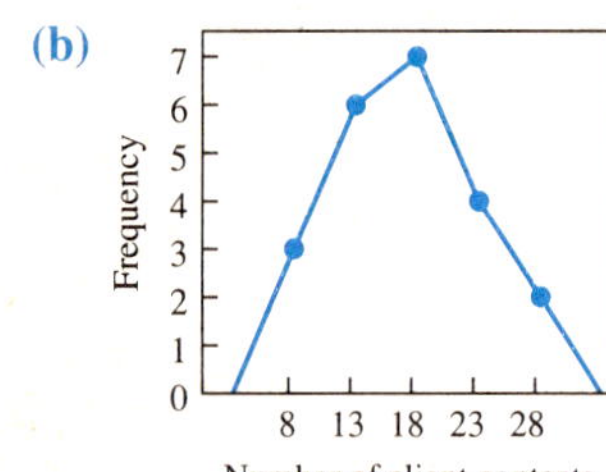

7. For the data above, how many uniform classes would be required if the first class had limits 7–9? 8

In Exercises 8–11, find the indicated measures for the following frequency distribution.

Value	4	6	8	10	12	14
Frequency	2	5	10	8	4	1

8. the mean 8.67 **9.** the median 8 **10.** the mode 8 **11.** the range 10

12. The following data are exam scores achieved by the students in a physics class. Arrange the data into a stem-and-leaf display with leaves ranked.

79	43	65	84	77	70	52	61	80	75	68	48	55	78	71
38	45	64	67	73	67	50	67	91	84	33	49	61	79	72

Stem	Leaves
3	3 8
4	3 5 8 9
5	0 2 5
6	1 1 4 5 7 7 7 8
7	0 1 2 3 5 7 8 9 9
8	0 4 4
9	1

Use the stem-and-leaf display shown here for Exercises 13–18.

Stem	Leaves
4	3 3 4
4	6 7 9 9
5	0 1 1 2 3 3 3 4
5	5 6 7 8 8 9
6	1 2 2 3
6	5 7
7	2 4
7	8
8	0
8	7

Compute the measures required in Exercises 13–17.

13. the median 55.5

14. the mode(s), if any 53

15. the range 44

16. the third decile 51

17. the fifteenth percentile 47

18. Construct a box plot for the above data, showing values for the five important quantities on the numerical scale.

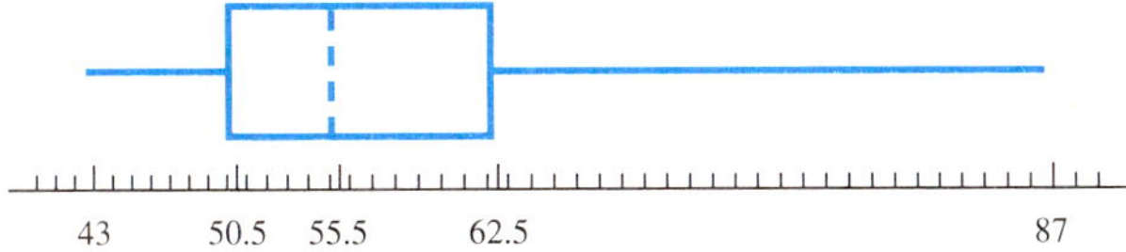

A certain training institute gives a standardized test to large numbers of applicants nationwide. The resulting scores form a normal distribution with mean 75 and standard deviation 10. Find the percent of all applicants with scores as follows. (Use the empirical rule.)

19. between 55 and 95 about 95%

20. greater than 105 or less than 45 about .3%

21. less than 65 about 16%

22. between 85 and 95 about 13.5%

In a certain young forest, the heights of the fir trees are normally distributed with mean 10.6 meters and standard deviation 3.1 meters. If a single tree is selected randomly, find the probability (to the nearest thousandth) that its height will fall in each of the following intervals.

23. less than 8.5 meters .248

24. between 14.3 and 16.2 meters .082

The 1994 Major League Baseball season effectively ended on August 11 when the Players Association went on strike. The table below gives statistics through that date for each of the three divisions of the two leagues. In each case, n = number of teams in the division, $\bar{x}$ = average (mean) number of games won, and s = standard deviation of number of games won. Refer to the table for Exercises 25–27.

American League			National League		
Eastern Division	**Central Division**	**Western Division**	**Eastern Division**	**Central Division**	**Western Division**
$n = 5$ $\bar{x} = 59$ $s = 7.3$	$n = 5$ $\bar{x} = 60.0$ $s = 7$	$n = 4$ $\bar{x} = 49.8$ $s = 2.2$	$n = 5$ $\bar{x} = 60.4$ $s = 10$	$n = 5$ $\bar{x} = 57.4$ $s = 8$	$n = 4$ $\bar{x} = 53.3$ $s = 4.6$

25. Overall, who had the lowest winning average, the Eastern teams, the Central teams, or the Western teams? **Western**

26. Overall, where were division teams the most "consistent" in number of games won, Eastern, Central, or Western? **Western**

27. Find (to the nearest tenth) the average number of games won for all 28 teams. **57.0**

Carry out the following for the paired data values shown here. (In Exercises 29–31, give all calculated values to two decimal places.)

x	1	4	6	7
y	9	7	8	1

28. Plot a scatter diagram.

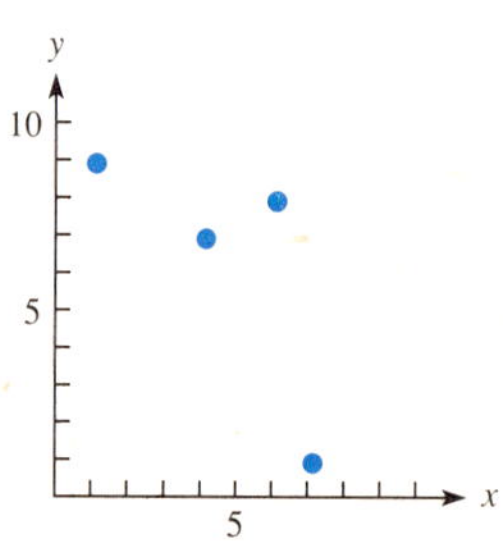

29. Find the equation for the least squares regression line. $y' = -.98x + 10.64$

30. Use your equation from Exercise 29 to predict y when $x = 3$. **7.70**

31. Find the sample correlation coefficient for the given data. **−.72**

32. Evaluate the strength of the linear relationship for the above data. How confident are you in your evaluation? Explain.

33. Relate the concepts of *inferential statistics* and *inductive reasoning.*

 CONCEPTUAL  WRITING  CHALLENGING SCIENTIFIC CALCULATOR GRAPHING CALCULATOR

CHAPTER 13 Consumer Mathematics

13.1 Exercises (Text Page 758)

Remember that answers obtained using Table 2 may lack accuracy and thus not agree exactly with answers given in the back of the text. Also, unless otherwise known, assume 12 months per year, 30 days per month, and 365 days per year whenever such equivalences are needed.

Find the simple interest owed for each of the following loans.

1. $1,200 at 10% for 1 year **$120**
2. $8,000 at 8% for 1 year **$640**
3. $450 at 9% for 9 months **$30.38**
4. $5,000 at 7.5% for 4 months **$125**
5. $2,675 at 9.2% for $2\frac{1}{2}$ years **$615.25**
6. $1,580 at 6.75% for 32 months **$284.40**

Find the future value of each of the following deposits if the account pays **(a)** *simple interest, and* **(b)** *interest compounded annually.*

7. $500 at 4% for 3 years
   (a) $560 (b) $562.43
8. $1,000 at 5% for 6 years
   (a) $1,300 (b) $1,340.10
9. $2,500 at 7% for 4 years
   (a) $3,200 (b) $3,276.99
10. $5,000 at 6% for 2 years
    (a) $5,600 (b) $5,618

Solve the following interest-related problems.

11. Ezra Mason was late on his property tax payment to the county. He owed $7,500 and paid the tax 4 months late. The county charges a penalty of 10% simple interest. Find the amount of the penalty.
    $250
12. Charles Paisley bought a new supply of police uniforms. He paid $625 for the uniforms and agreed to pay for them in 9 months at 13% simple interest. Find the amount of interest that he will owe.
    $60.94
13. Elanna Williams opened a security service on March 1. To pay for office furniture and guard dogs, Williams borrowed $12,500 at the bank and agreed to pay the loan back in 7 months at 12% simple interest. Find the *total amount* she must repay.
    $13,375
14. David Fontana is owed $180 by the Internal Revenue Service for overpayment of last year's taxes. The I.R.S. will repay the amount at 8% simple interest. Find the *total amount* Fontana will receive if the interest is paid for 9 months. **$190.80**

Find the missing final amount (future value) and/or interest earned.

	Principal	Rate	Compounded	Time	Final Amount	Compound Interest
15.	$ 975	8%	quarterly	4 years	$1,338.47	______
16.	$1,150	7%	semiannually	6 years	$1,737.73	______
17.	$ 480	6%	semiannually	9 years	______	$337.17
18.	$2,370	10%	quarterly	5 years	______	______
19.	$7,500	$5\frac{1}{2}$%	annually	25 years	______	______
20.	$3,450	12%	semiannually	10 years	______	______

15. $363.47 16. $587.73 17. $817.17 18. $3,883.52; $1,513.52
19. $28,600.44; $21,100.44 20. $11,064.62; $7,614.62

For each of the following deposits, find the future value *(final amount on deposit) when compounding occurs* **(a)** *annually,* **(b)** *semiannually, and* **(c)** *quarterly.*

	Principal	Rate	Time
21.	$1,000	8%	3 years
23.	$8,000	5%	5 years

	Principal	Rate	Time
22.	$ 4,000	6%	6 years
24.	$15,000	8%	6 years

21. (a) $1,259.71 (b) $1,265.32 (c) $1,268.24
22. (a) $5,674.08 (b) $5,703.04 (c) 5,718.01
23. (a) $10,210.25 (b) $10,240.68 (c) $10,256.30
24. (a) $23,803.11 (b) $24,015.48 (c) $24,126.56

Occasionally a bank may actually pay interest compounded continuously. For each of the following deposits, find the interest earned *if interest is compounded* **(a)** *semiannually,* **(b)** *quarterly,* **(c)** *monthly,* **(d)** *daily, and* **(e)** *continuously. (Recall that interest = future value − present value.)*

	Principal	Rate	Time
25.	$ 500	7.3%	4 years
26.	$1,550	8.5%	33 months (assume 1,003 days)

25. (a) $166.08 (b) $167.79 (c) $168.96 (d) $169.53 (e) $169.55
26. (a) $398.72 (b) $403.37 (c) $406.55 (d) $407.76 (e) $408.16

27. Describe the effect of interest being compounded more and more often. In particular, how good is continuous compounding?

Work the following interest-related problems.

28. Chris Siragusa takes out an 11% simple interest loan today which will be repaid 15 months from now with a payoff amount of $796.25. What amount is Chris borrowing? **$700**

29. What amount can you borrow today if it must be repaid in 5 months with simple interest at 10% and you know that at that time you will be able to repay no more than $1,500? **$1,440**

30. In the development of the future value formula for compound interest in the text at least four specific problem solving strategies were employed. Identify (name) as many of them as you can and describe their use in this case.

Find the present value for each of the following future amounts. (Table 2 may be applied in some cases, but may limit the accuracy of answers.)

31. $1,000 (6% compounded annually for 5 years) **$747.26**

32. $14,000 (4% compounded quarterly for 3 years) **$12,424.29**

33. $9,860 (8% compounded semiannually for 10 years) **$4,499.98**

34. $15,080 (5% compounded monthly for 4 years) **$12,351.59**

Ron and Diane want to establish an account that will help their 21-year-old son retire when he reaches age 61. For each of the following interest rates find the lump sum they must deposit today so that $500,000 will be available for retirement.

35. 8% compounded quarterly **$21,035.00**

36. 12% compounded quarterly **$4,415.85**

37. 8% compounded daily **$20,388.25**

38. 12% compounded daily **$4,118.12**

 CONCEPTUAL  WRITING  CHALLENGING  SCIENTIFIC CALCULATOR  GRAPHING CALCULATOR

Suppose a savings and loan pays a nominal rate of 7% on savings deposits. Find the effective annual yield *if interest is compounded as stated in Exercises 39–45. (Give answers to the nearest thousandth of a percent.)*

39. annually 7.000%

40. semiannually 7.123%

41. quarterly 7.186%

42. monthly 7.229%

43. daily 7.250%

44. 1,000 times per year 7.251%

45. 10,000 times per year 7.251%

46. Judging from Exercises 39–45, what do you suppose is the effective annual yield if a nominal rate of 7% is compounded continuously? Explain your reasoning.

On July 21, 1995, Schools Federal Credit Union in Sacramento posted the following savings rates for certificates of deposit (CDs) in several categories:

		Balance $10,000	Balance $50,000
12-month	Rate	6.00	6.10
	Yield	6.14	6.24
30-month	Rate	6.50	6.60
	Yield	6.66	6.77

47. If you put $40,000 into a 12-month CD, how much would you have in 1 year? $42,456

48. How often does compounding occur in these accounts? quarterly

Solve the following interest-related problems.

49. How long would it take to double your money in an account paying 9% compounded semiannually? (Give your answer in years plus days, ignoring leap years.)
7 years, 319 days

50. After what time period would the interest earned equal the original principal if the account pays 9% compounded daily? (Give your answer in years plus days, ignoring leap years.)
7 years, 256 days

51. At Greenway Savings interest is compounded monthly and the effective annual yield is 5.85%. What is the nominal rate? 5.70%

52. In Fair Oaks, one bank pays a nominal rate of 6.80% compounded daily on deposits. A second bank produces the same yield as the first but only compounds interest quarterly.

(a) What nominal rate does the second bank pay? 6.86%

(b) Which bank should Dave choose if he has $3,000 to deposit for 7 months? (Assume that the second bank pays no interest on funds deposited for less than an entire quarter.)
first bank

(c) Which bank should Carole choose if she has $5,000 to deposit for 1 year? no difference

Use Table 2 to estimate the number of years it would take for the general level of prices to double at the following annual inflation rates.

53. 3% 24 years **54.** 4% 18 years

55. 5% 15 years **56.** 6% 12 years

Use the rule of 70 to estimate the years to double for the following annual inflation rates.

57. 1% 70 years **58.** 2% 35 years

59. 8% 9 years **60.** 9% 8 years

Use the rule of 70 to estimate the annual inflation rate (to the nearest tenth of a percent) that would cause the general level of prices to double in the following time periods.

61. 6 years 11.7% **62.** 15 years 4.7%

Work the following inflation-related problems.

63. Derive a rule for estimating the "years to triple," that is, the number of years it would take for the general level of prices to triple for a given annual inflation rate.

$$\text{years to triple} = \frac{100 \ln 3}{\text{annual inflation rate}} \approx \frac{110}{\text{annual inflation rate}}$$

64. What would you call your rule of Exercise 63? Explain your reasoning.

The 1996 prices of several items are given below. Find the estimated future costs required to fill the blanks in the chart. (Give a number of significant figures consistent with the 1996 cost figures provided.)

Item	1996 Cost	2001 Cost 2% Inflation	2016 Cost 2% Inflation	2001 Cost 10% Inflation	2016 Cost 10% Inflation
65. House	$95,000	______	______	______	______
66. Meal Deal	$3.99	______	______	______	______
67. Gallon of gasoline	$1.34	______	______	______	______
68. Small car	$11,500	______	______	______	______

65. $105,000; $142,000; $157,000; $702,000 66. $4.41; $5.95; $6.58; $29.48
67. $1.48; $2.00; $2.21; $9.90 68. $12,700; $17,200; $19,000; $85,000

Work the following interest-related problems.

69. Mr. and Mrs. Ghardellini are close to retirement. They are selling a piece of land to a developer. The couple will be in a lower tax bracket in 4 years, and they wish to defer receipt of the money until then. Find the lump sum that the developer can deposit today, at 10% compounded quarterly, so that enough will be available to pay $200,000 to the couple in 4 years. **$134,724.99**

70. Radiology, Inc. will need $38,000 in 2 1/2 years when a new laser machine becomes available. What lump sum can they invest today, at 9% compounded semiannually, to have the required amount for the machine? **$30,493.14**

Extension Exercises (Text Page 763)

For each of the following ordinary annuity accounts, find **(a)** *the future value of the account (at the end of the accumulation period),* **(b)** *the total of all deposits, and* **(c)** *the total interest earned.*

Regular Deposit	Compounded	Annual Interest Rate	Accumulation Period
1. $1,000	yearly	8.5%	10 years
2. $2,000	yearly	10.0%	20 years
3. $50	monthly	6.0%	5 years
4. $75	monthly	7.2%	10 years
5. $20	weekly	5.2%	3 years
6. $30	weekly	7.8%	5 years

1. (a) $14,835.10 (b) $10,000 (c) $4,835.10
2. (a) $114,550 (b) $40,000 (c) $74,550
3. (a) $3,488.50 (b) $3,000 (c) $488.50
4. (a) $13,125.23 (b) $9,000 (c) $4,125.23
5. (a) $3,374.70 (b) $3,120 (c) $254.70
6. (a) $9,530.99 (b) $7,800 (c) $1,730.99

An annuity account into which regular deposits are made at the beginning *of each compounding period, rather than at the end, is called an* **annuity due.** *(It is usually assumed that a deposit is made at the end of compounding period number n so that the future value of an annuity due is the accumulation of n + 1 deposits rather than just n.)*

7. If the interest rate, regular deposit amount, compounding frequency, and accumulation period are all the same in both cases, would the future value of an annuity due be less or more than that of an ordinary annuity? Explain.

8. Modify the derivation of the formula for future value of an ordinary annuity (presented in the text) to show that the **future value of an annuity due** is given by

$$V = \frac{R\left[\left(1 + \frac{r}{m}\right)^{n+1} - 1\right]}{\frac{r}{m}}.$$

Use the formula of Exercise 8 to find the future value of each annuity due given below.

	Regular Deposit	Compounded	Annual Interest Rate	Accumulation Period	
9.	\$50	monthly	6.0%	5 years	\$3,555.94
10.	\$75	monthly	7.2%	10 years	\$13,278.98
11.	\$20	weekly	5.2%	3 years	\$3,398.08
12.	\$30	weekly	7.8%	5 years	\$9,575.28

13. Do you think it is practical for a financial institution (or insurance company) to guarantee to pay you a set interest rate for years into the future on your annuity account? Why or why not?

Interview some of your friends or acquaintances, teachers, or other employees of schools, hospitals, or other non-profit organizations and ask them about "tax deferred annuities" (TDAs) or "tax sheltered annuities" (TSAs). Explain the following features of these accounts.

14. their basic purpose

15. their tax advantages

16. their interest rates (as compared to those of regular savings accounts)

17. penalties for excessive deposits or for early withdrawals

18. the "annuitization period" (as opposed to the accumulation period)

13.2 Exercises (Text Page 769)

Suppose you buy appliances costing \$1,750 at a store charging 11% add-on interest, and you make a \$500 down payment.

1. Find the total amount you will be financing. \$1,250

2. Find the total interest if you will pay off the loan over a 2-year period. \$275

3. Find the total amount owed. \$1,525

4. The amount from Exercise 3 is to be repaid in 24 monthly installments. Find the monthly payment. \$63.54

Suppose you want to buy a new car that costs $12,000. You have no cash—only your old car, which is worth $3,000 as a trade-in.

5. How much do you need to finance to buy the new car? **$9,000**

6. The dealer says the interest rate is 9% add-on for 4 years. Find the total interest. **$3,240**

7. Find the total amount owed. **$12,240**

8. Find the monthly payment. **$255**

In the following exercises, find the total interest and the monthly payment using the add-on method of calculating interest. Round each answer to the nearest cent.

	Amount of Loan	Length of Loan	Interest Rate		Amount of Loan	Length of Loan	Interest Rate
9.	$3,700	3 years	9%	**13.**	$ 158	10 months	13%
10.	$1,746	24 months	10%	**14.**	$2,100	24 months	12%
11.	$ 400	1 year	13%	**15.**	$2,700	36 months	11%
12.	$2,896	18 months	9.2%	**16.**	$ 798	9 months	10.3%

9. $999; $130.53 10. $349.20; $87.30 11. $52; $37.67 12. $399.65; $183.09
13. $17.12; $17.51 14. $504; $108.50 15. $891; $99.75 16. $61.65; $95.52

Work the following problems. Give monetary answers to the nearest cent unless directed otherwise.

17. The Giordanos buy $7,000 worth of furniture for their new home. They pay $2,000 down. The store charges 10% add-on interest. The Giordanos will pay off the furniture in 30 monthly payments (2 1/2 years). Find the monthly payment. **$208.33**

18. Find the monthly payment required to pay off an auto loan of $6,780 over 3 years if the add-on interest rate is 11.3%. **$252.18**

19. The total purchase price of a new home entertainment system is $15,480. If the down payment is $3,500 and the balance is to be financed over 48 months at 12% add-on interest, what is the monthly payment? **$369.38**

20. What are the monthly payments Crystal Clymo pays on a loan of $790 for a period of 10 months if 9% add-on interest is charged? **$84.93**

▲ **21.** Jeff Klinger has misplaced the sales contract for his car and cannot remember the amount he originally financed. He does know that the add-on interest rate was 8.8% and the loan required a total of 36 monthly payments of $196.62 each. How much did Jeff borrow (to the nearest dollar)? **$5,600**

▲ **22.** Kelly Ng is making monthly payments of $318.35 to pay off a 2 1/2 year loan for $7,500. What is her add-on interest rate (to the nearest tenth of a percent)? **10.9%**

▲ **23.** How long (in years) will it take Matt Byerly to pay off a $10,000 loan with monthly payments of $245.83 if the add-on interest rate is 9.5%? **5 years**

▲ **24.** How many monthly payments must Jawann make on a $7,000 loan if he pays $354 a month and the add-on interest rate is 10.69%? **24**

Find the finance charge on each of the following open-end charge accounts. Assume interest is calculated on the unpaid balance of the account. Round to the nearest cent.

	Unpaid Balance	Monthly Interest Rate	
25.	$244.88	1.8%	**$4.41**
26.	$500.19	1.6%	**$8.00**
27.	$167.99	1.78%	**$2.99**
28.	$877.11	1.49%	**$13.07**

Complete each of the following tables, showing the unpaid balance at the end of each month. Assume a monthly interest rate of 1.5% on the unpaid balance.

29.

Month	Unpaid Balance at Beginning of Month	Finance Charge	Purchases During Month	Returns	Payment	Unpaid Balance at End of Month
February	$297.11	______	$86.14	0	$50	______
March	______	______	109.83	$15.75	60	______
April	______	______	39.74	0	72	______
May	______	______	56.29	18.09	50	______

Month	Unpaid Balance at Beginning of Month	Finance Charge	Purchases During Month	Returns	Payment	Unpaid Balance at End of Month
February	$297.11	$4.46	$ 86.14	0	$50	$337.71
March	337.71	5.07	109.83	$15.75	60	376.86
April	376.86	5.65	39.74	0	72	350.25
May	350.25	5.25	56.29	18.09	50	343.70

30.

Month	Unpaid Balance at Beginning of Month	Finance Charge	Purchases During Month	Returns	Payment	Unpaid Balance at End of Month
October	$554.19	______	$128.72	$23.15	$125	______
November	______	______	291.64	0	170	______
December	______	______	147.11	17.15	150	______
January	______	______	27.84	139.82	200	______

Month	Unpaid Balance at Beginning of Month	Finance Charge	Purchases During Month	Returns	Payment	Unpaid Balance at End of Month
October	$554.19	$ 8.31	$128.72	$ 23.15	$125	$543.07
November	543.07	8.15	291.64	0	170	672.86
December	672.86	10.09	147.11	17.15	150	662.91
January	662.91	9.94	27.84	139.82	200	360.87

31.

Month	Unpaid Balance at Beginning of Month	Finance Charge	Purchases During Month	Returns	Payment	Unpaid Balance at End of Month
August	$822.91	______	$155.01	$38.11	$100	______
September	______	______	208.75	0	75	______
October	______	______	56.30	0	90	______
November	______	______	190.00	83.57	150	______

Month	Unpaid Balance at Beginning of Month	Finance Charge	Purchases During Month	Returns	Payment	Unpaid Balance at End of Month
August	$822.91	$12.34	$155.01	$38.11	$100	$852.15
September	852.15	12.78	208.75	0	75	998.68
October	998.68	14.98	56.30	0	90	979.96
November	979.96	14.70	190.00	83.57	150	951.09

32.

Month	Unpaid Balance at Beginning of Month	Finance Charge	Purchases During Month	Returns	Payment	Unpaid Balance at End of Month
March	$1,522.83	______	$308.13	$74.88	$250	______
April	______	______	488.35	0	350	______
May	______	______	134.99	18.12	175	______
June	______	______	157.72	0	190	______

Month	Unpaid Balance at Beginning of Month	Finance Charge	Purchases During Month	Returns	Payment	Unpaid Balance at End of Month
March	$1,522.83	$22.84	$308.13	$74.88	$250	$1,528.92
April	1,528.92	22.93	488.35	0	350	1,690.20
May	1,690.20	25.35	134.99	18.12	175	1,657.42
June	1,657.42	24.86	157.72	0	190	1,650.00

Find the finance charge for each of the following charge accounts. Assume interest is calculated on the average daily balance of the account.

	Average Daily Balance	Monthly Interest Rate	
33.	$ 138.20	1 1/2%	$2.07
34.	$ 350.49	1 3/4%	$6.13
35.	$ 746.11	1.667%	$12.44
36.	$1,490.74	1.805%	$26.91

 CONCEPTUAL  WRITING ▲ CHALLENGING  SCIENTIFIC CALCULATOR  GRAPHING CALCULATOR

Find the average daily balance for each of the following credit card accounts. Assume one month between billing dates (using the proper number of days in the month). Then find the finance charge, if interest is 1.5% per month on the average daily balance.

37. Previous balance: $906.14

May 9	Billing date	
May 17	Payment	$200
May 30	Dinner	$ 46.11

$772.63; $11.59

38. Previous balance: $228.95

January 27	Billing date	
February 9	Cheese	$11.08
February 13	Returns	$26.54
February 20	Payment	$29
February 25	Repairs	$71.19

$221.44; $3.32

39. Previous balance: $312.78

June 11	Billing date	
June 15	Returns	$106.45
June 20	Jewelry	$115.73
June 24	Car rental	$ 74.19
July 3	Payment	$115

$312.91; $4.69

40. Previous balance: $714.58

August 17	Billing date	
August 21	Mail order	$ 14.92
August 23	Returns	$ 25.41
August 27	Beverages	$ 31.82
August 31	Payment	$108
September 9	Returns	$ 71.14
September 11	Plane ticket	$110
September 14	Cash advance	$100

$682.02; $10.23

Assume no purchases or returns are made in Exercises 41 and 42.

41. At the beginning of a 31-day billing period, Eleanora Garza has an unpaid balance of $720 on her credit card. Three days before the end of the billing period, she pays $600. Find her finance charge at 1.5% per month using the following methods.

(a) unpaid balance method **$10.80**

(b) average daily balance method **$9.93**

42. Jack Dominick's VISA bill dated April 14 shows an unpaid balance of $1,070. Five days before May 14, the end of the billing period, Dominick makes a payment of $900. Find his finance charge at 1.6% per month using the following methods.

(a) unpaid balance method **$17.12**

(b) average daily balance method **$14.72**

One version of the "90 Days Same as Cash" promotion is offered by a "major purchase card," which establishes an account charging 1.8583% interest per month on the account balance. Interest charges are added to the balance each month, becoming part of the balance on which interest is computed the next month. If you pay off the original purchase charge within 3 months, all interest charges are cancelled. Otherwise you are liable for all the interest. Suppose you purchase $2,400 worth of carpeting under this plan.

43. Find the interest charge added to the account balance at the end of

(a) the first month, **$44.60**

(b) the second month, **$45.43**

(c) the third month. **$46.27**

44. Suppose you pay off the account 1 day late (3 months plus 1 day). What total interest amount must you pay? (For purpose of calculating interest, ignore the extra day.) **$136.30**

45. Treating the 3 months as 1/4 year, find the equivalent simple interest rate for this purchase (to the nearest tenth of a percent). **22.7%**

Carole's bank card account charges 1.5% per month on the average daily balance as well as the following special fees:

Cash advance fee:	*2% (not less than $2 nor more than $10)*
Late payment fee:	*$15*
Over-the-credit-limit fee:	*$15*

In the month of June, Carole's average daily balance was $1,846. She was on vacation during the month and did not get her account payment in on time, which resulted in a late payment and also resulted in charges accumulating to a sum above her credit limit. She also used her card for six $100 cash advances while on vacation. Find the following based on account transactions in that month.

46. interest charges to the account **$27.69**

47. special fees charged to the account **$42**

Write out your responses to each of the following.

48. Is it possible to use a bank credit card for your purchases without paying anything for credit? If so, explain how.

49. Obtain applications or descriptive brochures for several different bank card programs, compare their features (including those in fine print), and explain which deal would be best for you, and why.

50. Research and explain the difference, if any, between a "credit" card and a "debit" card.

51. Many charge card offers include the option of purchasing credit insurance coverage, which would make your monthly payments if you became disabled and could not work and/or would pay off the account balance if you died. Find out the details on at least one such offer, and discuss why you would or would not accept it.

52. Make a list of "special incentives" offered by bank cards you are familiar with, and briefly describe the pros and cons of each one.

53. Early in 1995, one bank offered a card with a "low introductory rate" of 8.9%, good through the end of the year. And furthermore, you could receive back a percentage (up to 100%!) of all interest you pay, as shown in the table.

Use your card for:	2 years	5 years	10 years	15 years	20 years
Get back:	10%	25%	50%	75%	100%

(As soon as you take a refund, the time clock starts over.) Since you can eventually claim all your interest payments back, is this card a good deal? Explain why or why not.

54. Either recall a car-buying experience you have had, or visit a new-car dealer and interview a salesperson. Write a description of the procedure involved in purchasing a car on credit.

Daniel and Nora Onishi are considering two bank card offers that are the same in all respects except that Bank A charges no annual fee and charges monthly interest of 1.6% on the unpaid balance while Bank B charges a $40 annual fee and monthly interest of 1.3% on the unpaid balance. From their records, the Onishis have found that the unpaid balance they tend to carry from month to month is quite consistent and averages $900.

55. Estimate their total yearly cost to use the card if they choose the card from
(a) Bank A **$172.80**
(b) Bank B. **$180.40**

56. Which card is their better choice? **Bank A**

57. If Bank B raised its annual fee to $50 (and continued to charge 1.3% per month interest), then how high a monthly interest rate (to the nearest tenth of a percent) could Bank A charge and still offer the Onishis a more economical card? **1.7%**

58. Under the initial charges (1.6% per month for Bank A and 1.3% per month plus $40 per year for Bank B), what is the least average monthly unpaid balance (to the nearest cent) that would make Bank B's card a better deal for the Onishis? **$1,111.12**

13.3 Exercises (Text Page 780)

Find the APR (true annual interest rate), to the nearest half percent, for each of the following.

	Amount Financed	Finance Charge	Number of Payments	
1.	$1,200	$ 46	6	13.0%
2.	$4,800	$ 630	18	16.0%
3.	$5,000	$ 850	24	15.5%
4.	$4,900	$1,120	48	10.5%

CONCEPTUAL  WRITING CHALLENGING SCIENTIFIC CALCULATOR  GRAPHING CALCULATOR

Find the monthly payment for each of the following. Round to the nearest cent.

	Purchase Price	Down Payment	Finance Charge	Number of Payments	
5.	$2,000	$ 500	$ 100	24	**$66.67**
6.	$4,280	$ 280	$1,200	36	**$144.44**
7.	$3,950	0	$ 950	48	**$102.08**
8.	$9,000	$3,000	$1,500	60	**$125.00**

Find the APR (true annual interest rate), to the nearest half percent, for each of the following.

	Purchase Price	Down Payment	Add-on Interest Rate	Number of Payments	
9.	$ 4,190	$ 390	8%	18	**14.5%**
10.	$ 3,250	$ 750	9%	48	**16.0%**
11.	$ 7,480	$2,200	6%	12	**11.0%**
12.	$12,800	$4,500	7%	36	**13.0%**

For each of the following loans, find **(a)** *the finance charge, and* **(b)** *the APR.*

13. Frank Barry financed a $540 television with 18 monthly payments of $33.40 each. **(a) $61.20 (b) 14.0%**

14. Pat Peterson bought a horse trailer for $4,090. She paid $1,150 down and paid the remainder at $112.50 per month for 2 1/2 years. **(a) $435 (b) 11.0%**

15. Mike Karelius still owed $720 on his new refrigerator after the down payment. He agreed to pay monthly payments for a year at 6% add-on interest. **(a) $43.20 (b) 11.0%**

16. Phil Givant paid off a $12,000 car loan over 4 years with monthly payments of $304.35 each. **(a) $2,608.80 (b) 10.0%**

Each of the following loans was paid in full before its due date. **(a)** *Obtain the value of h from Table 3. Then* **(b)** *use the actuarial method to find the amount of unearned interest, and* **(c)** *find the payoff amount.*

	Regular Payment	APR	Remaining Number of Scheduled Payments After Payoff
17.	$285.20	10.5%	6
18.	$410.00	15.0%	18
19.	$595.80	12.0%	12
20.	$314.50	11.5%	24

17. (a) $3.08 (b) $51.13 (c) $1,945.27 18. (a) $12.29 (b) $807.73 (c) $6,982.27
19. (a) $6.62 (b) $443.92 (c) $7,301.48 20. (a) $12.42 (b) $833.89 (c) $7,028.61

Each of the following loans was paid off early. Find the unearned interest by **(a)** *the actuarial method, and* **(b)** *the rule of 78.*

	Amount Financed	Regular Monthly Payment	Total Number of Payments Scheduled	Remaining Number of Scheduled Payments After Payoff
21.	$ 1,650	$101.39	18	6
22.	$12,460	$319.03	48	12
23.	$ 9,850	$209.28	60	12
24.	$ 6,730	$230.02	36	18

21. (a) $22.44 (b) $21.49 22. (a) $209.19 (b) $189.26
23. (a) $130.92 (b) $115.37 24. (a) $425.37 (b) $398.16

Each of the following loans was paid in full before its due date. **(a)** *Obtain the value of h from the appropriate formula. Then* **(b)** *use the actuarial method to find the amount of unearned interest, and* **(c)** *find the payoff amount.*

	Regular Payment	APR	Remaining Number of Scheduled Payments After Payoff
25.	$148	9.1%	3
26.	$430	10.25%	5

25. (a) $1.52 (b) $6.65 (c) $585.35 26. (a) $2.58 (b) $54.07 (c) $2,525.93

Wei-Jen Luan needs to borrow $3,000 to supplement her down payment on a classic automobile. She can either borrow the $3,000 from a finance company at 7% add-on interest for 4 years or take a credit union loan that will require 48 monthly payments of $78.25 each. Use this information for the following exercises.

27. Find the APR (to the nearest half percent) for each loan and decide which one is Wei-Jen's better choice.
finance company APR: 12.5%; credit union APR: 11.5%; choose credit union

28. Suppose Wei-Jen takes the credit union loan. At the time of her fortieth payment she decides that this loan is a nuisance and decides to pay it off. If the credit union uses the rule of 78 for computing unearned interest, how much will she save by paying in full now? **$23.14**

29. What would Wei-Jen save in interest if she paid in full at the time of the fortieth payment and the credit union used the actuarial method for computing unearned interest? **$26.15**

30. Under the conditions of Exercise 29, what amount must Wei-Jen come up with to pay off her loan? **$678.10**

To convert an add-on interest rate to its corresponding APR, some people recommend using the formula

$$\text{APR} = \frac{2n}{n+1} \times r,$$

where r is the add-on rate and n is the total number of payments.

31. Apply the given formula to calculate the APR (to the nearest half percent) for the loan of Example 1 ($r = .08$, $n = 24$). **15.5%**

32. Compare your APR value in Exercise 31 to the value of Example 1. What do you conclude?

CONCEPTUAL  WRITING  CHALLENGING SCIENTIFIC CALCULATOR GRAPHING CALCULATOR

Exercises 33 and 34 also refer to Example 1 in the text.

33. Describe why, in the example, the APR and the add-on rate differ. Which one is more legitimate? Why?

34. In the example, the Hoangs paid off a $2,200 loan over 2 years (24 monthly payments), and their 8% add-on rate turned out to be equivalent to 14.5% APR. Compute the APR equivalent to 8% add-on given that they were paying over the following time periods.

(a) 3 years 14.5% **(b)** 4 years 14.5%
(c) 5 years 14.0%

A certain retailer's credit contract designates the rule of 78 for computing unearned interest, and also imposes a "prepayment penalty." In case of any payoff earlier than the due date, they will charge an additional 10% of the original finance charge. Find the least value of k (remaining payments after payoff) that would result in any net savings in each of the following cases.

35. 24 payments originally scheduled 8

36. 36 payments originally scheduled 12

The actuarial method of computing unearned interest assumes that, throughout the life of the loan, the borrower is paying interest at the rate given by APR for money actually being used by the borrower. When contemplating complete payoff along with the current payment, think of the k future payments as applying to a separate loan with the same APR, h being the finance charge per $100 of that *loan. Refer to the following formula.*

$$u = kR\left(\frac{h}{\$100 + h}\right)$$

37. Describe in words the quantity represented by $\dfrac{h}{\$100 + h}$.

38. Describe in words the quantity represented by kR.

39. Explain why the product of the two quantities above represents unearned interest.

Write out your responses to the following exercises.

40. Give reasons why a lender may be justified in imposing a prepayment penalty.

41. Discuss reasons that a borrower may want to pay off a loan early.

42. Find out what federal agency you can contact if you have questions about compliance with the Truth in Lending Act. (Any bank, or retailer's credit department, should be able to help you with this.)

13.4 Exercises (Text Page 791)

Find the monthly payment needed to amortize principal and interest for each of the following fixed-rate mortgages. You can use either the regular monthly payment formula or Table 4, as appropriate.

	Loan Amount	Interest Rate	Term	
1.	$50,000	13.0%	15 years	$632.62
2.	$30,000	11.0%	20 years	$309.66
3.	$48,500	9.8%	25 years	$433.90
4.	$85,000	7.5%	14 years	$818.67
5.	$27,750	15.5%	30 years	$362.00
6.	$95,450	8.9%	30 years	$761.15
7.	$132,500	7.6%	20 years	$1,075.53
8.	$105,000	5.5%	10 years	$1,139.53

Complete the first one or two months (as required) of the following amortization schedules for fixed-rate mortgages.

9. Mortgage: $43,500
Interest rate: 10.0%
Term of loan: 30 years

Amortization Schedule

Payment Number	Total Payment	Interest Payment	Principal Payment	Balance of Principal
1	(a) ______	(b) ______	(c) ______	(d) ______

(a) $381.74 (b) $362.50 (c) $19.24 (d) $43,480.76

10. Mortgage: $51,000
Interest rate: 9.5%
Term of loan: 20 years

Amortization Schedule

Payment Number	Total Payment	Interest Payment	Principal Payment	Balance of Principal
1	(a) ______	(b) ______	(c) ______	(d) ______

(a) $475.39 (b) $403.75 (c) $71.64 (d) $50,928.36

11. Mortgage: $58,500
Interest rate: 8.5%
Term of loan: 25 years

Amortization Schedule

Payment Number	Total Payment	Interest Payment	Principal Payment	Balance of Principal
1	(a) ______	(b) ______	(c) ______	(d) ______
2	(e) ______	(f) ______	(g) ______	(h) ______

(a) $471.06 (b) $414.38 (c) $56.68 (d) $58,443.32
(e) $471.06 (f) $413.97 (g) $57.09 (h) $58,386.23

12. Mortgage: $64,000
Interest rate: 11.0%
Term of loan: 15 years

Amortization Schedule

Payment Number	Total Payment	Interest Payment	Principal Payment	Balance of Principal
1	(a) ______	(b) ______	(c) ______	(d) ______
2	(e) ______	(f) ______	(g) ______	(h) ______

(a) $727.42 (b) $586.67 (c) $140.75 (d) $63,859.25
(e) $727.42 (f) $585.38 (g) $142.04 (h) $63,717.21

13. Mortgage: $28,000
Interest rate: 8.2%
Term of loan: 5 years

Amortization Schedule

Payment Number	Total Payment	Interest Payment	Principal Payment	Balance of Principal
1	(a) ______	(b) ______	(c) ______	(d) ______
2	(e) ______	(f) ______	(g) ______	(h) ______

(a) $570.42 (b) $191.33 (c) $379.09 (d) $27,620.91
(e) $570.42 (f) $188.74 (g) $381.68 (h) $27,239.23

14. Mortgage: $82,000
Interest rate: 12.5%
Term of loan: 22 years

Amortization Schedule

Payment Number	Total Payment	Interest Payment	Principal Payment	Balance of Principal
1	(a) ______	(b) ______	(c) ______	(d) ______
2	(e) ______	(f) ______	(g) ______	(h) ______

(a) $913.39 (b) $854.17 (c) $59.22 (d) $81,940.78
(e) $913.39 (f) $853.55 (g) $59.84 (h) $81,880.94

Find the total monthly payment, including taxes and insurance, on each of the following mortgages.

	Mortgage	Interest Rate	Term of Loan	Annual Taxes	Annual Insurance	
15.	$48,700	11%	30 years	$ 509	$176	$520.86
16.	$34,500	12%	20 years	$ 660	$105	$443.62
17.	$59,200	9%	25 years	$ 775	$287	$585.30
18.	$32,200	10%	10 years	$ 172	$165	$453.61
19.	$69,890	8.8%	25 years	$1,080.19	$423.74	$702.30
20.	$53,760	11.3%	15 years	$ 840.74	$209.77	$708.74

Suppose $70,000 is owed on a house after the down payment is made. The monthly payment for principal and interest at 10.5% for 30 years is 70 × $9.14739 = $640.32.

21. How many monthly payments will be made? 360

22. What is the total amount that will be paid for principal and interest? $230,515.20

23. The total interest charged is the total amount paid minus the amount financed. What is the total interest? $160,515.20

24. Which is more—the amount financed or the total interest paid? By how much? Interest exceeds amount financed by $90,515.20.

You may remember seeing home mortgage interest rates fluctuate widely in a period of not too many years. The following exercises show the effect of changing rates. Refer to Table 5, which compared the amortization of a $60,000, 30-year mortgage for rates of 4.5% and 14.5%. Give values of each of the following for **(a)** *a 4.5% rate, and* **(b)** *a 14.5% rate.*

25. monthly payments (a) $304.01 (b) $734.73

26. percentage of first monthly payment that goes toward principal (a) 26.0% (b) 1.3%

27. balance of principal after 1 year (a) $59,032.06 (b) $59,875.11

28. balance of principal after 20 years (a) $29,333.83 (b) $46,417.87

29. the first monthly payment that includes more toward principal than toward interest (a) payment 176 (b) payment 304

30. amount of interest included in final payment (a) $1.14 (b) $8.77

Suppose a $60,000 mortgage is to be amortized at 7.5% interest. Find the total amount of interest that would be paid for each of the following terms.

31. 10 years $25,465.20

32. 20 years $56,006.40

33. 30 years $91,030.80

34. 40 years $129,523.20

For each of the following adjustable-rate mortgages, find **(a)** *the initial monthly payment,* **(b)** *the monthly payment for the second adjustment period, and* **(c)** *the change in monthly payment at the first adjustment. (The "adjusted balance" is the principal balance at the time of the first rate adjustment. Assume no caps apply.)*

	Beginning Balance	Term	Initial Index Rate	Margin	Adjustment Period	Adjusted Index Rate	Adjusted Balance
35.	$75,000	20 years	6.5%	2.5%	1 year	8.0%	$73,595.52
36.	$44,500	30 years	7.2%	2.75%	3 years	6.6%	$43,669.14

35. (a) $674.79 (b) $746.36 (c) an increase of $71.57
36. (a) $388.88 (b) $370.20 (c) a decrease of $18.68

Jeff Walenski has a 1-year ARM for $50,000 over a 20-year term. The margin is 2% and the index rate starts out at 7.5% and increases to 10.0% at the first adjustment. The balance of principal at the end of the first year is $49,119.48. The ARM includes a periodic rate cap of 2% per adjustment period. (Use this information for Exercises 37–40.)

37. Find **(a)** the interest owed and **(b)** the monthly payment due for the first month of the first year. **(a) $395.83 (b) $466.07**

38. Find **(a)** the interest owed and **(b)** the monthly payment due for the first month of the second year. **(a) $470.73 (b) $531.09**

39. What is the monthly payment adjustment at the end of the first year? **$65.02**

40. If the index rate has dropped slightly at the end of the second year, will the third year monthly payments necessarily drop? Why or why not?

For Exercises 41–44, refer to the following table of closing costs for the purchase of a $95,000 house requiring a 20% down payment.

Title insurance premium	$200
Document recording fee	25
Loan fee (two points)	______
Appraisal fee	225
Prorated property taxes	545
Prorated fire insurance premium	190

Find each of the following.

41. the mortgage amount **$76,000**

42. the loan fee **$1,520**

43. the total closing costs **$2,705**

44. the total amount of cash required of the buyer at closing (including down payment) **$21,705**

On the basis of material in this section, or on your own research, give brief written responses to each of the following.

45. Suppose your ARM allows conversion to a fixed rate loan at each of the first five adjustment dates. Describe circumstances under which you would want to convert.

46. Describe each of the following types of mortgages.

Graduated payment
Balloon payment
Growing equity
Shared equity
Partnership

47. Should a home buyer always pay the smallest down payment that will be accepted? Explain.

48. Should a borrower always choose the shortest term available in order to minimize the total interest expense? Explain.

49. Under what conditions would an ARM probably be a better choice than a fixed-rate mortgage?

50. Why are second-year payments (slightly) less in Example 6 than in Example 5 even though the interest rate is exactly 12.7% in both cases?

51. Do you think that the discount in Example 6 actually makes the overall cost of the mortgage less? Explain.

52. Discuss the term "payment shock" mentioned at the end of Example 6.

53. Find out what is meant by the following terms and describe some of the features of each.

FHA-backed mortgage
VA-backed mortgage
Conventional mortgage

13.5 Exercises (Text Page 802)

Find each of the following from the stock table in the text. Give money answers in dollars.

1. high for the day for Alcan Aluminum (AL) **$36.625**

2. low for the day for Airborne Freight (ABF) **$21.125**

3. closing price for Singer Company (SEW) **$26.75**

4. change from the previous day for Sizzler International (SZ) **$.125 lower**

5. 52-week high for Agnico-Eagle Mines (AEM) **$14.75**

6. 52-week low for Sears (S) (*not* Sears preferred) **$21.50**

7. dividend for Aetna Life & Casualty (AET) **$2.76**

8. dividend for Snap-on Incorporated (SNA) **$1.08**

9. low for the day for Alex Brown (AB) **$45.50**

10. high for the day for Sherwin Williams (SHW) **$36.125**

11. sales for the day for Airtouch Communications (ATI) **558,400 shares**

12. sales for the day for Scott Paper (SPP) **489,400 shares**

13. yield for American Brands (AMB) **5.1%**

14. yield for Snyder Oil Corporation (SNY) (*not* the preferred stock) **2.2%**

Find the cost for each of the following stock purchases. Ignore any broker's fees.

	Stock	Symbol	Number of Shares	Transaction Price	
15.	Alcoa	AA	500	close	**$28,375**
16.	Allstate	ALL	300	low	**$9,637.50**
17.	Showboat	SBO	400	low	**$7,350**
18.	Seagram	VO	600	low	**$20,625**

Find the cost for each of the following stock purchases. Include typical discount broker's fees as described in the text, and any odd-lot differentials (charged only on the odd-lot portion of the order).

	Stock	Symbol	Number of Shares	Transaction Price
19.	Alaska Air Group	ALK	40	low
20.	Shoneys	SHN	50	high
21.	Sierra Pacific	SRP	425	close
22.	Albemarle Corporation	ALB	635	low

19. $773.75 **20. $666.09** **21. $9,369.06** **22. $10,174.30**

Find the amount received by the sellers of the following stocks. Use the sales expenses described in the text. (Note that each of Exercises 27–30 involves an odd-lot differential on a portion of the sale.)

23. 500 shares at 32 7/8 **$16,026.01**

24. 700 shares at 51 1/4 **$34,976.92**

25. 1,400 shares at 8 7/8 **$12,113.95**

26. 800 shares at 41 3/8 **$32,271.39**

27. 420 shares at 50 1/4 **$20,574.16**

28. 660 shares at 62 1/8 **$39,968.57**

29. 950 shares at 29 3/4 **$27,548.74**

30. 1,220 shares at 18 5/8 **$22,151.18**

Refer to the stock table in the text to work the following problems.

31. Ricardo Espinoza bought 70 shares of Albertsons (ABS) and 120 shares of Sears (S), both at the low price for the day. Find the total cost of this purchase including broker's commissions and any odd-lot differentials. **$5,881.94**

32. Betty Harrison purchased 20 shares of Alco Standard (ASN) and 40 shares of American Eagle (FLI), both at the closing price for the day. Find her total cost, including broker's commissions and any odd-lot differentials. **$2,111.32**

33. Margaret Decker bought 200 shares of Sears (preferred) at the low for the day and sold 140 shares of American Brands (AMB) at the high for the day. Considering all applicable expenses, what *net* amount did she take in or pay out for these combined transactions? **$11.75 net paid out**

34. Mehta Farak sold 125 shares of Allstate (ALL) and bought 90 shares of Alcoa (AA), both at the low price of the day. Including all applicable expenses, what *net* amount did Mehta take in or pay out for these transactions? **$1,276.81 net paid out**

Write a short essay in response to each of the following. Some research may be required.

35. Describe the concept of "dollar-cost averaging," and relate it to advantage number 1 of mutual funds as listed in the text.

36. Discuss advantage number 3 of mutual funds as listed in the text.

Chapter 13 Test (Text Page 804)

Find all monetary answers to the nearest cent. Use tables from the chapter if necessary.

Find the future value of each of the following deposits.

1. \$100 for 5 years at 6% simple interest **\$130**
2. \$50 for 2 years at 8% compounded quarterly **\$58.58**
3. Find the effective annual yield to the nearest hundredth of a percent for an account paying 6% compounded monthly. **6.17%**
4. Use the rule of 70 to estimate the years to double at an inflation rate of 12%. **6 years**
5. What amount deposited today in an account paying 10% compounded semiannually would grow to \$100,000 in 10 years? **\$37,688.95**
6. Darcy Randal's MasterCard statement shows an average daily balance of \$680. Find the interest due if the rate is 1.6% per month. **\$10.88**

Greg McRill buys a turquoise necklace for his wife on their anniversary. He pays \$4,000 for the necklace with \$1,000 down. The dealer charges add-on interest of 8% per year. McRill agrees to make payments for 24 months.

7. Find the total amount of interest he will pay. **\$480**
8. Find the monthly payment. **\$145**
9. Find the APR value for this loan (to the nearest half percent). **14.5%**
10. Find the unearned interest if he repays the loan in full with six payments remaining. Use the most accurate method available. **\$35.63**
11. Newark Hardware wants to include financing terms in their advertising. If the price of a floor waxer is \$150 and the finance charge with no down payment is \$7 over a 6-month period (six equal monthly payments), find the true annual percentage rate (APR). **16.0%**
12. Explain what a mutual fund is and what its advantages are.

Find the monthly payment required for each of the following home loans.

13. The mortgage is for \$50,000 at a fixed rate of 12% for 20 years. Amortize principal and interest only. **\$550.54**
14. The purchase price of the home is \$88,000. The down payment is 20%. Interest is at 10.5% fixed for a term of 30 years. Annual taxes are \$900 and annual insurance is \$600. **\$768.98**
15. If the lender in Exercise 14 charges two "points," how much does that add to the cost of the loan? **\$1,408**
16. Explain in general what "closing costs" are. Are they different from "settlement charges"?
17. To buy your home you obtain a 1-year ARM with 2.25% margin and a 2% periodic rate cap. The index starts at 7.85% but has increased to 10.05% by the first adjustment date. What interest rate will you pay during the second year? **12.1%**

Refer to the stock table in Section 5 of this chapter to find the following.

18. the 52-week high for Airlease (FLY) **\$17**
19. the sales for the day for Allied Products (ADP) **25,000 shares**
20. the cost to buy 1,500 shares of Sherwood (SHD) at the low price of the day, including a typical discount broker's commission as described in the chapter **\$12,684.38**

CHAPTER 14 Matrices and Their Applications

14.1 Exercises (Text Page 812)

Perform the indicated matrix operations. If a required result is not computable, state why not.

1. $\begin{bmatrix} 3 & 2 \\ 5 & 1 \end{bmatrix} + \begin{bmatrix} 8 & -1 \\ 4 & 3 \end{bmatrix}$

   $\begin{bmatrix} 11 & 1 \\ 9 & 4 \end{bmatrix}$

2. $\begin{bmatrix} -6 & 9 \\ 2 & 4 \end{bmatrix} + \begin{bmatrix} -2 & 5 \\ -1 & 3 \end{bmatrix}$

   $\begin{bmatrix} -8 & 14 \\ 1 & 7 \end{bmatrix}$

3. $\begin{bmatrix} 2 & 8 & -1 \\ 4 & 0 & 3 \end{bmatrix} - \begin{bmatrix} -1 & 5 & 2 \\ 0 & 4 & 3 \end{bmatrix}$

$\begin{bmatrix} 3 & 3 & -3 \\ 4 & -4 & 0 \end{bmatrix}$

4. $\begin{bmatrix} -1 & 2 & 4 \\ -2 & 3 & 0 \end{bmatrix} - \begin{bmatrix} 2 & -1 & 4 \\ -5 & 6 & 9 \end{bmatrix}$

$\begin{bmatrix} -3 & 3 & 0 \\ 3 & -3 & -9 \end{bmatrix}$

5. $\begin{bmatrix} -2 & 3 & 1 \\ 4 & 0 & -2 \\ 5 & -1 & 6 \end{bmatrix} + \begin{bmatrix} -1 & 1 & 1 \\ 4 & 2 & 3 \\ -1 & 4 & 7 \end{bmatrix}$

$\begin{bmatrix} -3 & 4 & 2 \\ 8 & 2 & 1 \\ 4 & 3 & 13 \end{bmatrix}$

6. $\begin{bmatrix} 3 & 4 & -1 \\ 7 & -8 & 2 \\ 9 & -1 & 3 \end{bmatrix} - \begin{bmatrix} -4 & 2 & -1 \\ 3 & -1 & 8 \\ 4 & 9 & 0 \end{bmatrix}$

$\begin{bmatrix} 7 & 2 & 0 \\ 4 & -7 & -6 \\ 5 & -10 & 3 \end{bmatrix}$

7. $\begin{bmatrix} -4 & 3 & 2 \\ -8 & 0 & 1 \\ 4 & 2 & 1 \end{bmatrix} - \begin{bmatrix} 3 & -4 \\ 2 & -3 \\ 8 & 9 \end{bmatrix}$

It is not computable because the matrices do not have the same order.

8. $\begin{bmatrix} -1 & 6 & 2 \\ 3 & 4 & 7 \\ 9 & 8 & 2 \end{bmatrix} + \begin{bmatrix} -4 & 9 & 2 & 0 \\ -3 & 1 & 4 & 0 \\ 8 & 9 & 7 & 0 \end{bmatrix}$

It is not computable because the matrices do not have the same order.

9. $\begin{bmatrix} 3 \\ 4 \\ 2 \end{bmatrix} + \begin{bmatrix} -1 \\ 9 \\ 8 \end{bmatrix} - \begin{bmatrix} 2 \\ 1 \\ 3 \end{bmatrix}$

$\begin{bmatrix} 0 \\ 12 \\ 7 \end{bmatrix}$

10. $\begin{bmatrix} 7 \\ 8 \\ 9 \end{bmatrix} + \begin{bmatrix} 1 \\ 9 \\ 3 \end{bmatrix} - \begin{bmatrix} -4 \\ 3 \\ 8 \end{bmatrix}$

$\begin{bmatrix} 12 \\ 14 \\ 4 \end{bmatrix}$

11. $\begin{bmatrix} -2 & 4 \\ 0 & 9 \end{bmatrix} + \begin{bmatrix} -8 & 1 \\ 3 & 6 \end{bmatrix} - \begin{bmatrix} 4 & 7 \\ -2 & 5 \end{bmatrix}$

$\begin{bmatrix} -14 & -2 \\ 5 & 10 \end{bmatrix}$

12. $\begin{bmatrix} -3 & 2 \\ 4 & 8 \end{bmatrix} + \begin{bmatrix} -9 & -1 \\ 6 & 3 \end{bmatrix} + \begin{bmatrix} -2 & 5 \\ 0 & 8 \end{bmatrix}$

$\begin{bmatrix} -14 & 6 \\ 10 & 19 \end{bmatrix}$

Let $A = \begin{bmatrix} -4 & 0 \\ 3 & -4 \end{bmatrix}$ *and let* $B = \begin{bmatrix} 2 & -1 \\ -4 & 0 \end{bmatrix}$. *Compute the following.*

13. $A + B$

$A + B = \begin{bmatrix} -2 & -1 \\ -1 & -4 \end{bmatrix}$

14. $A - B$

$A - B = \begin{bmatrix} -6 & 1 \\ 7 & -4 \end{bmatrix}$

15. $2A$

$2A = \begin{bmatrix} -8 & 0 \\ 6 & -8 \end{bmatrix}$

16. $3B$

$3B = \begin{bmatrix} 6 & -3 \\ -12 & 0 \end{bmatrix}$

17. $-4B$

$-4B = \begin{bmatrix} -8 & 4 \\ 16 & 0 \end{bmatrix}$

18. $-5A$

$-5A = \begin{bmatrix} 20 & 0 \\ -15 & 20 \end{bmatrix}$

▲ 19. $2A + 3B$

$2A + 3B = \begin{bmatrix} -2 & -3 \\ -6 & -8 \end{bmatrix}$

▲ 20. $-7A + 4B$

$-7A + 4B = \begin{bmatrix} 36 & -4 \\ -37 & 28 \end{bmatrix}$

▲ 21. $(-1)A + A$

$(-1)A + A = \begin{bmatrix} 0 & 0 \\ 0 & 0 \end{bmatrix}$

CONCEPTUAL  WRITING  ▲ CHALLENGING  SCIENTIFIC CALCULATOR GRAPHING CALCULATOR

A dietician prepares a diet specifying the allowable amounts of four main food groups: group I, meats; group II, fruits and vegetables; group III, breads and starches; and group IV, milk products. Amounts are given in appropriate units (1 ounce for meat, 1/2 cup for fruits and vegetables, 1 slice for bread, and 8 ounces for milk).

22. The numbers of units for breakfast for the four food groups respectively are 2, 1, 2, and 1; for lunch, 3, 2, 2, and 1; and for dinner, 4, 3, 2, and 1. Write a 3 × 4 matrix to display this information.

$$\begin{array}{l} \\ \text{Breakfast} \\ \text{Lunch} \\ \text{Dinner} \end{array} \begin{array}{c} \begin{array}{cccc} \text{I} & \text{II} & \text{III} & \text{IV} \end{array} \\ \begin{bmatrix} 2 & 1 & 2 & 1 \\ 3 & 2 & 2 & 1 \\ 4 & 3 & 2 & 1 \end{bmatrix} \end{array}$$

23. The amounts of fat, carbohydrates, and protein in the food groups respectively are as follows:

Fat: 5, 0, 0, 10 Carbohydrates: 0, 10, 15, 12 Protein: 7, 1, 2, 8.

Display this information in a 4 × 3 matrix.

$$\begin{array}{l} \\ \text{I} \\ \text{II} \\ \text{III} \\ \text{IV} \end{array} \begin{array}{c} \begin{array}{ccc} \text{Fat} & \text{Carbohydrates} & \text{Protein} \end{array} \\ \begin{bmatrix} 5 & 0 & 7 \\ 0 & 10 & 1 \\ 0 & 15 & 2 \\ 10 & 12 & 8 \end{bmatrix} \end{array}$$

24. Assume there are 8 calories per unit of fat, 4 calories per unit of carbohydrate, and 5 calories per unit of protein. Show these data in a 3 × 1 matrix.

$$\begin{array}{l} \\ \text{Fat} \\ \text{Carbohydrates} \\ \text{Protein} \end{array} \begin{array}{c} \text{Calories per unit} \\ \begin{bmatrix} 8 \\ 4 \\ 5 \end{bmatrix} \end{array}$$

Don and Connie run a business (D&C Creations) in California building and selling gold rush era tables and chairs. They maintain factories in Sloughhouse (F1) and Nevada City (F2) and ship their products to warehouses in Sacramento (W1), San Francisco (W2), and Reno (W3). The table here shows shipping costs per item from each factory to each warehouse.

Route	Cost Per Item
F1 to W1	$12
F1 to W2	20
F1 to W3	25
F2 to W1	16
F2 to W2	25
F2 to W3	30

25. Construct a 2 × 3 matrix, C, showing the data for shipping cost per item.

$$C = \begin{array}{l} \\ \text{F1} \\ \text{F2} \end{array} \begin{array}{c} \begin{array}{ccc} \text{W1} & \text{W2} & \text{W3} \end{array} \\ \begin{bmatrix} 12 & 20 & 25 \\ 16 & 25 & 30 \end{bmatrix} \end{array}$$

26. On April 1, the freight company that *D&C* utilizes adjusted per item shipping costs as shown here. Construct a 2 × 3 matrix, A, showing the shipping cost adjustments.

Route	Adjustment
F1 to W1	$3
F1 to W2	−2
F1 to W3	2
F2 to W1	4
F2 to W2	2
F2 to W3	−2

$$A = \begin{array}{l} \\ \text{F1} \\ \text{F2} \end{array} \begin{array}{c} \begin{array}{ccc} \text{W1} & \text{W2} & \text{W3} \end{array} \\ \begin{bmatrix} 3 & -2 & 2 \\ 4 & 2 & -2 \end{bmatrix} \end{array}$$

27. Show an appropriate matrix operation whose result gives *D&C*'s shipping costs per item after April 1.

$$C + A = \begin{bmatrix} 15 & 18 & 27 \\ 20 & 27 & 28 \end{bmatrix}$$

28. Each month, Don and Connie send to each of their warehouses 10 model-*A* tables, 12 model-*B* tables, 5 model-*C* tables, 15 model-*A* chairs, 20 model-*B* chairs, and 8 model-*C* chairs. Exhibit this information in a 2 × 3 matrix (with tables represented in the top row).

$$\begin{array}{l} \\ \text{Tables} \\ \text{Chairs} \end{array} \begin{array}{c} \begin{matrix} A & B & C \end{matrix} \\ \begin{bmatrix} 10 & 12 & 5 \\ 15 & 20 & 8 \end{bmatrix} \end{array}$$

29. The matrix here shows stock on hand at the *D&C* Reno warehouse as of September 1.

$$R = \begin{array}{l} \\ \text{Tables} \\ \text{Chairs} \end{array} \begin{array}{c} \begin{matrix} A & B & C \end{matrix} \\ \begin{bmatrix} 45 & 35 & 20 \\ 65 & 40 & 35 \end{bmatrix} \end{array}$$

If no stock is sent out of the warehouse during September, how much stock will be on hand October 1? Use the result of Exercise 28 and matrix methods.

$$\begin{bmatrix} 45 & 35 & 20 \\ 65 & 40 & 35 \end{bmatrix} + \begin{bmatrix} 10 & 12 & 5 \\ 15 & 20 & 8 \end{bmatrix} = \begin{bmatrix} 55 & 47 & 25 \\ 80 & 60 & 43 \end{bmatrix}$$

30. In Exercise 29, how many model-*B* chairs were on hand October 1 at the Reno warehouse? **60 model-*B* chairs**

31. In Exercise 29, how many model-*C* tables were on hand October 1 at the Reno warehouse? **25 model-*C* tables**

32. The number of each model held on September 1 at the Sacramento and San Francisco warehouses of *D&C* are:

$$S = \begin{bmatrix} 22 & 25 & 38 \\ 31 & 34 & 35 \end{bmatrix} \quad \text{and} \quad F = \begin{bmatrix} 30 & 32 & 28 \\ 43 & 47 & 30 \end{bmatrix}.$$

Find the total inventory in all three warehouses on September 1.

$$\begin{bmatrix} 45 & 35 & 20 \\ 65 & 40 & 35 \end{bmatrix} + \begin{bmatrix} 22 & 25 & 38 \\ 31 & 34 & 35 \end{bmatrix} + \begin{bmatrix} 30 & 32 & 28 \\ 43 & 47 & 30 \end{bmatrix} = \begin{bmatrix} 97 & 92 & 86 \\ 139 & 121 & 100 \end{bmatrix}$$

33. Suppose the Sacramento warehouse of *D&C* shipped the following numbers of items during September:

$$K = \begin{bmatrix} 5 & 10 & 8 \\ 11 & 14 & 15 \end{bmatrix}.$$

Find the stock on hand October 1, taking into account the numbers of items received and shipped during the month.

$$\begin{bmatrix} 22 & 25 & 38 \\ 31 & 34 & 35 \end{bmatrix} + \begin{bmatrix} 10 & 12 & 5 \\ 15 & 20 & 8 \end{bmatrix} - \begin{bmatrix} 5 & 10 & 8 \\ 11 & 14 & 15 \end{bmatrix} = \begin{bmatrix} 27 & 27 & 35 \\ 35 & 40 & 28 \end{bmatrix}$$

Matrix K shows the weights of four men and four women at the beginning of a diet designed to produce weight loss. Matrix M shows the weights after the diet.

$$K = \begin{matrix} \text{Men} \\ \text{Women} \end{matrix} \begin{bmatrix} 160 & 158 & 172 & 193 \\ 132 & 143 & 119 & 157 \end{bmatrix} \qquad M = \begin{matrix} \text{Men} \\ \text{Women} \end{matrix} \begin{bmatrix} 154 & 148 & 163 & 178 \\ 132 & 154 & 112 & 136 \end{bmatrix}$$

34. Show a matrix operation whose result gives the weight losses of all eight people on the diet.
$K - M = \begin{bmatrix} 6 & 10 & 9 & 15 \\ 0 & -11 & 7 & 21 \end{bmatrix}$

35. In Exercise 34, how much weight did the third man lose? **9 pounds**

36. In Exercise 34, what result did the second woman experience? **She gained 11 pounds.**

Find all real values of the variables that will make each of the following equalities true. In case the given equality is impossible, state why.

37. $\begin{bmatrix} x & y & z \\ 2 & 5 & 1 \end{bmatrix} = \begin{bmatrix} 7 & 2 \\ 4 & 5 \\ 2 & 1 \end{bmatrix}$
The given equality is impossible because the two matrices have different orders.

38. $\begin{bmatrix} 3 & 4 \\ z & x \end{bmatrix} = [3z \quad 4x]$
The given equality is impossible because the two matrices have different orders.

▲ **39.** $\begin{bmatrix} 5 & y \\ z & 2 \end{bmatrix} = \begin{bmatrix} 5 & z \\ y & 2 \end{bmatrix}$
$\{(y, z) \mid y = z\}$

▲ **40.** $[x \quad y \quad z] = [2x \quad 3y \quad 4z]$
$x = 0, y = 0, z = 0$

41. $\begin{bmatrix} 6 & y \\ k & m \end{bmatrix} = \begin{bmatrix} 6 & 3 \\ 8 & 1 \end{bmatrix}$
$k = 8, m = 1, y = 3$

42. $\begin{bmatrix} 5 & 7 \\ 1 & 8 \end{bmatrix} = \begin{bmatrix} a & b \\ 1 & c \end{bmatrix}$
$a = 5, b = 7, c = 8$

43. $\begin{bmatrix} x+2 & 5 & 9 \\ 1 & 3 & y \end{bmatrix} = \begin{bmatrix} 2x+1 & z-3 & 9 \\ 1 & 3 & 8 \end{bmatrix}$
$x = 1, y = 8, z = 8$

44. $\begin{bmatrix} 2 & m+5 & a \\ 1 & n-3 & -1 \end{bmatrix} = \begin{bmatrix} 2 & 2m+6 & 2a \\ 1 & 4n-12 & -1 \end{bmatrix}$
$a = 0, m = -1, n = 3$

▲ **45.** $\begin{bmatrix} -2 & 6 \\ x & 2 \end{bmatrix} + \begin{bmatrix} x & y \\ 3 & 5 \end{bmatrix} = \begin{bmatrix} 8 & 2 \\ 7 & 7 \end{bmatrix}$
The given equality is impossible. In the first row, first column of the matrix $\begin{bmatrix} x & y \\ 3 & 5 \end{bmatrix}$, **$x$ must be equal to 10. But in the second row, first column of the matrix** $\begin{bmatrix} -2 & 6 \\ x & 2 \end{bmatrix}$, **$x$ must be equal to 4.**

▲ **46.** $\begin{bmatrix} 5 & m & 3 \\ 2 & y & 5 \end{bmatrix} - \begin{bmatrix} 3 & 4 & a \\ 1 & 6 & a \end{bmatrix} = \begin{bmatrix} 2 & 5 & 2 \\ 1 & 4 & 1 \end{bmatrix}$
The given equality is impossible. In the first row, third column of the matrix $\begin{bmatrix} 3 & 4 & a \\ 1 & 6 & a \end{bmatrix}$, **$a$ must be equal to 1. But in the second row, third column of that matrix, a must be equal to 4.**

The dates in the calendar display for February 1998 form a 4×7 *matrix, as shown here.*

February 1998

Sun	Mon	Tue	Wed	Thu	Fri	Sat
1	2	3	4	5	6	7
8	9	10	11	12	13	14
15	16	17	18	19	20	21
22	23	24	25	26	27	28

◉ **47.** Will the month of March on a calendar ever form a matrix? Why or why not?
No. On a calendar, there are seven columns. In order for the days of a given month to form a matrix, the number of days in that month must be divisible by 7. March has 31 days, which is not divisible by 7; therefore the month of March on a calendar will never form a matrix.

48. What is the next year following 1998 in which February will again form a matrix? (*Hint:* You *may* want to consider "leap years" or "perpetual calendars." You can look up these concepts in this text or elsewhere.) **2009**

Example 10 in this section established the commutative property for addition of matrices. Use similar reasoning in Exercises 49–56 to determine whether each statement is true *or* false *for matrices A, B, and C of the same order and for scalars a and b.*

49. $A - B = B - A$ **false**

50. $A + (B + C) = (A + B) + C$ (associative property for addition) **true**

51. $A + B$ is a matrix. (closure property) **true**

52. If 0 is a matrix of the same order as A, containing only zero entries, then $0 + A = A$ and $A + 0 = A$. (identity property for addition) **true**

53. If $-A = (-1)A$, then $A + (-A) = 0$ and $-A + A = 0$. (inverse property for addition) **true**

54. $A - B = A + (-B)$, where $-B = (-1)B$. **true**

55. $a(B + C) = aB + aC$ **true**

56. $(a + b)C = aC + bC$ **true**

57. Does the set of all 2×2 matrices and the operation of addition form a *mathematical system?* Explain why or why not. (*Hint:* Look up the definition of "mathematical system.")

58. Does the system of Exercise 57 form a *group?* Explain why or why not. (*Hint:* Look up the definition of a mathematical "group.")

14.2 Exercises (Text Page 824)

Compute each of the following products, if possible. If a product is not computable, state why not.

1. $\begin{bmatrix} -3 & 2 \\ 4 & 1 \end{bmatrix} \cdot \begin{bmatrix} -2 & 0 \\ 1 & 3 \end{bmatrix}$

$\begin{bmatrix} 8 & 6 \\ -7 & 3 \end{bmatrix}$

2. $\begin{bmatrix} -5 & 2 \\ 1 & 4 \end{bmatrix} \cdot \begin{bmatrix} -1 & 4 \\ 3 & 0 \end{bmatrix}$

$\begin{bmatrix} 11 & -20 \\ 11 & 4 \end{bmatrix}$

3. $\begin{bmatrix} 0 & -2 \\ 5 & 1 \end{bmatrix} \cdot \begin{bmatrix} -3 & 6 \\ 1 & 4 \end{bmatrix}$

$\begin{bmatrix} -2 & -8 \\ -14 & 34 \end{bmatrix}$

4. $\begin{bmatrix} -3 & 5 \\ 1 & 0 \end{bmatrix} \cdot \begin{bmatrix} 8 & -2 \\ 1 & 7 \end{bmatrix}$

$\begin{bmatrix} -19 & 41 \\ 8 & -2 \end{bmatrix}$

5. $\begin{bmatrix} 1 & 3 \\ 4 & 1 \end{bmatrix} \cdot \begin{bmatrix} 2 & 1 & 0 \\ 5 & 2 & 3 \end{bmatrix}$

$\begin{bmatrix} 17 & 7 & 9 \\ 13 & 6 & 3 \end{bmatrix}$

6. $\begin{bmatrix} -5 & 6 \\ 3 & 4 \end{bmatrix} \cdot \begin{bmatrix} -3 & 5 & 0 \\ 0 & 1 & 5 \end{bmatrix}$

$\begin{bmatrix} 15 & -19 & 30 \\ -9 & 19 & 20 \end{bmatrix}$

7. $\begin{bmatrix} 3 & 2 \\ 5 & 1 \\ 0 & 4 \end{bmatrix} \cdot \begin{bmatrix} 2 & -1 \\ 1 & 3 \end{bmatrix}$

$\begin{bmatrix} 8 & 3 \\ 11 & -2 \\ 4 & 12 \end{bmatrix}$

8. $\begin{bmatrix} -1 & 0 \\ 4 & 1 \\ 2 & 0 \end{bmatrix} \cdot \begin{bmatrix} -5 & 0 \\ 4 & 2 \end{bmatrix}$

$\begin{bmatrix} 5 & 0 \\ -16 & 2 \\ -10 & 0 \end{bmatrix}$

9. $\begin{bmatrix} -5 & 1 \\ 6 & 2 \end{bmatrix} \cdot \begin{bmatrix} -3 & 2 \\ 4 & 1 \\ 5 & 8 \end{bmatrix}$

It is not computable because the number of columns in the first matrix is not equal to the number of rows in the second matrix.

10. $\begin{bmatrix} -7 & 3 \\ 9 & 2 \end{bmatrix} \cdot \begin{bmatrix} 4 & -1 \\ 2 & -5 \\ 3 & 6 \end{bmatrix}$

It is not computable because the number of columns in the first matrix is not equal to the number of rows in the second matrix.

11. $\begin{bmatrix} -5 & 1 & 3 \\ 2 & 0 & 4 \\ 3 & 0 & 2 \end{bmatrix} \cdot \begin{bmatrix} -2 & 0 & 0 \\ 4 & 1 & 0 \\ 0 & 1 & 0 \end{bmatrix}$

$\begin{bmatrix} 14 & 4 & 0 \\ -4 & 4 & 0 \\ -6 & 2 & 0 \end{bmatrix}$

12. $\begin{bmatrix} -3 & 4 & 0 \\ 2 & -1 & 5 \\ 0 & 4 & 1 \end{bmatrix} \cdot \begin{bmatrix} 0 & 4 & 0 \\ 2 & 0 & 3 \\ 0 & 5 & 1 \end{bmatrix}$

$\begin{bmatrix} 8 & -12 & 12 \\ -2 & 33 & 2 \\ 8 & 5 & 13 \end{bmatrix}$

13. Harry's Donuts, a small neighborhood bakery, sells four main items: sweet rolls, bread, cakes, and pies. Matrix A below shows the number of units of the main ingredients needed for these items.

$A =$

	Eggs	Flour	Sugar	Shortening	Milk
Sweet rolls	1	4	1/4	1/4	1
Bread	0	3	0	1/4	0
Cakes	4	3	2	1	1
Pies	0	1	0	1/3	0

The cost, in cents per egg or per cup, of each ingredient when purchased in large lots or in small lots is given by matrix B.

$B =$

	PURCHASE OPTION	
	Large lot	Small lot
Eggs	5	5
Flour	8	10
Sugar	10	12
Shortening	12	15
Milk	5	6

Use matrix multiplication to find a matrix representing the comparative costs per item for the two purchase options (buying in large lots or in small lots).

$AB =$

	PURCHASE OPTION	
	Large lot	Small lot
Sweet rolls	47.5	57.75
Bread	27	33.75
Cakes	81	95
Pies	12	15

Suppose a day's orders at Harry's (see Exercise 13) consist of 20 dozen sweet rolls, 200 loaves of bread, 50 cakes, and 60 pies.

14. Write this day's orders as a 1×4 matrix D, and use matrix multiplication to write as a matrix the amount of each ingredient needed to fill the day's orders.

$D = [20 \quad 200 \quad 50 \quad 60]$;
$DA = [220 \quad 890 \quad 105 \quad 125 \quad 70]$

15. Use matrix multiplication to find a matrix representing the costs to fill the day's orders under the two purchase options.

	Cost (large lot)	Cost (small lot)
$(DA)B =$ Day's orders	[11,120	13,555]

Refer to the expanded matrices of Example 3, part (a). Assume that Don and Connie have now expanded their product line to include desks, each requiring 10 units of wood, 7 units of hardware, and 4 units of paint.

16. Set up a matrix R relating furniture to materials for the complete line of $D\&C$ Creations.

$R =$

	Wood	Hardware	Paint
Table	8	5	3
Chair	6	4	2
Desk	10	7	4

17. Set up and compute an appropriate matrix product that will relate furniture to cost of materials.

$$RC = \begin{matrix} \text{Table} \\ \text{Chair} \\ \text{Desk} \end{matrix} \overset{\text{Cost per unit}}{\begin{bmatrix} 34 \\ 26 \\ 45 \end{bmatrix}}$$

▲ **18.** Don and Connie have found that the *labor* required for assembling and finishing is four hours per table, two hours per chair, and five hours per desk, and that their labor cost is \$9 per hour. Write a column matrix relating furniture to labor cost.

$$\begin{matrix} \text{Table} \\ \text{Chair} \\ \text{Desk} \end{matrix} \overset{\text{Labor cost}}{\begin{bmatrix} 36 \\ 18 \\ 45 \end{bmatrix}}$$

19. Combine your answers from Exercises 17 and 18 to obtain a 3×2 matrix relating furniture (table, chair, desk) to cost of resources (materials, labor).

$$\begin{matrix} \\ \text{Table} \\ \text{Chair} \\ \text{Desk} \end{matrix} \begin{matrix} \text{Materials} \quad \text{Labor} \\ \begin{bmatrix} 34 & 36 \\ 26 & 18 \\ 45 & 45 \end{bmatrix} \end{matrix}$$

▲ **20.** Write a column matrix relating furniture to selling price. Assume the selling price of each furniture item must cover its own material and labor costs, plus \$10 in overhead expenses, plus a profit (\$20 per table, \$10 per chair, and \$25 per desk).

$$\begin{matrix} \text{Table} \\ \text{Chair} \\ \text{Desk} \end{matrix} \overset{\text{Selling price}}{\begin{bmatrix} 100 \\ 64 \\ 125 \end{bmatrix}}$$

In Ohio, high school football teams are ranked by the Harbin Football Team Rating System. A team is granted one level 1 point for each team it defeats, and is also granted one level 2 point for each time that a team it defeats beats another team. Teams A, B, C, and D all played one another. The results are shown in matrix H, where an entry of 1 indicates a win. For example, A beat C only, B beat both A and D, and so on.

$$H = \begin{matrix} \\ A \\ B \\ C \\ D \end{matrix} \begin{matrix} A \quad B \quad C \quad D \\ \begin{bmatrix} 0 & 0 & 1 & 0 \\ 1 & 0 & 0 & 1 \\ 0 & 1 & 0 & 1 \\ 1 & 0 & 0 & 0 \end{bmatrix} \end{matrix}$$

21. Complete the table below for these four teams.

Team	Level 1 Points	Level 2 Points	Total Points
A	_____	_____	_____
B	_____	_____	_____
C	_____	_____	_____
D	_____	_____	_____

Team	Level 1 pts	Level 2 pts	Total pts
A	1	2	3
B	2	2	4
C	2	3	5
D	1	1	2

22. Let K be a 4×1 matrix with every entry a 1.

(a) Compute the product HK.

$$HK = \begin{bmatrix} 0 & 0 & 1 & 0 \\ 1 & 0 & 0 & 1 \\ 0 & 1 & 0 & 1 \\ 1 & 0 & 0 & 0 \end{bmatrix} \cdot \begin{bmatrix} 1 \\ 1 \\ 1 \\ 1 \end{bmatrix} = \begin{bmatrix} 1 \\ 2 \\ 2 \\ 1 \end{bmatrix}$$

⦿ **(b)** What does HK represent?

HK represents the number of Level 1 points earned by each team.

23. **(a)** Compute H^2K.

$$H^2K = \begin{bmatrix} 2 \\ 2 \\ 3 \\ 1 \end{bmatrix}$$

⦿ **(b)** What does H^2K represent?

H^2K represents the number of Level 2 points earned by each team.

24. **(a)** Compute $HK + H^2K$.

$$HK + H^2K = \begin{bmatrix} 3 \\ 4 \\ 5 \\ 2 \end{bmatrix}$$

⦿ CONCEPTUAL  WRITING ▲ CHALLENGING SCIENTIFIC CALCULATOR  GRAPHING CALCULATOR

(b) What does $HK + H^2K$ represent?

$HK + H^2K$ represents the total number of points earned by each team.

25. Is it true that $(H + H^2)K = HK + H^2K$? If so, verify it by direct computation.

$(H + H^2)K = \begin{bmatrix} 3 \\ 4 \\ 5 \\ 2 \end{bmatrix}$; **Yes,**

$(H + H^2)K = HK + H^2K$.

26. Is it true that $(I + H)HK = HK + H^2K$? If so, verify it by direct computation. (In this case, I is the 4×4 identity matrix.)

$(I + H)HK = \begin{bmatrix} 3 \\ 4 \\ 5 \\ 2 \end{bmatrix}$; **Yes,**

$(I + H)HK = HK + H^2K$.

Consider the matrices $A = \begin{bmatrix} 1 & 2 \\ 3 & 2 \end{bmatrix}$, $B = \begin{bmatrix} -1 & -3 \\ 2 & 2 \end{bmatrix}$, $C = \begin{bmatrix} 1 & 2 \\ -1 & -3 \end{bmatrix}$, and $I = \begin{bmatrix} 1 & 0 \\ 0 & 1 \end{bmatrix}$.

The following statements illustrate (but do not prove) several algebraic properties of the system of 2×2 *matrices. In each case, verify the given statement by direct computation.*

27. AB is computable and is of order 2×2. (closure property)

$AB = \begin{bmatrix} 3 & 1 \\ 1 & -5 \end{bmatrix}$

28. $AI = IA = A$. (identity property)

$AI = \begin{bmatrix} 1 & 2 \\ 3 & 2 \end{bmatrix}$; $IA = \begin{bmatrix} 1 & 2 \\ 3 & 2 \end{bmatrix}$; $A = \begin{bmatrix} 1 & 2 \\ 3 & 2 \end{bmatrix}$

29. $(AB)C = A(BC)$. (associative property)

$(AB)C = \begin{bmatrix} 2 & 3 \\ 6 & 17 \end{bmatrix}$; $A(BC) = \begin{bmatrix} 2 & 3 \\ 6 & 17 \end{bmatrix}$

30. $(A + B)C = AC + BC$. (the right distributive property for addition)

$(A + B)C = \begin{bmatrix} 1 & 3 \\ 1 & -2 \end{bmatrix}$; $AC + BC = \begin{bmatrix} 1 & 3 \\ 1 & -2 \end{bmatrix}$

31. $(A - B)C = AC - BC$. (the right distributive property for subtraction)

$(A - B)C = \begin{bmatrix} -3 & -11 \\ 1 & 2 \end{bmatrix}$;

$AC - BC = \begin{bmatrix} -3 & -11 \\ 1 & 2 \end{bmatrix}$

32. Explain why the verifications in Exercises 27–31 do not actually prove the corresponding properties for the system of 2×2 matrices.

In Exercises 33–36, determine whether $AB = BA$.

33. $A = \begin{bmatrix} 8 & 6 \\ 3 & 5 \end{bmatrix}$ $B = \begin{bmatrix} 10 & 14 \\ 7 & 3 \end{bmatrix}$

$AB = \begin{bmatrix} 122 & 130 \\ 65 & 57 \end{bmatrix}$; $BA = \begin{bmatrix} 122 & 130 \\ 65 & 57 \end{bmatrix}$;

Yes, $AB = BA$.

34. $A = \begin{bmatrix} -5/3 & -2/3 \\ -2 & -1 \end{bmatrix}$ $B = \begin{bmatrix} -3 & 2 \\ 6 & -5 \end{bmatrix}$

$AB = \begin{bmatrix} 1 & 0 \\ 0 & 1 \end{bmatrix}$; $BA = \begin{bmatrix} 1 & 0 \\ 0 & 1 \end{bmatrix}$;

Yes, $AB = BA$.

35. $A = \begin{bmatrix} 2 & 7 \\ -3 & 4 \end{bmatrix}$ $B = \begin{bmatrix} 6 & -4 \\ 1 & 2 \end{bmatrix}$

$AB = \begin{bmatrix} 19 & 6 \\ -14 & 20 \end{bmatrix}$; $BA = \begin{bmatrix} 24 & 26 \\ -4 & 15 \end{bmatrix}$;

No, AB is not equal to BA.

36. $A = \begin{bmatrix} 1 & 0 \\ 1 & 1 \end{bmatrix}$ $B = \begin{bmatrix} 1 & 1 \\ 0 & 1 \end{bmatrix}$

$AB = \begin{bmatrix} 1 & 1 \\ 1 & 2 \end{bmatrix}$; $BA = \begin{bmatrix} 2 & 1 \\ 1 & 1 \end{bmatrix}$;

No, AB is not equal to BA.

37. Find an original example of 2×2 matrices A and B such that $AB = BA$.

38. Let $M = \begin{bmatrix} 1 & 1 \\ 1 & 1 \end{bmatrix}$.

(a) Compute M^2.

$M^2 = \begin{bmatrix} 2 & 2 \\ 2 & 2 \end{bmatrix}$

(b) Compute M^3.

$M^3 = \begin{bmatrix} 4 & 4 \\ 4 & 4 \end{bmatrix}$

(c) Compute M^4.

$M^4 = \begin{bmatrix} 8 & 8 \\ 8 & 8 \end{bmatrix}$

(d) Without actually computing it, what is M^{10}?

$M^{10} = \begin{bmatrix} 512 & 512 \\ 512 & 512 \end{bmatrix}$

39. **(a)** Prove that $SI = S$ for every 3×3 matrix S.

Let $S = \begin{bmatrix} a & b & c \\ d & e & f \\ g & h & j \end{bmatrix}$; $I = \begin{bmatrix} 1 & 0 & 0 \\ 0 & 1 & 0 \\ 0 & 0 & 1 \end{bmatrix}$.

Then $SI = \begin{bmatrix} a \cdot 1 + b \cdot 0 + c \cdot 0 & a \cdot 0 + b \cdot 1 + c \cdot 0 & a \cdot 0 + b \cdot 0 + c \cdot 1 \\ d \cdot 1 + e \cdot 0 + f \cdot 0 & d \cdot 0 + e \cdot 1 + f \cdot 0 & d \cdot 0 + e \cdot 0 + f \cdot 1 \\ g \cdot 1 + h \cdot 0 + j \cdot 0 & g \cdot 0 + h \cdot 1 + j \cdot 0 & g \cdot 0 + h \cdot 0 + j \cdot 1 \end{bmatrix}$

$= \begin{bmatrix} a & b & c \\ d & e & f \\ g & h & j \end{bmatrix} = S.$

(b) Does part (a) establish that $I = \begin{bmatrix} 1 & 0 & 0 \\ 0 & 1 & 0 \\ 0 & 0 & 1 \end{bmatrix}$ is the identity for multiplication of 3×3 matrices? Explain why or why not.

40. In Example 6, how do you interpret the entry 3 in row 3, column 3 of the matrix M^2?

41. Refer to Example 6. Compute the matrix M^3 and interpret the meaning of its entries.

Assume that a fifth computer is introduced into the network referred to in Example 6. It is connected as shown here.

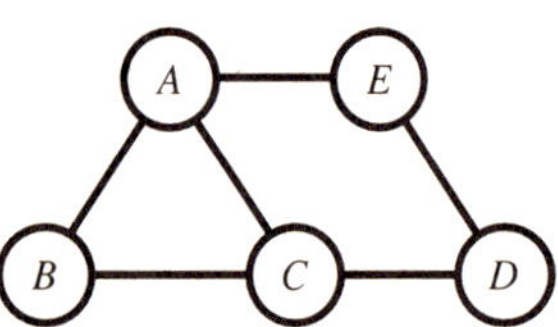

42. Write the communication matrix M for this new network.

$$M = \begin{bmatrix} 0 & 1 & 1 & 0 & 1 \\ 1 & 0 & 1 & 0 & 0 \\ 1 & 1 & 0 & 1 & 0 \\ 0 & 0 & 1 & 0 & 1 \\ 1 & 0 & 0 & 1 & 0 \end{bmatrix}$$

43. Compute M^2.

$$M^2 = \begin{bmatrix} 3 & 1 & 1 & 2 & 0 \\ 1 & 2 & 1 & 1 & 1 \\ 1 & 1 & 3 & 0 & 2 \\ 2 & 1 & 0 & 2 & 0 \\ 0 & 1 & 2 & 0 & 2 \end{bmatrix}$$

44. According to M^2, how many ways can computer E send data to C through a single intermediary?
2 ways

45. Without computing M^3, how many channels of communication would it show from B to E? What are those channels?
2 ways: B to C to A to E and B to C to D to E

46. How many channels of communication from B to E would be shown in the matrix M^4? 6 ways

The real numbers 2/5 and 5/2 are "multiplicative inverses" of each other because $(2/5) \cdot (5/2) = 1$, which is the identity for multiplication of real numbers. Similarly, two matrices A and B are multiplicative inverses of each other if $AB = I$ and $BA = I$, where I is the identity (of the same order in both cases) for multiplication of matrices. (In Section 4, we will address matrix inverses in more detail.)

47. In verifying that A and B are inverses, is it really necessary to compute both AB and BA? Why or why not?

48. What can you say about the orders of A and B if they are to be multiplicative inverses of each other?

CONCEPTUAL  WRITING  CHALLENGING  SCIENTIFIC CALCULATOR  GRAPHING CALCULATOR

Use the definition of matrix inverses (preceding Exercise 47) to determine whether each of the following pairs are inverses of each other.

49. $\begin{bmatrix} 1 & 0 \\ 0 & -1 \end{bmatrix}$ and $\begin{bmatrix} 1 & 0 \\ 0 & -1 \end{bmatrix}$

$\begin{bmatrix} 1 & 0 \\ 0 & -1 \end{bmatrix} \cdot \begin{bmatrix} 1 & 0 \\ 0 & -1 \end{bmatrix} = \begin{bmatrix} 1 & 0 \\ 0 & 1 \end{bmatrix}$;
Yes, they are inverses.

50. $\begin{bmatrix} 0 & 1 \\ 1 & 0 \end{bmatrix}$ and $\begin{bmatrix} 0 & 1 \\ 1 & 0 \end{bmatrix}$

$\begin{bmatrix} 0 & 1 \\ 1 & 0 \end{bmatrix} \cdot \begin{bmatrix} 0 & 1 \\ 1 & 0 \end{bmatrix} = \begin{bmatrix} 1 & 0 \\ 0 & 1 \end{bmatrix}$;
Yes, they are inverses.

51. $\begin{bmatrix} 2 & 1 \\ 4 & 3 \end{bmatrix}$ and $\begin{bmatrix} 3/2 & -1/2 \\ -2 & 1 \end{bmatrix}$

$\begin{bmatrix} 2 & 1 \\ 4 & 3 \end{bmatrix} \cdot \begin{bmatrix} 3/2 & -1/2 \\ -2 & 1 \end{bmatrix} = \begin{bmatrix} 1 & 0 \\ 0 & 1 \end{bmatrix}$;
$\begin{bmatrix} 3/2 & -1/2 \\ -2 & 1 \end{bmatrix} \cdot \begin{bmatrix} 2 & 1 \\ 4 & 3 \end{bmatrix} = \begin{bmatrix} 1 & 0 \\ 0 & 1 \end{bmatrix}$;
Yes, they are inverses.

52. $\begin{bmatrix} -3 & 2 \\ 6 & -5 \end{bmatrix}$ and $\begin{bmatrix} -5/3 & -2/3 \\ -2 & -1 \end{bmatrix}$

$\begin{bmatrix} -3 & 2 \\ 6 & -5 \end{bmatrix} \cdot \begin{bmatrix} -5/3 & -2/3 \\ -2 & -1 \end{bmatrix} = \begin{bmatrix} 1 & 0 \\ 0 & 1 \end{bmatrix}$;
$\begin{bmatrix} -5/3 & -2/3 \\ -2 & -1 \end{bmatrix} \cdot \begin{bmatrix} -3 & 2 \\ 6 & -5 \end{bmatrix} = \begin{bmatrix} 1 & 0 \\ 0 & 1 \end{bmatrix}$;
Yes, they are inverses.

53. $\begin{bmatrix} 3 & 1 & 4 \\ 1 & 0 & 1 \end{bmatrix}$ and $\begin{bmatrix} -1 & 1 \\ 0 & 1 \\ 1 & -1 \end{bmatrix}$

No, they are not inverses.

54. $\begin{bmatrix} -2 & 1/3 & 1 \\ 1/4 & 3 & 2 \end{bmatrix}$ and $\begin{bmatrix} 1 & 3 \\ 1/2 & 3/2 \end{bmatrix}$

No, they are not inverses.

55. $\begin{bmatrix} 5 & 6 \\ -10 & -13 \end{bmatrix}$ and $\begin{bmatrix} 13/5 & 6/5 \\ -2 & 1 \end{bmatrix}$

$\begin{bmatrix} 5 & 6 \\ -10 & -13 \end{bmatrix}\begin{bmatrix} 13/5 & 6/5 \\ -2 & 1 \end{bmatrix} = \begin{bmatrix} 1 & 12 \\ 0 & -25 \end{bmatrix}$;
No, they are not inverses.

56. $\begin{bmatrix} 1 & 1 \\ 1 & -1 \end{bmatrix}$ and $\begin{bmatrix} 1/2 & 1/2 \\ 1/2 & 1/2 \end{bmatrix}$

$\begin{bmatrix} 1 & 1 \\ 1 & -1 \end{bmatrix} \cdot \begin{bmatrix} 1/2 & 1/2 \\ 1/2 & 1/2 \end{bmatrix} = \begin{bmatrix} 1 & 1 \\ 0 & 0 \end{bmatrix}$;
No, they are not inverses.

57. $\begin{bmatrix} 1 & 3 & 3 \\ 1 & 4 & 3 \\ 1 & 3 & 4 \end{bmatrix}$ and $\begin{bmatrix} 7 & -3 & -3 \\ -1 & 1 & 0 \\ -1 & 0 & 1 \end{bmatrix}$

$\begin{bmatrix} 1 & 3 & 3 \\ 1 & 4 & 3 \\ 1 & 3 & 4 \end{bmatrix} \cdot \begin{bmatrix} 7 & -3 & -3 \\ -1 & 1 & 0 \\ -1 & 0 & 1 \end{bmatrix} = \begin{bmatrix} 1 & 0 & 0 \\ 0 & 1 & 0 \\ 0 & 0 & 1 \end{bmatrix}$; $\begin{bmatrix} 7 & -3 & -3 \\ -1 & 1 & 0 \\ -1 & 0 & 1 \end{bmatrix} \cdot \begin{bmatrix} 1 & 3 & 3 \\ 1 & 4 & 3 \\ 1 & 3 & 4 \end{bmatrix} = \begin{bmatrix} 1 & 0 & 0 \\ 0 & 1 & 0 \\ 0 & 0 & 1 \end{bmatrix}$;
Yes, they are inverses.

58. $\begin{bmatrix} -1 & 0 & 2 \\ 3 & 1 & 0 \\ 0 & 2 & -3 \end{bmatrix}$ and $\begin{bmatrix} -1/5 & 4/15 & -2/15 \\ 3/5 & 1/5 & 2/5 \\ 2/5 & 2/15 & -1/15 \end{bmatrix}$

$\begin{bmatrix} -1 & 0 & 2 \\ 3 & 1 & 0 \\ 0 & 2 & -3 \end{bmatrix} \cdot \begin{bmatrix} -1/5 & 4/15 & -2/15 \\ 3/5 & 1/5 & 2/5 \\ 2/5 & 2/15 & -1/15 \end{bmatrix} = \begin{bmatrix} 1 & 0 & 0 \\ 0 & 1 & 0 \\ 0 & 0 & 1 \end{bmatrix}$;
$\begin{bmatrix} -1/5 & 4/15 & -2/15 \\ 3/5 & 1/5 & 2/5 \\ 2/5 & 2/15 & -1/15 \end{bmatrix} \cdot \begin{bmatrix} -1 & 0 & 2 \\ 3 & 1 & 0 \\ 0 & 2 & -3 \end{bmatrix} = \begin{bmatrix} 1 & 0 & 0 \\ 0 & 1 & 0 \\ 0 & 0 & 1 \end{bmatrix}$; Yes, they are inverses.

59. Explain in your own words what a *row-column product* is, and how it is related to multiplication of matrices in general.

60. Explain why, in Example 7, the matrix product involves 90 products of entries.

Use a calculator to compute the following matrix products.

61. $$\begin{bmatrix} 23.8 & 10.6 & 12.8 & 21.9 \\ 15.8 & 24.5 & 11.6 & 14.8 \\ 13.8 & 10.6 & 13.3 & 14.7 \\ 16.4 & 12.8 & 11.1 & 15.9 \\ 21.3 & 12.3 & 13.2 & 32.1 \end{bmatrix} \cdot \begin{bmatrix} 8.23 & 7.45 & 8.25 & 6.27 & 9.91 & 7.29 \\ 2.29 & 6.52 & 3.35 & 5.94 & 6.37 & 6.73 \\ 2.58 & 9.88 & 1.28 & 7.53 & 1.23 & 3.28 \\ 8.45 & 5.27 & 6.37 & 4.38 & 4.39 & 3.46 \end{bmatrix}$$

$$\text{Product} = \begin{bmatrix} 438.227 & 488.299 & 387.747 & 404.496 & 415.265 & 362.598 \\ 341.127 & 470.054 & 321.549 & 396.768 & 391.883 & 369.323 \\ 296.377 & 380.795 & 260.023 & 314.025 & 285.172 & 266.426 \\ 327.277 & 399.097 & 293.671 & 332.085 & 327.514 & 297.122 \\ 508.767 & 538.464 & 438.303 & 446.607 & 446.589 & 392.418 \end{bmatrix}$$

62. $$\begin{bmatrix} 4.56 & 9.87 & 2.91 \\ 6.37 & 2.81 & 6.81 \\ 5.28 & 4.79 & 4.21 \\ 5.28 & 6.97 & 1.38 \\ 3.84 & 5.28 & 3.91 \\ 4.68 & 2.38 & 1.54 \\ 4.44 & 5.55 & 6.66 \\ 9.63 & 8.52 & 7.41 \end{bmatrix} \cdot \begin{bmatrix} 8.23 & 4.19 & 2.28 & 9.28 & 8.27 \\ 1.32 & 6.58 & 7.17 & 8.23 & 9.82 \\ 3.37 & 8.26 & 1.49 & 6.54 & 9.87 \end{bmatrix}$$

$$\text{Product} = \begin{bmatrix} 60.3639 & 108.0876 & 85.5006 & 142.5783 & 163.3563 \\ 79.084 & 101.4307 & 44.8182 & 126.7773 & 147.4888 \\ 63.9649 & 88.416 & 52.6556 & 115.9535 & 132.2561 \\ 57.3054 & 79.3846 & 64.0695 & 115.3867 & 125.7316 \\ 51.7495 & 83.1286 & 52.4387 & 104.661 & 122.1981 \\ 46.8478 & 47.99 & 30.0296 & 73.0894 & 77.275 \\ 66.3114 & 110.1342 & 59.8401 & 130.4361 & 156.954 \\ 115.473 & 157.6179 & 94.0857 & 207.9474 & 236.4432 \end{bmatrix}$$

14.3 Exercises (Text Page 833)

Let $A = \begin{bmatrix} -4 & 0 & 1 \\ 3 & -4 & 2 \\ -1 & 2 & 1 \end{bmatrix}$ *and* $B = \begin{bmatrix} 2 & -1 & 3 \\ -4 & 0 & 1 \\ 1 & 3 & -3 \end{bmatrix}$.

Apply the row operations indicated in each case, and write out the resulting matrix. (We work here with a square matrix rather than an augmented matrix just to practice row operations.)

1. Interchange rows 1 and 3 of A.

$$\begin{bmatrix} -1 & 2 & 1 \\ 3 & -4 & 2 \\ -4 & 0 & 1 \end{bmatrix}$$

2. Interchange rows 2 and 3 of B.

$$\begin{bmatrix} 2 & -1 & 3 \\ 1 & 3 & -3 \\ -4 & 0 & 1 \end{bmatrix}$$

3. Multiply each element of row 2 of A by -3.

$$\begin{bmatrix} -4 & 0 & 1 \\ -9 & 12 & -6 \\ -1 & 2 & 1 \end{bmatrix}$$

4. Double each element in row 1 of A.

$$\begin{bmatrix} -8 & 0 & 2 \\ 3 & -4 & 2 \\ -1 & 2 & 1 \end{bmatrix}$$

5. Multiply each element of row 2 of B by -4, and add the results to row 1.

$$\begin{bmatrix} 18 & -1 & -1 \\ -4 & 0 & 1 \\ 1 & 3 & -3 \end{bmatrix}$$

6. Multiply each element of row 3 of A by 5, and add the results to row 2.

$$\begin{bmatrix} -4 & 0 & 1 \\ -2 & 6 & 7 \\ -1 & 2 & 1 \end{bmatrix}$$

7. Multiply each element of row 1 of A by 3, and add the results to row 3.

$$\begin{bmatrix} -4 & 0 & 1 \\ 3 & -4 & 2 \\ -13 & 2 & 4 \end{bmatrix}$$

8. Multiply each element of row 3 of B by -1, and add the results to row 2.

$$\begin{bmatrix} 2 & -1 & 3 \\ -5 & -3 & 4 \\ 1 & 3 & -3 \end{bmatrix}$$

9. Form matrix $A + B$. Multiply each element of row 2 of $A + B$ by -3, and add the results to row 1.

$$\begin{bmatrix} 1 & 11 & -5 \\ -1 & -4 & 3 \\ 0 & 5 & -2 \end{bmatrix}$$

10. Form matrix $B - A$. Multiply each element of row 1 of $B - A$ by 5, and add the results to row 3.

$$\begin{bmatrix} 6 & -1 & 2 \\ -7 & 4 & -1 \\ 32 & -4 & 6 \end{bmatrix}$$

State which row operation must be applied to matrix A or B from above to get the following matrices.

11. to A; $\begin{bmatrix} 3 & -4 & 2 \\ -4 & 0 & 1 \\ -1 & 2 & 1 \end{bmatrix}$

Interchange rows 1 and 2.

12. to B; $\begin{bmatrix} 3 & 2 & 0 \\ -4 & 0 & 1 \\ 1 & 3 & -3 \end{bmatrix}$

Multiply each element of row 3 by 1, and add the results to row 1.

13. to A; $\begin{bmatrix} -4 & 0 & 1 \\ 6 & -8 & 4 \\ -1 & 2 & 1 \end{bmatrix}$

Double each element of row 2.

14. to A; $\begin{bmatrix} -4 & 0 & 1 \\ 3 & -4 & 2 \\ 1/2 & -1 & -1/2 \end{bmatrix}$

Multiply each element of row 3 by $-1/2$.

15. to B; $\begin{bmatrix} 2 & -1 & 3 \\ 0 & -2 & 7 \\ 1 & 3 & -3 \end{bmatrix}$

Multiply each element of row 1 by 2, and add the results to row 2.

16. to B; $\begin{bmatrix} -1 & -10 & 12 \\ -4 & 0 & 1 \\ 1 & 3 & -3 \end{bmatrix}$

Multiply each element of row 3 by -3, and add the results to row 1.

Find the sequence of row operations that will change each of the following matrices into the identity matrix, if possible.

17. $\begin{bmatrix} 1 & 2 \\ 0 & -1 \end{bmatrix}$

Multiply row 2 by -1. Multiply row 2 by -2 and add to row 1.

18. $\begin{bmatrix} 2 & 1 \\ 1 & -1 \end{bmatrix}$

Multiply row 1 by $1/2$. Multiply row 1 by -1 and add to row 2. Multiply row 2 by $-2/3$. Multiply row 2 by $-1/2$ and add to row 1.

19. $\begin{bmatrix} -2 & 2 \\ 4 & 1 \end{bmatrix}$

Multiply row 1 by $-1/2$. Multiply row 1 by -4 and add to row 2. Multiply row 2 by $1/5$. Multiply row 2 by 1 and add to row 1.

20. $\begin{bmatrix} 0 & -1 \\ -2 & 0 \end{bmatrix}$

Interchange rows 1 and 2. Multiply row 1 by $-1/2$. Multiply row 2 by -1.

21. $\begin{bmatrix} 6 & 3 & 0 \\ 0 & 4 & 0 \\ 0 & 0 & 2 \end{bmatrix}$

Multiply row 1 by $1/6$. Multiply row 2 by $1/4$. Multiply row 2 by $-1/2$ and add to row 1. Multiply row 3 by $1/2$.

22. $\begin{bmatrix} -2 & 2 & 0 \\ 0 & 1 & 0 \\ 1 & 1 & 1 \end{bmatrix}$

Multiply row 1 by $-1/2$. Multiply row 1 by -1 and add to row 3. Multiply row 2 by 1 and add to row 1. Multiply row 2 by -2 and add to row 3.

Write the augmented matrix for each system. Do not try to solve.

23. $2x + 3y = 11$
$x + 2y = 8$

$\left[\begin{array}{cc|c} 2 & 3 & 11 \\ 1 & 2 & 8 \end{array}\right]$

24. $3x + 5y = -13$
$2x + 3y = -9$

$\left[\begin{array}{cc|c} 3 & 5 & -13 \\ 2 & 3 & -9 \end{array}\right]$

25. $x + 5y = 6$
$3x - 4y = 1$

$\left[\begin{array}{cc|c} 1 & 5 & 6 \\ 3 & -4 & 1 \end{array}\right]$

26. $2x + 7y = 1$
$5x + y = -15$

$\left[\begin{array}{cc|c} 2 & 7 & 1 \\ 5 & 1 & -15 \end{array}\right]$

27. $2x + y + z = 3$
$3x - 4y + 2z = -7$
$x + y + z = 2$

$\left[\begin{array}{ccc|c} 2 & 1 & 1 & 3 \\ 3 & -4 & 2 & -7 \\ 1 & 1 & 1 & 2 \end{array}\right]$

28. $4x - 2y + 3z = 4$
$3x + 5y + z = 7$
$5x - y + 4z = 7$

$\left[\begin{array}{ccc|c} 4 & -2 & 3 & 4 \\ 3 & 5 & 1 & 7 \\ 5 & -1 & 4 & 7 \end{array}\right]$

29. $x + y = 6$
$2y + z = 2$
$z = 2$
$\left[\begin{array}{ccc|c} 1 & 1 & 0 & 6 \\ 0 & 2 & 1 & 2 \\ 0 & 0 & 1 & 2 \end{array}\right]$

30. $x = 6$
$y + 2z = 2$
$x - 3z = 2$
$\left[\begin{array}{ccc|c} 1 & 0 & 0 & 6 \\ 0 & 1 & 2 & 2 \\ 1 & 0 & -3 & 2 \end{array}\right]$

Use matrix row operations to solve the following systems of equations.

31. $x + y = 5$
$x - y = -1$
$\{(2, 3)\}$

32. $x + 2y = 5$
$2x + y = -2$
$\{(-3, 4)\}$

33. $x + y = -3$
$2x - 5y = -6$
$\{(-3, 0)\}$

34. $3x - 2y = 4$
$3x + y = -2$
$\{(0, -2)\}$

35. $2x - 3y = 10$
$2x + 2y = 5$
$\{(7/2, -1)\}$

36. $6x + y = 5$
$5x + y = 3$
$\{(2, -7)\}$

37. $2x - 5y = 10$
$3x + y = 15$
$\{(5, 0)\}$

38. $4x - y = 3$
$-2x + 3y = 1$
$\{(1, 1)\}$

39. $2x - 3y = 2$
$4x - 6y = 1$
$\emptyset$

40. $x + 2y = 1$
$2x + 4y = 3$
$\emptyset$

41. $x + y = -1$
$y + z = 4$
$x + z = 1$
$\{(-2, 1, 3)\}$

42. $x - z = -3$
$y + z = 9$
$x + z = 7$
$\{(2, 4, 5)\}$

43. $x + y - z = 6$
$2x - y + z = -9$
$x - 2y + 3z = 1$
$\{(-1, 23, 16)\}$

44. $x + 3y - 6z = 7$
$2x - y + z = 1$
$x + 2y + 2z = -1$
$\{(1, 0, -1)\}$

45. $-x + y = -1$
$y - z = 6$
$x + z = -1$
$\{(3, 2, -4)\}$

46. $x + y = 1$
$2x - z = 0$
$y + 2z = -2$
$\{(-1, 2, -2)\}$

▲ 47. $3x + 2y - w = 0$
$2x + z + 2w = 5$
$x + 2y - z = -2$
$2x - y + z + w = -2$
$\{(-1, 2, 5, 1)\}$

▲ 48. $x + 3y - 2z - w = 9$
$4x + y + z + 2w = 2$
$-3x - y + z - w = -5$
$x - y - 3z - 2w = 2$
$\{(0, 2, -2, 1)\}$

Use the Gauss-Jordan method to solve each problem.

49. Linda Ramirez is a building contractor. If she hires 7 day laborers and 2 concrete finishers, her payroll for the day is \$692, while 1 day laborer and 5 concrete finishers cost \$476. Find the daily wage charge for each type of worker.
A laborer charges \$76 per day and a concrete finisher charges \$80 per day.

50. A biologist wants to grow two types of algae, types A and B. She has available 15 gal of nutrient I and 26 gal of nutrient II. A vat of algae A needs 2 gal of nutrient I and 3 gal of nutrient II, while a vat of algae B needs 1 gal of I and 2 gal of II. How many vats of each type of algae should the biologist grow in order to use all the nutrients?
4 of *A*, 7 of *B*

51. To make his portrait bust of Millard Fillmore, Harry bought 2 lb of dark clay and 3 lb of light clay, paying \$13 for the clay. He later needed one more pound of dark clay and 2 lb of light clay, costing \$7 altogether. How much did he pay per pound for each type of clay? **Dark clay costs \$5 per lb, light clay costs \$1 per lb.**

52. The perimeter of a triangle is 21 cm. If two sides are of equal length, and the third side is 3 cm longer than one of the equal sides, find the lengths of the sides.
6 cm, 6 cm, 9 cm

53. The secretary of the local consumer group bought some decals at 8¢ each and some bumper stickers at 10¢ each to give to the members. He spent a total of $15.52. If he bought a total of 170 items, how many of each kind did he buy?
74 decals and 96 bumper stickers were purchased.

▲ **54.** The Matrix and the Patrix, a local musical group, is coming to play at a school festival. The reporter for the school newspaper doesn't know how many guitarists and how many members of the rhythm section there are in the group. Janet, one of the guitarists, says that, not counting herself, there are three times as many members of the rhythm group as guitarists. Steve, a member of the rhythm section, says that, not counting himself, the number of members of the rhythm section is one less than twice the number of guitarists. How many of each are there in the group?
6 rhythm, 3 guitarists

14.4 Exercises (Text Page 843)

Find the inverse of each matrix that has an inverse.

1. $\begin{bmatrix} 1 & 0 \\ 0 & -1 \end{bmatrix}$ $\begin{bmatrix} \mathbf{1} & \mathbf{0} \\ \mathbf{0} & \mathbf{-1} \end{bmatrix}$

2. $\begin{bmatrix} 0 & 1 \\ 1 & 0 \end{bmatrix}$ $\begin{bmatrix} \mathbf{0} & \mathbf{1} \\ \mathbf{1} & \mathbf{0} \end{bmatrix}$

3. $\begin{bmatrix} 2 & 1 \\ 4 & 3 \end{bmatrix}$ $\begin{bmatrix} \mathbf{3/2} & \mathbf{-1/2} \\ \mathbf{-2} & \mathbf{1} \end{bmatrix}$

4. $\begin{bmatrix} -3 & 2 \\ 6 & -5 \end{bmatrix}$ $\begin{bmatrix} \mathbf{-5/3} & \mathbf{-2/3} \\ \mathbf{-2} & \mathbf{-1} \end{bmatrix}$

5. $\begin{bmatrix} 1 & 1 \\ 1 & 1 \end{bmatrix}$ **no inverse**

6. $\begin{bmatrix} -1 & -1 \\ 1 & 1 \end{bmatrix}$ **no inverse**

7. $\begin{bmatrix} 1 & 2 & 0 \\ 0 & 1 & 0 \\ 1 & -1 & 1 \end{bmatrix}$ $\begin{bmatrix} \mathbf{1} & \mathbf{-2} & \mathbf{0} \\ \mathbf{0} & \mathbf{1} & \mathbf{0} \\ \mathbf{-1} & \mathbf{3} & \mathbf{1} \end{bmatrix}$

8. $\begin{bmatrix} 0 & 1 & 0 \\ 0 & 0 & -2 \\ 1 & -1 & 0 \end{bmatrix}$ $\begin{bmatrix} \mathbf{1} & \mathbf{0} & \mathbf{1} \\ \mathbf{1} & \mathbf{0} & \mathbf{0} \\ \mathbf{0} & \mathbf{-1/2} & \mathbf{0} \end{bmatrix}$

9. $\begin{bmatrix} 4 & 3 & 3 \\ -1 & 0 & -1 \\ -4 & -4 & -3 \end{bmatrix}$ $\begin{bmatrix} \mathbf{4} & \mathbf{3} & \mathbf{3} \\ \mathbf{-1} & \mathbf{0} & \mathbf{-1} \\ \mathbf{-4} & \mathbf{-4} & \mathbf{-3} \end{bmatrix}$

10. $\begin{bmatrix} 1 & 0 & 2 \\ -1 & 0 & -2 \\ 1 & 1 & 1 \end{bmatrix}$ **no inverse**

Use matrix algebra to find the unique solution of each system. If the system has no unique solution, state why.

11. $4x + y = 5$
$2x + y = 3$
(1, 1)

12. $3x - 7y = 31$
$2x - 4y = 18$
(1, −4)

13. $5x - y = 6$
$-10x + 2y = 15$
There is no unique solution because the coefficient matrix has no inverse.

14. $5x - y = 14$
$x + 8y = 11$
(3, 1)

15. $x + y - z = 6$
$2x - y + z = -9$
$x - 2y + 3z = 1$
(−1, 23, 16)

16. $4x + 2y - 3z = 6$
$x - 4y + z = -4$
$-x + 2z = 2$
(2, 2, 2)

17. $2x - y + 3z = 0$
$x + 2y - z = 5$
$2y + z = 1$
(2, 1, −1)

18. $x + y = 1$
$2x - z = 0$
$y + 2z = -2$
(−1, 2, −2)

19. $2x - 3y + z = 4$
$x + y - z = 1$
$-5y + 3z = 2$

There is no unique solution because the coefficient matrix has no inverse.

20. $x + 3y = 7$
$2x - y - 5z = 12$
$7y + 5z = 2$

There is no unique solution because the coefficient matrix has no inverse.

In each of Exercises 1 and 2, the given matrix was its own inverse. In each of the following cases, find the missing elements to make the given matrix its own inverse.

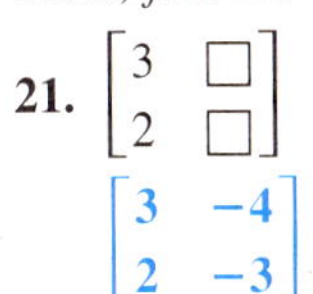

21. $\begin{bmatrix} 3 & \square \\ 2 & \square \end{bmatrix}$ $\begin{bmatrix} 3 & -4 \\ 2 & -3 \end{bmatrix}$

22. $\begin{bmatrix} \square & 8 \\ \square & 5 \end{bmatrix}$ $\begin{bmatrix} -5 & 8 \\ -3 & 5 \end{bmatrix}$

23. $\begin{bmatrix} 2 & \square \\ \square & -2 \end{bmatrix}$ **The missing elements can be any two numbers with a product of −3.**

24. $\begin{bmatrix} \square & 5 \\ -7 & \square \end{bmatrix}$ **The missing elements are 6 and −6 (in either order).**

Assuming that a and b are real numbers, determine whether each of the following statements is sometimes true, always true, *or* never true. *If sometimes true, give the conditions under which it is true.*

25. $\begin{bmatrix} a & 0 \\ 0 & a \end{bmatrix}$ and $\begin{bmatrix} b & 0 \\ 0 & b \end{bmatrix}$ are inverses of each other.

sometimes true (true whenever the numbers a and b are reciprocals of each other)

26. $\begin{bmatrix} 0 & a \\ a & 0 \end{bmatrix}$ and $\begin{bmatrix} 0 & b \\ b & 0 \end{bmatrix}$ are inverses of each other.

sometimes true (true whenever the numbers a and b are reciprocals of each other)

27. Recall that "row operation 2" is to multiply a given row by some nonzero constant. Show that the following product has the effect (on the second matrix) of multiplying its first row by k.

$$\begin{bmatrix} k & 0 \\ 0 & 1 \end{bmatrix} \cdot \begin{bmatrix} 2 & 4 \\ 3 & 5 \end{bmatrix}$$

$$\begin{bmatrix} k & 0 \\ 0 & 1 \end{bmatrix} \cdot \begin{bmatrix} 2 & 4 \\ 3 & 5 \end{bmatrix} = \begin{bmatrix} 2k + 0 \cdot 3 & 4k + 0 \cdot 5 \\ 0 \cdot 2 + 1 \cdot 3 & 0 \cdot 4 + 1 \cdot 5 \end{bmatrix} = \begin{bmatrix} 2k & 4k \\ 3 & 5 \end{bmatrix}$$

(The first row is multiplied by k.)

28. What 2×2 multiplier, on the left, will change

$$\begin{bmatrix} 2 & 4 \\ 3 & 5 \end{bmatrix} \quad \text{to} \quad \begin{bmatrix} 2 & 4 \\ 3k & 5k \end{bmatrix}$$

(that is, multiply the second row by the constant k)?

$$\begin{bmatrix} 1 & 0 \\ 0 & k \end{bmatrix}$$

29. "Row operation 3" is to add some multiple of one row to another row. Show that the following product has the effect (on the second matrix) of adding twice its first row elements to its corresponding second row elements.

$$\begin{bmatrix} 1 & 0 \\ 2 & 1 \end{bmatrix} \cdot \begin{bmatrix} 2 & 4 \\ 3 & 5 \end{bmatrix}$$

$$\begin{bmatrix} 1 & 0 \\ 2 & 1 \end{bmatrix} \cdot \begin{bmatrix} 2 & 4 \\ 3 & 5 \end{bmatrix} = \begin{bmatrix} 1 \cdot 2 + 0 \cdot 3 & 1 \cdot 4 + 0 \cdot 5 \\ 2 \cdot 2 + 1 \cdot 3 & 2 \cdot 4 + 1 \cdot 5 \end{bmatrix} = \begin{bmatrix} 2 & 4 \\ 2 \cdot 2 + 3 & 2 \cdot 4 + 5 \end{bmatrix}$$

▲ **30.** What 2×2 multiplier, on the left, will change

$$\begin{bmatrix} 2 & 4 \\ 3 & 5 \end{bmatrix} \quad \text{to} \quad \begin{bmatrix} 11 & 19 \\ 3 & 5 \end{bmatrix}$$

(that is, add three times the second row to the first row)?

$\begin{bmatrix} 1 & 3 \\ 0 & 1 \end{bmatrix}$

◉ *The matrix in Example 4 had no inverse. Here are several other 2×2 matrices that do not have inverses:*

$$\begin{bmatrix} 5 & 1 \\ 10 & 2 \end{bmatrix}, \quad \begin{bmatrix} 6 & -3 \\ -8 & 4 \end{bmatrix}, \quad \begin{bmatrix} 17 & 22 \\ 34 & 44 \end{bmatrix}.$$

▲ **31.** Describe the characteristic shared by the above matrices that causes them to not have inverses.

32. Find two additional 2×2 matrices, neither of which has an inverse.

Answers will vary. Examples are $\begin{bmatrix} 1 & 4 \\ 3 & 12 \end{bmatrix}$ **and** $\begin{bmatrix} -5 & 20 \\ 2 & -8 \end{bmatrix}$.

The ***determinant*** *of a 2×2 matrix is denoted by vertical bars (similar to absolute value signs), so that, for example, the determinant of* $\begin{bmatrix} 2 & 6 \\ 5 & 8 \end{bmatrix}$ *is denoted* $\begin{vmatrix} 2 & 6 \\ 5 & 8 \end{vmatrix}$. *The numerical* ***value*** *of a determinant is defined as follows:*

$$\begin{vmatrix} a & c \\ b & d \end{vmatrix} = ad - bc.$$

Evaluate each determinant.

33. $\begin{vmatrix} 7 & 4 \\ 5 & 3 \end{vmatrix}$ 1

34. $\begin{vmatrix} 2 & 8 \\ -3 & -6 \end{vmatrix}$ 12

35. $\begin{vmatrix} -4 & 8 \\ 5 & -10 \end{vmatrix}$ 0

36. $\begin{vmatrix} 3 & -2 \\ 6 & -4 \end{vmatrix}$ 0

▲ ◉ **37.** Extend the concept of a determinant to 3×3 matrices, and show how to evaluate a 3×3 determinant. (You may want to consult an algebra book.)

▲ ◉ **38.** Give an example of a 3×3 determinant with the value 0.

$\begin{vmatrix} 1 & 2 & 3 \\ 4 & 5 & 6 \\ 5 & 7 & 9 \end{vmatrix}$ **(Answers will vary.)**

◉ *A matrix is called* ***singular*** *if its determinant has the value 0.*

▲ **39.** Explain why a system of linear equations has no unique solution if its coefficient matrix is singular.

40. If A is a singular square matrix, what can you say about A^{-1}?

A^{-1} **does not exist.**

◉ CONCEPTUAL  ▲ CHALLENGING SCIENTIFIC CALCULATOR GRAPHING CALCULATOR

Use a calculator to find the inverse of each matrix. Give all elements to three decimal places.

41. $\begin{bmatrix} 1.2 & 3.7 \\ -4.8 & 5.5 \end{bmatrix}$ $\begin{bmatrix} .226 & -.152 \\ .197 & .049 \end{bmatrix}$

42. $\begin{bmatrix} 7/6 & 2/11 \\ 5/13 & 1/9 \end{bmatrix}$ $\begin{bmatrix} 1.861 & -3.046 \\ -6.443 & 19.542 \end{bmatrix}$

43. $\begin{bmatrix} 14.8 & 18.1 & 11.6 \\ 8.2 & 12.6 & 9.5 \\ 15.1 & 13.8 & 7.4 \end{bmatrix}$ $\begin{bmatrix} -.871 & .602 & .594 \\ 1.905 & -1.511 & -1.047 \\ -1.774 & 1.590 & .876 \end{bmatrix}$

44. $\begin{bmatrix} 125 & 391 & 416 \\ 277 & 143 & 209 \\ 610 & 157 & 504 \end{bmatrix}$ $\begin{bmatrix} -.002 & .007 & -.001 \\ .001 & .011 & -.005 \\ .002 & -.012 & .005 \end{bmatrix}$

45. In the real number system, how many numbers do not have multiplicative inverses? What are they? **one; 0**

46. In the system of 2 × 2 matrices, how many matrices do not have multiplicative inverses? What are they? **infinitely many; all the singular 2 × 2 matrices**

47. Assume the same basic wheat/oil economy described in the text. Find the gross production required for a net production of 750 metric tons of wheat and 1,000 metric tons of oil. Round final amounts to whole numbers.
1,174 metric tons of wheat and 1,565 metric tons of oil

48. Find the gross production required in Exercise 47 if production of 1 metric ton of wheat consumes 1/3 metric ton of oil (and no wheat), while the production of 1 metric ton of oil consumes 1/5 metric ton of wheat (and no oil). Round final amounts to whole numbers.
1,018 metric tons of wheat and 1,339 metric tons of oil

49. Consider an economy depending on three basic commodities: agriculture, manufacturing, and transportation. Suppose 1/4 unit of manufacturing and 1/2 unit of transportation are required to produce 1 unit of agriculture, 1/2 unit of agriculture and 1/4 unit of transportation are required to produce 1 unit of manufacturing, and 1/4 unit of agriculture and 1/4 unit of manufacturing are required to produce 1 unit of transportation. How many units of each commodity should be produced to satisfy a demand for 1,000 units of each commodity? Round final amounts to whole numbers.
3,077 units of agriculture; 2,564 units of manufacturing; 3,179 units of transportation

50. A simple economy depends on three commodities: oil, corn, and coffee. Production of one unit of oil requires .1 unit of oil, .2 unit of corn, and no units of coffee. To produce one unit of corn requires .2 unit of oil, .1 unit of corn, and .05 unit of coffee. To produce one unit of coffee requires .1 unit of oil, .05 unit of corn, and .1 unit of coffee. Find the gross production required to give a net production of 1,000 units each of oil, corn, and coffee. Round final amounts to whole numbers.
1,584 units of oil; 1,530 units of corn; 1,196 units of coffee

Extension Exercises (Text Page 847)

1. The message

THE THREE TENORS

is to be encoded by using the matrix method described in the text. The message is broken into groups of two letters (and spaces). Find the sequence of matrices derived from the given message.

$\begin{bmatrix} 20 \\ 8 \end{bmatrix} \begin{bmatrix} 5 \\ 27 \end{bmatrix} \begin{bmatrix} 20 \\ 8 \end{bmatrix} \begin{bmatrix} 18 \\ 5 \end{bmatrix} \begin{bmatrix} 5 \\ 27 \end{bmatrix} \begin{bmatrix} 20 \\ 5 \end{bmatrix} \begin{bmatrix} 14 \\ 15 \end{bmatrix} \begin{bmatrix} 18 \\ 19 \end{bmatrix}$

2. Use the multiplier matrix $M = \begin{bmatrix} -1 & 2 \\ 2 & -5 \end{bmatrix}$ to encode the sequence of matrices of Exercise 1, and list the resulting sequence of 2×1 matrices.
$\begin{bmatrix} -4 \\ 0 \end{bmatrix} \begin{bmatrix} 49 \\ -125 \end{bmatrix} \begin{bmatrix} -4 \\ 0 \end{bmatrix} \begin{bmatrix} -8 \\ 11 \end{bmatrix} \begin{bmatrix} 49 \\ -125 \end{bmatrix} \begin{bmatrix} -10 \\ 15 \end{bmatrix} \begin{bmatrix} 16 \\ -47 \end{bmatrix} \begin{bmatrix} 20 \\ -59 \end{bmatrix}$

3. A message that was encoded with the multiplier matrix of Exercise 2 was received as the following sequence of numbers.

39, −98, −19, 34, −7, 10, 3, −11, 36,
−99, 21, −62, 7, −18, 33, −87, 41, −109

What multiplier matrix should be used for decoding this message?
$\begin{bmatrix} -5 & -2 \\ -2 & -1 \end{bmatrix}$

4. Find the original message (of Exercise 3). **AT __ DODGER __ STADIUM**

5. Finish encoding the message SOME CODES ARE EASILY BROKEN given in the text.
$\begin{bmatrix} 103 \\ 118 \\ 116 \end{bmatrix} \begin{bmatrix} 95 \\ 122 \\ 98 \end{bmatrix} \begin{bmatrix} 42 \\ 46 \\ 47 \end{bmatrix} \begin{bmatrix} 103 \\ 130 \\ 104 \end{bmatrix} \begin{bmatrix} 114 \\ 119 \\ 141 \end{bmatrix} \begin{bmatrix} 65 \\ 66 \\ 84 \end{bmatrix} \begin{bmatrix} 120 \\ 132 \\ 145 \end{bmatrix} \begin{bmatrix} 87 \\ 89 \\ 105 \end{bmatrix} \begin{bmatrix} 63 \\ 74 \\ 68 \end{bmatrix} \begin{bmatrix} 176 \\ 203 \\ 203 \end{bmatrix}$

6. The person receiving the message of Exercise 5 (SOME CODES ARE EASILY BROKEN) also received the following encoded message (already converted to matrix form).

$$\begin{bmatrix} 99 \\ 119 \\ 107 \end{bmatrix} \begin{bmatrix} 116 \\ 134 \\ 135 \end{bmatrix} \begin{bmatrix} 84 \\ 85 \\ 102 \end{bmatrix} \begin{bmatrix} 98 \\ 125 \\ 102 \end{bmatrix} \begin{bmatrix} 45 \\ 51 \\ 51 \end{bmatrix} \begin{bmatrix} 81 \\ 84 \\ 102 \end{bmatrix} \begin{bmatrix} 153 \\ 173 \\ 180 \end{bmatrix}$$

Decode this second message, assuming it was encoded using the multiplier matrix given in the text. **OTHERS __ ARE __ DIFFICULT**

7. Suppose you are a special agent. The encoding multiplier matrix for today is
$M = \begin{bmatrix} 2 & 1 & 2 \\ 3 & 4 & 3 \\ 3 & 1 & 2 \end{bmatrix}$. You need to send the message

DESTROY FILES IMMEDIATELY

to another agent. What sequence of numbers would you transmit?
51, 89, 55, 88, 177, 108, 89, 201, 114, 40, 90, 49, 83, 192, 102, 49, 106, 62, 19, 51, 23, 69, 116, 89, 131, 264, 156

8. After your transmission of Exercise 7, you receive the response

45, 90, 51, 83, 172, 88, 51, 89, 55, 88, 177, 108, 63, 107, 88.

What is the message? **FILES __ DESTROYED**

9. Could the encoding matrix (the multiplier) used in the type of cryptography discussed here be of order 4×4? Explain why or why not.

10. What are the restrictions (if any) on the type of matrix used as the encoding multiplier?

11. Explain why the matrix type code discussed here would be difficult to break.

CONCEPTUAL WRITING CHALLENGING SCIENTIFIC CALCULATOR GRAPHING CALCULATOR

▲ **12.** The following message is written in a code in which the frequency of the symbols is the main key to the solution.

)?−−8))6*+8506*3×6;4?*7*&×*−6.48()6)985)?(8+2:;48)81&?(;46*3)6*;
48&(+8(*598+.8()8=8(5*−8−5(81?098;4&+)&15*50:)6)6*;?6;6&*0?−7

Find the frequency of each symbol. By comparing high frequency symbols with the high frequency letters in English (see the text above), try to decipher the message. (*Hint:* look for repeated two-symbol combinations and double letters for added clues. Try to identify vowels first.)
SUCCESS IN DEALING WITH UNKNOWN CIPHERS IS MEASURED BY THESE FOUR THINGS IN THE ORDER NAMED: PERSEVERANCE, CAREFUL METHODS OF ANALYSIS, INTUITION, LUCK.

14.5 Exercises (Text Page 854)

In the following games, assume that player A chooses rows and player B chooses columns. Eliminate all dominated strategies from each of the following games.

1. $\begin{bmatrix} 1 & 4 \\ 4 & -1 \\ 3 & 5 \\ -4 & 0 \end{bmatrix}$ $\begin{bmatrix} 4 & -1 \\ 3 & 5 \end{bmatrix}$

2. $\begin{bmatrix} 2 & 3 & 1 & -5 \\ -1 & 5 & 4 & 1 \\ 1 & 0 & 2 & -3 \end{bmatrix}$ $\begin{bmatrix} 2 & -5 \\ -1 & 1 \\ 1 & -3 \end{bmatrix}$

3. $\begin{bmatrix} 8 & 12 & -7 \\ -2 & 1 & 4 \end{bmatrix}$ $\begin{bmatrix} 8 & -7 \\ -2 & 4 \end{bmatrix}$

4. $\begin{bmatrix} 6 & 2 \\ -1 & 10 \\ 3 & 5 \end{bmatrix}$ $\begin{bmatrix} 6 & 2 \\ -1 & 10 \\ 3 & 5 \end{bmatrix}$

Where it exists, find the saddle point and the value of the game for each of the following.

5. $\begin{bmatrix} -6 & 2 \\ -1 & -10 \\ 3 & 5 \end{bmatrix}$ **(3, I); 3**

6. $\begin{bmatrix} 7 & 8 \\ -2 & 15 \end{bmatrix}$ **(1, I); 7**

7. $\begin{bmatrix} 3 & -4 & 1 \\ 5 & 3 & 2 \end{bmatrix}$ **(2, III); 2**

8. $\begin{bmatrix} 2 & 3 & 1 \\ -1 & 4 & -7 \\ 5 & 2 & 0 \\ 8 & -4 & -1 \end{bmatrix}$ **(1, III); 1**

9. $\begin{bmatrix} 1 & 4 & -3 & 1 & -1 \\ 2 & 5 & 0 & 4 & 10 \\ 1 & -3 & -2 & 5 & 2 \end{bmatrix}$ **(2, III); 0**

10. $\begin{bmatrix} -4 & 2 & -3 & -7 \\ 4 & 3 & 5 & 9 \end{bmatrix}$ **(2, II); 3**

Solve each of the following problems.

11. Hillsdale College has sold out all tickets for a jazz concert to be held in the stadium. If it rains, the show will have to be moved into the gym, which has a much smaller seating capacity. The dean must decide in advance whether to set up the seats and the stage in the gym or in the stadium, or both, just in case. The matrix below shows the net profit in each case.

		NATURE	
		Rain	No rain
	Set up stadium	−\$1,550	\$1,500
ACTIONS	Set up gym	\$1,000	\$1,000
	Set up both	\$750	\$1,400

(a) What action should the dean take if he is an optimist? **Set up stadium**
(b) If he is a pessimist? **Set up gym**
(c) If the weather forecaster predicts rain with a probability of .6, what action should the dean take to maximize expected profit? **Set up both**

12. A community is considering an anti-smoking campaign. The city manager has been asked to recommend one of three possible actions: a campaign for everyone in the community over age 10, a campaign for youths only, or no campaign at all. The two states of nature are a true cause-effect relationship between smoking and cancer or no such relationship. The cost to the community (including loss of life and productivity) in each case is shown below.*

	Cause-effect relationship	No such relationship
Campaign for all	\$100,000	\$800,000
Campaign for youth	\$2,820,000	\$20,000
No campaign	\$3,100,100	\$0

(a) What action should the manager recommend if she is an optimist? **No campaign**

(b) If she is a pessimist? **Campaign for all**

(c) If the Director of Public Health estimates that the probability of a true cause-effect relationship is .8, which action should be recommended to minimize expected costs? **Campaign for all**

13. The research department of Allied Manufacturing has developed a new process that they feel will result in an improved product. Management must decide whether or not to market the new product. The new product may be better than the old, or it may not be better. If the new product is better, and the company decides to market it, sales should increase by \$50,000. If it is not better and they replace the old product with the new product on the market, they will lose \$25,000 to competitors. If they decide not to market the new product, they will lose \$40,000 if it is better and research costs of \$10,000 if it is not.

(a) Prepare a payoff matrix.

	Better	**Not**
Market new	**50,000**	**−25,000**
Don't	**−40,000**	**−10,000**

(b) If management believes that the probability is .4 that the new product is better, find the expected profits under each possible course of action, and find the best action. **\$5,000 and −\$22,000; Market new**

14. Suppose Allied Manufacturing (of Exercise 13) decides to put their new product on the market with a big advertising campaign. At the same time, they find out that their major competitor, Bates Manufacturing, has also decided on a big advertising campaign for its version of the product. The payoff matrix below shows the increased sales in millions for Allied, which also represent decreases for Bates.

		BATES TV	BATES Radio
ALLIED	TV	1.0	−.7
	Radio	−.5	.5

* This problem is based on an article by B. G. Greenburg in the September 1969 issue of the *Journal of the American Statistical Association.*

Find the optimum strategy for Allied, and find the value of the game.
Allied should select strategy 1 with probability 10/27 and strategy 2 with probability 17/27. The value of the game is 1/18 million, which represents increased sales of about $55,556.

In each of the following games, find the optimum strategy for each player, and find the value of the game. Identify any fair games.

15. $\begin{bmatrix} 5 & 1 \\ 3 & 4 \end{bmatrix}$
I: 1/5; 2: 4/5;
I: 3/5; II: 2/5;
value = 17/5

16. $\begin{bmatrix} -4 & 5 \\ 3 & -4 \end{bmatrix}$
1: 7/16; 2: 9/16;
I: 9/16; II: 7/16;
value = −1/16

17. $\begin{bmatrix} -2 & 4 \\ 0 & 5 \end{bmatrix}$
The saddle point gives a value of 0. The game is fair. The first player should select strategy (row) 2, while the second player should select strategy (column) I.

18. $\begin{bmatrix} 0 & -4 \\ 4 & 0 \end{bmatrix}$
The saddle point gives a value of 0. The game is fair. Both players should always select their second strategy.

19. In the game of matching coins, two players each flip a coin. If both coins match (both show heads or both show tails), player A wins $1. If there is no match, player B wins $1, as in the matrix below. Find the optimum strategies for each player, and find the value of the game.

$$\begin{bmatrix} 1 & -1 \\ -1 & 1 \end{bmatrix}$$

Both A and B should choose either strategy with a probability of 1/2. The value of the game is 0, a fair game.

20. Explain the meaning of a "zero-sum" game. Give an example of a zero-sum game and an example of a non–zero-sum game.

21. Explain these terms: *maximax criterion, maximin criterion,* and *minimax criterion.*

22. Explain why, in a game matrix, a dominant row has entries that are *greater* than those of the dominated row, whereas a dominant column has entries that are *less* than those of the dominated column.

23. Explain what is meant by a "mixed" strategy.

Refer to the original payoff matrix in the text for the citrus farmer. Assume the farmer is neither an optimist nor a pessimist. Each of Exercises 24–26 gives a probability of having a freeze. In each case find **(a)** *the expected payoff for strategy 1 (use smudge pots) and* **(b)** *the expected payoff for strategy 2 (don't use them).*

24. 1/17
(a) $7,411.76 (b) $7,352.94

25. 1/18
(a) $7,388.89 (b) $7,388.89

26. 1/19
(a) $7,368.42 (b) $7,421.05

27. Discuss how your answers to Exercises 24–26 relate to the claim in the text that "pots should be used as long as you feel the probability of a freeze is greater than 1/18."

28. Refer to Example 3 in the text. Use E_{II} to find the farmer's expected gain for Nature's strategy II. Is the value the same as was obtained in the example using E_I?
$7,388.89; yes

In Example 3 it was stated that the value of a game (with no saddle point) comes out the same no matter which player's optimum strategy and expected gain are considered. Work the following exercises to consider the smudge pot question from Nature's viewpoint. Assume the original payoff matrix and suppose that Nature will randomly impose a freeze a certain fraction of the time, given by x.

29. Derive an expression for E_1, the expected payoff in case the farmer chooses to use smudge pots. **$7{,}000x + 7{,}000$**

30. Derive an expression for E_2, the expected payoff in case the farmer chooses not to use smudge pots. **$8{,}000 - 11{,}000x$**

31. Equate E_1 and E_2 from Exercises 29 and 30 to find Nature's optimum strategy. **Nature should impose a freeze 1/18 of the time.**

32. Find the value (expected payoff) of the game when Nature employs the strategy of Exercise 31. Is the value the same as that arrived at in Example 3? **\$7,388.89; yes**

Another citrus farmer facing the smudge pot decision has calculated the following payoff matrix. Refer to it for Exercises 33–37.

	Freeze	No freeze
Use smudge pots	\$8,000	\$7,000
Don't use them	−\$5,000	\$10,000

33. Which strategy would an optimist choose? **strategy 2**

34. Which strategy would a pessimist choose? **strategy 1**

35. If the farmer is neither an optimist nor a pessimist but is able to reasonably assess the probability of a freeze, how large must that probability be for the farmer's preferred strategy to be to use smudge pots? **greater than 3/16**

36. The payoff matrix above still has no saddle point, so the farmer's best plan is probably to use a mixed strategy. Use the procedure described just before Example 3 to decide what fraction of the time the farmer should use smudge pots. **Use pots 15/16 of the time.**

37. By using smudge pots the fraction of the time computed in Exercise 36, what expected gain does the farmer have? (Compare with Example 3.) **\$7,187.50**

38. For the citrus farmer problem, devise a new 2 × 2 payoff matrix for which an optimist and a pessimist would both choose not to use smudge pots. Describe conditions that might give rise to the payoffs in your matrix.

Chapter 14 Test (Text Page 858)

Let $A = \begin{bmatrix} 2 & -1 & 3 \\ 2 & 0 & 1 \\ 4 & 1 & 3 \end{bmatrix}$ *and* $B = \begin{bmatrix} -3 & 0 & 1 \\ 0 & 2 & -1 \\ -1 & 0 & 3 \end{bmatrix}$. *Perform the indicated matrix operations.*

1. $A + B$

$$\begin{bmatrix} -1 & -1 & 4 \\ 2 & 2 & 0 \\ 3 & 1 & 6 \end{bmatrix}$$

2. $A - B$

$$\begin{bmatrix} 5 & -1 & 2 \\ 2 & -2 & 2 \\ 5 & 1 & 0 \end{bmatrix}$$

3. AB

$$\begin{bmatrix} -9 & -2 & 12 \\ -7 & 0 & 5 \\ -15 & 2 & 12 \end{bmatrix}$$

4. BA
$$\begin{bmatrix} -2 & 4 & -6 \\ 0 & -1 & -1 \\ 10 & 4 & 6 \end{bmatrix}$$

5. $2A$
$$\begin{bmatrix} 4 & -2 & 6 \\ 4 & 0 & 2 \\ 8 & 2 & 6 \end{bmatrix}$$

6. $-3B + A$
$$\begin{bmatrix} 11 & -1 & 0 \\ 2 & -6 & 4 \\ 7 & 1 & -6 \end{bmatrix}$$

Apply the following row operations to matrix A from above.

7. Interchange rows 2 and 3.
$$\begin{bmatrix} 2 & -1 & 3 \\ 4 & 1 & 3 \\ 2 & 0 & 1 \end{bmatrix}$$

8. Multiply each element of row 2 of A by -3, and add the results to row 1.
$$\begin{bmatrix} -4 & -1 & 0 \\ 2 & 0 & 1 \\ 4 & 1 & 3 \end{bmatrix}$$

Find each of the following products, when possible. If not possible, state why not.

9. $\begin{bmatrix} -2 & 1 \\ 3 & 4 \end{bmatrix} \cdot \begin{bmatrix} 2 \\ 1 \end{bmatrix}$
$$\begin{bmatrix} -3 \\ 10 \end{bmatrix}$$

10. $\begin{bmatrix} 3 & 4 & 1 \\ 2 & 0 & 8 \end{bmatrix} \cdot \begin{bmatrix} 1 & 0 & 2 \\ 0 & 0 & 1 \\ 3 & 0 & 2 \end{bmatrix}$
$$\begin{bmatrix} 6 & 0 & 12 \\ 26 & 0 & 20 \end{bmatrix}$$

11. $\begin{bmatrix} 1 & 6 & 1 \end{bmatrix} \cdot \begin{bmatrix} 3 \\ 2 \\ 1 \end{bmatrix}$
$$\begin{bmatrix} 16 \end{bmatrix}$$

12. $\begin{bmatrix} 8 & 9 & 7 \\ 2 & 1 & 3 \end{bmatrix} \cdot \begin{bmatrix} 2 & 3 \\ 4 & 2 \end{bmatrix}$

There is no product, since the first matrix has a different number of columns than the second has rows.

Find the inverse of each of the following matrices, where possible.

13. $\begin{bmatrix} 2 & 1 \\ 5 & 3 \end{bmatrix}$
$$\begin{bmatrix} 3 & -1 \\ -5 & 2 \end{bmatrix}$$

14. $\begin{bmatrix} 1 & 0 & 1 \\ 2 & 3 & 1 \\ 1 & 3 & 2 \end{bmatrix}$
$$\begin{bmatrix} 1/2 & 1/2 & -1/2 \\ -1/2 & 1/6 & 1/6 \\ 1/2 & -1/2 & 1/2 \end{bmatrix}$$

Set up the augmented matrix for each of the following systems of equations. Do not solve the systems.

15. $9x - 4y = 20$
$5x + 10y = 12$
$$\left[\begin{array}{cc|c} 9 & -4 & 20 \\ 5 & 10 & 12 \end{array}\right]$$

16. $2x + y - 6z = 6$
$5x + 12z = 15$
$6y + 3z = 8$
$$\left[\begin{array}{ccc|c} 2 & 1 & -6 & 6 \\ 5 & 0 & 12 & 15 \\ 0 & 6 & 3 & 8 \end{array}\right]$$

Solve each system of equations by the Gauss-Jordan method (Exercises 17 and 18).

17. $5x + 6y = 15$
$3x + 7y = 9$
$\{(3, 0)\}$

18. $3x + 4y + z = 10$
$x - 3y - 5z = -3$
$2x + 7y = 1$
$\{(4, -1, 2)\}$

19. If the system of equations

$$\begin{aligned} x + y &= 1 \\ 2x - y &= -7 \end{aligned}$$

is written as $AX = B$, what exactly is represented by each of the following?
(a) A **(b)** X **(c)** B

(a) $\begin{bmatrix} 1 & 1 \\ 2 & -1 \end{bmatrix}$ (b) $\begin{bmatrix} x \\ y \end{bmatrix}$ (c) $\begin{bmatrix} 1 \\ -7 \end{bmatrix}$

20. Given the matrix equation $AX = B$, express X in terms of A and B. **$X = A^{-1}B$**

21. Use matrix algebra to solve the system of equations of Exercise 19. **$\{(-2, 3)\}$**

22. Explain the conditions under which the matrix product AB is computable.

The economy of a small country involves bananas and coal. Production of 1 unit of bananas consumes .3 unit of bananas and .1 unit of coal. Production of 1 unit of coal consumes .2 unit of bananas and .4 unit of coal.

23. Set up the technological matrix for this economy.

$A =$	Bananas	Coal
Bananas	.3	.2
Coal	.1	.4

24. Find the gross production required to meet a demand for net production of 400 units of bananas and 300 units of coal. **750 units of bananas, 625 units of coal**

25. Remove any dominated strategies from the following game.

$$\begin{bmatrix} 6 & 8 & -3 & 4 \\ 2 & 0 & 2 & -1 \end{bmatrix} \quad \begin{bmatrix} 8 & -3 \\ 0 & 2 \end{bmatrix}$$

26. Identify any saddle points in the following game.

$$\begin{bmatrix} -3 & 1 & 4 & 2 \\ 8 & 9 & 6 & 4 \\ 9 & 10 & 15 & 19 \end{bmatrix} \quad 9$$

27. Find the optimum strategy for player A and the value of the following game.

$$\begin{array}{cc} & \quad B \\ & \begin{array}{cc} \text{I} & \text{II} \end{array} \\ A \begin{array}{c} 1 \\ 2 \end{array} & \begin{bmatrix} -3 & -4 \\ -4 & 5 \end{bmatrix} \end{array}$$

Select strategy one 9/10 of the time and strategy two 1/10 of the time; value is $-31/10$.

CONCEPTUAL    SCIENTIFIC CALCULATOR GRAPHING CALCULATOR

28. Find two different pairs of values for the missing entries in the matrix

$$A = \begin{bmatrix} \square & -3 \\ 5 & \square \end{bmatrix}$$

that will cause A to be its own inverse.

4 and −4, or −4 and 4

Arturo Hogan makes silver jewelry inlaid with turquoise. Each ring requires 1.6 units of silver and .7 unit of turquoise, each bracelet requires 5.8 units of silver and 7.5 units of turquoise, and each necklace requires 26.3 units of silver and 18.8 units of turquoise. Arturo pays $5.50 per unit for silver and $2.10 per unit for turquoise.

29. Set up a matrix M relating jewelry items (ring, bracelet, necklace) to materials (silver, turquoise), and another matrix C relating materials to cost (per unit).

$$M = \begin{array}{l} \text{Ring} \\ \text{Bracelet} \\ \text{Necklace} \end{array} \overset{\begin{array}{cc} \text{Silver} & \text{Turquoise} \end{array}}{\begin{bmatrix} 1.6 & .7 \\ 5.8 & 7.5 \\ 26.3 & 18.8 \end{bmatrix}}; \quad C = \begin{array}{l} \text{Silver} \\ \text{Turquoise} \end{array} \overset{\text{Cost per unit}}{\begin{bmatrix} 5.50 \\ 2.10 \end{bmatrix}}$$

30. Perform an appropriate matrix operation to obtain a matrix relating jewelry items to cost (per item) for materials.

$$MC = \begin{bmatrix} 1.6 & .7 \\ 5.8 & 7.5 \\ 26.3 & 18.8 \end{bmatrix} \cdot \begin{bmatrix} 5.50 \\ 2.10 \end{bmatrix} = \begin{array}{l} \text{Ring} \\ \text{Bracelet} \\ \text{Necklace} \end{array} \overset{\text{Cost per item}}{\begin{bmatrix} 10.27 \\ 47.65 \\ 184.13 \end{bmatrix}}$$

APPENDIX The Metric System

Appendix Exercises (Text Page 864)

Perform the following conversions by multiplying or dividing by the appropriate power of 10.

1. 8 m to millimeters **8,000 mm**
2. 14.76 m to centimeters **1,476 cm**
3. 8,500 cm to meters **85 m**
4. 250 mm to meters **.25 m**
5. 68.9 cm to millimeters **689 mm**
6. 3.25 cm to millimeters **32.5 mm**
7. 59.8 mm to centimeters **5.98 cm**
8. 3.542 mm to centimeters **.3542 cm**
9. 5.3 km to meters **5,300 m**
10. 9.24 km to meters **9,240 m**
11. 27,500 m to kilometers **27.5 km**
12. 14,592 m to kilometers **14,592 km**

Use a metric ruler to perform the following measurements, first in centimeters, then in millimeters.

13. |———| **2.54 cm; 25.4 mm**

14. |————| **3.3 cm; 33 mm**

15. |——————| **5 cm; 50 mm**

16. Based on your measurement of the line segment in Exercise 13, one inch is about how many centimeters? how many millimeters? **2.54 cm; 25.4 mm**

Perform the following conversions by multiplying or dividing by the appropriate power of 10.

17. 6 L to centiliters **600 cl**
18. 4.1 L to milliliters **4,100 ml**
19. 8.7 L to milliliters **8,700 ml**
20. 12.5 L to centiliters **1,250 cl**
21. 925 cl to liters **9.25 L**
22. 412 ml to liters **.412 L**
23. 8,974 ml to liters **8.974 L**
24. 5,639 cl to liters **56.39 L**
25. 8,000 g to kilograms **8 kg**
26. 25,000 g to kilograms **25 kg**
27. 5.2 kg to grams **5,200 g**
28. 12.42 kg to grams **12,420 g**
29. 4.2 g to milligrams **4,200 mg**
30. 3.89 g to centigrams **389 cg**
31. 598 mg to grams **.598 g**
32. 7,634 cg to grams **76.34 g**

Use the formulas given in the text to perform the following conversions. If necessary, round to the nearest degree.

33. 86°F to Celsius **30°C**
34. 536°F to Celsius **280°C**
35. −114°F to Celsius **−81°C**
36. −40°F to Celsius **−40°C**
37. 10°C to Fahrenheit **50°F**
38. 25°C to Fahrenheit **77°F**
39. −40°C to Fahrenheit **−40°F**
40. −15°C to Fahrenheit **5°F**

Solve each of the following problems. Refer to geometry formulas as necessary.

41. One nickel weighs 5 g. How many nickels are in 1 kg of nickels? **200 nickels**

42. Sea water contains about 3.5 g salt per 1,000 ml of water. How many grams of salt would be in one liter of sea water? **3.5 g**

43. Helium weighs about .0002 g per milliliter. How much would one liter of helium weigh? **.2 g**

44. About 1,500 g sugar can be dissolved in a liter of warm water. How much sugar could be dissolved in one milliliter of warm water? **1.5 g**

45. Northside Foundry needed seven metal strips, each 67 cm long. Find the total cost of the strips, if they sell for $8.74 per meter. **$40.99**

46. Uptown Dressmakers bought fifteen pieces of lace, each 384 mm long. The lace sold for $54.20 per meter. Find the cost of the fifteen pieces. **$312.19**

47. Imported marble for desk tops costs $174.20 per square meter. Find the cost of a piece of marble 128 cm by 174 cm. **$387.98**

48. A special photographic paper sells for $63.79 per square meter. Find the cost to buy 80 pieces of the paper, each 9 cm by 14 cm. **$64.30**

49. An importer received some special coffee beans in a box 82 cm by 1.1 m by 1.2 m. Give the volume of the box, both in cubic centimeters and cubic meters. **1,082,400 cm^3; 1.0824 m^3**

50. A fabric center receives bolts of woolen cloth in crates 1.5 m by 74 cm by 97 cm. Find the volume of a crate, both in cubic centimeters and cubic meters. **1,076,700 cm^3; 1.0767 m^3**

51. A medicine is sold in small bottles holding 800 ml each. How many of these bottles can be filled from a vat holding 160 L of the medicine? **200 bottles**

52. How many 2-liter bottles of soda pop would be needed for a wedding reception if 80 people are expected, and each drinks 400 ml of soda? **16 bottles**

Perform the following conversions. Use a calculator and/or the table in the text as necessary.

53. 982 yd to meters **897.9 m**

54. 12.2 km to miles **7.581 mi**

55. 125 mi to kilometers **201.1 km**

56. 1,000 mi to kilometers **1,609 km**

57. 1,816 g to pounds **3.995 lb**

58. 1.42 lb to grams **644.7 g**

59. 47.2 lb to grams **21,428.8 g**

60. 7.68 kg to pounds **16.90 lb**

61. 28.6 L to quarts **30.22 qt**

62. 59.4 L to quarts **62.77 qt**

63. 28.2 gal to liters **106.7 L**

64. 16 qt to liters **15.14 L**

CONCEPTUAL  WRITING CHALLENGING SCIENTIFIC CALCULATOR GRAPHING CALCULATOR

Metric measures are very common in medicine. Since we convert among metric measures by moving the decimal point, errors in locating the decimal point in medical doses are not unknown. Decide whether the following doses of medicine seem reasonable *or* unreasonable.

65. Take 2 kg of aspirin three times a day. **unreasonable**

66. Take 4 L of Kaopectate every evening at bedtime. **unreasonable**

67. Take 25 ml of cough syrup daily. **reasonable**

68. Soak your feet in 6 L of hot water. **reasonable**

69. Inject 1/2 L of insulin every morning. **unreasonable**

70. Apply 40 g of salve to a cut on your finger. **unreasonable**

Select the most reasonable choice for each of the following.

71. length of an adult cow
(a) 1 m **(b)** 3 m **(c)** 5 m (b)

72. length of a Cadillac
(a) 1 m **(b)** 3 m **(c)** 5 m (c)

73. distance from Seattle to Miami
(a) 500 km **(b)** 5,000 km **(c)** 50,000 km (b)

74. length across an average nose
(a) 3 cm **(b)** 30 cm **(c)** 300 cm (a)

75. distance across this page
(a) 1.93 mm **(b)** 19.3 mm **(c)** 193 mm (c)

76. weight of this book
(a) 1 kg **(b)** 10 kg **(c)** 1,000 kg (a)

77. weight of a large automobile
(a) 1,300 kg **(b)** 130 kg **(c)** 13 kg (a)

78. volume of a 12-ounce bottle of beverage
(a) 35 ml **(b)** 355 ml **(c)** 3,550 ml (b)

79. height of a person
(a) 180 cm **(b)** 1,800 cm **(c)** 18 cm (a)

80. diameter of the earth
(a) 130 km **(b)** 1,300 km **(c)** 13,000 km (c)

81. length of a long freight train
(a) 8 m **(b)** 80 m **(c)** 800 m (c)

82. volume of a grapefruit
(a) 1 L **(b)** 4 L **(c)** 8 L (a)

83. the length of a pair of Levis
(a) 70 cm **(b)** 700 cm **(c)** 7 cm (a)

84. a person's weight
(a) 700 kg **(b)** 7 kg **(c)** 70 kg (c)

85. diagonal measure of the picture tube of a table model TV set
(a) 5 cm **(b)** 50 cm **(c)** 500 cm (b)

86. width of a standard bedroom door
(a) 1 m **(b)** 3 m **(c)** 5 m (a)

87. thickness of a standard audiotape cassette
(a) .9 mm **(b)** 9 mm **(c)** 90 mm (b)

88. length of an edge of a record album
(a) 300 mm **(b)** 30 mm **(c)** 3,000 mm (a)

89. the temperature at the surface of a frozen lake
(a) 0°C **(b)** 10°C **(c)** 32°C (a)

90. the temperature in the middle of Death Valley on a July afternoon
(a) 25°C **(b)** 40°C **(c)** 65°C (b)

91. surface temperature of desert sand on a hot summer day
(a) 30°C **(b)** 60°C **(c)** 90°C (b)

92. boiling water
(a) 100°C **(b)** 120°C **(c)** 150°C (a)

93. air temperature on a day when you need a sweater
(a) 30°C **(b)** 20°C **(c)** 10°C (c)

94. air temperature when you go swimming
(a) 30°C **(b)** 15°C **(c)** 10°C (a)

95. temperature when baking a cake
(a) 120°C **(b)** 170°C **(c)** 300°C (b)

96. temperature of bath water
(a) 35°C **(b)** 50°C **(c)** 65°C (a)

97. temperature in a sauna
(a) 25°C **(b)** 60°C **(c)** 90°C (b)

98. temperature in a freezer
(a) 32°C **(b)** 10°C **(c)** −5°C (c)

Acknowledgments

Literary Permissions

Chapter 4

74 (Table) "Months with a Friday the 13th" from "An Aid to the Superstitious" by G. L Ritter, S. R. Lowry, H. B. Woodruff, and T. L. Isenhour, *The Mathematics Teacher,* May 1977. Copyright © 1977 The National Council of Teachers of Mathematics, Inc. Reprinted by permission.

Chapter 9

226–228 (Exercises 1–25) Pages 1–4 from *National Council of Teachers of Mathematics Student Math Notes,* November 1991. Reprinted by permission of the National Council of Teachers of Mathematics.

Chapter 11

252 (Exercise 74) Problem #6 from *Mathematics Teacher,* January 1989, p. 38. Reprinted by permission.

252 (Exercise 75) Problem #2 from *Mathematics Teacher,* November 1989, p. 626. Reprinted by permission.

254 (Exercise 43) Problem #8 from *Mathematics Teacher,* December 1992, p. 736. Reprinted by permission.

Chapter 12

271 (Exercises 11–13) Adapted graph "The Growth of the Suburbs and the Black Migration to the Cities" from *The 1993 World Book Year Book.* Copyright © 1993 World Book, Inc. By permission of the publisher.

272 (Exercises 14–21) Adapted graph "U.S. Trade with Canada and Mexico" from *The 1993 World Book Year Book.* Copyright © 1993 World Book, Inc. By permission of the publisher.

273 (Exercises 22–23) Graph "The Decline in Federal Urban Aid ..." from *The 1993 World Book Year Book.* Copyright © 1993 World Book, Inc. By permission of the publisher.

274 (Exercises 26–29) Graph "Retirement Savings Net Worth" from *TSA Guide to Retirement Planning for California Educators,* Winter 1993. Reprinted by permission of John L. Kattman.

296–299 From *How to Lie with Statistics* by Darrell Huff, illustrated by Irving Geis. Copyright 1954 and renewed © 1982 by Darrell Huff and Irving Geis. Reprinted by permission of W. W. Norton & Company, Inc.

303 (Exercises 1–4) Adapted graph "Selected U.S. Economic Indicators" from *The 1993 World Book Year Book.* Copyright © 1993 World Book, Inc. By permission of the publisher.

Photo Credits

Unless otherwise acknowledged, all photographs are the property of Scott, Foresman and Company.

Chapter 1

21 Corbis-UPI/Bettmann
35 Hans Reinhard/Bruce Coleman Inc.

Chapter 5

95 Stadtmuseum, Aachen/Superstock, Inc.
96 Metropolitan Museum of Art/Harris Brisbane Dick Fund, 1943
98 Art Resource

Chapter 6

101 Times-Picayune

Chapter 9

210 Mount Rushmore National Park, U.S. Dept. of the Interior, National Park Service/Photo: Paul Harsled

Chapter 12

290 AP/Wide World